CAMBRIDGE
UNIVERSITY PRESS

Chemistry

for Cambridge IGCSE™

COURSEBOOK

Richard Harwood, Chris Millington & Ian Lodge

CAMBRIDGE
UNIVERSITY PRESS

Shaftesbury Road, Cambridge CB2 8EA, United Kingdom

One Liberty Plaza, 20th Floor, New York, NY 10006, USA

477 Williamstown Road, Port Melbourne, VIC 3207, Australia

314–321, 3rd Floor, Plot 3, Splendor Forum, Jasola District Centre, New Delhi – 110025, India

103 Penang Road, #05-06/07, Visioncrest Commercial, Singapore 238467

Cambridge University Press is part of the University of Cambridge.

It furthers the University's mission by disseminating knowledge in the pursuit of education, learning and research at the highest international levels of excellence.

www.cambridge.org
Information on this title: www.cambridge.org/9781108951609

First published 1998
Second edition 2002
Third edition 2010
Fourth edition 2014
Fifth edition 2021

20 19 18 17

Printed in Dubai by Oriental Press

A catalogue record for this publication is available from the British Library

ISBN 978-1-108-95160-9 Coursebook with Digital Access (2 Years)
ISBN 978-1-108-97040-2 Digital Coursebook (2 Years)
ISBN 978-1-108-94830-2 Coursebook eBook

Additional resources for this publication at www.cambridge.org/go

---

Exam-style questions and sample answers have been written by the authors. In examinations, the way marks are awarded may be different. References to assessment and/or assessment preparation are the publisher's interpretation of the syllabus requirements and may not fully reflect the approach of Cambridge Assessment International Education.

Contents

How to use this series

We offer a comprehensive, flexible array of resources for the Cambridge IGCSE™ Chemistry syllabus. We provide targeted support and practice for the specific challenges we've heard that students face: learning science with English as a second language; learners who find the mathematical content within science difficult; and developing practical skills.

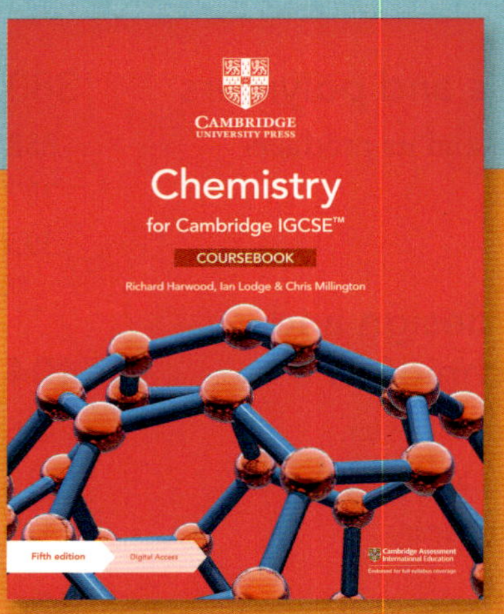

This coursebook provides coverage of the full Cambridge IGCSE Chemistry syllabus. Each chapter explains facts and concepts, and uses relevant real-world examples of scientific principles to bring the subject to life. Together with a focus on practical work and plenty of active learning opportunities, the coursebook prepares learners for all aspects of their scientific study. At the end of each chapter, examination-style questions offer practice opportunities for learners to apply their learning.

The digital teacher's resource contains detailed guidance for all topics of the syllabus, including common misconceptions identifying areas where learners might need extra support, as well as an engaging bank of lesson ideas for each syllabus topic. Differentiation is emphasised with advice for identification of different learner needs and suggestions of appropriate interventions to support and stretch learners. The teacher's resource also contains support for preparing and carrying out all the investigations in the practical workbook, including a set of sample results for when practicals aren't possible.

The teacher's resource also contains scaffolded worksheets and unit tests for each chapter. Answers for all components are accessible to teachers for free on the Cambridge GO platform.

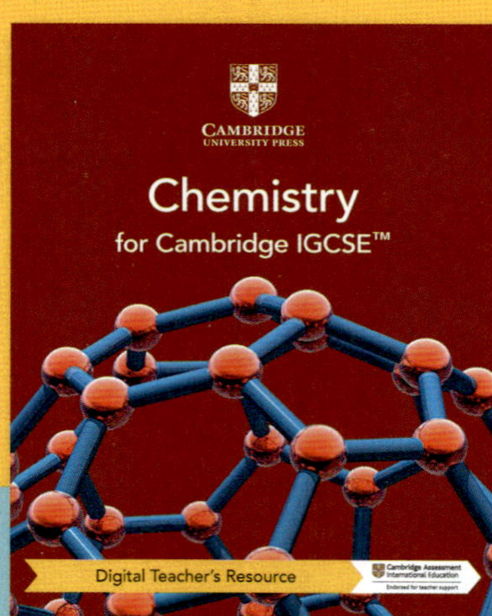

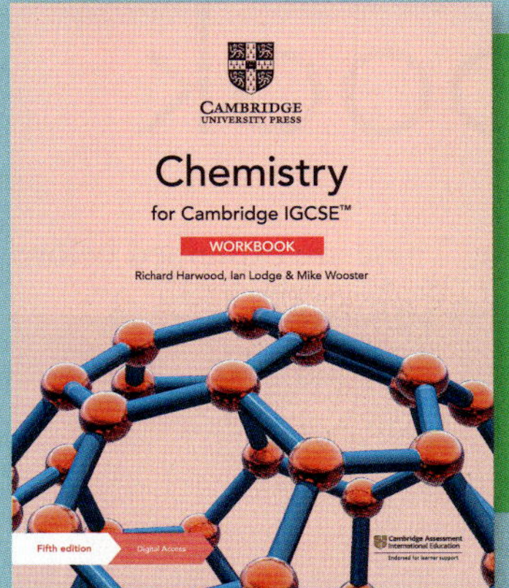

The skills-focused workbook has been carefully constructed to help learners develop the skills that they need as they progress through their Cambridge IGCSE Chemistry course, providing further practice of all the topics in the coursebook. A three-tier, scaffolded approach to skills development enables students to gradually progress through 'focus', 'practice' and 'challenge' exercises, ensuring that every learner is supported. The workbook enables independent learning and is ideal for use in class or as homework.

The practical workbook provides learners with additional opportunities for hands-on practical work, giving them full guidance and support that will help them to develop their investigative skills. These skills include planning investigations, selecting and handling apparatus, creating hypotheses, recording and displaying results, and analysing and evaluating data.

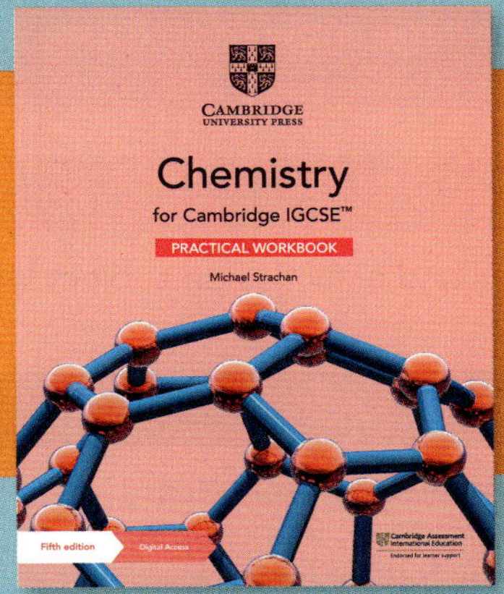

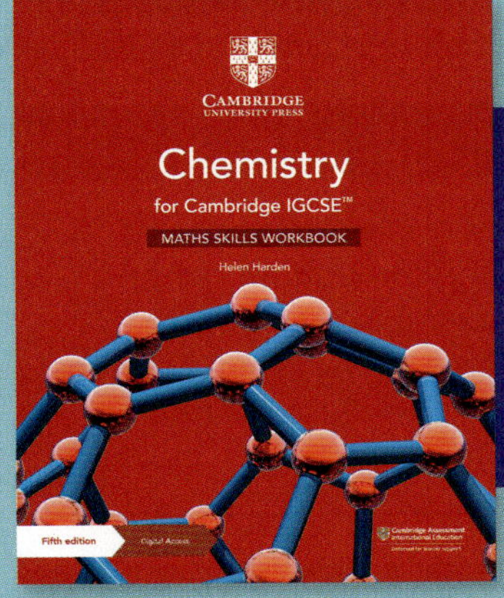

Mathematics is an integral part of scientific study, and one that learners often find a barrier to progression in science. The Maths Skills for Cambridge IGCSE Chemistry write-in workbook has been written in collaboration with the Association for Science Education, with each chapter focusing on several maths skills that students need to succeed in their Chemistry course.

Our research shows that English language skills are the single biggest barrier to students accessing international science. This write-in English language skills workbook contains exercises set within the context of Cambridge IGCSE Chemistry topics to consolidate understanding and embed practice in aspects of language central to the subject. Activities range from practising using the passive form of verbs in the context of electrolysis to the naming of chemical substances using common prefixes.

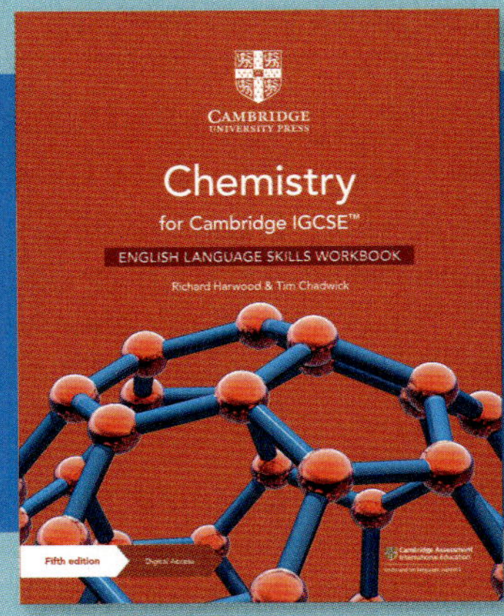

How to use this book

Throughout this book, you will notice lots of different features that will help your learning. These are explained below.

LEARNING INTENTIONS

These set the scene for each chapter, help with navigation through the coursebook and indicate the important concepts in each topic.

In the learning intentions table, the summary table and the exam-style questions, Supplement content is indicated with a large arrow and a darker background, as in the example here.

GETTING STARTED

This contains questions and activities on subject knowledge you will need before starting this chapter.

SCIENCE IN CONTEXT

This feature presents real-world examples and applications of the content in a chapter, encouraging you to look further into topics that may go beyond the syllabus. There are discussion questions at the end, which look at some of the benefits and problems of these applications, and encourage you to look further into the topics.

EXPERIMENTAL SKILLS

This feature focuses on developing your practical skills. They include lists of equipment required and any safety issues, step-by-step instructions so you can carry out the experiment, and questions to help you think about what you have learnt.

KEY WORDS

Key vocabulary is highlighted in the text when it is first introduced, and definitions are given in boxes near the vocabulary. You will also find definitions of these words in the Glossary at the back of this book.

Questions

Appearing throughout the text, questions give you a chance to check that you have understood the topic you have just read about. The answers to these questions are accessible to teachers for free on the Cambridge GO site.

ACTIVITY

Activities give you an opportunity to check and develop your understanding throughout the text in a more active way, for example by creating presentations, posters or role plays. When activities have answers, teachers can find these for free on the Cambridge GO site.

COMMAND WORDS

Command words that appear in the syllabus and might be used in exams are highlighted in the exam-style questions. In the margin, you will find the Cambridge International definition. You will also find these definitions in the Glossary at the back of the book.

Supplement content: Where material is intended for students who are studying the Supplement content of the syllabus as well as the Core, this is indicated using the arrow and the bar, as on the left here. You may also see just an arrow (and no bar), in boxed features such as the Key Words where part of the definition is Supplement.

WORKED EXAMPLE

Wherever you need to know how to use a formula to carry out a calculation, there are worked examples boxes to show you how to do this.

SELF/PEER ASSESSMENT

At the end of some activities and experimental skills boxes, you will find opportunities to help you assess your own work, or that of your classmates, and consider how you can improve the way you learn.

REFLECTION

These activities ask you to think about the approach that you take to your work, and how you might improve this in the future.

These boxes tell you where information in the book is extension content, and is not part of the syllabus.

SUMMARY

There is a summary of key points at the end of each chapter.

PROJECT

Projects allow you to apply your learning from the whole chapter to group activities such as making posters or presentations, or taking part in debates. They may give you the opportunity to extend your learning beyond the syllabus if you want to.

EXAM-STYLE QUESTIONS

Questions at the end of each chapter provide more demanding exam-style questions, some of which may require use of knowledge from previous chapters. The answers to these questions are accessible to teachers for free on the Cambridge GO site.

SELF-EVALUATION CHECKLIST

The summary checklists are followed by 'I can' statements which relate to the Learning intentions at the beginning of the chapter. You might find it helpful to rate how confident you are for each of these statements when you are revising. You should revisit any topics that you rated 'Needs more work' or 'Almost there'.

I can	See Topic...	Needs more work	Almost there	Confident to move on
Core				
Supplement				

Introduction

Chemistry is a laboratory science: its subject material and theories are based on experimental observation. However, its scope reaches out beyond the laboratory into every aspect of our lives – to our understanding of the nature of our planet, the environment we live in, the resources available to us and the factors that affect our health.

This is the fifth edition of our Cambridge IGCSE™ Chemistry Coursebook, and it provides everything that you need to support your course for Cambridge IGCSE Chemistry (0620/0971). It provides full coverage of the syllabus for examinations from 2023 onwards.

The chapters are arranged in the same sequence as the topics in the syllabus.

The various features that you will find in these chapters are explained in the How to use this book section.

Many of the questions you will meet during your course test whether you have a deep understanding of the facts and concepts you have learnt. It is therefore not enough just to learn words and diagrams that you can repeat in answer to questions; you need to ensure that you really understand each concept fully. Trying to answer the questions that you will find within each chapter, and at the end of each chapter, should help you to do this.

Although you will study your chemistry as a series of different topics, it is important to appreciate that all of these topics link up with each other. You need to make links between different areas of the syllabus to answer some questions.

As you work through your course, make sure that you keep reflecting on the work that you did earlier and how it relates to the current topic that you are studying. The reflection boxes throughout the chapters ask you to think about how you learn, which may help you to make the very best use of your time and abilities as your course progresses. You can also use the self-evaluation checklists at the end of each chapter to decide how well you have understood each topic in the syllabus, and whether or not you need to do more work on each one.

Practical skills are an important part of your chemistry course. You will develop these skills as you do experiments and other practical work related to the topics you are studying.

Note to teachers:
Guidance on safety has been included for each of the practical investigations in this coursebook. You should make sure that they do not contravene any school, education authority or government regulations. You and your school are responsible for safety matters.

Chapter 1
States of matter

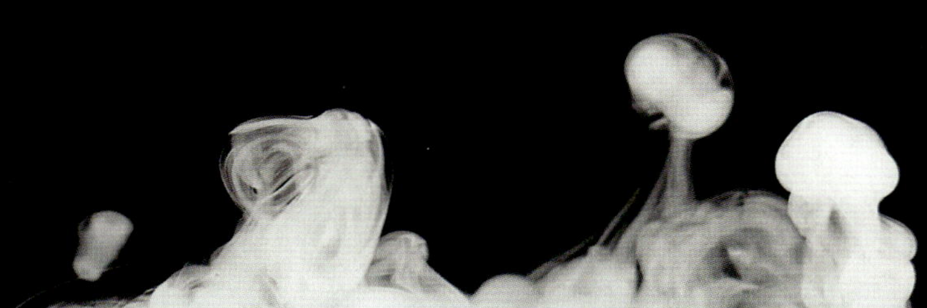

IN THIS CHAPTER YOU WILL:

- learn that matter can exist in three different states: solid, liquid or gas
- understand that substances can change state depending on the physical conditions
- understand that matter is made up of very small particles such as atoms or molecules
- see how changes in temperature produce changes of state by affecting the motion of the particles involved
- learn how to describe the structure of the physical states in terms of the arrangement and movement of particles
- describe how changes in temperature and pressure affect the volume of a gas
- explain diffusion in terms of the movement of particles

- think about how the movement of particles (kinetic particle theory) helps explain how changes of state happen
- understand the effects of changes in temperature and pressure on the volume of a gas
- learn how the molecular mass of particles in a gas affects the rate of diffusion.

GETTING STARTED

You will know about solids, liquids and gases from general life experience and your science courses. However, the ideas concerning the ways in which one state of matter changes into another are more complex.

Spray a small amount of air freshener at the front of a room. How does the smell spread around the room? Shake the can. Can you hear the liquid in it? Or try placing some liquid perfume in a dish in front of you and see how long it takes for someone to notice the scent some distance away. Discuss these observations in terms of changes of state and the movement of the molecules involved.

Look at the flowchart (Figure 1.1). Working in groups, can you add more detail to improve and extend the flowchart?

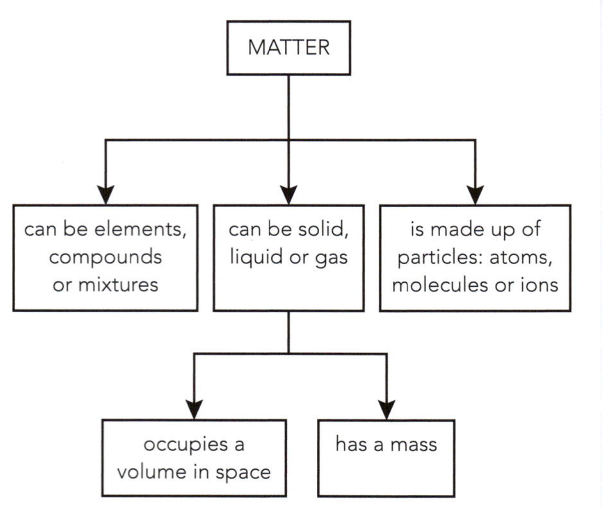

Figure 1.1: Flowchart of the nature of matter.

LORD OF THE RINGS

Saturn is perhaps the most beautiful of the planets of our solar system. Saturn has fascinated astronomers, even the famous Galileo, because of its mysterious rings. Galileo himself was confused by the rings, thinking at first that the rings were planets. The *Pioneer*, *Voyager* and *Cassini-Huygens* space probes have sent back a lot of information to Earth about the structure of the moons and rings of Saturn. Figure 1.2 is an image captured by the Hubble Space Telescope. The photograph shows the rings at close to their maximum tilt (angle) to the Earth, which shows their magnificence.

Each ring around Saturn is made up of a stream of icy particles, following each other nose-to-tail, like cars in a traffic jam, around the planet. The particles can be of widely varying sizes. The rings resemble a snowstorm, in which tiny snowflakes are mixed with snowballs that can be as large as a house. The ice that surrounds one of the most spectacular planets of our solar system is made of water, which is the same substance (with the same chemical formula, H_2O) that covers so much of the Earth's surface.

Figure 1.2 also shows the pastel colours of the clouds of ammonia and methane in Saturn's atmosphere. However, the bulk of the planet is made of hydrogen and helium gases. Deep in the centre of these lightweight gases is a small rocky core, surrounded by a liquid layer of the gases. The hydrogen is liquid because of the high pressure in the inner regions of the planet nearest the core. The liquid hydrogen shows metallic properties, producing the planet's magnetic field. A study of Saturn's physical structure emphasises how substances that we know on Earth can exist in unusual physical states in different environments.

Discussion questions

1 Why are the planets Jupiter and Saturn called 'gas giants'? What progression do we see in the physical nature of the planets as we move away from the Sun?

2 Why does hydrogen only exist as a liquid under such extreme conditions of temperature and pressure?

Figure 1.2: Saturn and its rings. A photograph taken by the Hubble Space Telescope.

1.1 States of matter

There are many different kinds of **matter**. The word is used to cover all the substances and materials of which the universe is composed. Samples of all of these materials have two properties in common: they each occupy space (they have volume) and they have mass.

Chemistry is the study of how matter behaves, and of how one kind of substance can be changed into another. Whichever chemical substance we study, we find that the substance can exist in three different forms (or physical states) depending on the conditions. These three different **states of matter** are known as *solid*, *liquid* and *gas*. Changing the temperature and/or pressure can change the state in which a substance exists (Figure 1.3).

Each of the different physical states have certain general characteristics that are true whatever chemical substance is being considered. These are summarised in Table 1.1.

Table 1.1 highlights a major difference between solids and the other two physical states. Liquids and gases are able to flow, but a solid has a fixed shape and volume. Liquids and gases are **fluids**. This means that liquids and gases can be poured, or pumped, from one container to another. The three physical states also show differences in the way they respond to changes in temperature and pressure. All three show an increase in volume (an expansion) when the temperature is increased and a decrease in volume (a contraction) when the temperature is lowered. The effect is much bigger for a gas than for a solid or a liquid.

The volume of a gas at a fixed temperature can easily be reduced by increasing the pressure on the gas. Gases are easily compressed ('squashed'). Liquids are only slightly compressible, and the volume of a solid is unaffected by changing the pressure.

KEY WORDS

matter: anything that occupies space and has mass

states of matter: solid, liquid and gas are the three states of matter in which any substance can exist, depending on the conditions of temperature and pressure

fluid: a gas or a liquid; they are able to flow

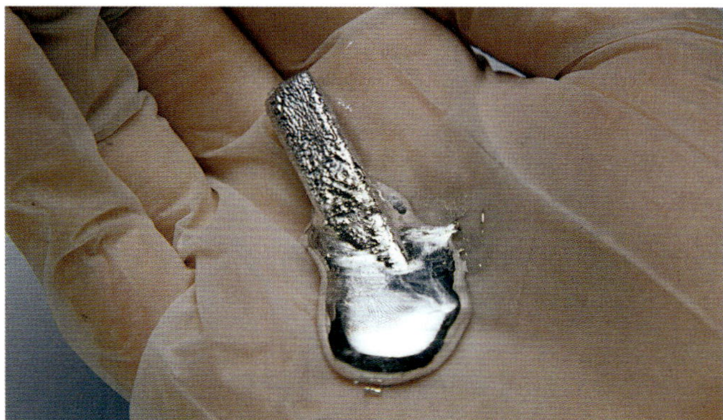

Figure 1.3: Gallium metal melts with the warmth of the hand.

Physical state	Volume	Density	Shape	Fluidity
solid	has a fixed volume	high	has a definite shape	does not flow
liquid	has a fixed volume	moderate to high	no definite shape – takes the shape of the container	generally flows easily
gas	no fixed volume – expands to fill the container	low	no definite shape – takes the shape of the container	flows easily

Table 1.1: Differences in the properties of the three states of matter.

Changes in state

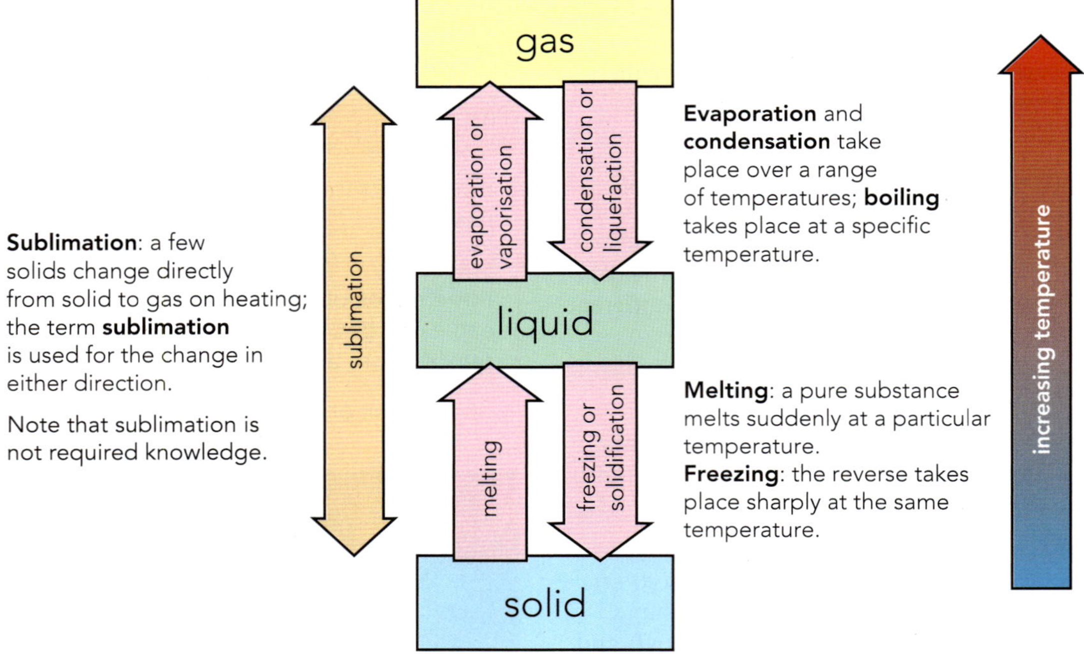

Figure 1.4: Changes of physical state and the effect of increasing temperature at atmospheric pressure.

Large increases, or decreases, in temperature and pressure can cause changes that are more dramatic than expansion or contraction. They can cause a substance to change its physical state. The changes between the three states of matter are shown in Figure 1.4. At atmospheric pressure, these changes can occur by raising or lowering the temperature of the substance.

KEY WORDS

melting point (m.p): the temperature at which a solid turns into a liquid – it has the same value as the freezing point; a pure substance has a sharp melting point

Melting and freezing

The temperature at which a substance turns to a liquid is called the **melting point (m.p.)**. This always happens at one particular temperature for each substance (Figure 1.5). The process is reversed at precisely the same temperature if a liquid is cooled down. It is then called the *freezing point* (f.p.). The melting point and freezing point of any given substance are both the same temperature. For example, the melting and freezing of pure water take place at 0 °C.

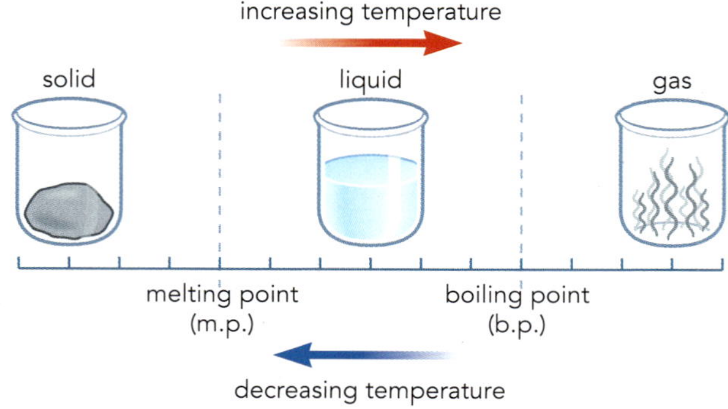

Figure 1.5: Relationship between the melting and boiling points of a substance.

Sublimation

A few solids, such as solid carbon dioxide, do not melt when they are heated at normal pressures. Instead, they turn directly into gas. Solid carbon dioxide is often called 'dry ice' because the surface of the block is dry (Figure 1.6). This is different to a normal ice cube, which has a thin film of liquid water on the surface.

This change of state is called sublimation: the solid *sublimes* (see Figure 1.4). Sublimation is a direct change of state from solid to gas, or gas to solid; the liquid phase is bypassed. As with melting, this also happens at one particular temperature for each pure solid.

Figure 1.6: Solid carbon dioxide sublimes. The white smoke is composed of water droplets condensed from the air; there is no liquid film on the solid pieces.

Evaporation, boiling and condensation

If a liquid is left with its surface exposed to the air, it evaporates. When liquids change into gases in this way, the process is called **evaporation**. Evaporation takes place from the surface of the liquid. The larger the surface area, the faster the liquid evaporates. The warmer the liquid is, the faster it evaporates. The hot climate around the Dead Sea means that water evaporates easily and the sea has a high salt concentration (Figure 1.7).

KEY WORD

evaporation: a process occurring at the surface of a liquid, involving the change of state from a liquid into a vapour at a temperature below the boiling point

Figure 1.7: An aerial view showing large surface salt formations in the southern part of the Dead Sea.

Eventually, at a certain temperature, a liquid becomes hot enough for gas to form within the liquid and not just at the surface. Bubbles of gas appear inside the liquid (Figure 1.8a). This process is known as **boiling**. It takes place at a specific temperature, known as the **boiling point** for each pure liquid (Figure 1.5).

Water evaporates fairly easily and has a relatively low boiling point (100 °C). Water is quite a **volatile** liquid. Ethanol, with a boiling point of 78 °C, is more volatile than water. It has a higher **volatility** than water and evaporates more easily.

The reverse of evaporation is **condensation**. This is usually brought about by cooling. However, we saw earlier that the gas state is the one most affected by changes in pressure. It is possible, at normal temperatures, to condense a gas into a liquid by increasing the pressure, without cooling.

We can see these different processes in action if we look closely at a kettle as water boils (Figure 1.8b). Colourless, invisible water vapour escapes from the kettle. Water vapour is present in the clear region we can see at the mouth of the kettle. The visible cloud of steam is made up of droplets of liquid water formed by condensation as the vapour cools in the air.

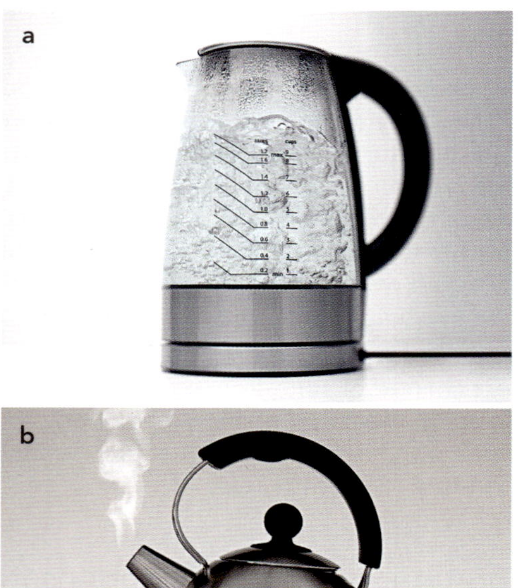

Figure 1.8 a: Water boiling in a glass kettle; bubbles are formed throughout the liquid. **b:** A boiling kettle produces colourless, invisible water vapour that then condenses to produce a cloud of steam.

For a beaker of boiling water, the bubbles form when there are enough high-energy water molecules to give a pocket of gas with a pressure equal to atmospheric pressure. The boiling point of a liquid can change if the surrounding pressure changes. The value given for the boiling point is usually stated at the pressure of the atmosphere at sea level (*atmospheric pressure* or *standard pressure*). If the surrounding pressure falls, the boiling point falls. The boiling point of water at standard pressure is 100 °C. On a high mountain, the boiling point is lower than 100 °C. If the surrounding pressure is increased, the boiling point rises.

KEY WORDS

boiling: the process of change from liquid to gas at the boiling point of the substance; a condition under which gas bubbles are able to form within a liquid – gas molecules escape from the body of a liquid, not just from its surface

boiling point: the temperature at which a liquid boils, when the pressure of the gas created above the liquid equals atmospheric pressure

volatile: term that describes a liquid that evaporates easily; it is a liquid with a low boiling point because there are only weak intermolecular forces between the molecules in the liquid

volatility: the property of how easily a liquid evaporates

condensation: the change of a vapour or a gas into a liquid; during this process heat is given out to the surroundings

Pure substances

A **pure substance** consists of only one substance without any contaminating impurities. A pure substance melts and boils at definite temperatures. Table 1.2 shows the precise melting points and boiling points of some common substances at atmospheric pressure.

Substance	Physical state at room temperature (25 °C)	Melting point / °C	Boiling point / °C
oxygen	gas	−219	−183
nitrogen	gas	−210	−196
ethanol (alcohol)	liquid	−117	78
water	liquid	0	100
sulfur	solid	115	444
common salt (sodium chloride)	solid	801	1465
copper	solid	1083	2600
carbon dioxide	gas	−78[a]	

[a] *Sublimes at atmospheric pressure*

Table 1.2: Melting and boiling points of some common chemical substances.

KEY WORDS

pure substance: a single chemical element or compound – it melts and boils at definite precise temperatures

The values for the melting point and boiling point of a pure substance are precise and predictable. This means that we can use them to test the purity of a sample. These values can also be used to check the identity of an unknown substance. The melting point of a solid can be measured using an electrically heated melting-point apparatus or by the apparatus described later in Figure 1.9.

A substance's melting and boiling points in relation to room temperature (standard taken as 25 °C) determine whether it is usually seen as a solid, a liquid or a gas. For example, if the melting point is below 25 °C and the boiling point is above 25 °C, the substance will be a liquid at room temperature.

Effect of impurities

Seawater is impure water. This fact can be easily demonstrated if you put some seawater in a dish and heat it until all of the water evaporates. A solid residue of salt is left behind in the dish (you can see this effect in Figure 1.7, which shows solid salt formations on the surface of the Dead Sea).

Impurities often affect the value of the melting or boiling point of a substance. An impure substance sometimes melts or boils over a *range* of temperatures, not at the precise point of the pure substance.

Seawater freezes at a temperature below the freezing point of pure water (0 °C) and boils at a temperature above the boiling point of pure water (100 °C). Other substances that contain impurities show differences in their freezing and boiling points when compared with the known values for the pure substance.

Questions

1 State the names for the following physical changes:
   - **a** liquid to solid
   - **b** liquid to gas at a precise temperature
   - **c** gas to liquid.

2 The melting and boiling points of three pure substances are given in Table 1.3.

Substance	Melting point / °C	Boiling point / °C
ethanol	−117	78
methane	−182	−164
mercury	−30	357

Table 1.3: Melting and boiling points of ethanol, methane and mercury.

   - **a** All three substances have negative values for their melting point. Which of them has the lowest melting point?
   - **b** Which two substances are liquids at room temperature? Explain your answer.
   - **c** What effect does the presence of an impurity have on the freezing point of a liquid?

3 **a** What do you understand by the word *volatile* when used in chemistry?

b Put these three liquids in order of volatility, with the most volatile first: water (b.p. 100 °C), ethanoic acid (b.p. 128 °C) and ethanol (b.p. 78 °C).

c Table 1.4 shows the melting and boiling points of four substances A–D. In which of these four substances are the particles arranged in a **lattice** (a regular structure) at room temperature?

Substance	Melting point / °C	Boiling point / °C
A	−115	79
B	80	218
C	−91	−88
D	−23	77

Table 1.4: Melting and boiling points of four unknown substances.

4 Iodine is often seen as an example of a substance that changes directly from a solid to a gas. However, data books that give the standard physical measurements for substances show values for the melting point (114 °C) and boiling point (184 °C) of iodine at atmospheric pressure.

a Explain why iodine seems to miss out the liquid stage if crystals are heated strongly in a boiling tube.

b Suggest how you could demonstrate that iodine can melt to form a liquid at atmospheric pressure.

KEY WORD

lattice: a regular three-dimensional arrangement of atoms, molecules or ions in a crystalline solid

Heating and cooling curves

The melting point of a solid can also be measured using the apparatus shown in Figure 1.9. A powdered solid is put in a narrow melting-point tube so that it can be heated easily. An oil bath can be used so that melting points above 100 °C can be measured. We can follow the temperature of the sample before and after melting.

On heating, the temperature rises until the solid starts to melt. However, close observation shows that the temperature stays constant until all the solid has melted. The temperature then rises as the liquid warms further.

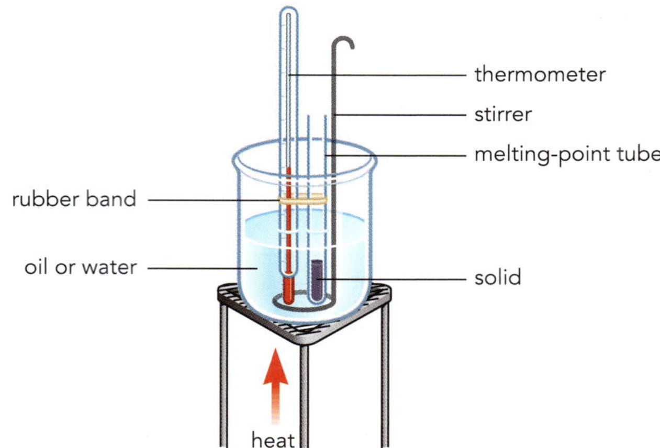

Figure 1.9: Apparatus for measuring the melting point of a solid. A water-bath can be used for melting points below 100 °C and an oil bath for melting points above 100 °C.

It is possible to continue to heat the liquid in the same apparatus until its boiling point is reached. Again, the temperature stays the same until all the liquid has completely evaporated.

We can perform this experiment in reverse. Similar apparatus can be used to produce a cooling curve, but the thermometer must be placed in a test-tube containing the solid being studied. The solid is then melted completely and the liquid heated. Heating is then stopped. The temperature is noted every minute as the substance cools. This produces a cooling curve (Figure 1.10). The level (horizontal) part of the curve occurs where the liquid freezes, forming the solid.

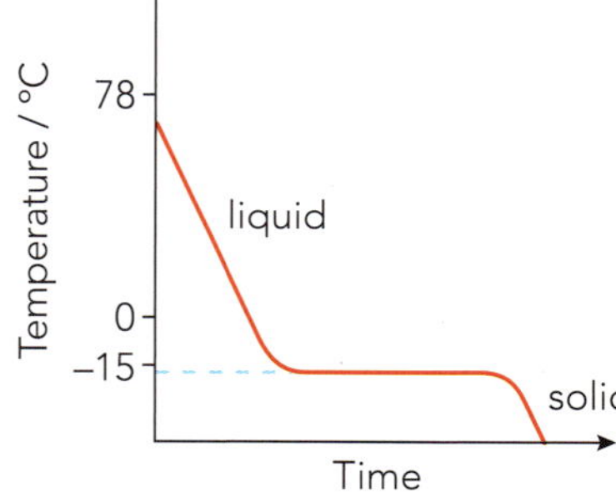

Figure 1.10: A cooling curve. The temperature stays constant while the liquid solidifies. A cooling mixture of ice and salt could be used to lower the temperature below 0 °C.

These experiments show that heat energy is needed to change a solid into a liquid, or a liquid into a gas. During the reverse processes, heat energy is given out.

EXPERIMENTAL SKILLS 1.1

Plotting a cooling curve

In this experiment, you are going to plot cooling curves for two substances, A and B. This experiment investigates the energy changes taking place as a liquid cools down below its freezing point.

You will need:

- two beakers (250 cm^3)
- Bunsen burner
- tripod
- gauze
- heat-resistant mat
- stopwatch, stopclock or other timer
- two boiling tubes labelled A and B
- two stirring thermometers (−10 to 110 °C).

Substance A is paraffin wax (choose a low m.p. type, m.p. around 55 °C). Substance B is either octadecanoic acid (stearic acid) m.p. 70 °C or phenyl salicylate (salol) m.p. 43 °C.

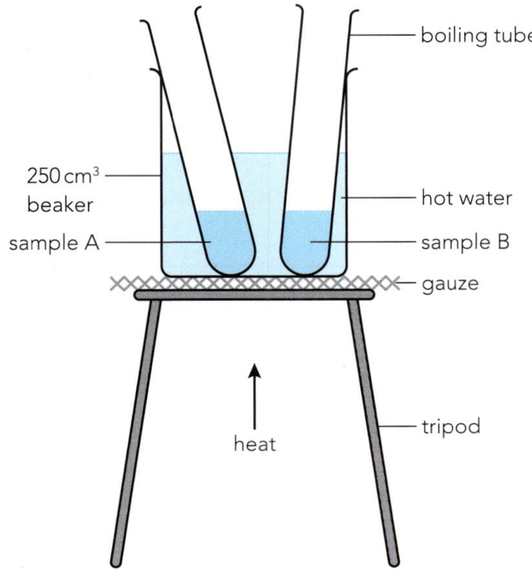

Figure 1.11: Samples A and B are melted in a water-bath.

Safety

It is important that you follow the safety rules set out by your teacher for all practicals. In particular, pay attention to the type of Bunsen burner flame needed as well as the concentrations and volumes of chemicals used. Wear eye protection throughout.

Getting started

Before starting the experiment, make sure you are familiar with the scale on the thermometer you are using. You will need to be able to read it confidently. You can check on your reading of the thermometer as you are heating up the water-bath.

Make sure you and your partner are clear as to the tasks you each have.

Method

1. Fill a 250 cm^3 beaker three-quarters full of water and heat using a Bunsen burner to make a water-bath. Place a thermometer in the water. Heat the water until it is at 90 °C.
2. Put boiling tubes containing a sample of each solid A and B into the water-bath (Figure 1.11).
3. When the solid has melted, place a thermometer in each tube. There should be enough liquid to cover the bulb at the base of the thermometer.
4. Remove the tubes from the water-bath and stand them in an empty beaker for support.
5. Look at the thermometer and record the temperature in each tube. Then start the timer.
6. Look at the thermometer and record the temperature in each tube every minute until the temperature reaches 40 °C.
7. Plot a graph for each set of readings with time on the x-axis and temperature on the y-axis.

Questions

1. Which of the two substances is a pure substance? Explain your answer.
2. Explain any ways in which your method could be improved to give more reliable results.

CONTINUED

Self-assessment

Complete the self-assessment checklist below to assess your graph drawing skills.

For each point, award yourself:

2 marks if you did it really well

1 mark if you made a good attempt at it and partly succeeded

0 marks if you did not try to do it, or did not succeed

Checkpoint	Marks awarded
Have you drawn the axes with a ruler, using most of the width and height of the grid?	
Have you used a good scale for the *x*-axis and the *y*-axis, going up in 0.25 s, 0.5 s, 1 s or 2 s? (Note that the axes do not necessarily need to start at the origin (0,0).)	
Have you labelled the axes correctly, giving the correct units for the scales on both axes?	
Have you plotted each point precisely and correctly?	
Have you used a small, neat cross or encircled dot for each point?	
Have you drawn a single, clear best-fit line through each set of points?	
Have you ignored any anomalous results when drawing the line through each set of points?	
Total (out of 14):	

Your total score will reflect how clear and well-presented your graph is. Drawing graphs is an important skill in chemistry as you need be able to deduce reliable information from your graph.

Take a look at where you gave yourself 2 marks and where you gave yourself less than 2 marks. What did you do well, and what aspects will you focus on next time? Having thought about your assessment, talk it through with your teacher to gain further advice on areas that would help you improve your presentation of graphical data.

Questions

5 Sketch a cooling curve for water from 80 °C to −20 °C, noting what is taking place in the different regions of the graph.

6 Energy is needed to overcome the forces of attraction holding the particles in position in a solid. Energy is absorbed during melting. Figure 1.12 shows how energy is involved in the different changes of state. Complete Figure 1.12 by providing labels for the positions A, B and C.

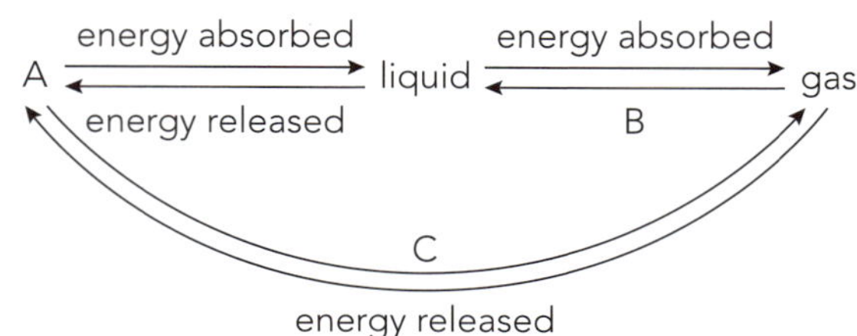

Figure 1.12: Energy changes involved in different changes of state.

7 As an alternative to following the cooling of a substance, it is possible to draw a heating curve. Figure 1.13 shows the heating curve for substance X.

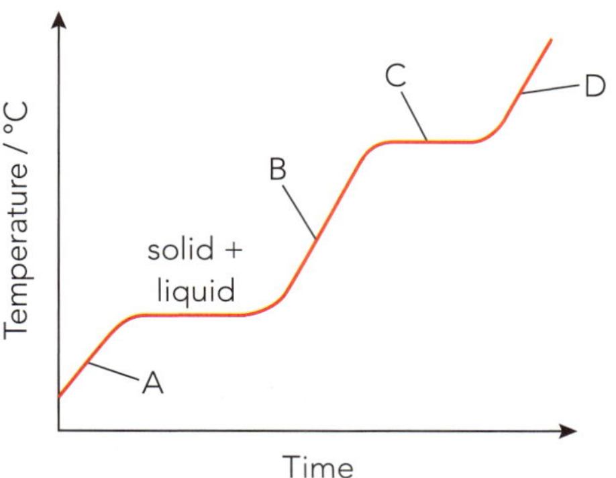

Figure 1.13: Heating curve for substance X.

What physical state, or states, will X be in at points A, B, C and D on the curve?

1.2 Kinetic particle theory of matter

Existence of atoms and molecules

Elements and compounds mix and react to produce the world around us. They produce massive objects such as the 'gas giants' (the planets Jupiter and Saturn) that we met at the start of this chapter. They also give rise to the tiny highly structured crystals of solid sugar or salt. How do the elements organise themselves to give this variety? How can an element exist in the three different states of matter simply through a change in temperature?

Our modern understanding is based on the idea that all matter is divided into very small particles known as **atoms**. The key ideas in our understanding are that:

- each element is composed of its own type of atom
- atoms of different elements can combine to make the molecules of a compound.

This idea that all substances consist of very small particles begins to explain the structure of the three different states of matter. The different levels of freedom of movement of the particles explains some of the different features of the three states. Figure 1.14 illustrates the basic features of the three states we discussed earlier (see Table 1.1).

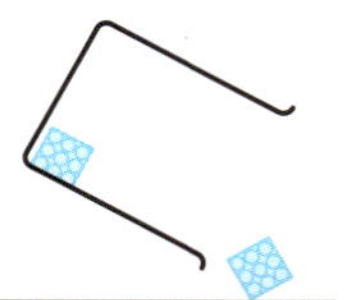
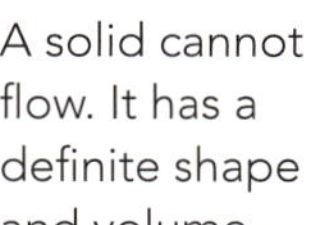

A solid cannot flow. It has a definite shape and volume.

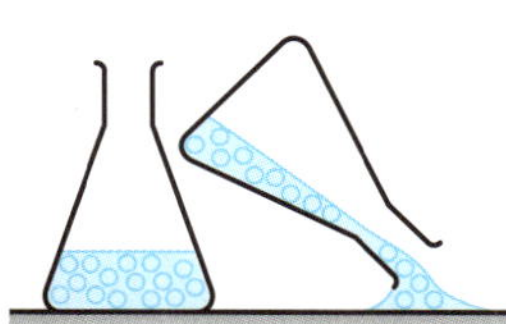

A liquid can flow; it is a fluid. It has a definite volume but takes the shape of its container.

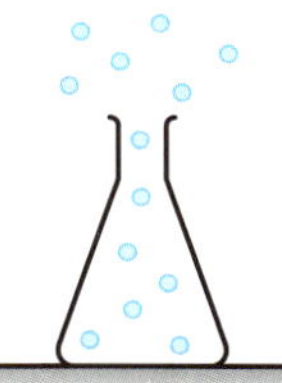

A gas is a fluid and spreads throughout its container. It has no definite volume.

Figure 1.14: The basic differences between the physical properties of the three states of matter.

Main points of the kinetic particle theory

The **kinetic particle theory** of matter describes the three different states, and the changes between them, in terms of the movement of particles. The major points of the theory are:

- All matter is made up of very small particles (different substances contain different types of particles, such as atoms, **molecules** or ions).
- Particles are moving all the time (the higher the temperature, the higher the average energy of the particles).
- The freedom of movement and the arrangement of the particles is different for the three states of matter.
- The pressure of a gas is produced by the atoms or molecules of the gas hitting the walls of the container. The more often the particles collide with the walls, the greater the pressure.

KEY WORDS

atom: the smallest particle of an element that can take part in a chemical reaction

kinetic particle theory: a theory which accounts for the bulk properties of the different states of matter in terms of the movement of particles (atoms or molecules) – the theory explains what happens during changes in physical state

molecule: a group of atoms held together by covalent bonds

Figure 1.15 is a more detailed summary of the organisation of the particles in the three states of matter and explains the changes involved in the different changes in state.

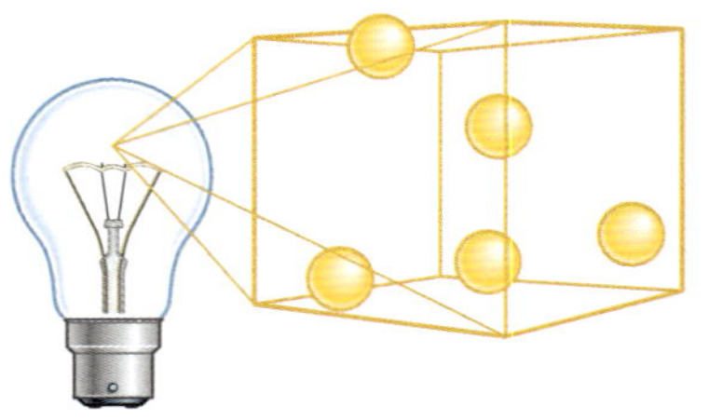

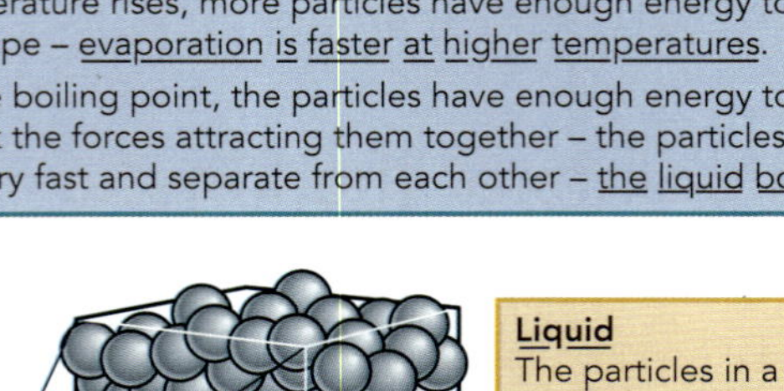

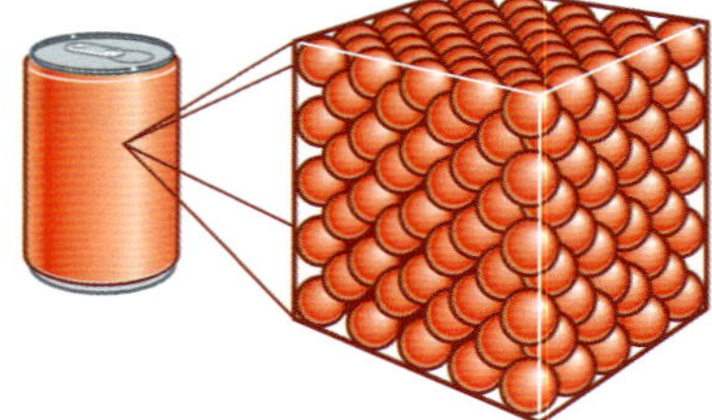

Figure 1.15: Applying the kinetic particle theory to changes in physical state.

The highly structured, ordered microscopic arrangements in solids can produce the regular crystal structures seen in this state. In a solid, the particles are packed close together. The particles cannot move freely. They simply vibrate about fixed positions in their regular arrangement (lattice).

In a liquid, the particles are still close together. However, they can now move about past each other. The separation between particles is much greater in a gas. In a gas, the particles are very far apart and move randomly.

The ability of the particles to move in the liquid and gas phases produces their fluid properties. The particles are very widely separated in a gas, but close together in a liquid or solid. The space between the particles is called the **intermolecular space**. In a gas, the intermolecular space is large and can be reduced by increasing the external pressure. Therefore, gases are easily compressible. In liquids, this space is very much smaller. As a result, liquids are not very compressible.

Changing the external pressure on a sample of a gas produces a change in volume that can easily be seen.

- An increase in external pressure produces a contraction in volume. The gas is compressed.
- A decrease in external pressure produces an increase in volume. The gas expands.

The volume of a gas is also altered by changes in temperature.

- An increase in the temperature of a gas produces an increase in volume. The gas expands.
- A decrease in temperature produces a contraction of the volume of a gas.

KEY WORDS

intermolecular space: the space between atoms or molecules in a liquid or gas. The intermolecular space is small in a liquid, but relatively very large in a gas.

The movement of particles in a liquid also helps to explain evaporation from the surface of a liquid. Some of the particles are moving faster than other particles. At the surface, these faster moving particles may have enough energy to escape into the gaseous state (Figure 1.16).

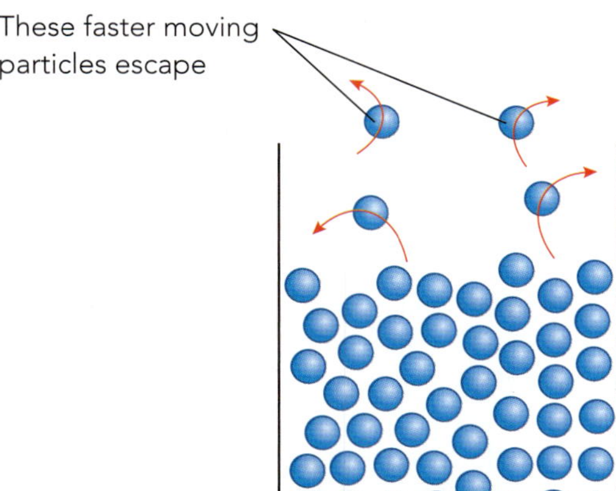

Figure 1.16: Faster moving particles leaving the surface of a liquid, causing evaporation.

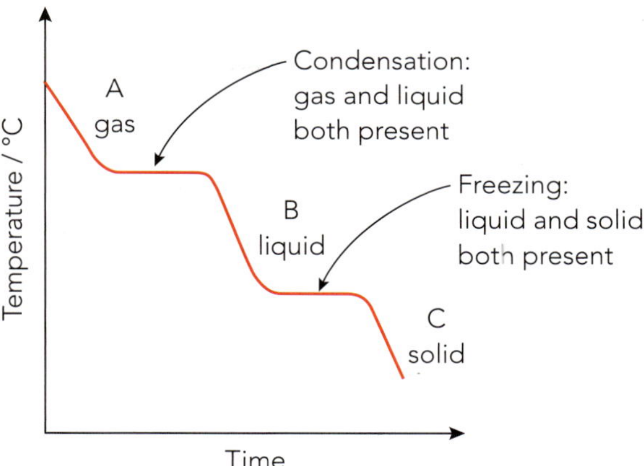

Figure 1.17: The energy changes taking place during the cooling of a gas to a solid.

The fact that the space between the rapidly moving particles in a gas is much greater than in the other two states of matter explains why the volume of a gas is much more easily changed by conditions of temperature and pressure. If the temperature is raised then the gas particles move faster and there is less chance of interaction between them. The gas particles move faster and more freely and occupy a greater volume. The opposite is true if the temperature is lowered. The particles are moving more slowly. They are more likely to interact with each other and move together to occupy a smaller volume.

Changes in pressure also affect the volume of a gas sample. An increase in pressure pushes the particles closer together meaning that the moving particles are more likely to interact with each other and move closer together. The opposite is true when the external pressure is lowered. The particles occupy a greater space and interactions between the particles are less likely.

The interpretation of a cooling curve

The way the particles in the three states are arranged and interact with each other also helps to explain the energy changes involved when a substance is heated or cooled. Figure 1.17 summarises the energy changes that take place at the different stages of a cooling-curve experiment.

The cooling of the gas gives rise to a sequence of changes during which the particles move less rapidly and interact more strongly with each other. The substance passes through the liquid state, eventually becoming a solid. Over the course of the experiment the temperature falls. However, the graph shows two periods during which the temperature remains constant. These regions are the time when first condensation, and then freezing takes place.

In region A (Figure 1.17), the temperature is falling. The energy of the particles decreases. The particles move more slowly and interact with each other more strongly. The particles begin to come together to form the liquid. As the **intermolecular forces** increase between the particles, energy is given out. This results in the temperature staying constant until the gas is completely condensed to liquid.

Once the liquid is formed the temperature starts to fall again (region B). The liquid cools. The particles in the liquid slow down and eventually the solid begins to form. The forces holding the solid together form and energy is given out. While the solid is forming this release of energy keeps the temperature constant. The temperature stays the same until freezing is complete.

After the solid has formed the temperature falls again (region C). The particles in the solid vibrate less strongly as the temperature falls.

KEY WORDS

intermolecular forces: the weak attractive forces that act between molecules

The key points about the processes taking place during condensation and freezing are:

- as the particles come closer together, new forces of interaction take place
- this means that energy is given out during these changes
- therefore, the temperature remains unchanged until the liquid or solid is totally formed.

As energy is given out during these changes, condensation and freezing are **exothermic changes** (Chapter 7).

Carrying out the experiment in the opposite direction, starting from the solid, would give a heating curve. In this case, the temperature stays constant during melting and boiling. At these stages, energy has to be put in to overcome the forces between the particles. The energy put in breaks these interactions and the particles are able to move more freely and faster. As energy has to be put in during these changes, melting, evaporation and boiling are **endothermic changes** (Chapter 7).

What are the forces that hold a solid or liquid together? They must be attractive forces between the particles. They are the forces that act when a substance condenses or freezes. Their formation releases energy. However, their nature depends on the substance involved. For substances such as water or ethanol they act between the molecules present, and so are intermolecular forces.

KEY WORDS

exothermic changes: a process or chemical reaction in which heat energy is produced and released to the surroundings.
ΔH for an exothermic change has a negative value.

endothermic changes: a process or chemical reaction that takes in heat from the surroundings.
ΔH for an endothermic change has a positive value.

1.3 Mixtures of substances and diffusion

The chemical world is very complex, owing to the vast range of pure substances available and to the variety of ways in which these pure substances can mix with each other. Each **mixture** must be made from at least two parts, which may be either solid, liquid or gas. There are a number of different ways in which the three states can be combined. In some, the states are completely mixed to become one single state or phase. Technically, the term **solution** is used for this type of mixture composed of two or more substances.

Solid salt dissolves in liquid water to produce a liquid mixture. This is called a salt solution (Figure 1.18). The solid has completely disappeared into the liquid. In general terms, the solid that dissolves in the liquid is called the **solute**. The liquid in which the solid dissolves is called the **solvent**. In other types of mixture, the states remain separate. One phase is broken up into small particles, droplets or bubbles, within the main phase. The most obvious example of this type of mixture is a **suspension** of fine particles of a solid in a liquid, such as we often get after a **precipitation reaction** (Chapters 12 and 22).

KEY WORDS

mixture: two or more substances mixed together but not chemically combined – the substances can be separated by physical means

solution: is formed when a substance (solute) dissolves into another substance (solvent)

solute: the solid substance that has dissolved in a liquid (the solvent) to form a solution

solvent: the liquid that dissolves the solid solute to form a solution; water is the most common solvent but liquids in organic chemistry that can act as solvents are called *organic solvents*

suspension: a mixture containing small particles of an insoluble solid, or droplets of an insoluble liquid, spread (suspended) throughout a liquid

precipitation reaction: a reaction in which an insoluble salt is prepared from solutions of two soluble salts

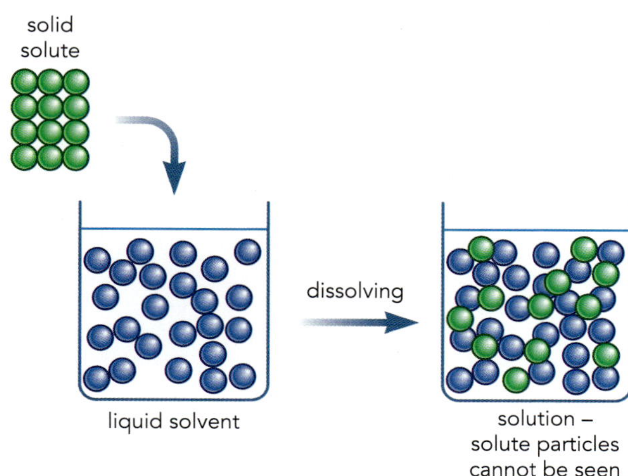

Figure 1.18: When solute dissolves in a solvent, the solute particles are completely dispersed in the liquid.

Solutions

We most often think of a solution as being made of a solid dissolved in a liquid. Two-thirds of the Earth's surface is covered by a solution of various salts in water. The salts are totally dispersed in the water and cannot be seen. However, other substances that are not normally solid are dissolved in seawater. For example, the dissolved gases, oxygen and carbon dioxide, are important for life to exist in the oceans.

A closer look at solutions

Water is the commonest solvent in use, but other liquids are also important. Most of these other solvents are organic liquids, such as ethanol, propanone and trichloroethane. These organic solvents are important because they will often dissolve substances that do not dissolve in water. If a substance dissolves in a solvent, it is said to be **soluble**; if it does not dissolve, it is **insoluble**.

Less obvious, but quite common, are solutions of one liquid in another. Alcohol mixes (dissolves) completely with water. Alcohol and water are completely **miscible**: this means that they make a solution.

Alloys are similar mixtures of metals, though we do not usually call them solutions. They are made by mixing the liquid metals together (dissolving one metal in the other) before solidifying the alloy.

Solubility of solids in liquids

If we try to dissolve a substance such as copper(II) sulfate in a fixed volume of water, the solution becomes more concentrated as we add more solid. A *concentrated* solution contains a high proportion of solute. A *dilute* solution contains a small proportion of solute. If we keep adding more solid, a point is reached when no more will dissolve at that temperature. This is a **saturated solution**. To get more solid to dissolve, the temperature must be increased. The **concentration** of solute in a saturated solution is the **solubility** of the solute at that temperature.

The solubility of most solids increases with temperature. The process of crystallisation depends on these observations. When a saturated solution is cooled, the solution can hold less solute at the lower temperature and some solute crystallises out.

KEY WORDS

soluble: a solute that dissolves in a particular solvent

insoluble: a substance that does not dissolve in a particular solvent

miscible: if two liquids form a completely uniform mixture when added together, they are said to be miscible

alloys: mixtures of elements (usually metals) designed to have the properties useful for a particular purpose, e.g. solder (an alloy of tin and lead) has a low melting point

saturated solution: a solution that contains as much dissolved solute as possible at a particular temperature

concentration: a measure of how much solute is dissolved in a solvent to make a solution. Solutions can be dilute (with a high proportion of solvent), or concentrated (with a high proportion of solute)

solubility: a measure of how much of a solute dissolves in a solvent at a particular temperature

Solubility of gases in liquids

Unlike most solids, gases become less soluble in water as the temperature rises. The solubility of gases from the air in water is quite small, but the amount of dissolved oxygen is enough to support fish and other aquatic life.

The solubility of gases increases with pressure. Sparkling drinks contain carbon dioxide dissolved under pressure. They 'fizz' when the pressure is released by opening the container. They go 'flat' if the container is left to stand open, and more quickly if left to stand in a warm place.

Diffusion in fluids

Some of the earliest evidence for the kinetic model of the states of matter came from observations on **diffusion**, where particles spread to fill the space available to them.

The main ideas involved in diffusion are:

- particles move from a region of higher concentration towards a region of lower concentration; eventually, the particles are evenly spread. Their concentration is the same throughout.
- the rate of diffusion in liquids is much slower than in gases.
- diffusion does not take place in solids as the particles cannot move from place to place.

KEY WORD

diffusion: the process by which different fluids mix as a result of the random motions of their particles

Dissolving

A potassium manganate(VII) crystal is placed at the bottom of a dish of water. The dish is then left to stand. At first the water around the crystal becomes purple as the solid dissolves (Figure 1.19). Particles move off from the surface of the crystal into the water. Eventually, the crystal dissolves completely and the purple colour spreads through the liquid. The whole solution becomes purple. The particles from the solid become evenly spread through the water.

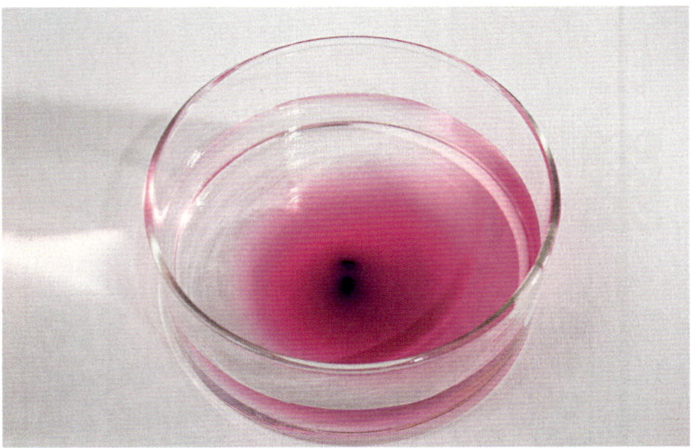

Figure 1.19: The diffusion of potassium manganate(VII) in water.

Whether a solid begins to break up like this in a liquid depends on the particular solid and liquid involved. But the spreading of the solute particles throughout the liquid is an example of diffusion. Diffusion in solution is also important when the solute is a gas. This is especially important in breathing. Diffusion contributes to the movement of oxygen from the lungs to the blood, and of carbon dioxide from the blood to the lungs.

Diffusion of gases

A few drops of liquid bromine are put into a gas jar and the lid is replaced. The liquid bromine evaporates easily. Liquid bromine is highly volatile. After a short time, the brown gas begins to spread throughout the jar. The jar becomes full of brown gas. Bromine vaporises easily and its gas will completely fill the container (Figure 1.20). Gases diffuse to fill all the space available to them. Diffusion is important for our 'sensing' of the world around us. It is the way that smells reach us.

The atoms or molecules in gases move at high speeds. We are being bombarded constantly by nitrogen and oxygen molecules in the air, which are travelling at about 1800 km/hour. However, these particles collide very frequently with other particles in the air (many millions of collisions per second), so their path is not direct. These frequent collisions slow down the overall rate of diffusion from one place to another. The pressure of a gas is the result of collisions of the fast-moving particles with the walls of the container.

Figure 1.20: Bromine vapour diffuses throughout the container to fill the space available.

Not all gases diffuse at the same rate. The speed at which a gas diffuses depends on the mass of the particles involved. At the same temperature, molecules that have a lower mass move, on average, faster than those with a higher mass. This is shown by the experiment in Figure 1.21. The ammonia and hydrochloric acid fumes react when they meet, producing a white smoke ring of ammonium chloride. This smoke ring is made of fine particles of solid ammonium chloride. The fact that the ring is not formed halfway along the tube shows that ammonia, the lighter molecule of the two, diffuses faster.

The important points derived from the kinetic particle theory relevant here are:

- heavier gas particles move more slowly than lighter particles at the same temperature
- larger molecules diffuse more slowly than smaller ones
- the rate of diffusion is inversely related to the mass of the particles
- the average speed of the particles increases with an increase in temperature.

Figure 1.21: Ammonia and hydrochloric acid fumes diffuse at different rates.

EXPERIMENTAL SKILLS 1.2

Investigating diffusion in liquids

This experiment helps to demonstrate the process of diffusion in a liquid. Diffusion is shown by the formation of an insoluble precipitate where the ions meet in a solution.

You will need:

- Petri dish
- tweezers
- white tile
- silver nitrate, one crystal
- potassium iodide, one crystal
- distilled or deionised water
- test-tubes
- silver nitrate solution
- potassium iodide solution
- dropping pipettes.

Safety

Wear eye protection throughout. Use tweezers to handle the crystals. Be careful with chemicals. Never ingest them and always wash your hands after handling them. Note that silver nitrate is corrosive, oxidising and can stain the skin. Silver nitrate is also hazardous to the aquatic environment. Waste silver nitrate solution must not be poured down the drain.

Getting started

Before starting, try the reaction between potassium iodide and silver nitrate solutions in a test-tube. Add 1 cm^3 of aqueous silver nitrate to a similar volume of potassium iodide solution. Note the formation of the precipitate, particularly its colour.

Method

1 Put a Petri dish on a white tile or piece of white paper. Fill the Petri dish nearly to the top with deionised water.

2 Using tweezers, put a crystal of silver nitrate at one side of the dish and a crystal of potassium iodide at the other side (Figure 1.22).

3 Look at the crystals. Notice that as crystals begin to dissolve in the water, a new compound is formed within the solution.

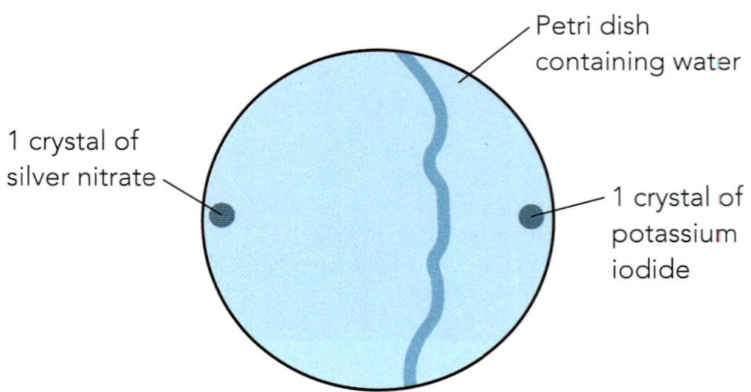

Figure 1.22: Experiment to investigate diffusion through water.

Questions

1 What is the precipitate formed in this reaction?

2 Write a word equation to show the reaction taking place.

3 What factors control where the solid is formed in the Petri dish?

4 Why does the solid not form exactly in the middle of the dish?

Questions

8 A small amount of liquid bromine is placed in a gas jar, which is then sealed with a lid. Evaporation of the liquid bromine takes place.

$Br_2(l) \rightarrow Br_2(g)$

Use the ideas of the kinetic theory to explain why, after about an hour, the gaseous bromine molecules have spread to evenly occupy the whole container.

9 A teacher carried out a class demonstration on diffusion similar to that using ammonia ($M_r = 17$) and hydrochloric acid ($M_r = 36.5$) (Figure 1.21). However, they replaced the ammonia with methylamine ($M_r = 31$), which reacts in a similar way to ammonia (note that M_r is the relative molecular mass of the substance).

a Where would you predict the position of the smoke ring to be in this experiment? Explain your answer.

b Suggest other gases similar to hydrochloric acid that could replace it in this demonstration (use textbooks or the internet to find a possible acid).

10 Experiments comparing the rate of diffusion of different gases can be done using the apparatus shown in Figure 1.23.

A cylinder of **porous pot** is used through which gas molecules are able to pass. Any change in pressure in the cylinder pot shows itself in a change of liquid levels in the side tube. When there is air both inside and outside the pot, the liquid levels are the same.

Explain why the levels of liquid change when hydrogen is placed *outside* the porous pot cylinder (Figure 1.23 b).

KEY WORDS

porous pot: an unglazed pot that has channels (pores) through which gases can pass

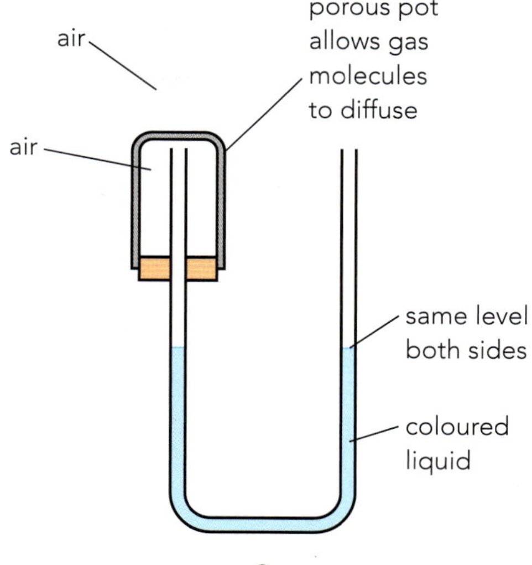

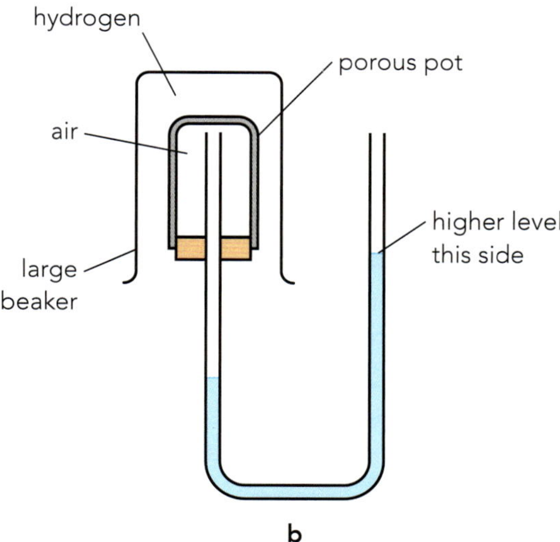

Figure 1.23: Gas diffusion through a porous pot **a:** with air inside and outside the pot, **b:** with hydrogen outside, air inside, the pot.

ACTIVITY 1.1

The kinetic model of matter

Modelling the arrangement of the particles in a solid, liquid or gas is one way to help understand the properties of the different states of matter.

Working in a small group, create a model or visual representation that explains the movement of the particles in the different states. Think about:

- What could you use to represent the particles? (Balls or marbles in a tray or dish, circular pieces of card on a plate, groups of people, symbols perhaps?)
- How will you arrange the particles to demonstrate solids, liquids and gases?
- How could you represent the movement of the particles?

Your model, diagram or display should answer three of the following questions:

- Why can three states of matter exist?
- Why is it that it takes time for a solid to melt?
- Why do solids not diffuse over a normal time period?
- What is different about substances that means that they each have different melting points?
- Different substances also have different boiling points. Is the reason for this similar to why they have different melting points?
- Why is it that you can feel a liquid cool when it evaporates in your hand?

After you have taken time to answer the questions, each group should choose one of the questions to demonstrate how your model works to the rest of the class.

REFLECTION

To understand some the ideas introduced in this chapter, you need to be able to think about the behaviour of particles smaller than you can see.

- What strategies could you use to help you to visualise particles such as atoms and molecules?
- Are there any experiments which give you clues to the existence of sub-microscopic particles?
- How useful do you find the different approaches?

SUMMARY

There are three different physical states in which a substance can exist: solid, liquid or gas.
The structures of solids, liquids and gases can be described in terms of particle separation, arrangement and motion.
Different changes in state can take place, including melting and freezing, evaporation and condensation, and boiling.
Changes of state can be produced by changing conditions of temperature and/or pressure.
Pure substances have precise melting and boiling points.
The kinetic particle model describes the idea that the particles of a substance are in constant motion and that the nature and amount of motion of these particles differs in a solid, liquid or gas.

CONTINUED

Changing physical state involves energy being absorbed or given out, the temperature of the substance staying constant while the change takes place (as illustrated by the experimental construction of cooling curves).
Changes in temperature or the external pressure produce changes in the volumes of gases which can be explained in terms of the kinetic particle theory.
Diffusion in liquids and gases is the spreading of particles to fill all of the space available.
The rate of diffusion of a gas is dependent on molecular size, with molecules of lower mass diffusing more quickly than those of higher mass.

PROJECT

The 'Goldilocks principle'

How we experience the world around us depends upon the physical conditions and states in which substances exist. This is particularly true in the case of water. The Earth is the only body in our solar system where water exists in all three states of matter.

Work in a group of three or four. Use the internet to search for some information on the topics listed here. Then select one to research in detail.

The presence of water: What is distinctive about the physical conditions on Earth that mean that life could begin, and continue to exist, here? Why is water so important when thinking about how life began? Is Earth the only planet to have water and therefore life? Have other planets had water in their past? Recent space probes have been sent to try to find water on Mars and the moons of Jupiter and Saturn (Figure 1.24). Research the various missions to find out whether there are other planets in our solar system where life may have existed.

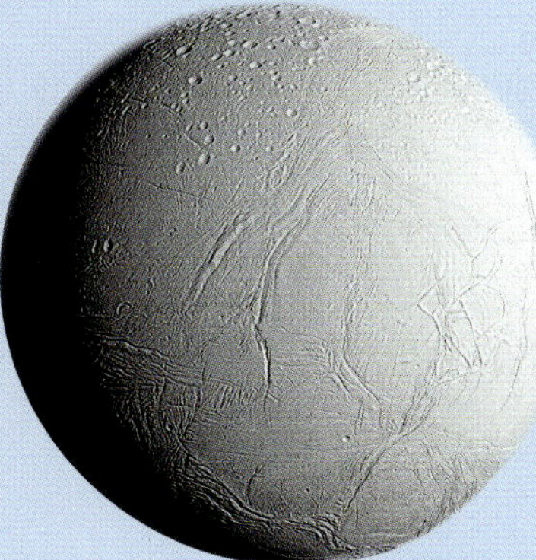

Figure 1.24: Saturn's moon Enceladus has a global ocean of liquid salty water beneath its crust.

The 'Goldilocks Zone': Earth orbits the Sun at just the right distance for liquid water to exist on its surface. It is neither too hot nor too cold for this. Research this situation, which is known as the 'Goldilocks Zone', and its meaning. Then think how it applies to the orbits of Venus, Earth and Mars.

Exo-planets and life beyond our solar system: The *Kepler* and *CHEOPS* probes have searched for planets outside our solar system (exo-planets) where life may have evolved. Research these missions and find out the characteristics of the other solar systems and planets they were hoping to find.

Decide how you will share out the tasks between the members of your group. Then bring your research together as an illustrated talk delivered to the whole class. A good illustrated talk should include the following:

- a clear structure
- a strong introduction that includes details of the question(s) you have investigated
- a short summary of the different areas you researched: make sure your points are in a sensible order
- a list of the key conclusions at the end
- the key information presented in a graphic format (e.g. as a table, chart, pie chart) instead of just text: illustrations will make your presentation much easier for your audience to understand and help them to remember your key points.

EXAM-STYLE QUESTIONS

1 A group of friends sit sharing a special meal together. When the food was put on the table, they could all smell the appetising food. How did the smell reach them when the dishes were placed on the table?

A decomposition
B diffusion
C distillation
D decolourisation **[1]**

2 The figure shows one of the changes of physical state.

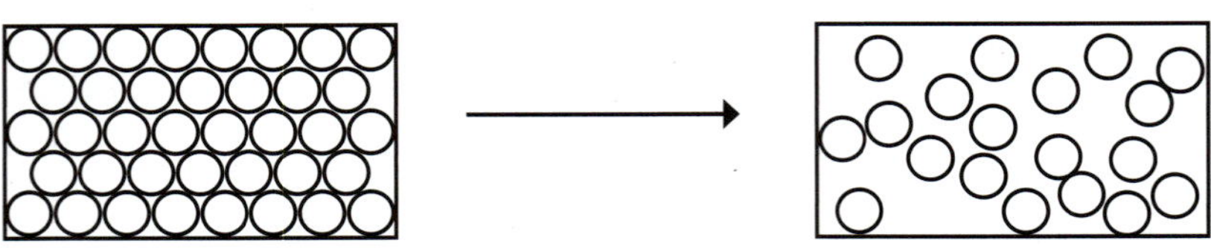

Which change of state is shown?

A boiling
B condensation
C melting
D evaporation **[1]**

3 The figure shows ice cubes floating on the surface in a glass of fizzy drink.

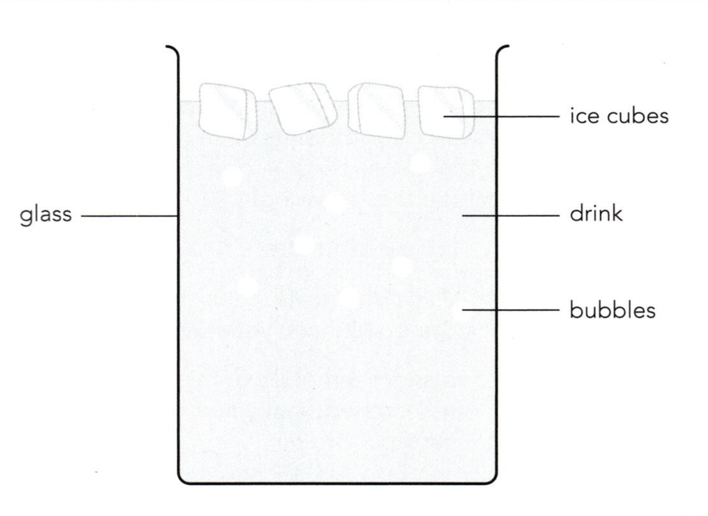

In which of **A–D** are the particles close together but free to move past each other?

A bubbles **B** glass **C** drink **D** ice cubes **[1]**

CONTINUED

4 Which of A–D in the figure shows the process of diffusion?

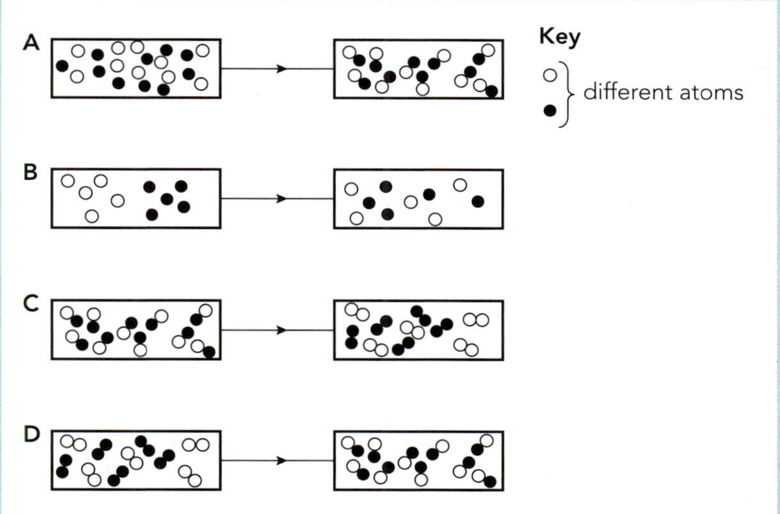

[1]

5 An experiment on the diffusion of ammonia and hydrogen chloride gases is carried out in a glass tube. The gases are given off by solutions held at each end of the tube.

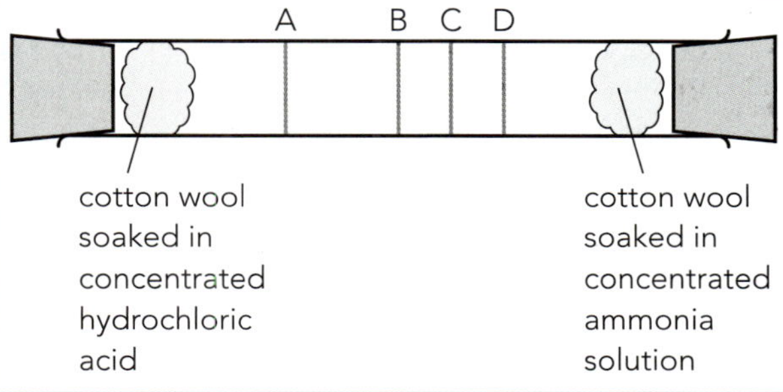

When the two gases meet, they react to produce a white solid, ammonium chloride.

Which line (A–D) shows where the white solid is formed? **[1]**

CONTINUED

6 The figure shows the arrangement of particles in each state of matter.

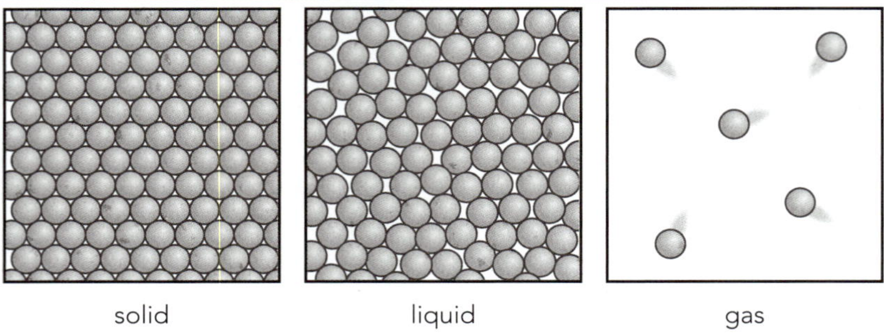

a In a gas, the particles are moving rapidly and randomly. **Describe** the movement of the particles in a liquid. **[2]**

b How does the movement of the particles in a solid change when it is heated? **[1]**

c What name is given to the process which happens when liquid water changes to water vapour at room temperature? **[1]**

d What is meant by the term *freezing*? **[1]**

[Total: 5]

7 A teacher opens a bottle of perfume at the front of her laboratory. She notices a smell of flowers. A few minutes later, students at the front of the lab notice the smell too. Those students at the back do not notice it until later.

a What two processes must take place for the smell from the perfume to reach the back of the lab? **[2]**

Later in the day, when the room had cooled, the teacher tries the same experiment with a different class. The smell is the same but it takes longer to reach the back of the lab.

b **Explain** this observation by reference to the particles of perfume. **[2]**

[Total: 4]

COMMAND WORDS

describe: state the points of a topic / give characteristics and main features

explain: set out purposes or reasons/make the relationships between things evident/provide why and/or how and support with relevant evidence

CONTINUED

8 The figure shows the change in temperature as a substance cools down.

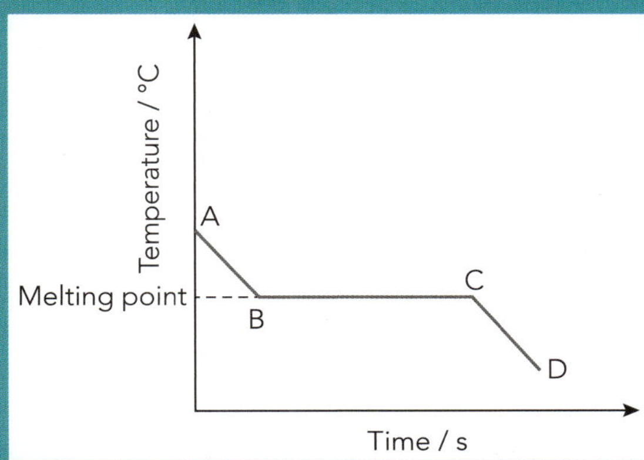

a What is happening to the substance between **C** and **D**? [2]

b What is happening to the particles of the substance between A and **B**? [2]

c Why does the temperature not change between **B** and **C**? [1]

[Total: 5]

9 Ammonia gas (M_r = 17) is a base that changes universal indicator to purple. Hydrogen chloride gas (M_r = 36.5) is an acid that changes universal indicator to red.

The figure shows an experiment done with these two gases. After two minutes, the universal indicator paper changed to purple.

a Why did the universal indicator change to purple and not red? [3]

CONTINUED

b A further experiment is shown in the figure, measuring the rates of diffusion of ammonia and hydrogen chloride.

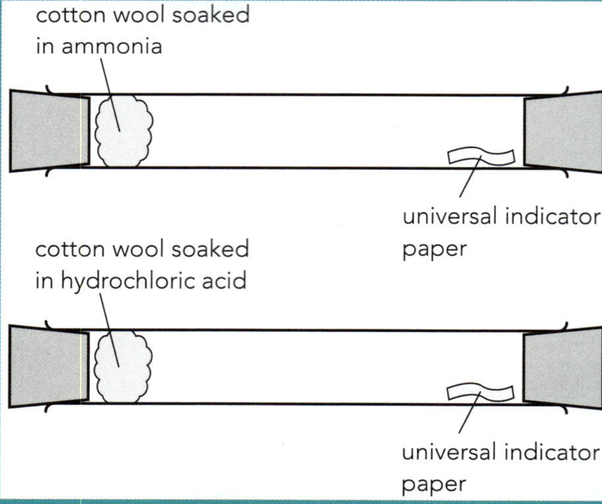

In the ammonia tube, the colour changed in 34 s. Estimate approximately how long it took for the indicator paper in the HCl tube to change colour. Choose your answer (A–D) and give a reason.

A 34 s

B 100 s

C 50 s

D 25 s

[2]

[Total: 5]

SELF-EVALUATION CHECKLIST

After studying this chapter, think about how confident you are with the different topics. This will help you to see any gaps in your knowledge and help you to learn more effectively.

I can	See Topic...	Needs more work	Almost there	Confident to move on
state the major differences between the three states of matter	1.1			
describe the changes of state observed with increasing or decreasing temperature	1.1			
describe the effect of changes in temperature on the motion of particles in the different states of matter	1.2			
interpret the shape of a cooling curve for a substance in terms of the kinetic particle theory	1.2			
state the effects of changing temperature and pressure on the volume of a gas	1.2			
explain, in terms of the kinetic particle theory, the effects of changing temperature and pressure on the volumes of gases	1.2			
understand how solids and gases can dissolve in liquids and the terms used to describe this	1.3			
describe diffusion in gases and liquids	1.3			
describe the effect of relative molecular mass on the rate of diffusion of a gas	1.3			

Chapter 2
Atomic structure

IN THIS CHAPTER YOU WILL:

- learn how atoms are the particles that make up the different elements
- study the nuclear model of the atom
- learn about the relative charge and mass of the subatomic particles: the proton, neutron and electron
- discover how the structure of any atom is defined by its proton number and mass number
- learn how the isotopes of an element have the same proton number but different numbers of neutrons
- learn how the electrons in an atom are organised in shells around the nucleus
- see how the electronic configuration of the atoms of an element relates to its position in the Periodic Table

> learn that the isotopes of an element all have the same chemical properties

> establish how to calculate the relative atomic mass of an element.

GETTING STARTED

Your earlier science courses will have introduced you to the world of the very small. In biology, you may have used a microscope to look at some microscope slides to see the detail of leaves and small flies. In chemistry, we have seen that matter is made up of very small particles such as atoms or molecules.

We have also seen that the chemical formula for water is H_2O. From your previous studies, discuss the following questions to begin your study of the sub-microscopic world:

1 What do you understand that the formula of water means? What do the letters 'H' and 'O' mean in the formula? What is the difference between an atom and a molecule?

2 Are you aware of there being any particles that are smaller than an atom?

SEEING IS BELIEVING

Scientists have used new imaging techniques to 'see' atoms. Over the past few decades, the use of scanning tunnelling microscopes (STM) – a form of 'atomic microscopy' – has opened up the manipulation of the atomic world. In 1990, scientists at the Zurich laboratories of the multinational technology company IBM were able to create an 'atomic logo' using individual xenon atoms. The ability to move and position individual atoms has led to other images of interactions between atoms. One such image is the atomic 'corral' of 48 iron atoms arranged in a ring, which appeared in the international press (Figure 2.1).

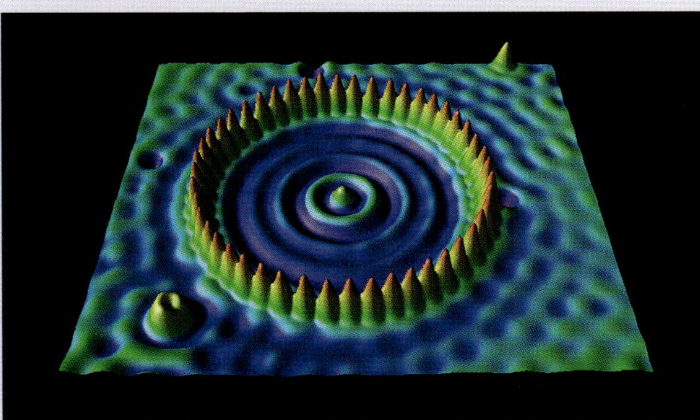

Figure 2.1: An atomic 'corral' of 48 iron atoms in a ring on a copper surface. The iron atoms are viewed using a scanning tunnelling microscope (IBM, published 1993).

New forms of atomic microscope have been developed, including the atomic force microscope and the quantum microscope. More recent advances using both scanning tunnelling and atomic force microscopes have led to the building of specific individual molecules. One such molecule was built to celebrate the 2012 Olympic Games in London. The single molecule was created using a combination of clever synthetic chemistry and state-of-the-art imaging technique. As you would expect, this molecule, called olympicene, was made up of five rings and was about 100 000 times thinner than a human hair. Further advances in microscopy and image processing are anticipated that will allow us to observe digitally and 'see' the reactions between individual atoms.

Discussion questions

1 Which atoms do you think form the basic structure of olympicene? Think of the element that forms the basis of life.

2 Olympicene was made up of five hexagons arranged with shared sides as in a honeycomb. Predict the structure of the molecule. How many atoms do you think make up the skeleton structure of the five rings of olympicene?

2.1 Atoms and elements

Every substance around us is made up of atoms. They are the incredibly small particles from which all the material world is built (Figure 2.2).

We talked of atoms, and the molecules they can form, when discussing the kinetic particle theory of matter in Chapter 1. A substance made up of just one type of atom is called an **element**. Elements cannot be broken down into anything simpler by chemical reactions.

There are now 118 known elements, but most of the known mass of the universe consists of just two elements, hydrogen (92%) and helium (7%), with all the other elements contributing only 1% to the total (Figure 2.3). How a certain number of these elements concentrated together to form the Earth is of great interest and significance. There are 94 elements found naturally on Earth but just eight account for more than 98% of the mass of the Earth's crust. Two elements, silicon and oxygen, which are bound together in silicate rocks, make up almost three-quarters of the crust. Only certain elements are able to form the complex **compounds** that are found in living things. For example, the human body contains 65% oxygen, 18% carbon, 10% hydrogen, 3% nitrogen, 2% calcium and 2% of other elements.

KEY WORDS

element: a substance that cannot be further divided into simpler substances by chemical methods; all the atoms of an element contain the same number of protons

compound: a substance formed by the chemical combination of two or more elements in fixed proportions

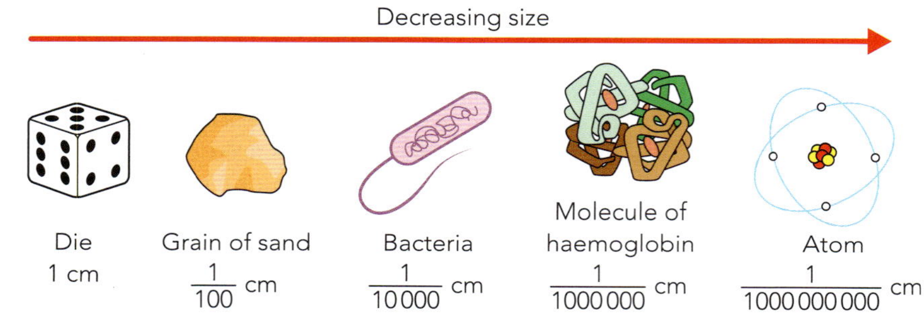

Figure 2.2: A sense of perspective on the size of the atom.

Figure 2.3: The Horsehead nebula showing huge clouds made up predominantly of hydrogen and helium.

Structure of the atom

Our modern understanding of the atom is based on the atomic theory put forward by the English chemist John Dalton in 1807. Dalton's idea was that atoms were the basic building blocks of the elements. He thought of them as indivisible particles that could join together to make molecules. Although certain parts of the theory have had to change as a result of what we have discovered since Dalton's time, his theory was one of the great leaps of understanding in chemistry. It meant that we could explain many natural processes. Whereas Dalton only had theories for the existence of atoms, modern techniques (e.g. scanning tunnelling microscopy) can now directly reveal the presence of individual atoms.

Research since Dalton's time has shown that atoms are made up of several subatomic particles. The **electron** was discovered in 1897, followed soon after by the **proton**. Crucial experiments then showed that an atom is mostly space occupied by the negatively charged electrons, surrounding a very small, positively charged **nucleus**. The nucleus is at the centre of the atom and contains almost all the mass of the atom. By 1932, when the **neutron** was discovered, it was clear that atoms consisted of three **subatomic particles** – protons, neutrons and electrons. These particles are universal – all atoms are made from them. The atom remains the smallest particle that shows the chemical characteristics of a particular element. Note that the term subatomic particles, while a useful description, is not an essential term to learn.

KEY WORDS

electron: a subatomic particle with negligible mass and a relative charge of −1; electrons are present in all atoms, located in the shells (energy levels) outside the nucleus

proton: a subatomic particle with a relative atomic mass of 1 and a charge of +1 found in the nucleus of all atoms

nucleus: (of an atom) the central region of an atom that is made up of the protons and neutrons of the atom; the electrons orbit around the nucleus in different 'shells' or 'energy levels'

neutron: an uncharged subatomic particle present in the nucleus of atoms – a neutron has a mass of 1 relative to a proton

subatomic particles: very small particles – protons, neutrons and electrons – from which all atoms are made

ACTIVITY 2.1

Understanding atomic structure – a timeline of discovery

The discovery of the nature of the subatomic particles that make up all atoms took place in a relatively short space of time around the beginning of the 20th century. Working in a group, investigate this key period in the history of science and produce a timeline showing how our understanding of the model of the atom developed. You should include words and phrases such as:

indivisible, subatomic particles, protons, neutrons, electrons, 'plum pudding' model, nuclear model, nucleus, orbiting electrons.

Key scientists to research include J. J. Thompson, Hantaro Nagaoka, Ernest Rutherford, James Chadwick and Niels Bohr.

On your timeline, try to answer the following questions:

1 What was remarkable about the structure of the atom suggested by the Geiger and Marsden gold foil experiments carried out in Rutherford's laboratory?

2 What is it about the nature of the neutron that made it the last of the particles to be discovered?

Once your timeline is complete, work individually to answer this question:

3 What do you think was the most important discovery? Justify your answer.

CONTINUED

Peer assessment

Once your group has completed a timeline and you have all answered the questions as a group and individually (question 3), exchange your group's timeline and answers with another group. Now review the timeline and the answers prepared by the other group. When you have finished your group review, share your feedback and comments with each other.

Points to consider when preparing your own timeline and in your review of the other group's work:

- Are all the major discoveries and experiments included?
- Has the work and discoveries of any scientists other than those listed in the introduction been included?
- Are the events in the correct time sequence?
- Is the timeline presented in a clear way and can the information be quickly and easily understood?

Characteristics of protons, neutrons and electrons

The three subatomic particles are found in distinct regions of the atom. The protons and neutrons are located in the small central nucleus. The electrons are present in the space surrounding the nucleus. The electrons are held within the atom by an electrostatic force of attraction between them and the positive charge of the protons in the nucleus.

The key characteristics of these three subatomic particles are listed in Table 2.1. You will note that protons and neutrons have almost the same mass. Both are given a relative mass of 1. Electrons have virtually no mass (1/1840 or 0.00054 of the mass of a proton). An important feature of these different particles is their electric charge. Protons and electrons have equal and opposite charges (+1 and −1, respectively), while neutrons are electrically neutral (have no charge).

Although atoms contain electrically charged particles, the atoms themselves are electrically neutral (they have no overall charge). This must mean that in any atom there are an equal number of protons and electrons. In this way, the total positive charge on the nucleus (due to the protons) is balanced by the total negative charge of the orbiting electrons.

Subatomic particle	Relative mass	Relative charge	Location in atom
proton	1	+1	in nucleus
neutron	1	0	in nucleus
electron	$\frac{1}{1840}$ (negligible)	−1	outside nucleus

Table 2.1: Properties of the subatomic particles.

The simplest atom is the hydrogen atom, which has one proton in its nucleus. It is the only atom that has no neutrons; it consists of one proton and one electron.

The next simplest atom is that of helium. This has two protons and two neutrons in the nucleus, and two orbiting electrons (Figure 2.4).

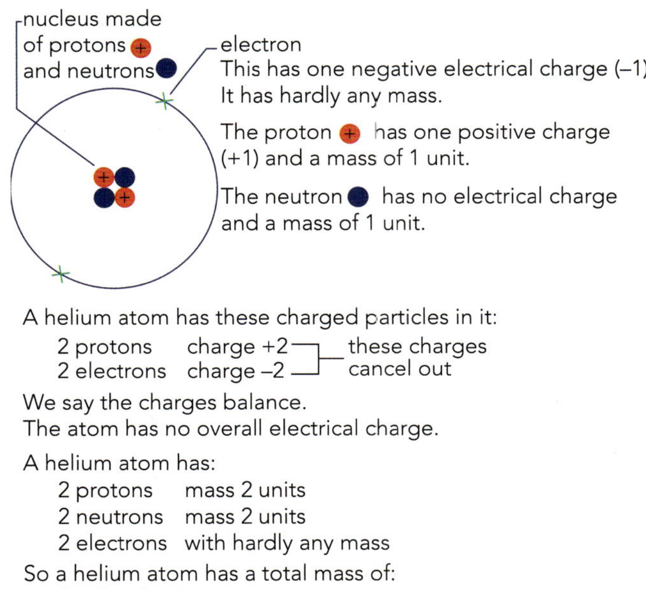

Figure 2.4: Structure of a helium atom.

Lithium is the next simplest atom after helium. A lithium atom has three protons, four neutrons and three electrons. The subatomic arrangement of larger atoms gets more complicated with the addition of more protons and electrons. The number of neutrons required to hold the nucleus together increases as the atomic size increases. An atom of gold consists of 79 protons, 118 neutrons and 79 electrons.

Proton (atomic) number and mass (nucleon) number

Only hydrogen atoms have one proton in their nuclei. Only helium atoms have two protons. Indeed, only gold atoms have 79 protons. The number of protons in the nucleus of an atom determines which element it is. This important number is known as the **proton number** or **atomic number** of an atom. The proton number is given the symbol Z.

We have seen that protons alone do not make up all the mass of an atom. The neutrons in the nucleus also contribute to the total mass. Because a proton and a neutron have the same relative mass, the mass of a particular atom depends on the total number of protons and neutrons present. Protons and neutrons are known as nucleons and the total number of protons and neutrons present is called the **mass number** or **nucleon number** of an atom. The mass number is given the symbol A.

KEY WORDS

proton number (or atomic number) (Z): the number of protons in the nucleus of an atom

mass number (or nucleon number) (A): the total number of protons and neutrons in the nucleus of an atom

An important way of representing a particular atom of an element is to combine the chemical symbol of the element (discussed in more detail in Chapter 4) with the proton and mass numbers of the atom. The symbol Z representing the proton number and symbol A representing the mass number of an atom can be written alongside the symbol for that element, in the general format ${}^{A}_{Z}\mathrm{X}$. Figure 2.5 shows how a helium atom's structure is written in this way.

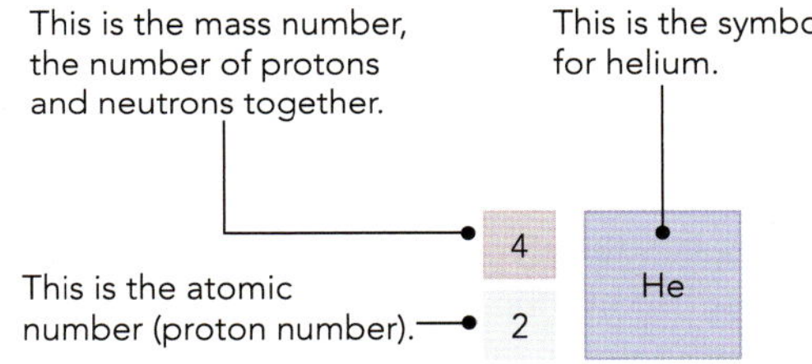

Figure 2.5: Representing the structure of a helium atom using the format ${}^{A}_{Z}\mathrm{X}$.

Using this format, an atom of lithium is represented as ${}^{7}_{3}\mathrm{Li}$. The atoms of carbon, oxygen and uranium are represented as ${}^{12}_{6}\mathrm{C}$, ${}^{16}_{8}\mathrm{O}$ and ${}^{238}_{92}\mathrm{U}$.

When the proton and mass numbers are known for the atoms of an element, we can work out the following:

- proton number (Z) = number of protons in the nucleus
- mass number (A) = number of protons + number of neutrons

It is also possible to establish two other important relationships:

- number of electrons = number of protons = atomic number
- number of neutrons = mass number – atomic number = A–Z

Atom	Symbol	Proton number (Z)	Mass number (A)	Inside the nucleus:		Outside the nucleus: electrons (Z)
				Protons (Z)	Neutrons ($A - Z$)	
hydrogen	H	1	1	1	0	1
helium	He	2	4	2	2	2
lithium	Li	3	7	3	4	3
beryllium	Be	4	9	4	5	4
carbon	C	6	12	6	6	6
oxygen	O	8	16	8	8	8
sodium	Na	11	23	11	12	11
calcium	Ca	20	40	20	20	20
gold	Au	79	197	79	118	79
uranium	U	92	238	92	146	92

Table 2.2: Subatomic composition and structure of atoms of different elements.

Table 2.2 shows the numbers of protons, neutrons and electrons in particular atoms in various elements. Note that the rules apply even to the smallest or largest atoms.

An atom of any element is defined by the number of the different constituent subatomic particles present in that atom. The most important number is the proton number of that atom as it defines which element it belongs to. We shall see later in this chapter, and in Chapter 4, that the Periodic Table of the elements is arranged in order of increasing proton number. For example, magnesium is the twelfth atom in the table, so it must have 12 protons and 12 electrons in its atoms.

Questions

1 What are the relative masses of a proton, neutron and electron, given that a proton has a mass of 1?

2 How many protons, neutrons and electrons are there in an atom of phosphorus, which has a proton number of 15 and a mass number of 31?

3 Explain the terms *atom* and *element*, and include a clear description of the relationship between the two terms.

4 Explain why neutrons are important in making the nucleus of an atom stable. You need to consider the charges on the different subatomic particles present.

2.2 Isotopes

Measuring the mass of atoms

A single atom cannot be weighed on a balance. However, the mass of one atom can be compared to that of another using a **mass spectrometer**. Since we are comparing the masses of atoms, the values we obtain are relative values and we need to set a standard against which other atoms are measured. The element carbon has been chosen as the standard. Carbon was chosen as the standard because, as we shall see in Chapters 18–20, there are far more compounds containing carbon than any other element. The masses of all other atoms are compared to the mass of a carbon atom. This gives a series of values of the **relative atomic mass (A_r)** for the different elements.

KEY WORDS

mass spectrometer: an instrument in which atoms or molecules are ionised and then accelerated; the ions are then separated according to their mass

relative atomic mass (A_r): the average mass of naturally occurring atoms of an element on a scale where the carbon-12 atom has a mass of exactly 12 units

However, analysing the masses of various elements showed up a complication. Pure samples of many elements (e.g. carbon, hydrogen and chlorine) are found to contain atoms that have different masses. This is despite the fact that the atoms have the same numbers of protons and electrons. The different masses observed are the result of the presence of different numbers of neutrons in the nuclei of atoms of the same element. When this occurs, the atoms of the same element are called **isotopes**.

Because there are several isotopes of carbon, the standard against which all atomic masses are measured has to be defined precisely. The isotope carbon-12 is used as this standard. One atom of carbon-12 is given the mass of 12 precisely. From this we obtain 1 atomic mass unit (a.m.u.) = $\frac{1}{12}$ × mass of one atom of carbon-12.

Table 2.3 gives some examples of the values obtained for other elements. It shows that carbon-12 atoms are 12 times as heavy as hydrogen atoms, which are the lightest atoms of all. Calcium atoms are 40 times as heavy as hydrogen atoms.

Element	Atomic symbol	Relative atomic mass (A_r)
carbon	C	12
hydrogen	H	1
oxygen	O	16
calcium	Ca	40
copper	Cu	64
gold	Au	197

Table 2.3: Relative atomic masses of some elements.

KEY WORD

isotopes: atoms of the same element that have the same proton number but a different nucleon number; they have different numbers of neutrons in their nuclei. Some isotopes are radioactive because their nuclei are unstable (radioisotopes).

Characteristics of isotopes

The difference between isotopes of the same element is just the number of neutrons in the atoms. The atoms have the same number of protons and electrons. The isotopes of an element are defined by their difference in mass number. The isotopes are referred to using their mass number. For example, the isotopes of carbon are carbon-12, carbon-13 and carbon-14. Table 2.4 gives the details of the isotopes of several elements.

Element	Isotopes		
Hydrogen	hydrogen (99.99%)	deuterium (0.01%)	tritium[(a)]
	$^{1}_{1}H$	$^{2}_{1}H$	$^{3}_{1}H$
	1 proton	1 proton	1 proton
	0 neutrons	1 neutron	2 neutrons
	1 electron	1 electron	1 electron
Carbon	carbon-12 (98.9%)	carbon-13 (1.1%)	carbon-14[(a)] (trace)
	$^{12}_{6}C$	$^{13}_{6}C$	$^{14}_{6}C$
	6 protons	6 protons	6 protons
	6 neutrons	7 neutrons	8 neutrons
	6 electrons	6 electrons	6 electrons
Neon	neon-20 (90.5%)	neon-21 (0.3%)	neon-22 (9.2%)
	$^{20}_{10}Ne$	$^{21}_{10}Ne$	$^{22}_{10}Ne$
	10 protons	10 protons	10 protons
	10 neutrons	11 neutrons	12 neutrons
	10 electrons	10 electrons	10 electrons
Chlorine	chlorine-35 (75%)	chlorine-37 (25%)	
	$^{35}_{17}Cl$	$^{37}_{17}Cl$	
	17 protons	17 protons	
	18 neutrons	20 neutrons	
	17 electrons	17 electrons	

[(a)]*Tritium and carbon-14 atoms are radioactive isotopes because their nuclei are unstable.*

Table 2.4: Several elements that exist as mixtures of isotopes.

Many elements have naturally occurring isotopes. Hydrogen, the simplest element, has two naturally occurring isotopes: hydrogen and deuterium (Figure 2.6). A third isotope, tritium, can be made artificially.

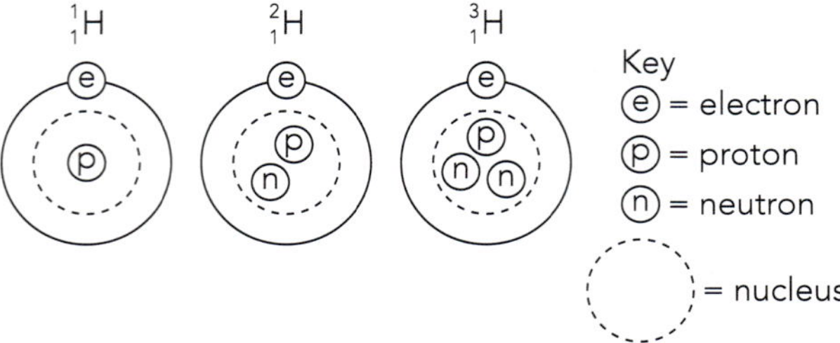

Figure 2.6: Isotopes of hydrogen: hydrogen ($^1_1$H), deuterium ($^2_1$H) and tritium ($^3_1$H).

Questions

5 Which of the diagrams A, B or C represents the nucleus of an atom of the carbon-14 isotope (Figure 2.7)?

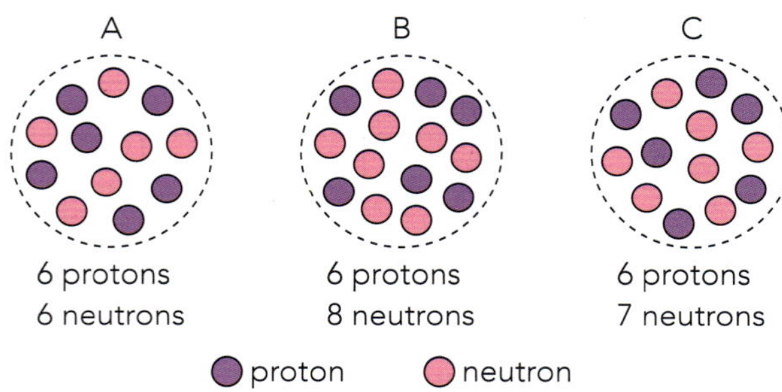

Figure 2.7: The nuclei of carbon isotopes.

6 An atom of neon has 10 protons, 10 electrons and 11 neutrons. How many nucleons does it contain?

7 What is the difference in terms of subatomic particles between an atom of chlorine-35 and an atom of chlorine-37?

8 By considering the subatomic particles present, explain why an atom of nitrogen is electrically neutral.

The isotopes of an element have the same chemical properties because they contain the same number of electrons and therefore the same electronic configuration. It is the number of electrons in an atom that determines the way in which it forms bonds and reacts with other atoms. However, some physical properties of the isotopes *are* different. The masses of the atoms differ and therefore other properties, such as density and rate of diffusion, also vary.

A visual illustration of how the presence of isotopes affects physical properties is the difference in density between ordinary ice and heavy-water ice (frozen deuterium oxide, D_2O). Frozen deuterium oxide has a 10.6% greater density than ordinary ice. Normal ice cubes float in water, but a heavy-water ice cube will sink (Figure 2.8).

Figure 2.8: A heavy-water ice cube (frozen D_2O) sinks in water (left). An ordinary ice cube floats (right).

Tritium and carbon-14 illustrate another difference in physical properties that can occur between isotopes, as they are *radioactive*. The imbalance of neutrons and protons in their nuclei causes them to be unstable so the nuclei break up spontaneously (without any external energy being supplied), emitting certain types of radiation. They are known as radioisotopes.

Calculating relative atomic mass

Most elements exist naturally as a mixture of isotopes. Therefore, the value we use for the atomic mass of an element is an average mass. This takes into account the proportions (abundance) of all the naturally occurring isotopes. If a particular isotope is present in a high proportion, it will make a large contribution to the average. This average value for the mass of an atom of an element is known as the relative atomic mass (A_r).

The existence of isotopes also explains why the accurate values for the relative atomic masses of most elements are not whole numbers. However, to make calculations more straightforward, in this book the values are rounded to the nearest whole number. This is also true in the Periodic Table in this book (see Appendix). There is one exception, chlorine, where this would be misleading. Chlorine has two isotopes, chlorine-35 and chlorine-37, in an approximate ratio of 3 : 1 (or 75.0% : 25.0%). If the ratio was 1 : 1 (or 50% : 50%), then the relative atomic mass of chlorine would be 36. The actual value is 35.5.

This value for the relative atomic mass of chlorine can be calculated by finding the total mass of 100 atoms.

Mass of 100 atoms of chorine = (35 × 75.0) + (37 × 25.0) = 3550

Then the average mass of one atom = $\frac{3550}{100}$ = 35.5

Thus, for chlorine A_r (Cl) = 35.5

WORKED EXAMPLE 2.1

Calculating relative atomic mass

Lithium has two isotopes, $^{6}_{3}Li$ and $^{7}_{3}Li$. The relative abundance of these two isotopes is shown in Figure 2.9.

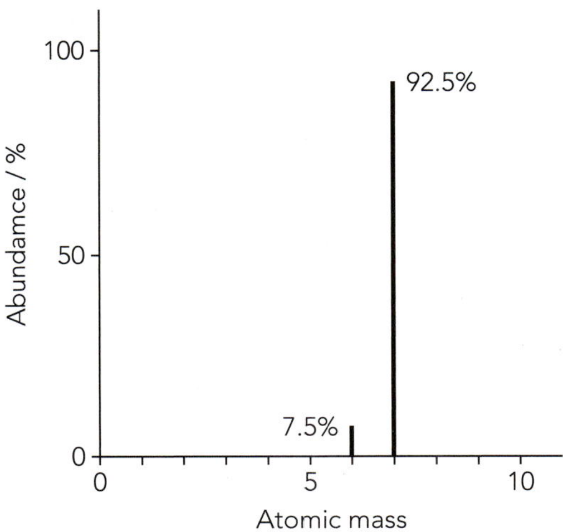

Figure 2.9: Isotopes of lithium.

Calculate the relative atomic mass of lithium.

Step 1. Look at Figure 2.9. The graph shows you that the lithium sample consists of 7.5% lithium-6 and 92.5% lithium-7.

Step 2. Calculate the mass of 100 atoms by multiplying the mass of each isotope by its relative abundance.

Total mass of 100 atoms of lithium = (6 × 7.5) + (7 × 92.5) = 692.5

Step 3. Calculate the average mass of one lithium atom by dividing the total mass by 100.

Average mass of one lithium atom 692.5 / 100 = 6.925

Step 4. Round your answer to required level of accuracy.

Relative atomic mass of lithium = 6.9 (to two significant figures)

The isotopes of magnesium and their abundances are given in Table 2.5.

Isotope	Symbol	Abundance / %
magnesium-24	24 / 12 Mg	78.6
magnesium-25	25 / 12 Mg	10.1
magnesium-26	26 / 12 Mg	11.3

Table 2.5: Isotopes of magnesium.

Calculate the relative atomic mass of magnesium.

Step 1. Review data provided.

Step 2. Calculate total mass of 100 atoms of magnesium.

= (24 × 78.6) + (25 × 10.1) + (26 × 11.3) = 2432.7

Step 3. Calculate the average mass of a magnesium atom.

$= \frac{2432.7}{100} = 24.327$

Step 4. Round your answer to the required level of accuracy asked for.

Relative atomic mass of magnesium = 24.3 (to three significant figures) *(Note that the answer to both of these worked examples is given to the same number of significant figures as the lowest accuracy of the data given.)*

Here are two further examples for you to try for yourself.

1 Iridium has two isotopes. These isotopes are iridium-191 and iridium-193. A natural sample of iridium consists of 37.3% iridium-191. Calculate the relative atomic mass of iridium from the data.

2 Bromine is an element with two isotopes: bromine-79 and bromine-81. Usually, the relative atomic mass of bromine is given as 80. What does this value for the A_r of bromine tell you about the approximate proportions of the two isotopes in a natural sample?

2.3 Electronic configuration of elements

Electrons in shells

The aurora borealis is a spectacular display seen in the sky in the far north (a similar phenomenon – the aurora australis – occurs in the southern hemisphere). It is caused by radiation from the Sun moving the electrons in atoms of the gases of the atmosphere.

Similar colour effects can be created in a simpler way in the laboratory by heating the compounds of some metals in a Bunsen flame (Figure 2.10; see also the flame tests in Chapter 22).

Figure 2.10: Some flame test colours for different elements.

These flame test colours are also seen in fireworks. Different metal compounds are used in various types of firework to produce the spectacular effects we see at displays (Figure 2.11). The colours produced are due to electrons in the atom moving between two different **electron shells**.

In 1913, Niels Bohr, working with Ernest Rutherford, developed a theory to explain how electrons were arranged in atoms. This theory helps to explain how the colours in the flame test are produced.

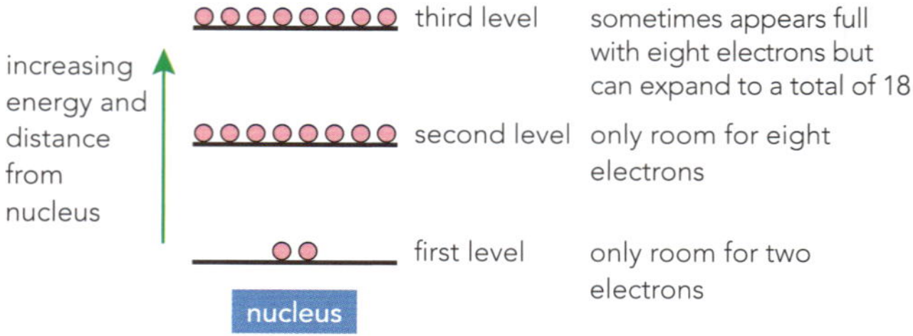

Figure 2.12: The different energy levels (shells) in an atom (not to scale). Electrons fill the shells starting from the one closest to the nucleus.

A simplified version of Bohr's **atomic theory** of the arrangement of electrons in an atom can be summarised as follows (see Figures 2.12 and 2.13):

- electrons are in orbit around the central nucleus of the atom
- the electron orbits are called **electron shells** (or energy levels) and have different energies
- shells that are further from the nucleus have higher energies
- the shells are filled starting with the one with lowest energy (closest to the nucleus)
- the first shell can hold only two electrons
- the second and subsequent shells can hold eight electrons to give a stable (noble gas) arrangement of electrons.

KEY WORDS

electron shells (energy levels): (of electrons) the allowed energies of electrons in atoms – electrons fill these shells (or levels) starting with the one closest to the nucleus

atomic theory: a model of the atom in which electrons can only occupy certain shells (or energy levels) moving outwards from the nucleus of an atom

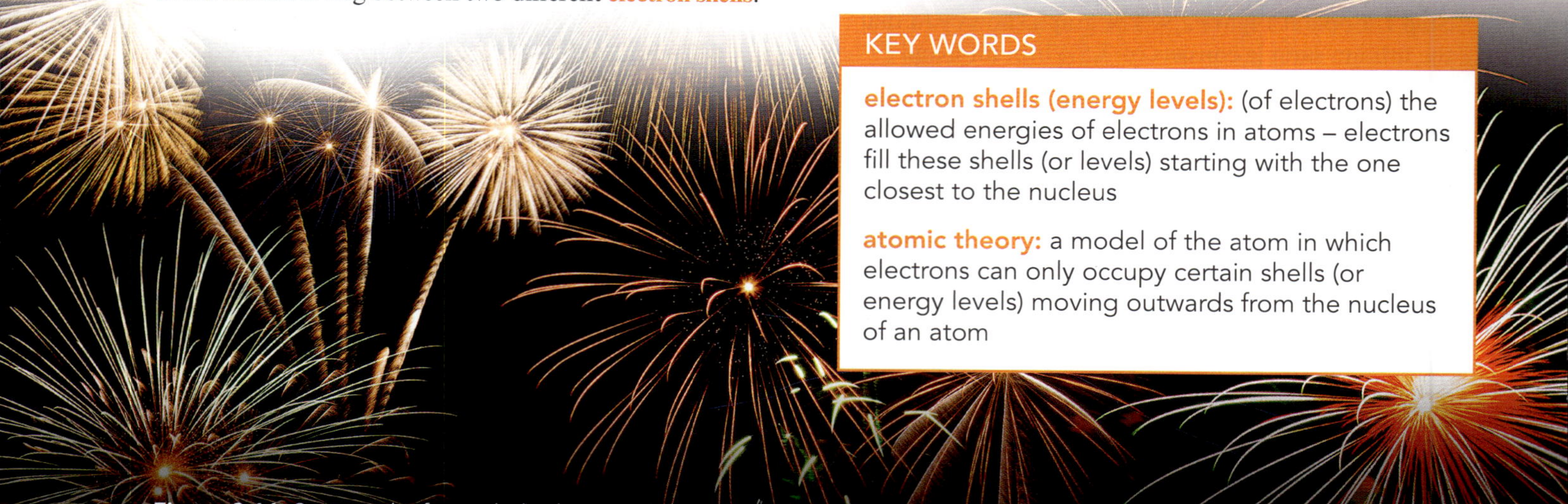

Figure 2.11: Spectacular firework displays are the product of electrons changing energy level in the different metal compounds present in the fireworks.

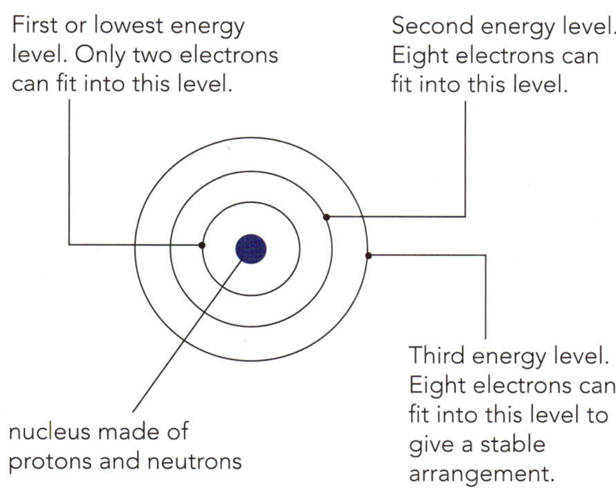

Figure 2.13: Bohr's atomic theory of the arrangement of electrons in an atom.

Further evidence was found that supported these ideas of how the electrons are arranged in atoms. The number and arrangement of the electrons in the atoms of the first 20 elements in the Periodic Table (see the Appendix) are shown in Table 2.6.

For any atom, the combination of the arrangement of electrons in shells with the numbers of protons and neutrons makes it possible to draw a full subatomic representation. Figure 2.14 shows such a representation for an atom of carbon-12 (carbon atoms being possibly the most versatile atoms in the universe; see Chapters 18–20).

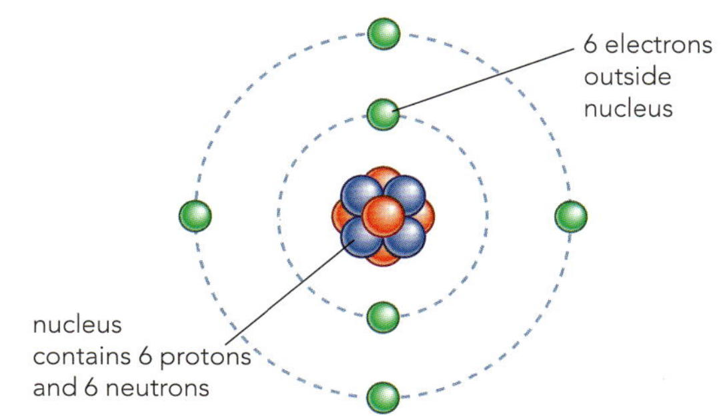

Figure 2.14: A visualisation of the subatomic structure of carbon-12.

Element	Symbol	Atomic number (Z)	First shell	Second shell	Third shell	Fourth shell	Electronic configuration
hydrogen	H	1	•				1
helium	He	2	••				2
lithium	Li	3	••	•			2,1
beryllium	Be	4	••	••			2,2
boron	B	5	••	•••			2,3
carbon	C	6	••	••••			2,4
nitrogen	N	7	••	•••••			2,5
oxygen	O	8	••	••••••			2,6
fluorine	F	9	••	•••••••			2,7
neon	Ne	10	••	••••••••			2,8
sodium	Na	11	••	••••••••	•		2,8,1
magnesium	Mg	12	••	••••••••	••		2,8,2
aluminium	Al	13	••	••••••••	•••		2,8,3
silicon	Si	14	••	••••••••	••••		2,8,4
phosphorus	P	15	••	••••••••	•••••		2,8,5
sulfur	S	16	••	••••••••	••••••		2,8,6
chlorine	Cl	17	••	••••••••	•••••••		2,8,7
argon	Ar	18	••	••••••••	••••••••		2,8,8
potassium	K	19	••	••••••••	••••••••	•	2,8,8,1
calcium	Ca	20	••	••••••••	••••••••	••	2,8,8,2

Table 2.6: Electronic configurations of the first 20 elements.

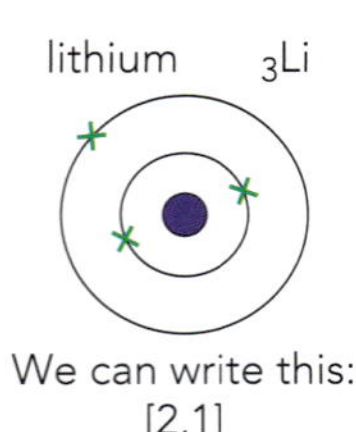

We can write this:
[2,1]

We can write this:
[2,8,1]

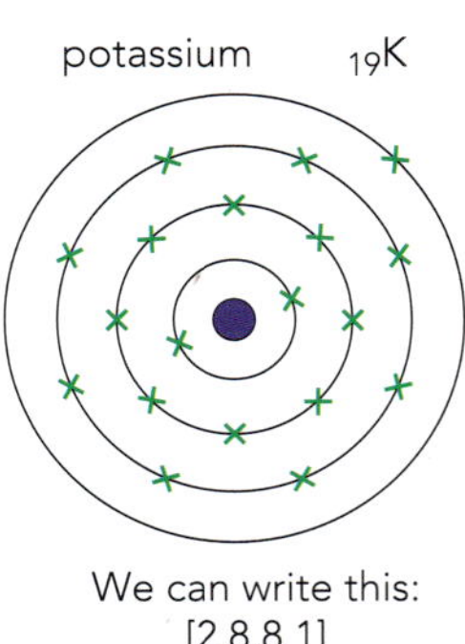

We can write this:
[2,8,8,1]

Figure 2.15: Different ways of showing electron structure.

The electron arrangement or **electronic configuration** can be given simply in terms of numbers: 2,8,4 for silicon, for example. You should make sure that you remember how to work out the electron arrangements of the first 20 elements and can draw them in rings (shells) as in Figure 2.15.

KEY WORDS

electronic configuration: a shorthand method of describing the arrangement of electrons within the electron shells (or energy levels) of an atom; also referred to as electronic structure

EXPERIMENTAL SKILLS 2.1

A chemical rainbow!

How are the spectacular colours of fireworks produced? The essential ingredients of fireworks are the explosive mixture and the metal compounds (also known as salts – see Chapters 11 and 12) that produce the distinctive colours. This experiment explores how different metal salts produce different colours as a result of different electron shifts within the atoms present.

You will need:

- a Bunsen burner
- a heat-resistant mat
- a nichrome wire loop
- dilute hydrochloric acid solution (0.1 mol dm^{-3})
- a hand-held spectroscope (if available)
- several different solutions of metal salts (concentration 0.5 mol dm^{-3}): NaCl(aq), KCl(aq), LiCl(aq), $CuCl_2$(aq) and $CaCl_2$(aq); for example. A 0.1 mol dm^{-3} $BaCl_2$(aq) solution can also be used.

Safety

Wear eye protection throughout. Be careful with chemicals. Never ingest them and always wash your hands after handling them. All the salt solutions are low hazard at this concentration, except $CuCl_2$(aq), which is harmful and a skin/eye irritant.

Getting started

You will need a roaring flame on your Bunsen burner to carry out these tests. Practise getting this hottest flame using your burner by varying the air supply. If available, you should practise using a hand-held spectroscope before you take your practical observations. Be aware that the energy of the light given out by the salts (and its frequency) is related to the colour produced.

Method

1. In a darkened room, place the Bunsen burner on a heat-resistant mat and light the Bunsen burner.
2. Heat a nichrome loop in in a roaring Bunsen flame and then dip the loop into the dilute hydrochloric acid solution. Repeat this process several times to clean the loop and remove any contaminants.

CONTINUED

3 Dip the loop into one of the salt solutions and then hold it in the hot part of the roaring Bunsen flame.

4 Observe the colour produced in the flame; repeat using a hand-held spectroscope, noting the colour of the major lines produced.

5 Rinse the loop with the dilute hydrochloric acid solution and flame it several times to clean away any remaining salt solution.

6 Continue to test the other salt solutions in the same way.

Questions

1 Is it the metal or non-metal present in the salts that is responsible for the colour emitted in the flame?

2 Is the flame colour seen the result of a chemical or physical change?

3 Table 2.7 shows the general relationship between the colour, frequency and energy of the light released. The bigger the change in energy level of the electrons, the greater the energy of the light released.

Colour	Frequency / THz	
Red	405–480	
Orange	480–510	
Yellow	510–530	
Green	530–600	
Cyan (blue-green)	600–620	energy increasing ↓
Blue	620–680	
Violet	680–790	

Which of the metal salt solutions produces light of the greatest energy? Put the salt solutions in order of the energy change involved, from the least to the greatest.

Table 2.7: Frequency (in terahertz, THz) of the different colours of the visible spectrum.

Electronic configuration and the Periodic Table

Electron arrangement is key to understanding and predicting the properties and characteristics of an element. The properties and characteristics of each element, when compared to other elements, are the basis for how all known elements are arranged in the Periodic Table.

You can directly link the electronic configuration of an element to its position in the Periodic Table (Figure 2.16).

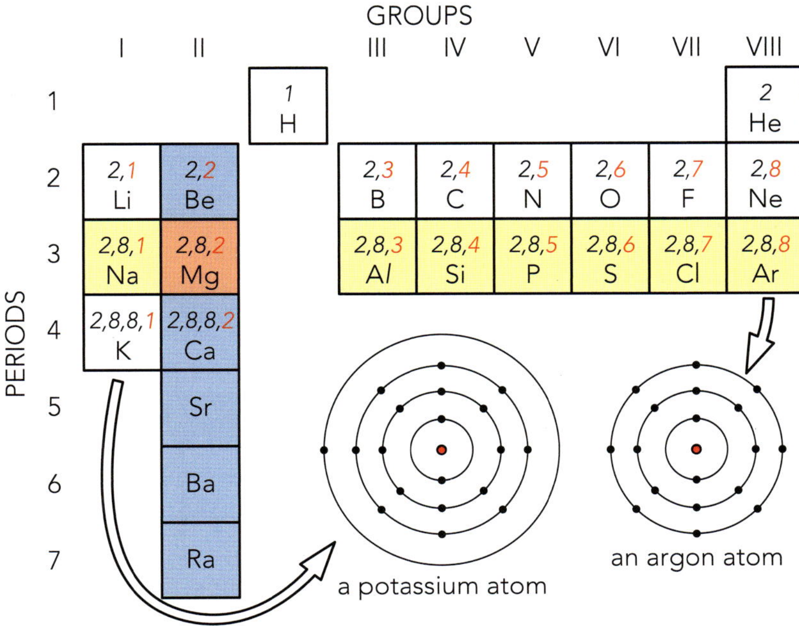

Figure 2.16: The relationship between an element's position in the Periodic Table and the electronic configuration of its atoms.

Group and period number

Figure 2.16 indicates that the number of outer electrons in an atom is the same as the **group number** for the element in the Periodic Table. The number of occupied electron shells in an atom tells you the **period (row) number** of the element in the table.

Elements in the same group have the same number of outer electrons. For the elements in Groups I to VII, the number of the group is the number of electrons in the outer shell. As you move across a period in the table, a shell of electrons is being filled.

The properties of an element are closely determined by the number of outer electrons in the atom. The elements in Group II have two outer electrons. The elements in Period 3 have three shells of electrons. A magnesium atom has two electrons in its third (outer) shell and is in Group II. A potassium atom has one electron in its fourth, outer shell, and is in Group I and Period 4 (Figure 2.16). An argon atom has an outer shell containing eight electrons – a very stable arrangement – and is in Group VIII (the noble gases).

KEY WORDS

group number: the number of the vertical column that an element is in on the Periodic Table

period (row) number: the horizontal row of the Periodic Table that an element is in

The noble gas electronic configuration

Noble gases (Group VIII) are very unreactive gases. The atoms of the noble gas elements all have a very stable electron arrangement. That means their atoms exist naturally as single atoms. Noble gas atoms do not make chemical bonds with the atoms of other elements by sharing or transferring their outer electrons. Noble gas atoms all have a 'full' outer shell of electrons (Table 2.8). This means that they usually have eight electrons in their outer shell.

When the atoms of elements other than the noble gases combine together to form molecules, the atoms involved often achieve the same stable arrangement of electrons characteristic of the noble gases. We will explore the formation of chemical bonds in detail in Chapter 3.

KEY WORDS

noble gases: elements in Group VIII – a group of stable, very unreactive gases

Questions

9 a What are the maximum numbers of electrons that can fill the first and the second shells (energy levels) of an atom?
   b What is the electron arrangement of a calcium atom, which has an atomic number of 20?
   c How many electrons are there in the outer shells of the atoms of the noble gases argon and neon?

10 Carbon-12 and carbon-14 are different isotopes of carbon. How many electrons are there in an atom of each isotope?

11 The electronic configurations of the atoms of four elements are:

   A 2,4   B 2,8,18,7
   C 2,8,4   D 2,8,8

   a Which two elements are in Group IV of the Periodic Table? Explain your answer.
   b Which element is a noble gas? Explain your answer.
   c Which of these elements is in Group VII of the Periodic Table? Explain your answer.
   d Which two elements are in the third period (row) of the Periodic Table? Explain your answer.
   e What is the proton number of element C? Explain your answer.

Element	Atomic number (*Z*)	Period number	Electronic configuration	Number of outer electrons*
Helium	2	1	2	2
Neon	10	2	2,**8**	8
Argon	18	3	2,8,**8**	8

**The other noble gases in Group VIII (krypton, xenon, and radon) also have eight outer electrons.*

Table 2.8: Electronic configurations of noble gas elements.

REFLECTION

Do you find diagrams help you to visualise the structure, movement and interaction of small objects such as atoms and molecules?

If diagrams do not help, can you think of other ways that could? Discuss your ideas with a partner.

SUMMARY

The elements are the basic building units of the material world – they cannot be chemically broken down into anything simpler.
The atoms of the elements are made up of different combinations of the subatomic particles (protons, neutrons and electrons) and an element is made up of atoms that all have the same number of protons.
These subatomic particles have particular electrical charges and relative masses.
In any atom, the protons and neutrons are bound together in a central nucleus, and the electrons 'orbit' the nucleus in different energy levels (or shells).
The number of protons in an atom is defined as the proton (atomic) number (Z) of the element and the total number of protons and neutrons in as atom is defined as the mass (nucleon) number (A).
Isotopes of the same element can exist and differ only in the number of neutrons in their nuclei.
The chemical properties of all the isotopes of an element are the same as they have the same electronic configuration.
The relative atomic mass of an element can be calculated from data on the abundance of the different isotopes of that element.
The electrons in atoms are arranged in different shells (or energy levels) that are at different distances from the nucleus of the atom.
Each shell (or energy level) has a maximum number of electrons that it can contain and the electrons fill the shells closest to the nucleus first.
The electronic configuration of an element determines the group number and period number of that element in the Periodic Table.

PROJECT

Modelling the structure of an atom

You will have seen many different representations of atoms in the media. Often images show electrons orbiting like planets (Figure 2.17a). After completing this chapter, you will have learnt much more about atomic structure, and understand that electrons are arranged around a central nucleus in energy levels (Figure 2.17b).

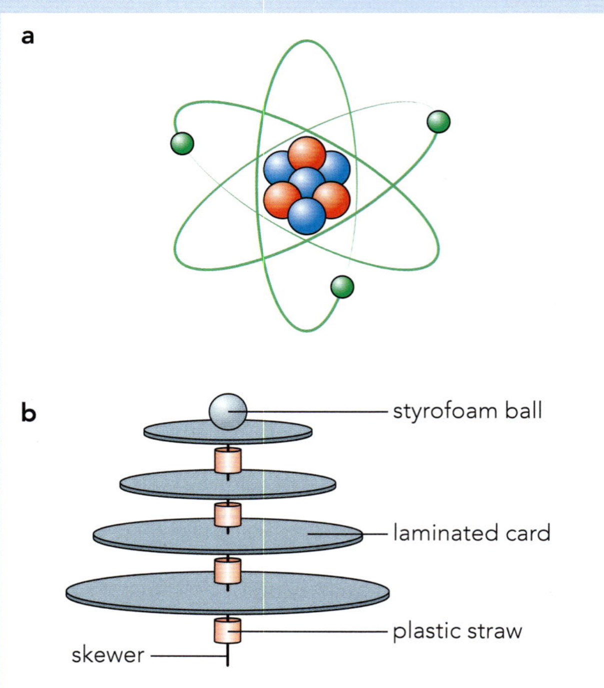

Figure 2.17 a: A planetary model of an atomic showing three electrons orbiting a nucleus. **b:** Model of the electron energy levels in an atom.

Now it is your turn to try to represent an atom by creating a 3D model.

Working in groups, construct a model of either:

- a planetary model of a simple atom, or
- a model showing the energy levels that electrons can occupy in an atom – representing the energy levels like trays with a certain number of electrons on each tray (see Figure 2.15 for the basis of such a model).

Having constructed your model, take a moment to assess how well it represents an atom. Be critical of your model and discuss the limitations of the model with your group.

How well does your model represent what you have learnt in this chapter?

EXAM-STYLE QUESTIONS

1 The structure of an atom is defined by two numbers: the proton (atomic) number and the mass (nucleon) number. What is the electronic configuration of an atom with proton number 5 and mass number 11?

A 2,8,1 **B** 3,2 **C** 2,3 **D** 1,8,2 **[1]**

2 Cadmium is an element that has several isotopes. One of these isotopes is $^{112}_{48}Cd$. Which particle in the table is another isotope of cadmium?

	Protons	Neutrons
A	48	64
B	48	112
C	112	48
D	64	48

[1]

3 Elements in the Periodic Table are arranged in order of their proton number.

a **Define** the meaning of the term *proton number*. **[1]**

b Argon, proton number 18, has a larger mass number than potassium, proton number 19. **Explain** why this is. **[2]**

c Complete this table showing the properties of the three particles that make up atoms.

Particle	Charge	Mass	Position in atom
proton	+		
neutron		1	in the nucleus
electron	–		

[5]

d An atom of which element has only two of the three particles listed in the table above? **[1]**

e Argon and helium are both noble gases. In what way are their electron structures similar? **[1]**

[Total: 10]

4 $^{12}_{6}C$ and $^{14}_{6}C$ are two isotopes of the element carbon.

a In what way does the structure of their atoms differ? **[3]**

b Carbon reacts with oxygen to produce carbon dioxide. What is the electronic configuration of an oxygen atom? **[1]**

c Germanium (A_r =32) is an element in the same group as carbon.

i How many electrons does germanium have in its outer electron shell? **[1]**

ii **State** which period of the Periodic Table germanium is found in. **[1]**

[Total: 6]

COMMAND WORDS

define: give precise meaning

explain: set out purposes or reasons/make the relationships between things evident/provide why and/or how and support with relevant evidence

state: express in clear terms

CONTINUED

5 The element copper has two main isotopes $^{63}_{29}Cu$ and $^{65}_{29}Cu$. Both isotopes have the same chemical properties.

a Explain why the chemical properties of these two isotopes are the same. [2]

b The Periodic Table gives the approximate relative atomic mass of copper as 64.
Natural copper consists of 70.0% ^{63}Cu and 30.0% ^{65}Cu. **Calculate** the accurate relative atomic mass of copper. [2]

[Total: 4]

COMMAND WORD

calculate: work out from given facts, figures or information

SELF-EVALUATION CHECKLIST

After studying this chapter, think about how confident you are with the different topics. This will help you to see any gaps in your knowledge and help you to learn more effectively.

I can	See Topic...	Needs more work	Almost there	Confident to move on
understand that matter is made up of elements and compounds	2.1			
describe the structure of the atom as a central nucleus surrounded by electrons	2.1			
define the properties of protons, neutrons and electrons, and the meaning of the proton (atomic number) number and nucleon (mass) number of an atom	2.1			
define isotopes as atoms of an element with the same proton number but different nucleon numbers	2.2			
describe isotopes of an element as having the same chemical properties as they have the same electronic configuration	2.2			
calculate the relative atomic mass of an element from the relative masses and abundancies of its isotopes	2.2			
describe the electronic configuration of an atom and the significance of it in terms of the position of the element in the Periodic Table	2.3			
determine the electronic configuration of atoms with proton numbers 1 to 20	2.3			

> Chapter 3

Chemical bonding

IN THIS CHAPTER YOU WILL:

- outline the differences between elements, compounds and mixtures
- understand that a covalent bond is formed by sharing a pair of outer electrons between two atoms and describe bonding in simple molecules
- understand the formation of ions and that an ionic bond is the strong electrostatic force of attraction between oppositely charged ions
- relate the properties of simple molecular and ionic compounds to their structure and bonding
- outline the giant covalent structures of diamond and graphite and relate these to their uses

> describe bonding in more complex molecules

> describe the formation of ionic bonds between metals and non-metals

> explain the properties of simple molecular and ionic compounds in terms of their structure and bonding

> describe the giant covalent structure of silicon(IV) oxide

> learn that metallic bonding is the electrostatic attraction between the positive ions in the metallic lattice and a 'sea' of delocalised electrons between them

CONTINUED

- explain physical properties of metals in terms of metallic bonding.

GETTING STARTED

In groups, use a molecular model-building kit or simple construction materials such as plasticine and drinking straws to build models of two or three simple molecules that you have encountered in your earlier science courses. It would be useful to build models of compounds such as:

- water
- ammonia
- methane.

Use the models to discuss what you remember of the type of bonding in the molecules and how the bonds are made. It would be useful also to recall the shapes of these different molecules if you have covered that previously.

EXPLOITING THE PROPERTIES OF GRAPHENE

Two different forms of carbon are familiar to us in everyday life. When we write with a pencil or make a charcoal sketch, we are sliding layers of graphite onto the paper. Diamonds are used as precious gemstones in jewellery and can be a very special gift for someone. Diamond and graphite are two structural forms of this particularly important element, carbon, and we will look at them in more detail later in this chapter.

However, interest in different structural types of carbon has grown dramatically in recent years as other new forms have been discovered. The discovery of the fullerenes (such as C_{60}) and carbon nanotubes has led to many new technological applications. Even more recently (in 2004), a further new form of carbon – graphene – has been isolated at the University of Manchester in the UK. Andre Geim and Konstantin Novoselov were awarded the Nobel Prize in Physics in 2010 and the potential for this new material has generated immense interest and research investment.

Graphene is essentially a single-layered material made up of individual sheets of graphite. The first samples of graphene were isolated in experiments aimed at seeing how thin a piece of graphite could be made by polishing or by cleaning graphite with 'sticky tape'. The layers of graphite were peeled off for surface science experiments.

High-quality graphene is strong, has a low density, and is almost transparent. Graphene has many interesting properties:

- it is an excellent conductor of heat and electricity (300 times better than copper)
- the introduction of graphene into plastics could make those plastics electrically conducting
- it behaves as a *semiconductor* and may one day replace silicon in computer chips and other technology applications, it is 200 times stronger than steel and is incredibly flexible.

Of the different forms of carbon, graphene is the most chemically reactive. Membranes of graphene oxide have been shown to filter water and clean it, suggesting possible uses in desalination and water purification. This application is highly significant for countries where water supply is difficult, as current methods of large-scale desalination by distillation or reverse osmosis are economically expensive.

Discussion questions

1 How many different structural forms of the element carbon do you know of?

2 Graphite is used to strengthen everything from tennis rackets to the chassis of Formula 1 cars, but do you think that it is equally strong in all directions?

3.1 Non-metallic substances and covalent bonding

Elements, compounds and mixtures

Elements are the building blocks of the universe and each element has its own distinctive properties. Some elements are very reactive. Other elements, particularly the noble gases (Group VIII of the Periodic Table) are very stable and inert. We now know of over 100 elements, although only 94 occur naturally, and they are essentially divided into **metals** and **non-metals**. The atoms of an element are all the same in terms of their content of protons and electrons. The atoms of any one element all have the same atomic number and are distinct from those of any other atom (Chapter 2).

Elements combine to make all the **compounds** around us. The simplest compounds that we are all familiar with, to the complex molecules that determine our biology, are all made from these elements. Each compound is made of particular elements, always combined in the same proportions. Simple compounds such as water, ammonia and methane begin to show the variety that can be achieved when the atoms of elements combine together. Water is formed from hydrogen and oxygen. Each water molecule contains two hydrogen atoms bonded to an oxygen atom (Figure 3.1).

Figure 3.1: A water molecule consists of two hydrogen atoms (white) bonded to an oxygen atom (red).

> **KEY WORDS**
>
> **element:** a substance that cannot be further divided into simpler substances by chemical methods; all the atoms of an element contain the same number of protons
>
> **metals:** a class of chemical elements (and alloys) that all have a characteristic shiny appearance and are good conductors of heat and electricity
>
> **non-metals:** a class of chemical elements that are typically poor conductors of heat and electricity
>
> **compound:** a substance formed by the chemical combination of two or more elements in fixed proportions

When different elements react together to form a compound their characteristic properties are lost, being replaced by those of that compound. Importantly, the compound formed has properties that bear no direct relationship to those of the elements that make it up. A clear example of this is sodium chloride (common salt) (Figure 3.2). Sodium is a highly reactive metal. On contact with water, sodium fizzes violently and may even cause a flame as the reaction takes place. The toxic gas chlorine was once used as a chemical weapon, and is still used to kill harmful microbes in water treatment. Yet the compound of these two elements, sodium chloride, is a crystalline solid that is not only harmless enough to flavour food, but is in fact essential for life.

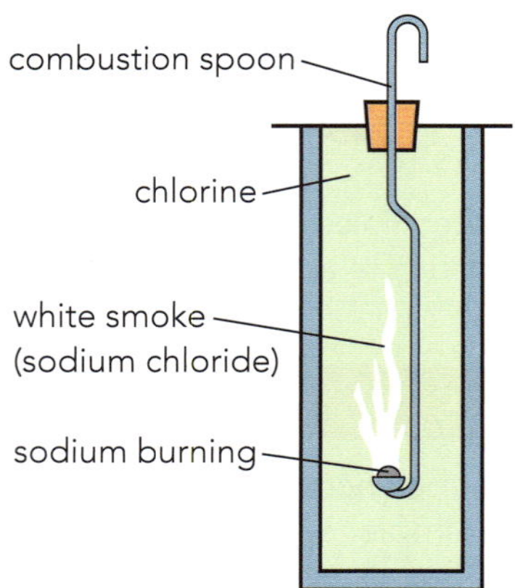

Figure 3.2: Burning sodium in chlorine gas produces sodium chloride (common or table salt).

The formulae of the compounds sodium chloride and water are NaCl and H_2O, respectively. Here you begin to see how formulae are constructed using the chemical symbol of each element. Both formulae indicate the ratio of the atoms involved in making the compound. We will look at **chemical formulae** in more detail in Chapter 4.

Chemical bonding involves the outer electrons of each atom (Chapter 2). As we examine a range of substances, we shall see that, whatever type of bonding holds the structure together, it is the outer electrons that are used. The diversity of the material world is produced by the different ways in which atoms can join together. Generally speaking, there are two types of compound (Figure 3.3):

- molecular compounds where the atoms are bonded together: water, ammonia and methane are examples of simple molecular compounds
- ionic compounds where many ions (charged atoms) are held together in a regular structure: sodium chloride is an example of an ionic compound.

A mixture contains two or more elements or compounds that are not chemically bonded together. The parts of a mixture are not present in fixed amounts. Sodium chloride can be mixed with water to form a type of mixture known as a solution. However, the amount of salt dissolved can be varied to give a dilute or concentrated solution. Mixtures can be separated by one or other of the physical separation methods described in Chapter 21. Other mixtures include air, which is a mixture of elements such as nitrogen, oxygen and argon, together with compounds such as carbon dioxide (Chapter 17), and alloys such as brass (Chapter 14).

A mixture of iron and sulfur powders can contain various amounts of the two elements. The elements still keep their characteristic properties and can be separated using a magnet (Figure 3.4b). However, when a mixture of iron and sulfur is heated, a black compound iron sulfide is formed, which is not attracted to a magnet (Figure 3.4a).

KEY WORDS

chemical formula: a shorthand method of representing chemical elements and compounds using the symbols of the elements

chemical bonding: the strong forces that hold atoms (or ions) together in the various structures that chemical substances can form – metallic bonding, covalent bonding and ionic bonding

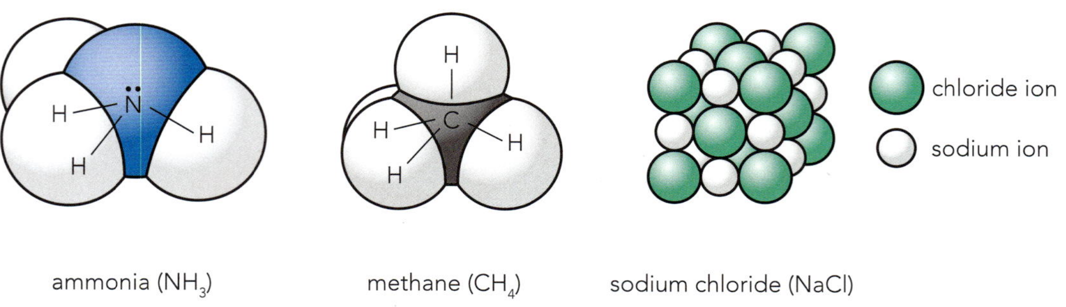

Figure 3.3: Ammonia and methane are examples of molecular compounds. Sodium chloride is an ionic compound.

Figure 3.4 a: A mixture of iron and sulfur can be separated with a magnet. **b:** Iron powder and sulfur react together to form the black compound, iron sulfide, which is not magnetic.

Table 3.1 summarises how compounds and mixtures compare.

Compounds	Mixtures
A compound is a single substance.	A mixture contains two or more substances.
The composition is always the same.	The composition can be varied.
The formation involves a chemical reaction.	No chemical change takes place when made.
The properties are very different from the elements present in the compound.	The properties of the substances making the mixture are still present.
Can only be broken down by chemical reactions.	The substances present can be separated by physical methods.

Table 3.1: A comparison of the nature of chemical compounds and mixtures of substances.

Covalent bonding in simple molecular elements and compounds

In Chapter 2, we learnt that each element contains a defined type of atom (with a defined number of protons). However, elements are not simply made up of separate atoms individually arranged. Elements such as oxygen (O_2) and hydrogen (H_2) consist of **diatomic molecules**. Indeed, the only elements that are made up of individual atoms moving almost independently of each other are the noble gases (Group VIII). The noble gases are the elements whose electron arrangements are most stable and so their atoms do not combine with each other.

All other elements do form structures; their atoms are linked by some type of bonding.

For non-metallic elements the type of bonding involved is **covalent bonding**. Covalent bonding is also the bonding used in compounds between one non-metal and another.

Simple molecular elements

Hydrogen normally exists in the form of diatomic molecules (H_2). Two atoms bond together by sharing their electrons. The orbits overlap with each other and a molecule is formed (Figure 3.5).

> **KEY WORDS**
>
> **diatomic molecules:** molecules containing two atoms, e.g. hydrogen, H_2
>
> **covalent bonding:** chemical bonding formed by the sharing of one or more pairs of electrons between two atoms

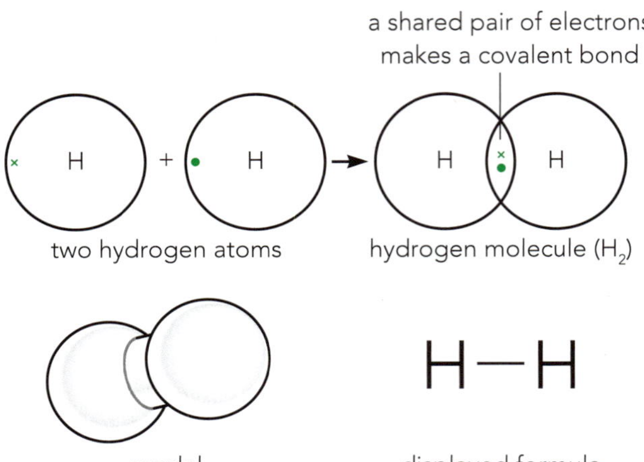

Figure 3.5: The hydrogen molecule is formed by sharing the electrons between the atoms. A space-filling model can be used to show the atoms overlapping.

Through this sharing, each atom gains a share in two electrons. This is the number of electrons in the outer shell of helium, the nearest noble gas to hydrogen. (Remember that the electron arrangement of helium is very stable; helium atoms do *not* form He_2 molecules.) Sharing electrons like this is known as covalent bonding. It has been shown that in a hydrogen molecule, the electrons are more likely to be found between the two nuclei. The forces of attraction between the shared electrons and the nuclei are greater than any repulsive forces. The molecule is held together by the bond.

The main features of covalent bonding are:

- the bond is formed by the sharing of a pair of electrons between two atoms, leading to noble gas electronic configurations
- each atom contributes one electron to each bond
- molecules are formed from atoms linked together by covalent bonds.

Many non-metallic elements form diatomic molecules. However, elements other than hydrogen form bonds in order to gain a share of *eight* electrons in their outer shells. This is the number of electrons in the outer shell of all the noble gases apart from helium. Thus, the halogens (Group VII) form covalent molecules (Figure 3.6). In forming these molecules, each atom gains a share in eight electrons in its outer shell. A **dot-and-cross diagram** can be drawn showing the outer electrons only (Figure 3.6a), because the inner electrons are not involved in the bonding.

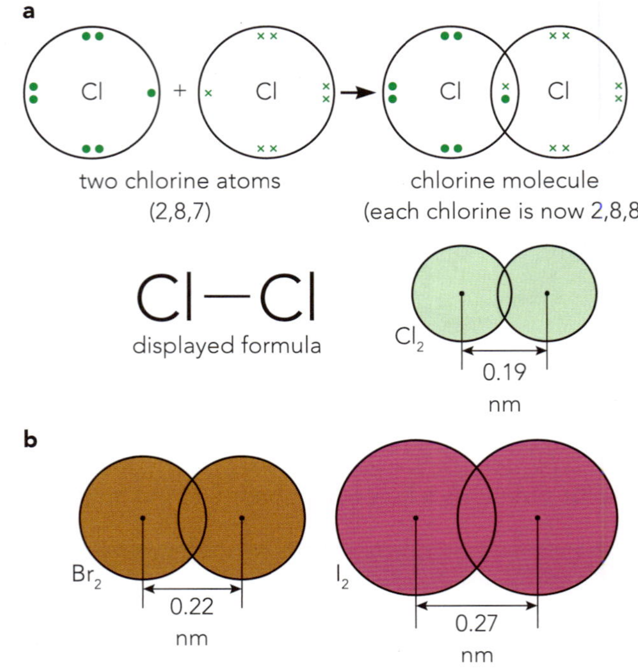

Figure 3.6 a: The formation of the covalent bond in chlorine molecules (Cl_2). **b:** Molecules of Br_2 and I_2 are formed in the same way. They are larger because the original atoms are bigger.

KEY WORDS

dot-and-cross diagram: a diagram drawn to represent the bonding in a molecule, or the electrons in an ion; usually, only the outer electrons are shown and they are represented by dots or crosses depending on which atom they are from

Molecules of hydrogen and the halogens are each held together by a single covalent bond. Such a single bond uses two electrons, one from each atom. The bond can be drawn as a single line between the two atoms. Note that, when we draw diagrams showing the overlap of the outer shells, we can show the outer electrons only because the inner electrons are not involved in the bonding. Each atom gains a share in eight electrons in its outer shell.

When molecules of oxygen (O_2) or nitrogen (N_2) are formed, more electrons have to be used in bonding if the atoms are to gain a share of eight electrons. These molecules are held together by a double bond (O_2) or a triple bond (N_2) (Figure 3.7).

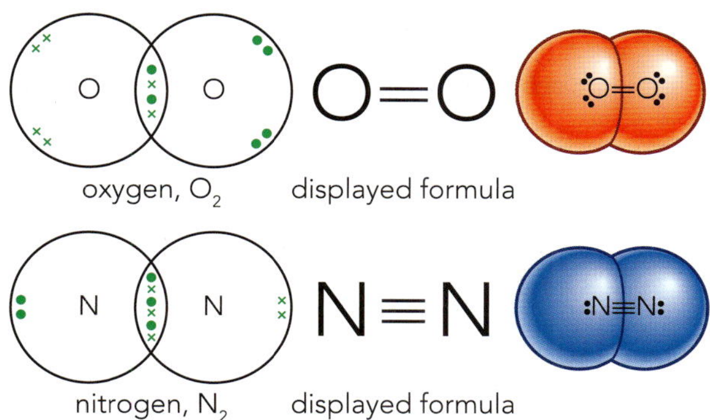

Figure 3.7: Structures of oxygen (O_2) and nitrogen (N_2) molecules involve multiple covalent bonding. An oxygen molecule contains a double bond; a nitrogen molecule contains a triple bond.

Simple covalent compounds

In covalent compounds, bonds are again made by sharing electrons between atoms. In simple molecules, the atoms combine to achieve a more stable arrangement of electrons, most often that of a noble gas. The formation of hydrogen chloride (HCl) molecules involves the two atoms sharing a pair of electrons (Figure 3.8).

The examples shown in Figure 3.9 illustrate different ways of representing this sharing. They also show how the formula of the compound corresponds to the numbers of each atom in a molecule.

In each case, the atoms achieve a share in the same number of electrons as the noble gas nearest to that element in the Periodic Table. In all but the case of hydrogen, this means a share of eight electrons in their outer shell.

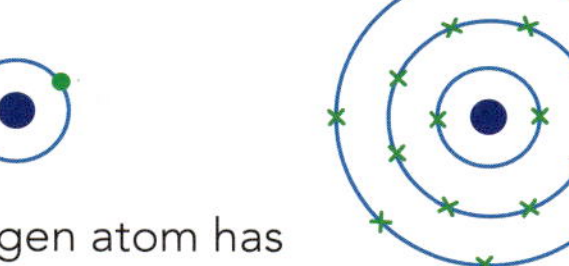

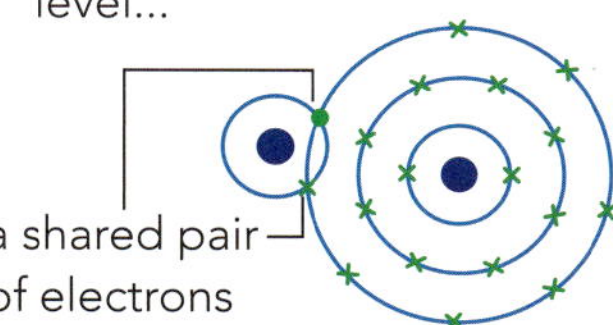

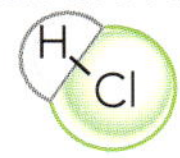

Figure 3.8: Hydrogen and chlorine atoms share a pair of electrons to form a molecule of hydrogen chloride.

Earlier we saw that multiple covalent bonds can exist in molecules of the elements, oxygen and nitrogen. Multiple covalent bonds can also exist in compounds. The carbon dioxide molecule is held together by double bonds between the atoms (Figure 3.10). Figure 3.10 also shows some other examples of bonding in organic compounds that you will meet again in Chapters 18–20.

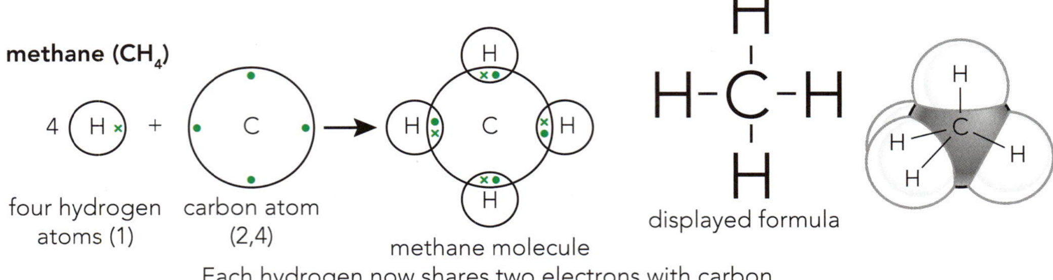

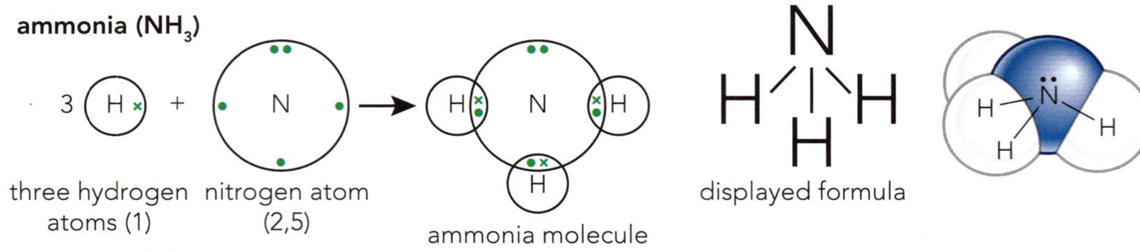

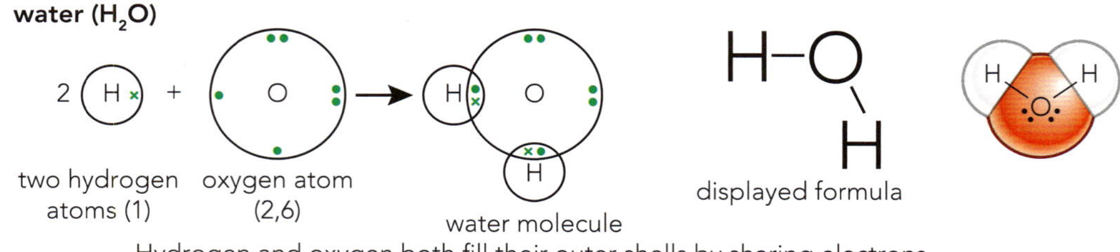

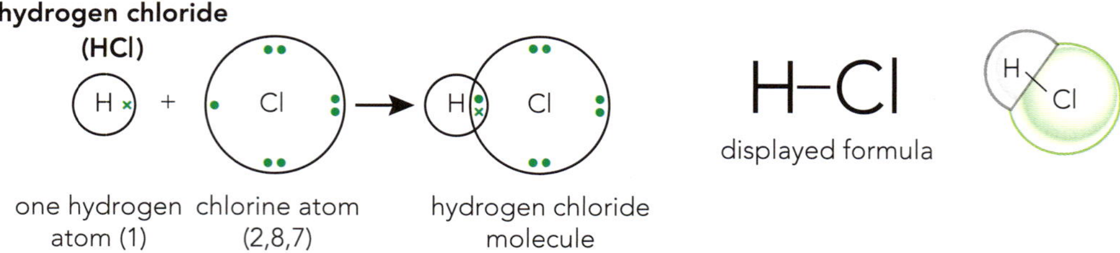

Figure 3.9: Examples of the formation of simple covalent molecules. Again, only the outer electrons of the atoms are shown.

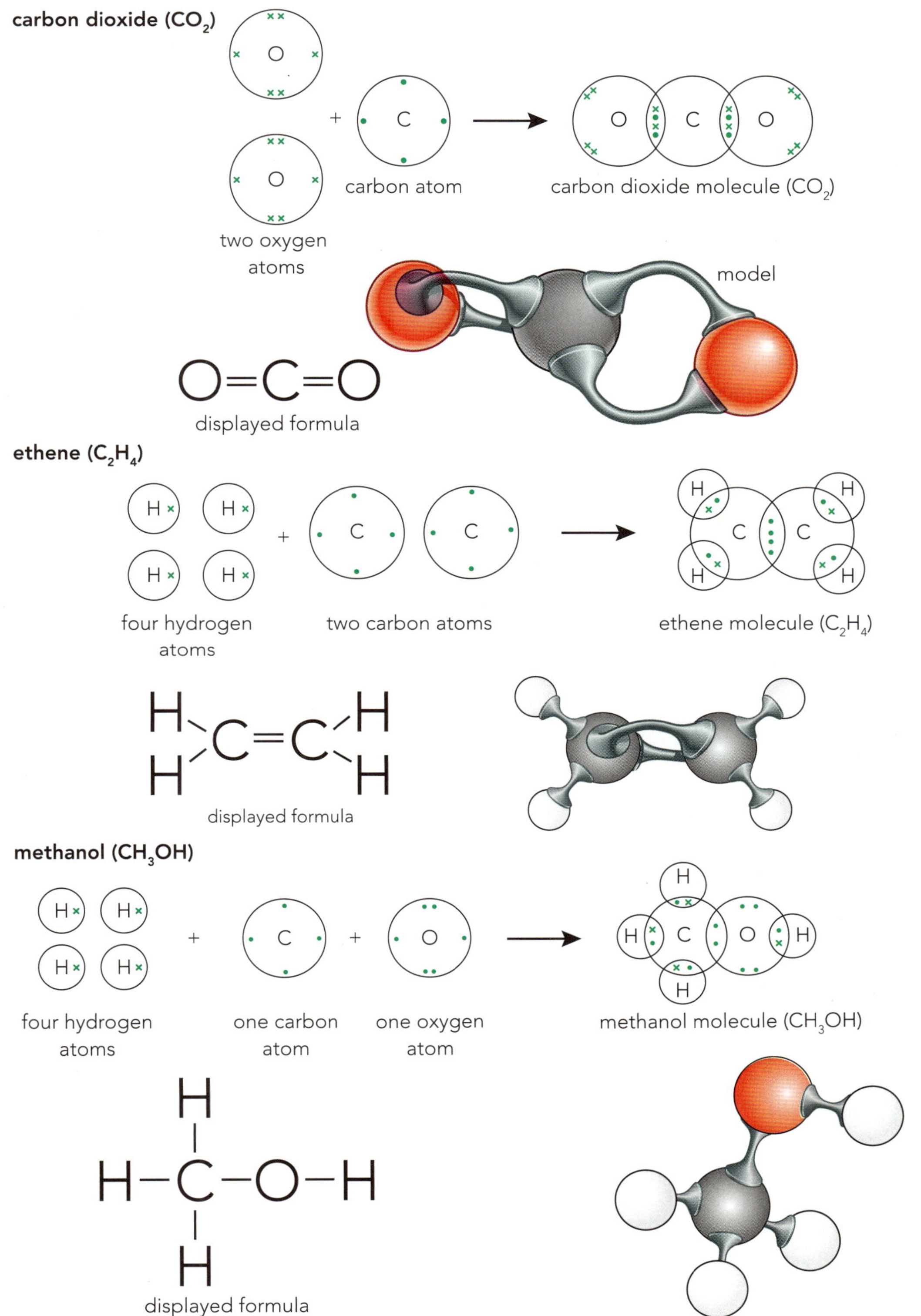

Figure 3.10: Formation of the carbon dioxide, ethene and methanol molecules, showing the outer electrons only. Ball-and-stick models can be used to show the structure.

Physical properties of simple covalent compounds

Knowledge of how atoms combine to make different types of structures helps us begin to understand why substances have different physical properties. The properties of simple molecular compounds are:

- they have low melting points and boiling points and are often liquids or gases at room temperature
- they show poor electrical conductivity.

When considering the properties of simple molecular compounds, it is important to be aware of the two levels of interactive force involved. The covalent bonds within the molecule are strong and difficult to break. However, the forces between the molecules, the intermolecular forces, are only weak and are relatively easily broken. If you think back to the discussion of changes of state in Chapter 1, you will understand that the presence of weak intermolecular forces gives rise to low melting and boiling points. Because the forces between the molecules are weak only a small amount of energy is needed to enable the molecules to move away from each other.

Table 3.2 shows key physical properties of simple covalent compounds.

Properties of simple covalent compounds	Reasons for these properties
Often liquids or gases at room temperature and have low melting points and boiling points	They are made of simple covalent molecules. The forces between the molecules (intermolecular forces) are only very weak. Not much energy is needed to move the molecules further apart.
Do not conduct electricity	There are no free electrons or ions present to carry the electrical current.

Table 3.2: The physical properties of simple molecular compounds.

Questions

1 In a pure sample of any element, what is the most important feature of all the atoms present?

2 A gas is made up of simple molecules that have the formula NOC1. Which of the diagrams A–D in Figure 3.11 correctly represents the molecules of NOCl?

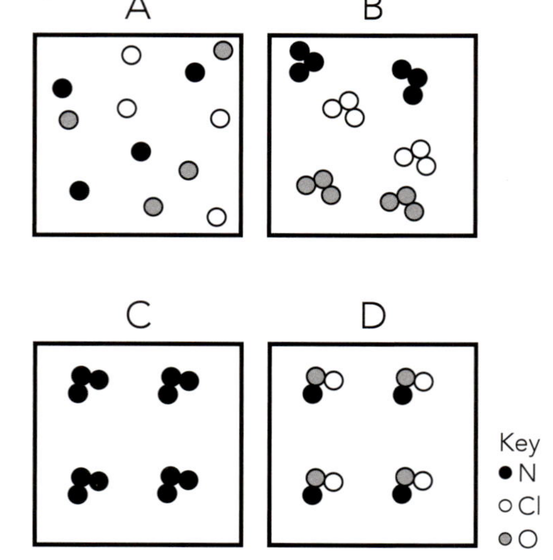

Figure 3.11: Representations of various gases.

3 Methane is the simplest hydrocarbon (a compound of carbon and hydrogen *only*) and has the formula CH_4 (Figure 3.12).

```
     H
     |
 H — C — H
     |
     H
```

Figure 3.12: The structure of a methane molecule.

What is the total number of electrons involved in the bonding in methane?

A 10 **B** 2 **C** 8 **D** 4

4 Covalent bonding involves electrons being shared between the atoms bonded together. Methane is made up of covalently bonded molecules. Which of the diagrams A–D in Figure 3.13 represents the bonding in methane?

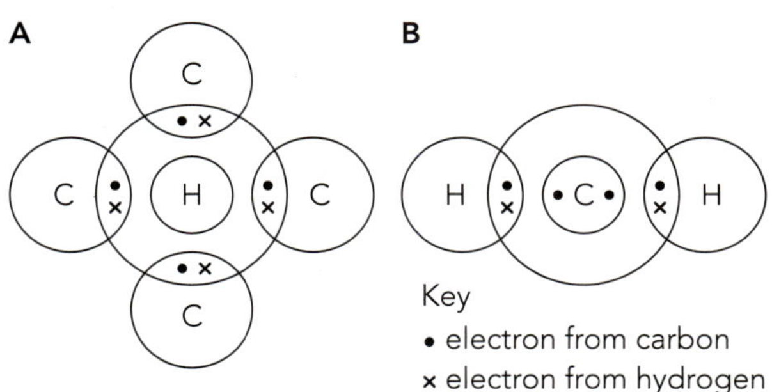

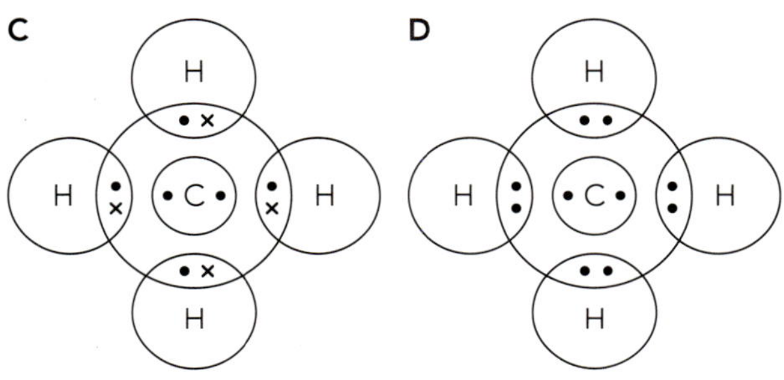

Figure 3.13: The possible bonding in methane.

3.2 Ions and ionic bonding

Two major types of bond hold compounds together. The first is covalent bonding, which, as we have seen, involves the sharing of electrons, usually between atoms of non-metals.

However, compounds between metals and non-metals involve a second type of bonding known as **ionic bonding**. Electrons are transferred from one atom to another, forming **ions**. These ionic compounds are held together by **electrostatic forces** of attraction between the oppositely charged ions.

KEY WORDS

ionic bonding: a strong electrostatic force of attraction between oppositely charged ions

ions: charged particles made from an atom, or groups of atoms (compound ions), by the loss or gain of electrons

electrostatic forces: strong forces of attraction between particles with opposite charges – such forces are involved in ionic bonding

Formation of ions

An ion is an electrically charged particle. Ions are formed when atoms lose or gain electrons. We saw in Chapter 2 that atoms with a noble gas electronic configuration are particularly stable and unreactive. Noble gas atoms, except helium, have eight electrons in their outer shells. Most atoms do not have this arrangement of outer electrons and so are more reactive. One way of gaining this noble gas structure is to completely transfer electrons from one atom to another.

The metals in Group I all have one electron in their outer energy level. Figure 3.14 shows the loss of an outer electron from a sodium atom. This results in the formation of a positive sodium ion (Na^+).

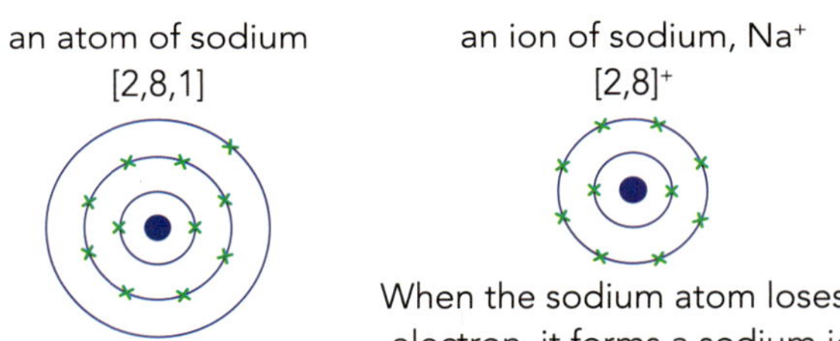

Figure 3.14: A sodium atom loses an electron to become a sodium ion.

The sodium ion has a single positive charge because it now has just 10 electrons in total, but there are still 11 protons in the nucleus of the atom.

Halogen atoms in Group VII of the Periodic Table have seven outer electrons. They can achieve a stable noble gas configuration by gaining an outer electron (Figure 3.15). The chloride ion (Cl^-) formed has a negative charge because it has one more electron (18) than there are protons in the nucleus (17).

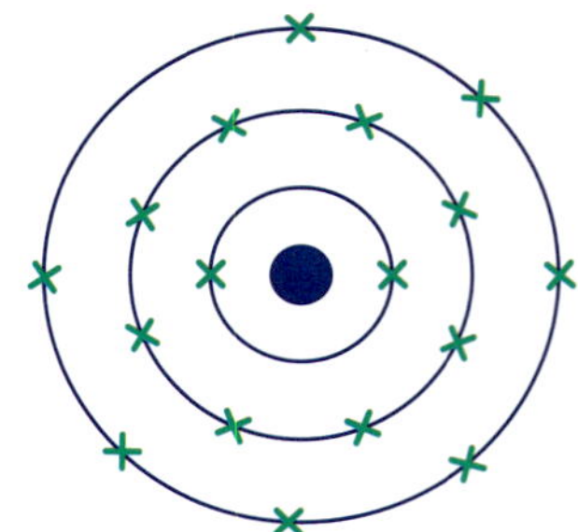

The chlorine atom [2,8,7] needs to gain an electron to make it more stable.

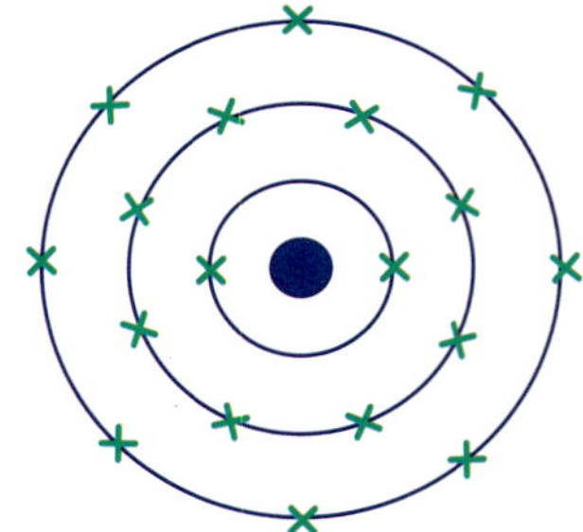

This is an ion of chlorine $[2,8,8]^-$.

Figure 3.15: A chlorine atom gains an electron to become a chloride ion.

The ease with which the atoms of Group I and Group VII can lose or gain outer electrons results in these elements being strongly reactive.

Note that positive ions are also known as **cations** and negative ions are also known as **anions**. This naming relates to the movement of these ions in electrolysis (Chapter 6).

KEY WORDS

cation: a positive ion that would be attracted to the cathode in electrolysis

anion: a negative ion that would be attracted to the anode in electrolysis

Ionic compounds

Compounds formed between a metal and a non-metal generally involve ionic bonding. This involves the transfer of electrons from one atom to another. This transfer of electrons results in the formation of positive and negative ions. In the case of sodium chloride, the ions formed are sodium (Na^+) and chloride (Cl^-) ions (Figure 3.16).

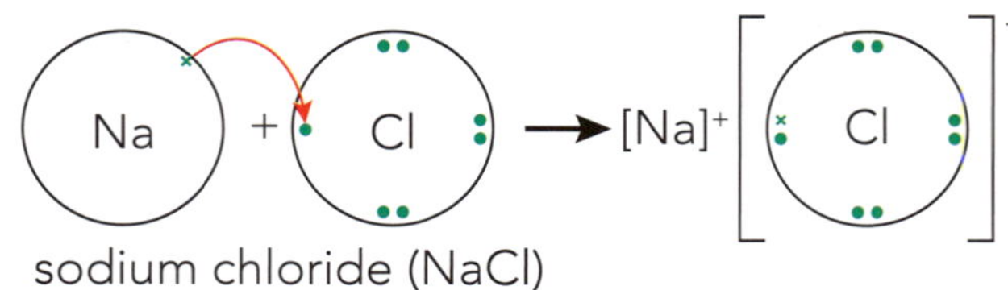

Figure 3.16: Transfer of electrons from a sodium atom to a chlorine atom to form ions.

The sodium ion then has the stable electron arrangement (2,8) of a neon atom – the element just before it in the Periodic Table. The electron released is transferred to a chlorine atom (Figure 3.16). The chloride ion formed (electron arrangement 2,8,8) has the electron arrangement of an argon atom. The positive and negative ions in sodium chloride are held together by the electrostatic attraction between opposite charges.

The important features of ionic bonding are:

- The electrons involved in the formation of ions are those in the outer shell of the atoms.
- Metal atoms lose their outer electrons to become positive ions (cations). In doing so, they achieve the more stable electron arrangement of the nearest noble gas to them in the Periodic Table.
- Generally, atoms of non-metals gain electrons to become negative ions (anions). Again, in doing so, they achieve the stable electron arrangement of the nearest noble gas to them in the Periodic Table.

The type of diagram we have used here is a dot-and-cross diagram. When drawing such a diagram we usually only show the outer electrons as these are the ones that have taken part in the bonding. Similar diagrams can be drawn for the ionic bonding in other compounds between Group I metals and Group VII non-metals – the alkali metals and the halogens. Try drawing diagrams like the one in Figure 3.16 for compounds such as lithium fluoride or potassium bromide. You will see that there is a great similarity in the diagrams.

More complex ionic compounds than those formed between the alkali metals and the halogens involve the transfer of a greater number of electrons. Diagrams showing the bonding on such compounds require care when drawing. Figure 3.17 shows two examples of such compounds. In the first case, of magnesium oxide (MgO), two electrons are being transferred from a magnesium atom to an oxygen atom. A magnesium atom has two electrons in its outer shell. An oxygen atom has six outer electrons. By transferring two electrons from magnesium to oxygen to complete its outer shell, both the magnesium and oxide ions formed have the configuration of a neon atom.

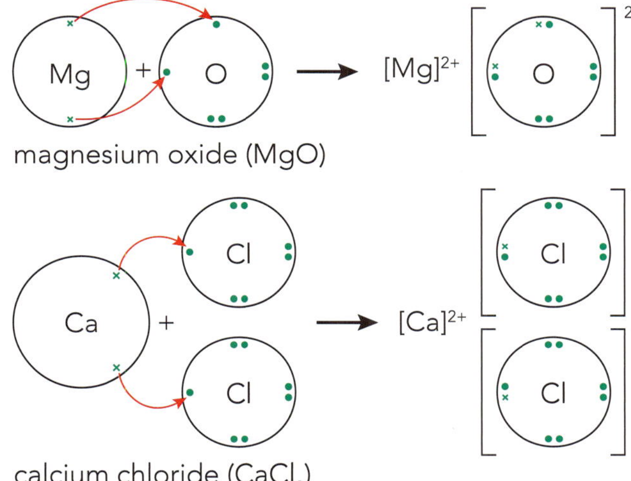

Figure 3.17: Diagrams showing the formation of ionic bonds in magnesium oxide and calcium chloride. Only the outer electrons are shown.

The case of calcium chloride is slightly more complicated. The calcium atom (2,8,8,2) has two electrons to transfer. So, two chlorine atoms are needed. Each chlorine atom accepts one electron. In calcium chloride there are two chloride ions (Cl^-) formed for each calcium ion (Ca^{2+}) produced.

In summary, the following features are common to the formation of ionic bonds between metallic and non-metallic elements:

- metal atoms always lose their outer electrons to form positive ions (cations)
- the number of positive charges on a metal ion is equal to the number of electrons lost
- non-metal atoms, with the exception of hydrogen, always gain electrons to become negative ions (anions).
- the number of negative charges on a non-metal ion is equal to the number of electrons gained
- in both cases, the ions formed have a more stable electronic configuration, usually that of the nearest noble gas to them in the Periodic Table
- ionic bonds result from the electrostatic attraction between oppositely charged ions.

Physical properties of ionic compounds

The physical properties of ionic compounds are different to those of covalent simple molecular compounds (Topic 3.1) because the type of bonding involved is different. The key general physical properties of ionic compounds are:

- they have high melting points and boiling points, and so are crystalline solids at room temperature
- they are often soluble in water
- they conduct electricity when molten or dissolved in water (not when solid).

Ionic compounds (such as sodium chloride and magnesium oxide) are solids at room temperature. The ions arrange themselves into a regular lattice structure (Figure 3.18). In this regular arrangement, each ion is surrounded by ions of the opposite charge. The whole structure is held together by the electrostatic forces of attraction that occur between particles of opposite charge.

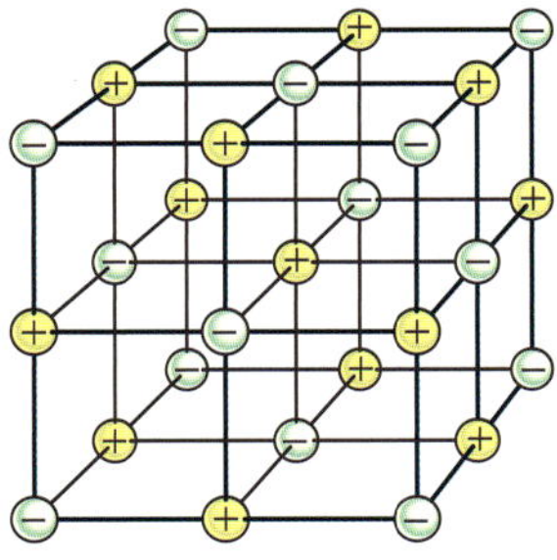

Figure 3.18: An ionic lattice where each ion is surrounded by ions of opposite charge.

Table 3.3 outlines the relationship between the general physical properties of ionic compounds and their structure and bonding. We will explore further the structural features of ionic crystals in a later section of this chapter.

Properties of typical ionic compounds	Reasons for these properties
They have high melting points and boiling points.	Ions are attracted to each other by strong electrostatic forces. Large amounts of energy are needed to separate them.
They are crystalline solids at room temperature.	There is a regular arrangement of the ions in a lattice. Ions with opposite charge are next to each other.
They are often soluble in water (not usually soluble in organic solvents, e.g. ethanol, methylbenzene).	Water is attracted to charged ions and therefore many ionic solids dissolve.
They conduct electricity when molten or dissolved in water (not when solid).	In the liquid or solution, the ions are free to move about. They can move towards the electrodes when a voltage is applied.

Table 3.3: General physical properties of ionic compounds.

Questions

5 The boiling point of a substance relates to the type of bonding present in the substance. Two elements X and Y combine to form a liquid with the relatively low boiling point of 120 °C. Which of the alternatives A–D in Table 3.4 correctly describes this substance?

	Type of element		Type of bonding
	X	Y	
A	metal	metal	covalent
B	non-metal	non-metal	ionic
C	non-metal	non-metal	covalent
D	metal	non-metal	ionic

Table 3.4: Possible descriptions of the make-up of a liquid.

6 Draw dot-and-cross diagrams of the ionic bonding in the following compounds and give their formula:

- a lithium fluoride
- b sodium oxide
- c magnesium chloride.

7 Figure 3.19 shows the structure of a molecule of methanoic acid.

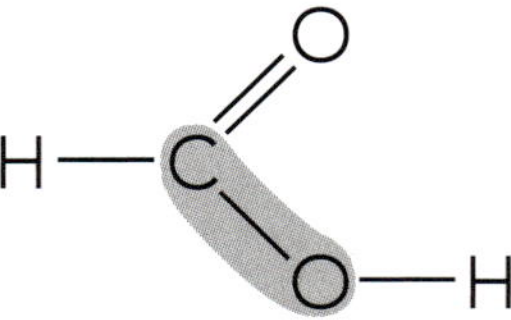

Figure 3.19: Methanoic acid.

Table 3.5 lists four descriptions of the bonds formed between the carbon and oxygen atoms in the shaded region of the structure. Which of the descriptions A–D is correct?

	Carbon	Oxygen
A	shares 1 electron with hydrogen and 3 electrons with oxygen	shares 1 electron with hydrogen and 1 electron with carbon
B	shares 2 electrons with hydrogen and 2 electrons with oxygen	shares 1 electron with hydrogen and 2 electrons with carbon
C	shares 1 electron with hydrogen and 3 electrons with oxygen	shares 1 electron with hydrogen and 2 electrons with carbon
D	shares 2 electrons with hydrogen and 2 electrons with oxygen	shares 1 electron with hydrogen and 1 electron with carbon

Table 3.5: Bonding in methanoic acid.

ACTIVITY 3.1

Part 1: Mind-mapping chemical bonding

Mind-mapping is a technique that allows you to visually organise your thoughts and share information on a complex topic. Figure 3.20 shows part of a mind map on atoms, ions and molecules. A section of the map has been drawn based on the ideas behind atomic theory.

1 Copy the current map and complete the 'atomic theory' group by filling in the missing words.

2 Draw new clouds on the unoccupied branches: the upper branch labelled 'ionic bonding' and the middle branch labelled 'covalent bonding'.

3 Add boxes around each new cloud giving the key details of that type of bonding. Include as many ideas as you can.

4 Using another coloured pen, can you make a new link between one of the lower boxes and the middle cloud?

You could also try to create your own mind map using online tools and resources available on the internet.

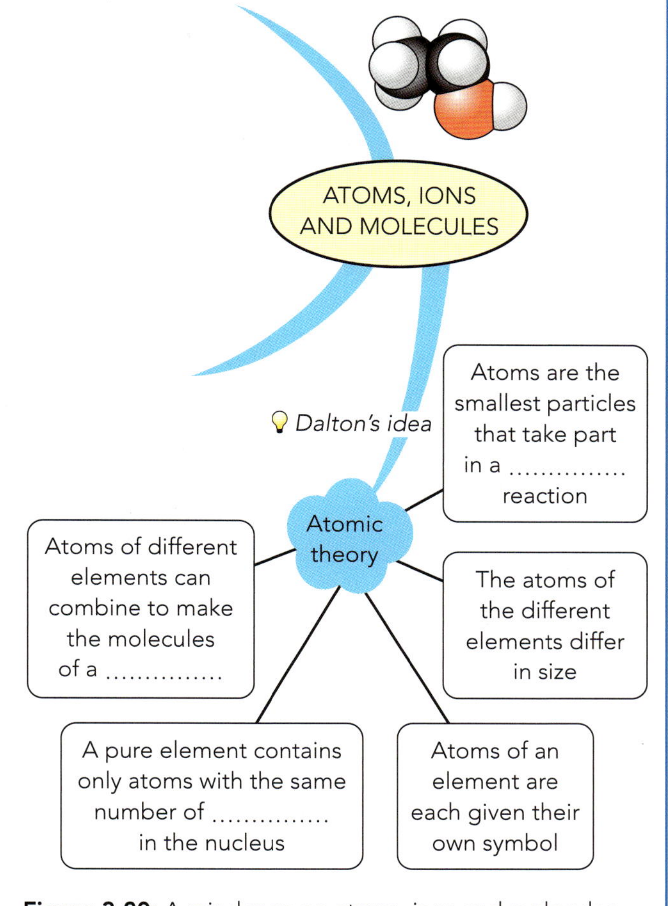

Figure 3.20: A mind map on atoms, ions and molecules.

3.3 Giant structures

Giant ionic lattice structures

Ionic compounds form lattices consisting of positive and negative ions (**giant ionic lattice**). In an ionic lattice, the nearest neighbours of an ion are always of the opposite charge. Thus, in sodium chloride, each sodium (Na^+) ion is surrounded by six chloride (Cl^-) ions (Figure 3.21), and each Cl^- ion is surrounded by six Na^+ ions. Overall, there are equal numbers of Na^+ and Cl^- ions, so the charges balance. The actual arrangement of the ions in other compounds depends on the numbers of ions involved and on their sizes. However, it is important to remember that all ionic compounds are electrically neutral.

Ionic crystals are hard but much more brittle than other types of crystal lattice. This is a result of the structure of the layers. In an ionic crystal, pushing one layer against another brings ions of the same charge next to each other. The repulsions force the layers apart (Figure 3.22).

Water can also disrupt an ionic lattice. Many ionic compounds dissolve in water. Water molecules are able to interact with both positive and negative ions. When an ionic crystal dissolves, each ion becomes surrounded by water molecules. This breaks up the lattice and keeps the ions apart (Figure 3.23). For those ionic compounds that do not dissolve in water, the forces between the ions must be very strong.

Ions in solution are able to move, so the solution can carry an electric current. Ionic compounds can conduct electricity when dissolved in water. This is also true when they are melted because here, again, the ions are able to move through the liquid and carry the current.

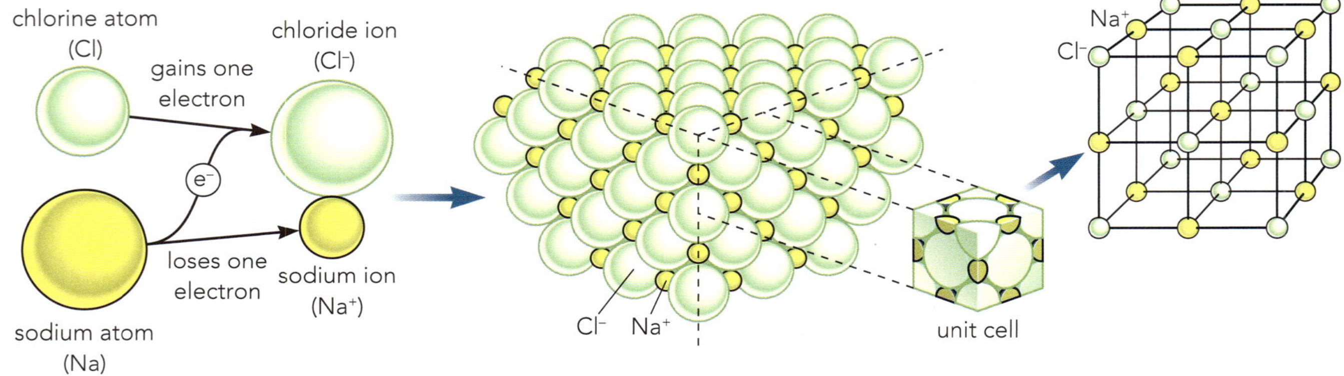

Figure 3.21: The arrangement of positive and negative ions in a sodium chloride crystal.

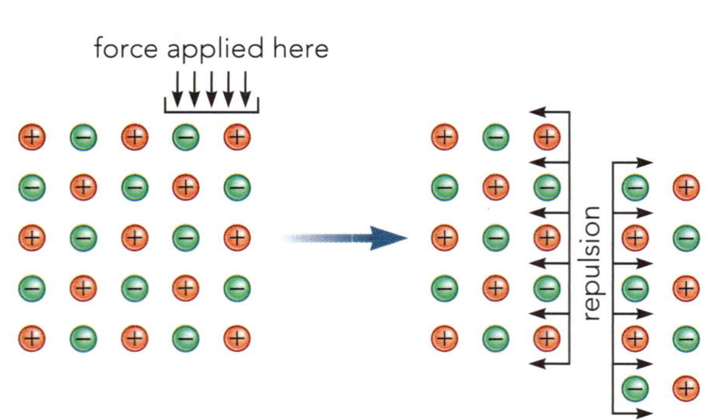

Figure 3.22: In ionic crystals, when one layer is forced to slide against another, repulsions cause the crystal to fracture.

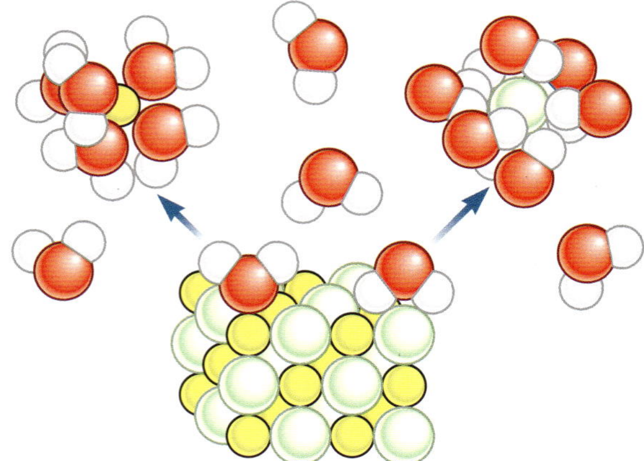

Figure 3.23: Water molecules (red and white) surround metal (yellow) and non-metal (green) ions. This helps ionic substances (e.g. sodium chloride, NaCl) to dissolve in water.

Ionic crystal lattices are called 'giant' structures because the structure repeats itself in all directions. The forces involved are the same in all directions holding the whole structure together. There are two other types of **giant structure**: **giant covalent structures** and **giant metallic lattices**.

KEY WORDS

giant ionic lattice (structure): a lattice held together by the electrostatic forces of attraction between positive and negative ions

giant structures: these are lattices where the structure repeats itself in all directions; the forces involved are the same in all directions holding the whole structure together

giant covalent structures: a substance where large numbers of atoms are held together by covalent bonds forming a strong lattice structure

giant metallic lattice: a regular arrangement of positive metal ions held together by the mobile 'sea' of electrons moving between the ions

Giant covalent structures

Giant molecular crystals are held together by strong covalent bonds. This type of structure is shown by some elements (e.g. carbon, in the form of diamond and graphite) and also by some compounds (e.g. silica, SiO_2).

Diamond and graphite

The properties of diamond are due to the fact that the strong covalent bonds extend in all directions through the whole crystal. Each carbon atom is attached to four others – the atoms are arranged tetrahedrally (Figure 3.24) and all the atoms are bonded to each other.

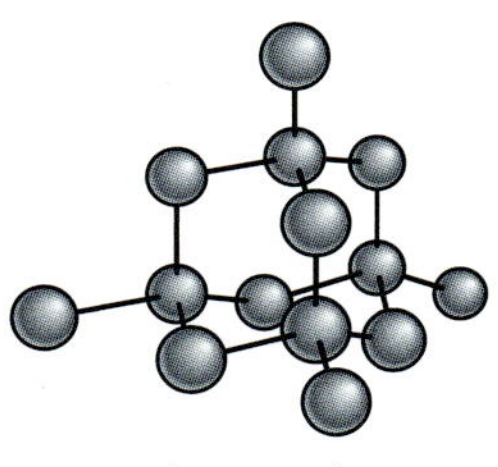

Figure 3.24: The tetrahedral structure of diamond.

Diamond has a very high melting point and, because the bonding extends throughout the whole structure, it is very hard and is used in cutting tools. The bonds are rigid, however, and these structures are much more brittle than giant metallic lattices. All the outer electrons of the atoms in these structures are used to form covalent bonds. There are no electrons free to move. Diamond is therefore a typical non-metallic element. It does not conduct electricity (Table 3.6).

	Diamond		Graphite	
	Properties	Uses	Properties	Uses
Appearance	Colourless, transparent crystals that sparkle in light	In jewellery and ornamental objects	Dark grey, shiny solid	
Hardness	The hardest natural substance	In drill bits, diamond saws and glass-cutters	Soft – the layers can slide over each other – and solid has a slippery feel	In pencils, and as a lubricant
Electrical conductivity	Does not conduct electricity		Conducts electricity	As electrodes and for the brushes in electric motors

Table 3.6: A comparison of the properties and uses of diamond and graphite.

Graphite is a different form of carbon that does conduct electricity (Table 3.6). The carbon atoms are arranged in a different way in the molecular structure of graphite. They are arranged in flat layers of linked hexagons (Figure 3.25).

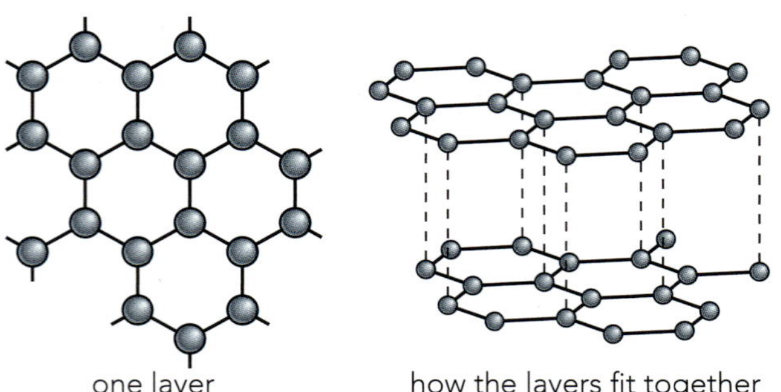

Figure 3.25: Layered structure of graphite.

Each graphite layer is a two-dimensional giant molecule. Within these layers, each carbon atom is bonded to three others by strong covalent bonds. Between the layers there are weaker forces of attraction. The layers are able to slide over each other easily. This means that graphite feels slippery and can be used as a lubricant. Graphite is used in pencils. When we write with a pencil, thin layers of graphite are left stuck to the paper.

The most distinctive property, however, results from the fact that there are free electrons not used for covalent bonding by the atoms in the layers. These electrons can move between the layers, carrying charge, so that graphite can conduct electricity in a similar way to metals.

The first samples of graphene (see the Science in Context section) were isolated in experiments aimed at seeing how thin a piece of graphite could be made by polishing or by cleaning graphite with 'sticky tape' (Figure 3.26). The layers of graphite were peeled off for surface science experiments.

Figure 3.26: A kit for preparing graphene. Items donated to the Nobel Museum in Stockholm (a piece of graphite, a graphene transistor and a sticky tape dispenser).

Silicon(IV) oxide

Sand and quartz are examples of silica (silicon(IV) oxide, SiO_2). The silicon(IV) oxide found in quartz has a structure similar to diamond (Figure 3.27). In this structure, each silicon atom is bonded to four oxygen atoms, but each oxygen is only bonded to two silicon atoms. This means that the formula for the structure is SiO_2.

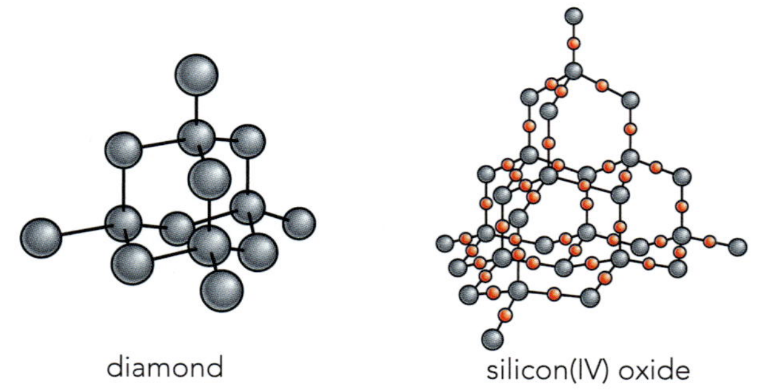

Figure 3.27: Comparison of the structures of diamond and silicon(IV) oxide.

Both diamond and silicon(IV) oxide structures have a rigid, tetrahedral arrangement of atoms. All the atoms in these structures are held together throughout by strong covalent bonds. As a result, both diamond and silicon(IV) oxide show similar physical properties. They are both very hard and have high melting points. Silicon(IV) oxide, like diamond, does not conduct electricity as there are no electrons free to move through the structure.

Metallic bonding

Giant metallic lattices

Metal atoms have relatively few electrons in their outer shells. When they are packed together, each metal atom loses its outer electrons into a **'sea' of delocalised electrons**. Having lost electrons, the atoms present are no longer electrically neutral. They become positive ions because they have lost electrons but the number of protons in the nucleus has remained unchanged. Therefore, the structure of a metal is made up of positive ions packed together. These ions are surrounded by electrons, which can move freely between the ions. These free electrons are delocalised (not restricted to orbiting one positive ion) and form a kind of electrostatic 'glue' holding the structure together in what is called **metallic bonding** (Figure 3.28).

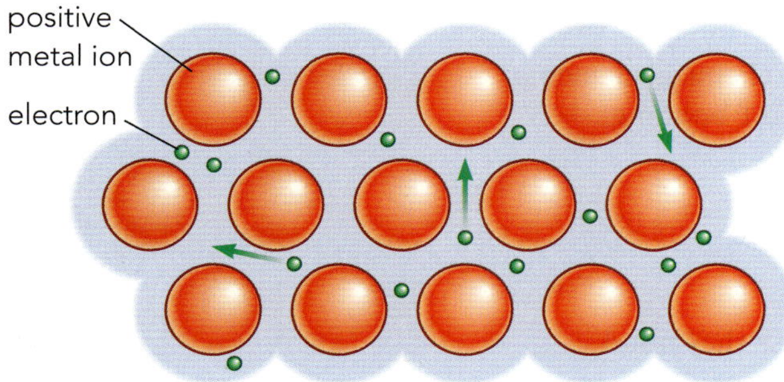

Figure 3.28: In metallic bonding, the metal ions are surrounded by a 'sea' of delocalised electrons that are free to move about.

KEY WORDS

'sea' of delocalised electrons: term used for the free, mobile electrons between the positive ions in a metallic lattice

metallic bonding: an electrostatic force of attraction between the mobile 'sea' of electrons and the regular array of positive metal ions within a solid metal

Physical properties of metals

The key physical properties of metals can be explained using the model of metallic structure and bonding:

- Most metals have high melting and boiling points. A large amount of energy is needed to overcome the strong and extensive force of attraction between the positive metal ions and the 'sea' of delocalised electrons moving within the lattice. These attractive forces can only be overcome when the temperature is high.
- Metals are good conductors of electricity. In an electrical circuit, metals can conduct electricity because the mobile electrons can move through the structure, carrying the current. This type of bonding is present in alloys as well. Alloys such as solder and brass, for example, will conduct electricity.
- Metals are easily bent and shaped (**malleable**) or stretched into wires (**ductile**). The positive ions in a metal are arranged in layers. When a force is applied, the layers can slide over each other. The attractive forces in metallic bonding act in all directions to hold the structure together. This means that when the layers slide over each other new bonds are easily formed (Figure 3.29). This movement of layers leaves the metal with a different shape.

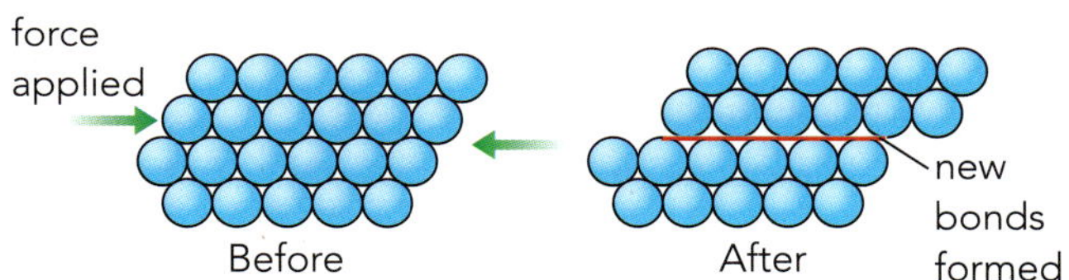

Figure 3.29: When a force is applied to a metal, the layers can slide over each other without the structure being broken.

KEY WORDS

malleable: a word used to describe the property that metals can be bent and beaten into sheets

ductile: a word used to describe the property that metals can be drawn out and stretched into wires

REFLECTION

This chapter involves drawing different diagrams and visual representations of molecules and structures. Do you find drawing these diagrams useful? Do these visualisations help you to understand the topic? Could you use diagrams in other topics in chemistry to help develop your understanding?

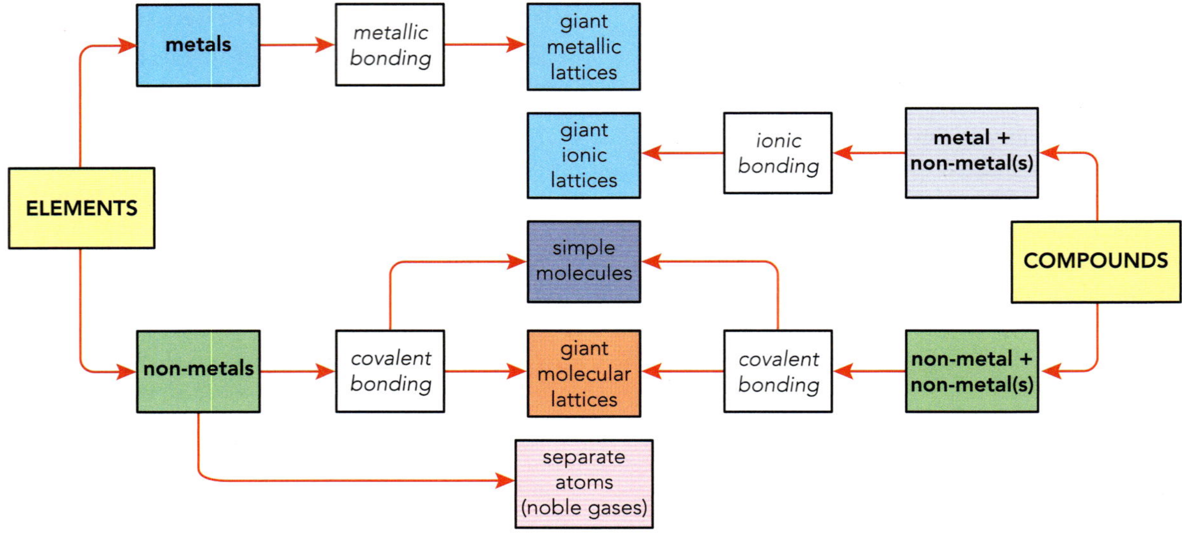

Figure 3.30: Summary of chemical bonding in elements and compounds.

Metallic bonding represents the final type of bonding present in chemical elements and compounds. We have discussed the following different types of bonding:

- covalent bonding that occurs in elements and compounds
- ionic bonding that occurs in metallic compounds
- metallic bonding present in metallic elements.

These different types of bonding and the nature of the substances they bond together are summarised in Figure 3.30.

Questions

8 Which of the following elements exists as a giant covalent structure?

A carbon **B** iodine **C** helium **D** oxygen

9 Carbon and silicon are both in Group IV of the Periodic Table. Using Table 3.7, which row (A–D) contains the text that best describes the bonding structure of carbon dioxide and silicon(IV) oxide?

	Carbon dioxide	Silicon(IV) oxide
A	Molecules attracted to each other by strong covalent bonds	Giant molecule with strong covalent bonds
B	Molecules held together by weak intermolecular forces	Molecules attracted to each other by strong covalent bonds
C	Molecules held together by weak intermolecular forces	Giant molecule with strong covalent bonds
D	Molecules attracted to each other by strong covalent bonds	Molecules attracted to each other by strong covalent bonds

Table 3.7: Bonding in carbon dioxide and silicon(IV) oxide.

10 Metals such as copper are bonded together by the attraction between metal ions and a 'sea' of delocalised electrons surrounding them. Which of the properties of metals (A–D) is not explained by this type of bonding?

A electrical conductivity
B malleability
C melting point
D reaction with acids

ACTIVITY 3.2

Part 2: Mind-mapping chemical bonding

The types of bonding and particles described in the mind map you devised in Activity 3.1 (Part 1) produce several different giant structures present in solid substances. Look back at the mind map you created and see how you can extend it to include these structures.

Peer assessment

Swap your mind map with a partner and use the following checklist to ask them questions:

- Have they completed the ideas about atomic theory?
- Have they included the key information on ionic bonding?
- Have they included the key ideas on covalent bonding?
- Have they made links between the different ideas in a different colour?
- Have they included ideas about giant structures?
- Were there ideas you could gain from looking at your partner's mind map?

SUMMARY

There are differences between chemical elements, compounds and mixtures.
A covalent bond is formed between two atoms by the sharing of a pair of outer electrons and this is illustrated using dot-and-cross diagrams.
Covalent bonding involves sharing more electrons in more complex molecules including those containing multiple bonds.
Compounds formed between Group I metals and Group VII elements involve ionic bonding using the forces of electrostatic attraction between positive metal ions (cations) and negative non-metal ions (anions), and dot-and-cross diagrams show electron transfer.
An ionic bond is a strong electrostatic attraction between oppositely charged ions.
Dot-and-cross diagrams are used to describe the formation of ionic bonds between metals and non-metals.
The physical properties of simple molecular compounds and ionic compounds can be related to their structure and bonding.
The physical properties of simple molecular compounds and ionic compounds can be explained in terms of their structure and bonding.
Diamond and graphite have giant covalent structures and their very strong structures make them useful and versatile.
Silicon(IV) oxide has a giant covalent structure similar to that of diamond, and the two substances therefore have similar physical properties.
Metallic bonding is the electrostatic attraction between positive metal ions in the metallic lattice and a 'sea' of delocalised electrons between them.
Metallic bonding explains why metals are malleable and ductile, and good electrical conductors.

PROJECT

Choice materials

Being able to link the type of bonding to the properties that a chemical has is important when selecting which material to use for a specific role (Figure 3.31). In the design of new electronic components, you need excellent conductors, while for sports equipment you might need high tensile strength and in panels for spacecraft you would need high melting points.

You have been asked to design a training tool for new chemists so that they can learn how to predict the properties of a given compound/structure and link this to their understanding of dot-and-cross diagrams.

Figure 3.31: An electric motor, showing the gold rotor (centre), the small graphite brushes on either side of the rotor and the rotating copper coil. The bonding present in each of these components is suited to its role in the motor.

CONTINUED

- Take a piece of A4 card and cut it into quarters.
- Select four different chemicals: an ionic compound, a simple covalent compound, a giant molecular compound and a metal.
- Draw diagrams of each material on one side of each card (for the ionic and simple covalent this must include dot-and-cross diagrams, for the metal it should include the lattice of ions surrounded by delocalised electrons and for the giant molecular it should just be a part of the lattice).
- Draw equally spaced lines on the reverse of each card to make five spaces and add the following details:
  1. name of compound
  2. type of bonding
  3. melting point: state whether it is high or low
  4. electrical conductivity: state whether it conducts as a solid, liquid, in solution or no electrical conductivity
  5. other properties: list any other properties e.g. lubricant, malleable, etc. (if none just leave as blank).

For example, a card on ionic bonding for calcium chloride is shown in Figure 3.32.

Important information for designing the card:

You must have the numbers 1–5 on the cards for each property.

Take one of the quarters of paper (one of the four compounds) and cut it into five segments, each giving one of the properties. Repeat for the other three compounds. Then, mix all of the properties and ask your partner to try to sort them out and put them into the correct number sequence. When completed you can ask them to draw the structure diagrams for each compound. These can be checked by flipping the card over.

1) **Name:** Calcium chloride

2) **Type of bonding:** Ionic

3) **Melting point:** High

4) **Electrical conductivity:** Conducts when molten or in solution but not as a solid

5) **Other properties:**

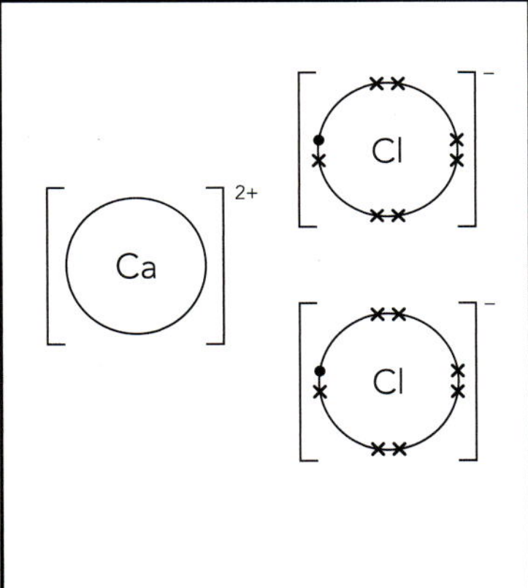

Figure 3.32: Front and reverse side of a card for an ionic compound.

EXAM-STYLE QUESTIONS

1 Some non-metallic elements form covalent simple molecular structures involving a number of atoms. Phosphorus is one of these, forming the molecule P_4.

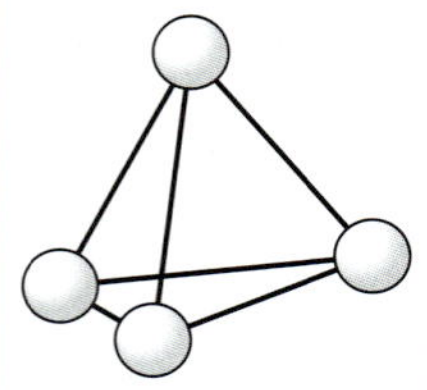

Which of A–D indicates the total number of electrons shared in the bonds in this molecule?

A 12 **B** 8 **C** 4 **D** 2 **[1]**

2 There are various different types of structure that compounds can form, depending on the nature of their bonding. A substance has an ionic structure that can be represented as shown in the figure below.

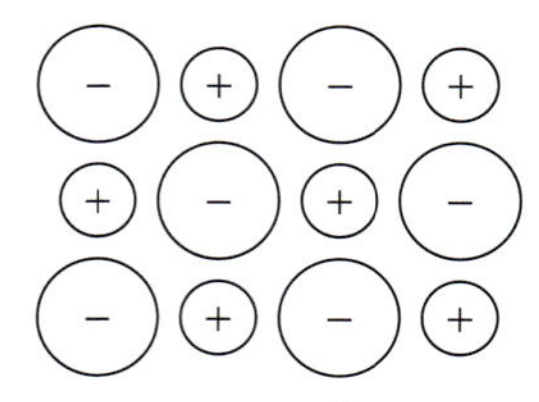

a What could this substance be?

A iodine

B water

C potassium bromide

D diamond **[1]**

b Give reasons for your answer to **a**. **[2]**

[Total: 3]

3 When potassium reacts with chlorine the ionic compound potassium chloride is formed. The electronic structures of the two ions formed are the same.

$[K\ 2,8,8]^+$ $[Cl\ 2,8,8]^-$

a Explain what has happened during the reaction to form these two ions. **[2]**

b When two chlorine atoms react with each other ions are not formed. Sketch a dot-and-cross diagram to show the bond between two chlorine atoms. Show outer shell electrons only. **[2]**

c What name is given to this type of bond? **[1]**

d Give a difference in physical properties between the chlorine and potassium chloride. **[1]**

[Total: 6]

COMMAND WORDS

give: produce an answer from a given source or recall/memory

explain: set out purposes or reasons/make the relationships between things evident/ provide why and/or how and support with relevant evidence

sketch: make a simple freehand drawing showing the key features, taking care over proportions

CONTINUED

4 Calcium chloride is an ionic compound. When it is dissolved in water, the solution conducts electricity.

a Explain why a solution of calcium chloride conducts electricity but solid calcium chloride does not. [2]

b In what other way can calcium chloride be made to conduct electricity? [1]

c Calcium chloride has a melting point of 782 °C. Water melts at 0 °C. Use this difference to **contrast** the types of attractive forces involved. [2]

[Total: 5]

5 Carbon exists in two forms, diamond and graphite. Both forms have very high melting points. Graphite conducts electricity but diamond does not.

a Explain why graphite has a high melting point and conducts electricity. [2]

b Graphite is soft and can be used as a lubricant, but diamond is very hard. Explain this difference. [2]

c Metals and ionic solids both form giant lattices containing ions.

i Give a reason why ionic solids are brittle but metals are malleable. [2]

ii Why do metals conduct electricity when solid but ionic solids do not? [2]

[Total: 8]

COMMAND WORD

contrast: identify/comment on differences

SELF-EVALUATION CHECKLIST

After studying this chapter, think about how confident you are with the different topics. This will help you see any gaps in your knowledge and help you to learn more effectively.

I can	See Topic...	Needs more work	Almost there	Confident to move on
describe the difference between elements, compounds and mixtures	3.1			
describe how a covalent bond is formed when a pair of outer electrons are shared between two atoms, using dot-and-cross diagrams to illustrate the sharing of electrons	3.1			
describe the electronic configuration in more complex molecules, using dot-and-cross diagrams	3.1			
outline some of the properties of simple molecular compounds	3.1			
explain some of the properties of simple molecular compounds (including reference to weak intermolecular forces)	3.1			
describe the formation of ionic bonds between the ions of Group I metals and Group VII halogens by electron transfer	3.2			
outline some of the physical properties of ionic compounds	3.2			
explain some of the physical properties of ionic compounds	3.2			
describe the giant lattice structures of ionic compounds as a regular arrangement of alternating cations (positive) and anions (negative)	3.3			
describe the giant molecular structures of diamond and graphite and their key properties	3.3			
describe the giant covalent structure of silicon(IV) oxide and the similarity of its properties to those of diamond	3.3			
describe the nature of metallic bonding as the electrostatic attraction between a giant metallic lattice and a 'sea' of delocalised electrons	3.3			
explain the key properties of metals in terms of their structure and metallic bonding	3.3			

> Chapter 4

Chemical formulae and equations

IN THIS CHAPTER YOU WILL:

- understand how to write the chemical formulae of elements and compounds
- define the molecular formula as the number and type of different atoms in one molecule
- deduce the formula of simple covalent compounds
- explore how to write word and balanced symbol equations, including state symbols
- describe the relative atomic mass (A_r) and understand how to use A_r to calculate the relative molecular (or formula) mass (M_r)

> deduce the formula of ionic compounds

> understand how the empirical formula of a compound is the simplest whole number ratio of the different atoms or ions present

> understand how to write ionic equations and deduce symbol equations for chemical reactions.

GETTING STARTED

Knowledge of the symbols of the elements and their use in chemical formulae is important in building confidence in chemistry.

- Think about the symbols of elements and any formulae of compounds you have met when learning science: write a list of as many as you can think of.
- Based on this list, create five questions to ask a partner.
- Working in pairs, take it in turns to ask each other your questions.

How did you do? Discuss with your partner how much you know about the elements and formulae included in your questions.

WHAT'S IN A NAME?

The 'language' of chemistry is intended to be above cultural, linguistic and national boundaries, and meant to be understood internationally. For this reason, a system of chemical symbols and formulae is used. The system of symbols and formulae used today was invented by the Swedish chemist, Jöns Jacob Berzelius (1779–1848). The agreed symbols are meant to be used by people of all languages. Each chemical element has a unique chemical symbol based on its name, but not necessarily its name in English. For example, tungsten has the chemical symbol 'W' after the German 'Wolfram'.

The use of symbols has an advantage over simply using the name of an element. The names of elements can show a language dependency, particularly in the name ending. So, for example, magnesium changes to *magnésium* in French, *magnesio* in Spanish, *magnesion* in Greek and *magnij* in Russian. In Japanese, *katakana* reproduces the sound of the English 'magnesium'. The use of symbols means that **chemical formulae** can be written that are truly universal.

The last element to be discovered in a natural source was francium (Fr) isolated by Marguerite Perey in 1939. The elements discovered after francium have all been synthesised in particle accelerators. Indeed, the recent authentication of four new elements has led to some new names (and symbols) completing the bottom row (or period) of the modern Periodic Table:

- **element 113** – nihonium, (Nh): this element was first discovered at the RIKEN Nishima Centre for Accelerator Science in Japan, and the name for the element comes from the Japanese name for the country of its discovery
- **element 115** – moscovium, (Mc): is named after the Moscow region in Russia where the Joint Institute for Nuclear Research is located
- **element 117** – tennessine, (Ts): was named in recognition of the research work on new elements carried out at Vanderbilt University and the Oak Ridge National Laboratory, which are in the state of Tennessee in the USA
- **element 118** – oganesson, (Og): is named after the Armenian nuclear physicist Yuri Oganessian (Figure 4.1), who has played a leading role in the search for new elements.

Figure 4.1: An Armenian stamp issued in 2017 to celebrate Yuri Oganessian after whom element 118 is named.

Discussion questions

1. What are the names of the elements with the symbols Lv, Fl and Cn shown in the stamp (Figure 4.1)? Which places, or people, were these elements named after? What is significant about the names of the elements Cm and Mt?
2. How many neutrons are there thought to be in an atom of oganesson? Why do you think there is uncertainty about the number in brackets next to the symbol for the element? What is the scientific purpose of studying the atoms of elements that only exist for a very short time?

4.1 Chemical names and formulae

Formulae of elements

When we were discussing the difference between elements and compounds at the beginning of Chapter 3, we briefly commented that each element has its own **chemical symbol**. The symbol for an element is one of the most important pieces of information presented in the Periodic Table. You will rapidly become familiar with the symbols that you most often use. All chemical symbols are either a single capital letter, such as hydrogen (H), oxygen (O) and nitrogen (N), or made up of two letters, such as aluminium (Al), argon (Ar) or magnesium (Mg). It is important to remember that the second letter of any symbol is always a small (lower case) letter. For some elements the symbol is taken from the Latin name of the element (Table 4.1).

Element	Latin name	Chemical symbol
silver	argentum	Ag
gold	aurum	Au
copper	cuprum	Cu
sodium	natrium	Na
potassium	kalium	K
iron	ferrum	Fe
lead	plumbum	Pb

Table 4.1: Latin names and chemical symbols for some important elements.

The use of symbols means that we can very quickly represent an element and its structure. For the elements whose structures are made up of individual atoms (the noble gases), the formula of the element is simply its symbol (Figure 4.2).

Where elements exist as giant structures, whether held together by metallic or covalent bonding, the formula is simply the symbol of the element (e.g. Cu, Mg, Fe, Na, K, C, Si, Ge). For convenience, the same applies to elements such as phosphorus (P) or sulfur (S) where the molecules contain more than three atoms (strictly speaking they should be P_4 and S_8).

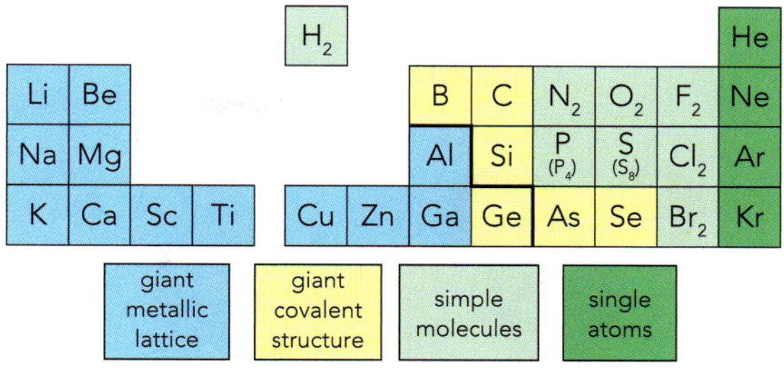

Figure 4.2: Formulae of the elements reflect their structure and position in the Periodic Table.

Naming chemical compounds

Chemical compounds consist of elements bonded together. However, many compounds are not simply made up of one element bonded to another. The naming system must take this into account. One important aspect of this is that there are some ions that are made up of more than two elements. Some of these ions form part of some very important substances (e.g. calcium carbonate and potassium nitrate).

Simple and compound ions

The ionic compounds mentioned in earlier chapters are all composed of two simple ions bonded together (e.g. Na^+, K^+, Mg^{2+}, Cl^-, O^{2-}). However, in many important ionic compounds the metal ion is combined with a negative ion containing a group of atoms (e.g. SO_4^{2-}, NO_3^-, CO_3^{2-}). These **compound ions**, or groups, are made up of atoms covalently bonded together. These groups have a negative charge because they have gained electrons to make a stable structure. Examples of such ions are shown in Figure 4.3. In addition to these negative compound ions, there is one important compound ion that is positively charged, the ammonium ion, NH_4^+ (Figure 4.3).

KEY WORDS

chemical symbol: a letter or group of letters representing an element in a chemical formula

compound ion: an ion made up of several different atoms covalently bonded together and with an overall charge (can also be called a molecular ion; negatively charged compound ions containing oxygen can be called oxyanions)

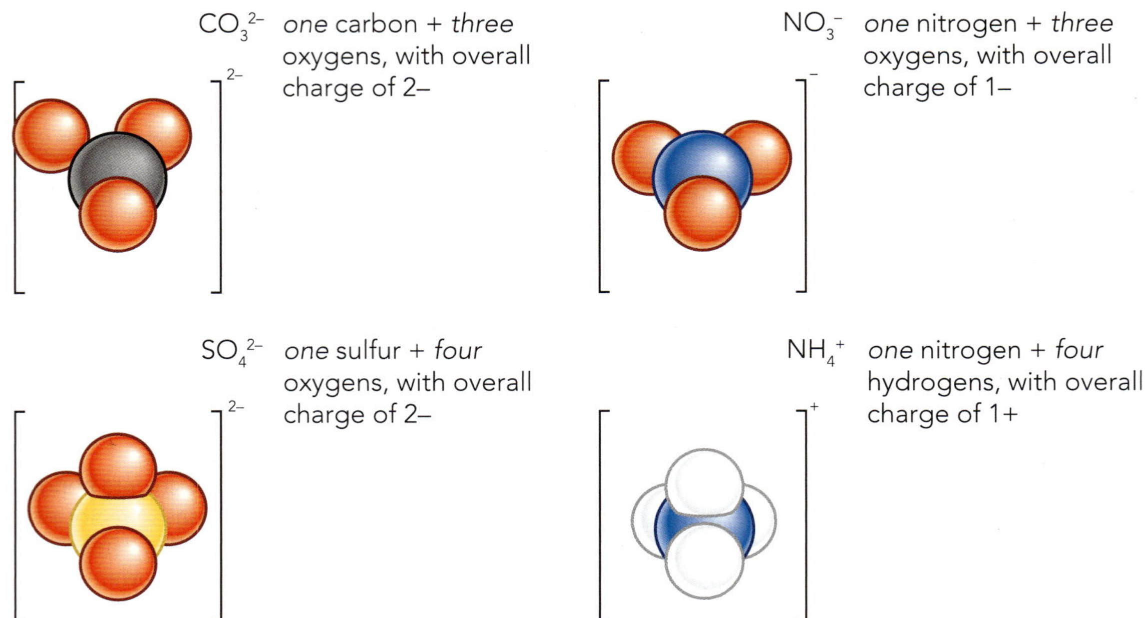

Figure 4.3: Three examples of negatively charged compound ions and a positively charged compound ion. The numbers of atoms and the overall charge carried by each group of atoms are shown.

Rules for naming compounds

Compounds are built from atoms and ions. Giving a name to a compound is a way of classifying the compound and it is useful to know there are some rules to naming most compounds. As we commented earlier, the name of a compound should then tell you which atoms or ions are present. There are some basic generalisations that are useful:

- If there is a metal in the compound, it is named first.
- Where the metal can form more than one ion, then the name indicates which ion is present. For example, iron(II) chloride contains the Fe^{2+} ion, while iron(III) chloride contains the Fe^{3+} ion.
- Compounds containing only two elements have names ending in *-ide*. For example, sodium chloride (NaCl), calcium bromide ($CaBr_2$) and magnesium nitride (Mg_3N_2). The important exception to this is the hydroxides, which contain the hydroxide (OH^-) ion.
- Compounds containing a compound ion (usually containing oxygen) have names that end with *-ate*. For example, calcium carbonate ($CaCO_3$), potassium nitrate (KNO_3), magnesium sulfate ($MgSO_4$) and sodium ethanoate (CH_3COONa).
- The names of some compounds use prefixes to tell you the number of that particular atom in the molecule. This is useful if two elements form more than one compound. For example, carbon *mon*oxide (CO) and carbon *di*oxide (CO_2), nitrogen *di*oxide (NO_2) and *di*nitrogen *tetr*oxide (N_2O_4), and sulfur *di*oxide (SO_2) and sulfur *tri*oxide (SO_3).

The names for the important mineral acids follow a logical system but are best simply learnt at this stage, e.g. sulfuric acid (H_2SO_4). Some common and important compounds have historical names that do not fit with the above rules. Examples of these include water (H_2O), ammonia (NH_3) and methane (CH_4).

Formulae of covalent compounds

Atoms of each element have a combining power, sometimes referred to as their *valency*. The combining power of an atom in a covalent molecule is the number of bonds that atom makes. This idea of an atom having a valency can be used to work out the formulae of covalent compounds. Note that the term valency is not required knowledge.

The valency of an element in the main groups of the Periodic Table can be deduced from the group number of the element. The relationship between valency and group number is:

- For elements in Groups I–IV
  valency = group number
- For elements in Groups V–VII
  valency = 8–the group number
- For elements in Group VIII
  valency = 0

This trend in valency with the group number can be seen by looking at typical compounds of the elements of Period 3 (Table 4.2). You can see that the valency rises to a value of 4 and then decreases to 0 as we cross the period (e.g. carbon is in Group IV, so its valency is 4, and oxygen is in Group VI, so its valency is 8 − 6 = 2).

The 'cross-over' method for working out chemical formulae can be applied to simple covalent compounds where the molecules have a central atom, e.g. water, methane, carbon dioxide and ammonia (Worked example 4.1).

Group	I	II	III	IV	V	VI	VII	VIII
Valency (combining power)	1	2	3	4	3	2	1	0
Typical compound	NaCl	$MgCl_2$	$AlCl_3$	$SiCl_4$	PH_3	H_2S	HCl	—

Table 4.2: Typical compounds of the elements of Period 3 showing the change in valency across the period.

WORKED EXAMPLE 4.1

Work out the formula of carbon dioxide

A method involving the symbols and valencies of the elements can be used:

1 Write down the symbols of the elements.

2 Write the valencies of the elements beneath their symbols (Figure 4.4).

3 If necessary, simplify the ratio.

Formula of carbon dioxide

Write down the symbols

Write down the valencies

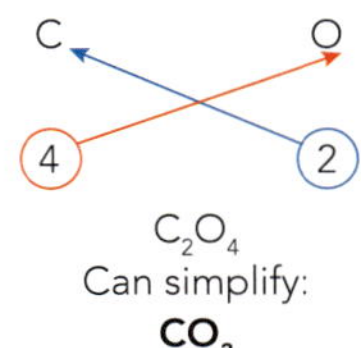

Formula C_2O_4
Can simplify:
CO_2

Figure 4.4: A method for finding the formula of carbon dioxide.

Now apply this 'cross-over' method to finding the formula of:

a methane (the simplest compound of carbon and hydrogen)

b nitrogen chloride (a simple molecular compound)

If we are given a diagram of a molecule showing all the atoms and bonds, we can easily work out its **molecular formula**. We do this by simply counting the number of each type of atom present. Figure 4.5 shows diagrams of molecules with the molecular formulae BrF_3, NH_3 and C_3H_6.

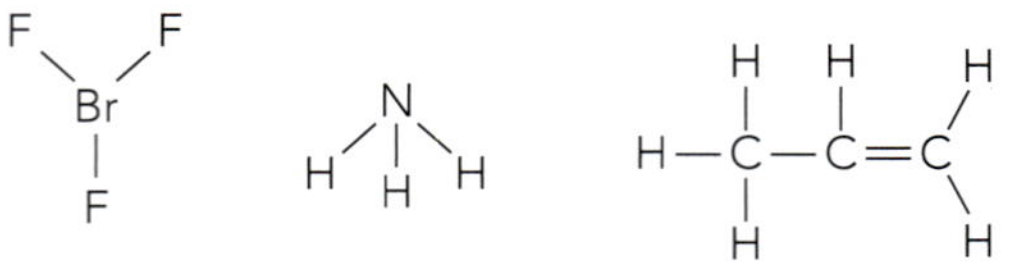

Figure 4.5: Diagrams showing the structures of BrF_3, NH_3 and C_3H_6.

The molecular formula of a compound is the formula you should use in chemical equations because it shows the actual number of each type of atom present in a molecule.

KEY WORDS

molecular formula: a formula that shows the actual number of atoms of each element present in a molecule of the compound

Empirical and molecular formulae of covalent compounds

If you look closely at the third structure in Figure 4.5, it has the molecular formula C_3H_6. We do not simplify the formula of this compound down to CH_2 as that would not represent the actual molecule. The method for working out the formulae of simple covalent molecules by crossing over the valencies does not work for large numbers of covalent molecules, particularly those that do not have a single central atom, e.g. H_2O_2, C_2H_6, C_3H_6, etc. (Figures 4.5 and 4.6).

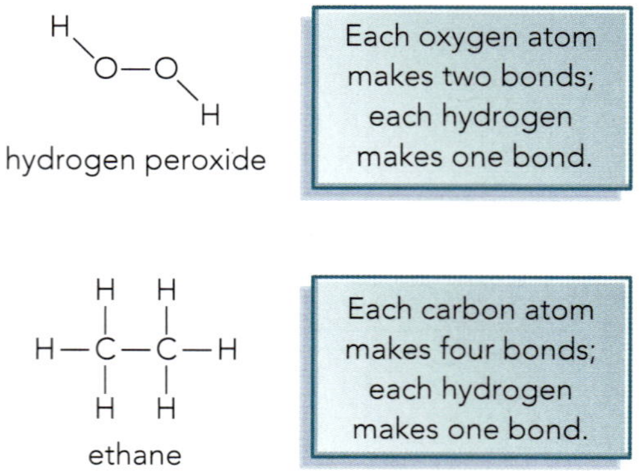

hydrogen peroxide

Each oxygen atom makes two bonds; each hydrogen makes one bond.

ethane

Each carbon atom makes four bonds; each hydrogen makes one bond.

Figure 4.6: Structures of hydrogen peroxide (H_2O_2) and ethane (C_2H_6), showing the bonds made.

The molecular formulae of compounds such as hydrogen peroxide and ethane still obey the valency rules but it is important to know the structure of the molecules. The numbers in the formula represent the actual number of atoms of each element present in a molecule of the compound.

When chemists were first working out the formulae of compounds, they were calculating the formulae from data on the reacting masses of the elements that formed a compound. These calculations from practical experiments gave the ratio of the elements present in a compound. We will consider this type of calculation in Chapter 5. For the third compound (propene) in Figure 4.5, such experiments and data would give a ratio for C:H of 1:2. This result indicates a formula of CH_2 for this compound. Other evidence on the structure of the molecule tells us that the molecular formula of propene is C_3H_6. This type of formula based on the simplest ratio of the elements present is known as the **empirical formula**. The empirical formula of a compound is the simplest whole number ratio of the different atoms in that compound. In the case of propene, the empirical formula is not the same as the molecular formula.

The two types of formula are numerically related; the molecular formula must be a whole number multiple of the empirical formula for a compound. Compounds such as ethane (Figure 4.6) have two different formulae:

- the molecular formula of ethane is C_2H_6
- the empirical formula of ethane is CH_3

KEY WORDS

empirical formula: a formula for a compound that shows the simplest ratio of atoms present

Some non-metallic compounds form giant covalent structures. The most important example of this is silicon(IV) oxide (Chapter 3). The formula for any giant covalent compound is always the simplest whole number ratio of the atoms present in the structure. Such compounds are always represented by their empirical formula. The formula of silicon(IV) oxide is therefore SiO_2 as the atoms are present in a 1:2 ratio in the crystal structure. The formula of any giant covalent compound can be worked out by the cross-over method using the valencies of the elements. You could prove this for yourself for silicon(IV) oxide.

Formulae of ionic compounds

Ionic compounds are solids at room temperature. Such compounds have giant ionic lattice structures (Chapter 3) and their formulae are simply the whole number ratio of the positive to negative ions in the structure. Thus, in magnesium chloride, there are two chloride ions (Cl^-) for each magnesium ion (Mg^{2+}).

ions present	Mg^{2+}	Cl^-
total charge	2+	2–

The chemical formula is $MgCl_2$. The overall structure must be neutral. The positive and negative charges must balance each other. Because they are giant ionic structures, ionic compounds are always represented by their empirical formulae and the cross-over method can be used to work these formulae out.

The size of the charge on an ion is a measure of its valency (Table 4.3). Mg^{2+} ions can combine with Cl^- ions in a ratio of 1 : 2, but Na^+ ions can only bond in a 1 : 1 ratio with Cl^- ions. Table 4.3 gives a summary of some simple and compound ions.

In ionic compounds the valency of each ion is equal to its charge. This idea can be used to ensure that you always use the correct formula for an ionic compound. Follow the examples of aluminium oxide and calcium oxide in Figure 4.7 (Worked example 4.2), and make sure you understand how this works.

Valency (combining power)	Simple metal ions	Simple non-metallic ions		Compound ions	
	(+ve cation)	(+ve cation)	(−ve anion)	(+ve cation)	(−ve anion)
1	sodium, Na^+	hydrogen, H^+	hydride, H^-	ammonium, NH_4^+	hydroxide, OH^-
	potassium, K^+		chloride, Cl^-		nitrate, NO_3^-
	silver, Ag^+		bromide, Br^-		hydrogencarbonate, HCO_3^-
	copper(I), Cu^+		iodide, I^-		
2	magnesium, Mg^{2+}		oxide, O^{2-}		sulfate, SO_4^{2-}
	calcium, Ca^{2+}		sulfide, S^{2-}		carbonate, CO_3^{2-}
	zinc, Zn^{2+}				
	iron(II), Fe^{2+}				
	copper(II), Cu^{2+}				
3	aluminium, Al^{3+}		nitride, N^{3-}		phosphate, PO_4^{3-}
	iron(III), Fe^{3+}				

Table 4.3: Some common simple and compound ions.

WORKED EXAMPLE 4.2

Working out the formulae of aluminium oxide and calcium oxide

Formula for aluminium oxide

Write down the correct symbols — Al, O

Write down the charges on the ions — 3+, 2−

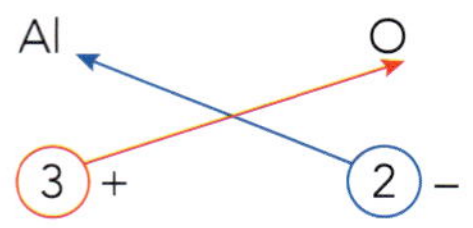

Formula Al_2O_3

Formula for calcium oxide

Write down the correct symbols — Ca, O

Write down the charges on the ions — 2+, 2−

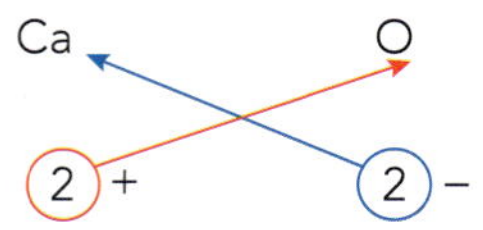

Ca_2O_2

Simplify the ratio:

Formula **CaO**

Figure 4.7: Working out the formulae of aluminium oxide and calcium oxide.

Now find the formulae of:

a magnesium iodide

b aluminium bromide

The same rules apply when writing the formulae of compounds containing compound ions because each of them has an overall charge (Table 4.3). It is useful to put the formula of the compound ion in brackets, showing that the formula of this ion cannot be changed. For example, the formula of the carbonate ion is always CO_3^{2-}. Worked example 4.3 shows how the formulae for sodium carbonate and ammonium sulfate can be determined.

WORKED EXAMPLE 4.3

Working out the formulae of sodium carbonate and ammonium sulfate

Formula of sodium carbonate

Write down the correct symbols — Na, (CO_3)

Write down the charges on the ions — 1+, 2−

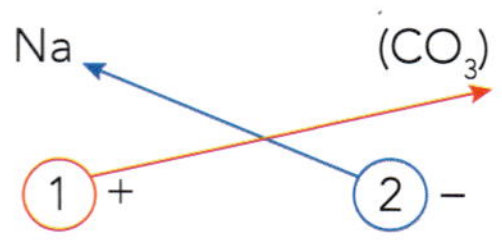

Formula Na_2CO_3

The brackets are not needed if there is only one ion present

CONTINUED

Formula of ammonium sulfate

Write down the correct symbols (NH_4) (SO_4)

Write down the charges on the ions 1+ 2−

Formula $(NH_4)_2SO_4$

Figure 4.8: Working out the formulae of sodium carbonate and ammonium sulfate.

Now find the formulae of:

a potassium phosphate

b ammonium nitrate

Ionic compounds can be represented by diagrams of their structure. Figure 4.9 shows the structure of calcium chloride.

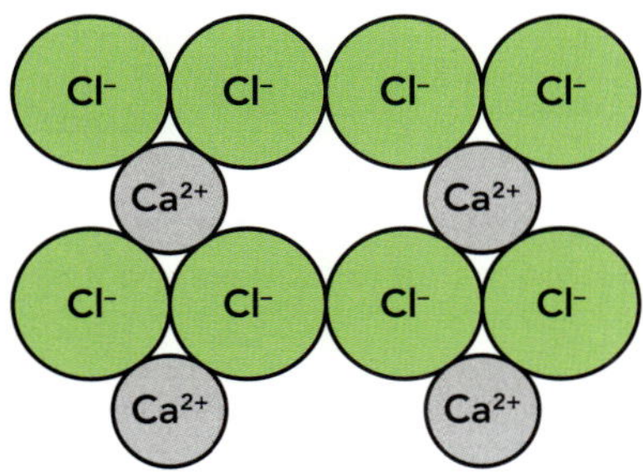

Figure 4.9: Representing the structure of calcium chloride ($CaCl_2$) diagrammatically.

The diagram shows four positive ions and eight negative ions. So Ca^{2+} ions = 4 and Cl^- ions = 8. If we divide through by 4, we get the simplest ratio, which is Ca^{2+} ions = 1 and Cl^- ions = 2. So, the formula of calcium chloride is $CaCl_2$.

The formula of an ionic compound is always the simplest whole number ratio of the ions present in the compound (the empirical formula). That is why we should always simplify the ratio by cancelling down, as we did for the example of calcium chloride. Table 4.4 summarises the formulae of some important ionic compounds.

Name	Formula	Ions present		Ratio
sodium chloride	$NaCl$	Na^+	Cl^-	1 : 1
ammonium nitrate	NH_4NO_3	NH_4^+	NO_3^-	1 : 1
potassium sulfate	K_2SO_4	K^+	SO_4^{2-}	2 : 1
calcium hydrogencarbonate	$Ca(HCO_3)_2$	Ca^{2+}	HCO_3^-	1 : 2
copper(II) sulfate	$CuSO_4$	Cu^{2+}	SO_4^{2-}	1 : 1
magnesium nitrate	$Mg(NO_3)_2$	Mg^{2+}	NO_3^-	1 : 2
aluminium chloride	$AlCl_3$	Al^{3+}	Cl^-	1 : 3

Table 4.4: Formulae of some ionic compounds.

ACTIVITY 4.1

Jigsaw compounds

Work in groups to make a set of jigsaw cards for the common ions that form compounds (Figure 4.10).

Step 1: Use the selection of ionic compounds and ions in Table 4.4.

Step 2: In order for the jigsaw pieces to make a compound, the shapes need to fit together to make a rectangle. On a piece of paper, draw seven rectangles to represent the seven compounds.

Step 3: Divide each rectangle into cards representing the ions needed for each compound. For the positive ions, create 'slots' in the card. For the negative ions, create 'tabs'. The 'tabs' on the negative ions should fit into the 'slots' in the positive ion cards. The number of slots or tabs on each card represents the charge of the ion. Use Figure 4.10 as a guide. More than one card of the same ion may be needed to complete the compound. In addition to the cards based on the ions in Table 4.4 you should make cards for CO_3^{2-}, O_2^- and Fe^{3+} ions (see Figure 4.10).

Step 4: Make sure your cards are labelled with the formula of the ion it represents. Then your rectangle of cards will give you the formula of the ionic compound concerned.

Step 5: Cut out your cards to be used as a jigsaw.

- Having made your set of cards, use them to work out the formulae of the following compounds: copper carbonate, iron(III) oxide and ammonium sulfate.
- Then challenge each other in the group to make a compound you name.
- Using some of your examples, compare this method to the cross-over method described earlier in this chapter.

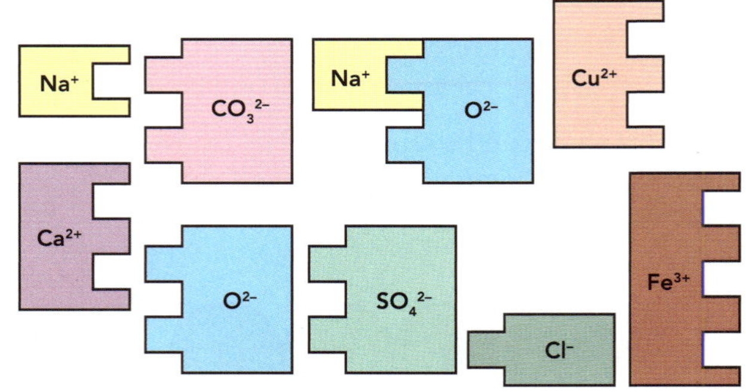

Figure 4.10: Examples of jigsaw cards designed to help with matching ionic charge to form compounds.

Self-assessment

Think about the following questions, and give yourself a ☺ 😐 or ☹

How confident did you feel …

- working out and writing the formulae of ionic compounds?
- working out the formula of copper carbonate?
- working out the formula of iron(III) oxide?
- working out the formula of ammonium sulfate?

Do you find this method of finding the formula of an ionic compound more helpful than the cross-over method?

4.2 Chemical equations for reactions

Word equations

The 'volcano reaction', in which ammonium dichromate is decomposed, gives out a large amount of energy and produces nitrogen gas (Figure 4.11). Other reactions produce gases much less violently. The neutralisation of an acid solution with an alkali produces no change that you can see. However, a reaction has happened.

Figure 4.11: Decomposition of ammonium dichromate (the 'volcano experiment') produces heat, light and an apparently large amount of powder.

We can write out descriptions of chemical reactions, but these would be quite long. To understand and group similar reactions together, it is useful to have a shorter way of describing them. The simplest way to do this is in the form of a **word equation**. This type of equation links together the names of the substances that react (**reactants**) with those of the new substances formed (**products**). The word equation for burning magnesium in oxygen would be:

$$\underset{\text{reactants}}{\text{magnesium + oxygen}} \rightarrow \underset{\text{product}}{\text{magnesium oxide}}$$

The reaction between hydrogen and oxygen is another highly exothermic reaction. The hydrogen–oxygen fuel cell is based on this reaction (Chapter 6). The word equation for this reaction is:

$$\text{hydrogen + oxygen} \rightarrow \text{water}$$

Note that, although a large amount of energy is produced in this reaction, it is *not* included in the equation. An equation includes only the chemical substances involved, and energy is not a chemical substance.

KEY WORDS

word equation: a summary of a chemical reaction using the chemical names of the reactants and products

reactants: (in a chemical reaction) the chemical substances that react together in a chemical reaction

products: (in a chemical reaction) the substance(s) produced by a chemical reaction

This type of equation gives us some information. But equations can be made even more useful if we write them using chemical formulae.

Constructing balanced symbol equations

From investigations of a large number of different chemical reactions, a very important point about all reactions has been discovered. It is summed up in a law, known as the law of conservation of mass. This law of conservation of mass states that the total mass of all the products of a chemical reaction is always equal to the total mass of all the reactants.

This important law becomes clear if we consider what is happening to the atoms and molecules involved in a reaction. During a chemical reaction, the atoms of one element are not changed into those of another element. Nor do atoms disappear from the mixture or indeed appear from nowhere. A reaction involves the breaking of some bonds between atoms, and then the making of new bonds between atoms to give the new products. During a chemical reaction, some of the atoms present 'change partners'. During the reaction, atoms separate from those they were joined to and become bonded to atoms from the other reactants.

Look more closely at the reaction between hydrogen and oxygen molecules shown in Figure 4.12.

hydrogen + oxygen → water

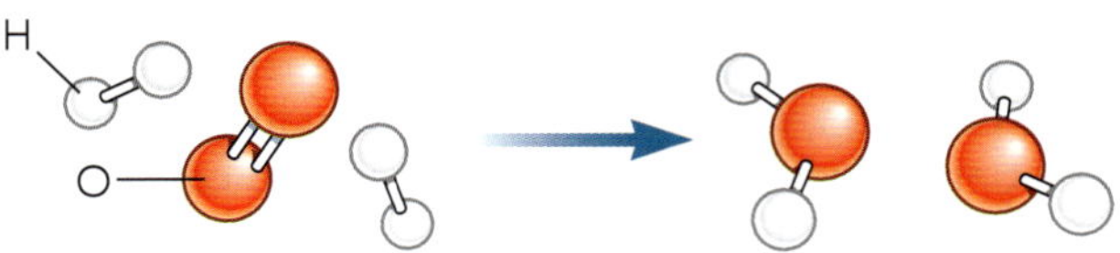

Figure 4.12: A chemical reaction involves atoms 'changing partners' and new substances being formed.

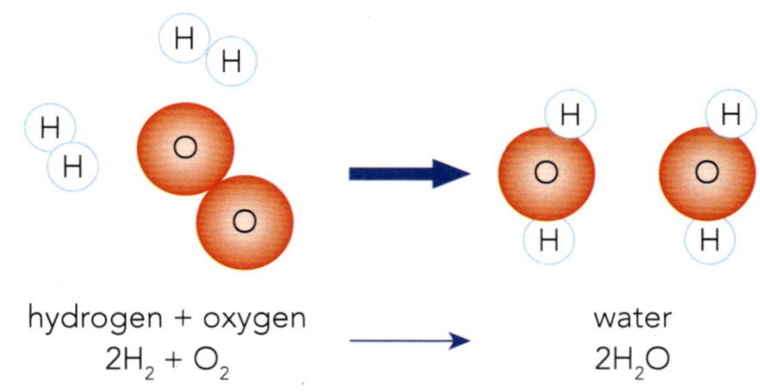

Figure 4.13: Summary of the reaction between hydrogen and oxygen.

Each molecule of water (H_2O) contains only one oxygen atom (O). It follows that one molecule of oxygen (O_2) has enough oxygen atoms to produce two molecules of water (H_2O). Therefore, two molecules of hydrogen (H_2) will be needed to provide enough hydrogen atoms (H) to react with each oxygen molecule. The numbers of hydrogen and oxygen atoms are then the same on both sides of the equation.

The symbol equation for the reaction between hydrogen and oxygen is therefore written:

$$2H_2 + O_2 \rightarrow 2H_2O$$

This is a **balanced chemical equation**. The numbers of each type of atom are the same on both the reactant side and the product side of the equation: four hydrogen atoms and two oxygen atoms on each side (Figure 4.13).

A balanced equation gives us more information about a reaction than we can get from a simple word equation. Worked example 4.4 shows a step-by-step approach to working out the balanced equation for a reaction.

KEY WORDS

balanced chemical (symbol) equation: a summary of a chemical reaction using chemical formulae – the total number of any of the atoms involved is the same on both the reactant and product sides of the equation

WORKED EXAMPLE 4.4

What is the balanced equation for the reaction between magnesium and oxygen?

Step 1. Make sure you know what the reactants and products are. For example, magnesium burns in air (oxygen) to form magnesium oxide.

Step 2. From this you can write out the word equation:

magnesium + oxygen → magnesium oxide

Step 3. Write out the equation using the formulae of the elements and compounds:

$Mg + O_2 \rightarrow MgO$

Remember that oxygen exists as diatomic molecules. This equation is not balanced: there are two oxygen atoms on the left, but only one on the right.

Step 4. Balance the equation:

$2Mg + O_2 \rightarrow 2MgO$

Now work through the steps to find the balanced symbol equation for the reaction that takes place when burning sodium is lowered into a gas jar of chlorine gas. Sodium chloride is produced in this reaction. Remember that chlorine is a diatomic gas.

We cannot alter the formulae of the substances involved in the reaction. These are fixed by the bonding in the substance itself. We can only put multiplying numbers in front of each formula where necessary.

Chemical reactions do not only involve elements reacting together. In most reactions, compounds are involved. For example, potassium metal is very reactive and gives hydrogen gas when it comes into contact with water. potassium reacts with water to produce potassium hydroxide and hydrogen (Figure 4.14).

Figure 4.14: Potassium reacts strongly with water to produce hydrogen.

All the alkali metals react with water in the same way. So, if you know one of these reactions, you know them all. The general equation is:

alkali metal + water → metal hydroxide + hydrogen

Therefore:

potassium + water → potassium hydroxide + hydrogen

Then:

$$K + H_2O \rightarrow KOH + H_2$$

This symbol equation needs to be balanced. An even number of hydrogen atoms is needed on the product side, because on the reactant side the hydrogen occurs as H_2O. Therefore, the amount of KOH must be doubled. Then the number of potassium atoms and water molecules must be doubled on the left:

$$2K + 2H_2O \rightarrow 2KOH + H_2$$

This equation is now balanced. Check for yourself that the numbers of the three types of atom are the same on both sides.

Questions

1. Write word equations for the reactions described in a–c.
   - **a** Iron rusts because it reacts with oxygen in the air to form a compound called iron(III) oxide.
   - **b** Sodium hydroxide neutralises sulfuric acid to form sodium sulfate and water.
   - **c** Sodium reacts strongly with water to give a solution of sodium hydroxide; hydrogen gas is also given off.
2. Balance the following symbol equations:
   - **a** $\_\_Cu + O_2 \rightarrow \_\_CuO$
   - **b** $N_2 + \_\_H_2 \rightarrow \_\_ NH_3$
   - **c** $\_\_Na + O_2 \rightarrow \_\_Na_2O$
   - **d** $\_\_NaOH + H_2SO_4 \rightarrow Na_2SO_4 + \_\_H_2O$
   - **e** $\_\_Al + \_\_Cl_2 \rightarrow \_\_AlCl_3$
   - **f** $\_\_Fe + \_\_H_2O \rightarrow \_\_Fe_3O_4 + \_\_H_2$
3. Change these word equations into balanced symbol equations.
   - **a** hydrogen + chlorine → hydrogen chloride
   - **b** copper + oxygen → copper oxide
   - **c** magnesium + zinc chloride → magnesium chloride + zinc

Using state symbols

So far, our equations have told us nothing about the physical state of the reactants and products. Chemical equations can be made more useful by including symbols that give us this information. These are called **state symbols**. They show clearly whether a gas is given off or a solid precipitate is formed during a reaction. The four symbols used are shown in Table 4.5.

Symbol	Meaning
s	solid
l	liquid
g	gas
aq	aqueous solution (dissolved in water)

Table 4.5: State symbols used in chemical equations.

The examples A–C show how state symbols can be used (the points of particular interest are shown in bold type). Note that when water itself is produced in a reaction, it has the symbol (l) for liquid, not (aq).

A magnesium + nitric acid
→ magnesium nitrate + hydrogen

$$Mg(s) + 2HNO_3(aq) \rightarrow Mg(NO_3)_2(aq) + \mathbf{H_2(g)}$$

B hydrochloric acid + sodium hydroxide
→ sodium chloride + water

$$HCl(aq) + NaOH(aq) \rightarrow NaCl(aq) + \mathbf{H_2O(l)}$$

C copper(II) sulfate + sodium hydroxide
→ copper(II) hydroxide + sodium sulfate

$$CuSO_4(aq) + 2NaOH(aq) \rightarrow \mathbf{Cu(OH)_2(s)} + Na_2SO_4(aq)$$

KEY WORDS

state symbols: symbols used to show the physical state of the reactants and products in a chemical reaction

Ionic equations

Equations **B** and **C** in the previous section involve mixing solutions that contain ions. Only some of the ions present actually change their status – by changing either their bonding or their physical state. The other ions present are simply *spectator ions* to the change. These spectator ions are present during a chemical reaction but do not take part in it. Note that use of this term is not specifically required at this level.

Equation B involves the neutralising of hydrochloric acid with sodium hydroxide solution:

$$HCl(aq) + NaOH(aq) \rightarrow NaCl(aq) + H_2O(l)$$

Writing out all the ions present, we get:

$$[H^+(aq) + \cancel{Cl^-(aq)}] + [\cancel{Na^+(aq)} + OH^-(aq)] \rightarrow [\cancel{Na^+(aq)} + \cancel{Cl^-(aq)}] + H_2O(l)$$

The use of state symbols clearly shows which ions have not changed during the reaction. They have been crossed out (~~like this~~) and can be left out of the equation. This leaves us with the essential **ionic equation** for all neutralisation reactions (Chapter 11):

$$H^+(aq) + OH^-(aq) \rightarrow H_2O(l)$$

Applying the same principles to a precipitation reaction again gives us a clear picture of which ions are reacting (Figure 4.15).

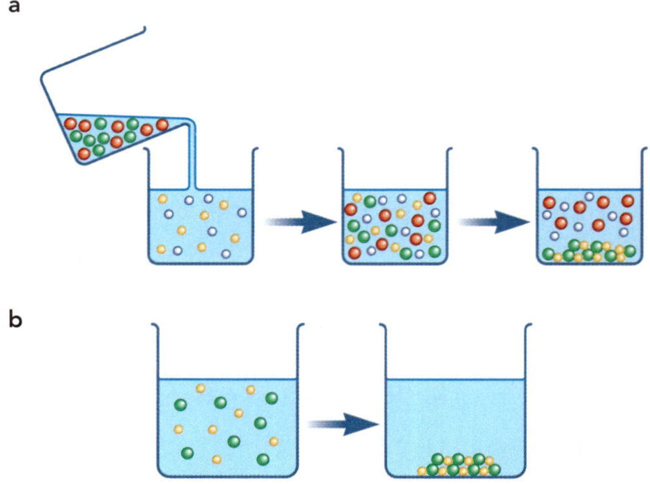

Figure 4.15: A precipitation reaction in which two solutions containing ions are mixed: **a:** the overall reaction, and **b:** the actual reaction with the spectator ions not shown.

The equation:

$$CuSO_4(aq) + 2NaOH(aq) \rightarrow Cu(OH)_2(s) + Na_2SO_4(aq)$$

for the precipitation of copper(II) hydroxide becomes:

$$Cu^{2+}(aq) + 2OH^-(aq) \rightarrow Cu(OH)_2(s)$$

This is the essential ionic equation for the precipitation of copper(II) hydroxide; the spectator ions (sulfate and sodium ions) have been left out.

KEY WORDS

ionic equation: the simplified equation for a reaction involving ionic substances: only those ions which actually take part in the reaction are shown

Questions

4 Explain the meaning of the symbols (s), (l), (aq) and (g) in the following equation, with reference to each reactant and product:

$Na_2CO_3(s) + 2HCl(aq)$
$\rightarrow 2NaCl(aq) + H_2O(l) + CO_2(g)$

5 Write an ionic equation, including state symbols, for each of the following reactions:

  a silver nitrate solution + sodium chloride solution → silver chloride + sodium nitrate solution

  b sodium sulfate solution + barium nitrate solution → sodium nitrate solution + barium sulfate

6 Write full symbol equations for the following reactions, followed by the ionic equation for the reaction (include state symbols in both types of equation).

  a dilute hydrochloric acid + potassium hydroxide solution → potassium chloride solution + water

  b dilute hydrochloric acid + copper carbonate → copper chloride solution + water + carbon dioxide

4.3 Relative masses of atoms and molecules

Relative atomic mass of elements

The mass of a single hydrogen atom is incredibly small when measured in grams (g):

mass of one hydrogen atom = 1.7×10^{-24} g

= 0.0000000000000000000000017g

It is much more useful and convenient to measure the masses of atoms relative to each other. To do this, a standard atom has been chosen, against which all others are then compared.

This standard atom is an atom of the carbon-12 isotope, the 'mass' of which is given the value of exactly 12 (Figure 4.16).

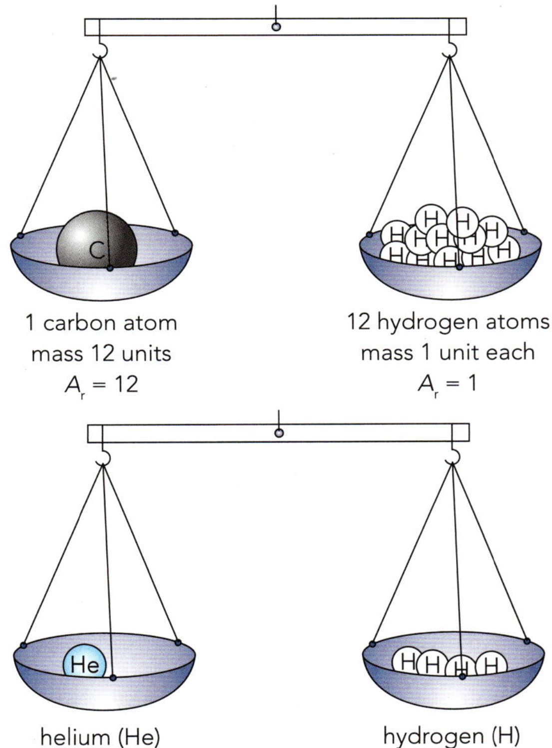

Figure 4.16: Relative mass of atoms. Twelve hydrogen atoms have the same mass as one atom of carbon-12. A helium atom has the same mass as four hydrogens.

The use of the mass spectrometer first showed the existence of isotopes. Isotopes are atoms of the same element that have different masses because they have different numbers of neutrons in the nucleus. We discussed the existence and structure of isotopes in Chapter 2. In that chapter we also discussed the calculation of the relative atomic mass (A_r) of an element that has isotopes. The majority of elements do in fact have several isotopes (Figure 4.17).

The relative atomic mass of an element is the average mass of an atom of the element, taking into account the different natural isotopes of that element. So most relative atomic masses are not whole numbers. But in this book and the Periodic Table used, with the exception of chlorine, they are rounded to the nearest whole number to make our calculations easier.

It is important to note that the mass of an ion will be the same as that of the parent atom. The mass of the electron(s) gained or lost in forming the ion can be ignored in comparison to the total mass of the atom.

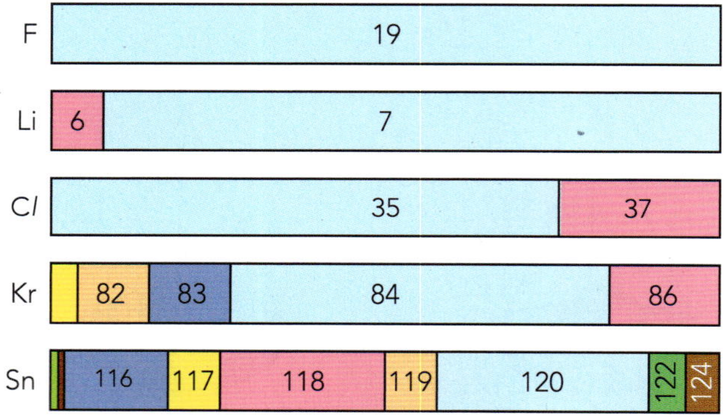

Figure 4.17: Many different elements have more than one isotope. These bars show the proportions of different isotopes for some elements. Fluorine is rare in having just one isotope.

Relative molecular (formula) mass of compounds

Atoms combine to form molecules or groups of ions. The total masses of these molecules or groups of ions provide useful information on the way the elements have combined with each other. The formula of an element or compound is taken as the basic unit (the **formula unit**) that summarises the composition of the substance. The masses of the atoms or ions in the formula are added together. The mass of a simple covalent molecular substance found in this way is called the **relative molecular mass (M_r)**. Here we illustrate the method by calculating the relative molecular masses of three simple substances.

Hydrogen gas is made up of H_2 molecules (H–H). Each molecule contains two hydrogen atoms. So, its relative molecular mass is twice the relative atomic mass of hydrogen:
$M_r(H_2) = 2 \times 1 = 2$

Water is a liquid made up of H_2O molecules (H–O–H). Each molecule contains two hydrogen atoms and one oxygen atom. So, its relative molecular mass is twice the relative atomic mass of hydrogen plus the relative atomic mass of oxygen:
$M_r(H_2O) = (2 \times 1) + 16 = 18$

Sodium chloride is an ionic solid. It contains one chloride ion for each sodium ion present. In this case there are no molecules of sodium chloride. The formula unit of sodium chloride is therefore Na^+Cl^-. So, its relative formula mass is the relative atomic mass of sodium plus the relative atomic mass of chlorine:
$M_r(NaCl) = 23 + 35.5 = 58.5$

If the substance is ionic or a giant covalent compound (such as silicon(IV) oxide) then we calculate the mass of the formula unit. We call this the **relative formula mass (M_r)**.

The practical result of these definitions can be seen by looking at further examples (Table 4.6).

The percentage by mass of a particular element in a compound can be found from calculations of relative formula mass. Figure 4.18 shows how this works for the simple case of sulfur dioxide (SO_2), whose mass is made up of 50% each of the two elements.

KEY WORDS

formula unit: this unit of an element or compound is the molecule or group of ions defined by the chemical formula of the substance

relative molecular mass (M_r): the sum of all the relative atomic masses of the atoms present in a molecule

relative formula mass (M_r): the sum of all the relative atomic masses of the atoms present in a 'formula unit' of a substance

Substance	Formula	Atoms in formula	Relative atomic masses, A_r	Relative molecular (or formula) mass, M_r
hydrogen	H_2	2H	H = 1	2 × 1 = 2
carbon dioxide	CO_2	1C	C = 12	1 × 12 = 12
		2O	O = 16	2 × 16 = 32
				44
calcium carbonate	$CaCO_3$ (one Ca^{2+} ion, one CO_3^{2-} ion)	1Ca	Ca = 40	1 × 40 = 40
		1C	C = 12	1 × 12 = 12
		3O	O = 16	3 × 16 = 48
				100
ammonium sulfate	$(NH_4)_2SO_4$ (two NH_4^+ ions, one SO_4^{2-} ion)	2N	N = 14	2 × 14 = 28
		8H	H = 1	8 × 1 = 8
		1S	S = 32	1 × 32 = 32
		4O	O = 16	4 × 16 = 64
				132
hydrated magnesium sulfate	$MgSO_4.7H_2O$ (one Mg^{2+} ion, one SO_4^{2-} ion, seven H_2O molecules)	1Mg	Mg = 24	1 × 24 = 24
		1S	S = 32	1 × 32 = 32
		4O	O = 16	4 × 16 = 64
		14H	H = 1	14 × 1 = 14
		7O	O = 16	7 × 16 = 112
				246

Table 4.6: Relative formula masses of some elements and compounds.

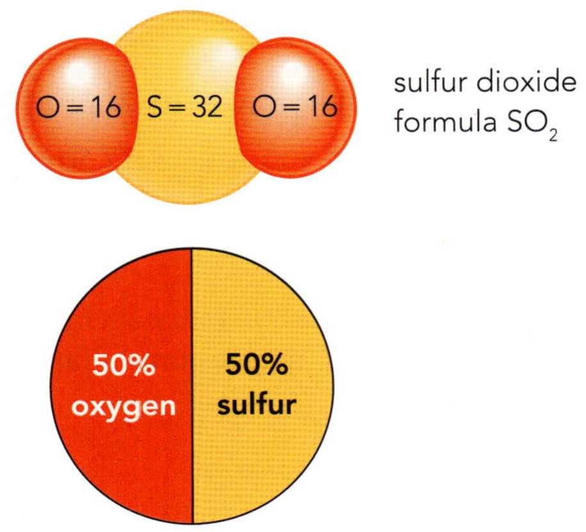

Figure 4.18: Percentage composition by mass of sulfur dioxide.

We will look further at the percentage composition of more complex compounds in Chapter 5.

Compound formation and chemical formulae

The idea that compounds are made up of elements combined in fixed amounts can be shown experimentally. Samples of the same compound made in different ways always contain the same elements. Also, the masses of the elements present are always in the same ratio.

Several different groups in a class can prepare magnesium oxide by heating a coil of magnesium in a crucible (Figure 4.19). The crucible must first be weighed empty, and then re-weighed with the magnesium in it. The crucible is then heated strongly. Air is allowed in by occasionally lifting the lid very carefully but solid must not be allowed to escape as a white smoke. The crucible and products are then allowed to cool before re-weighing. The crucible may be heated and re-weighed until there is no further change in mass (known as 'heating to constant mass'). This process is a way of making sure the reaction has completely finished.

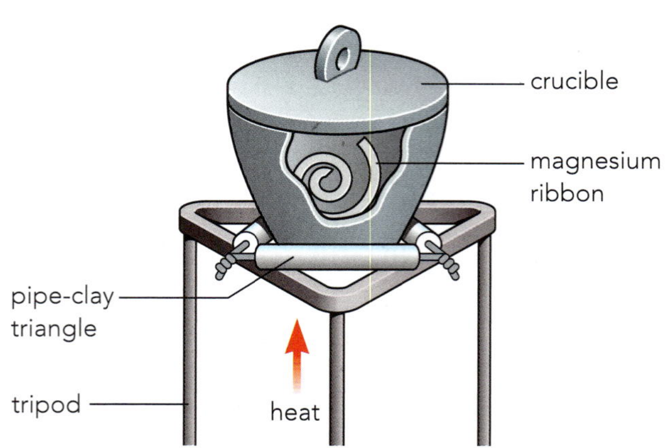

Figure 4.19: Heating magnesium in a crucible.

The increase in mass is due to the oxygen that has now combined with the magnesium. The mass of magnesium used and the mass of magnesium oxide produced can be found from the results (Worked example 4.5).

The results of the various experiments in the class can be plotted on a graph. The mass of oxygen combined with the magnesium (y-axis) is plotted against the mass of magnesium used (x-axis). Figure 4.20 shows some results obtained from this experiment.

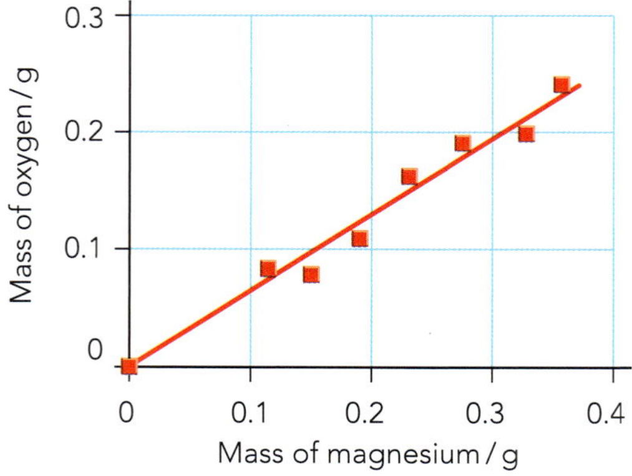

Figure 4.20: Graph of the results obtained from heating magnesium in air. The graph shows the mass of oxygen (from the air) that reacts with various masses of magnesium.

WORKED EXAMPLE 4.5

How much magnesium oxide is produced from a given mass of magnesium?

Here are some results obtained from this experiment:

a mass of empty crucible + lid = 8.52 g

b mass of crucible + lid + magnesium = 8.88 g

c mass of crucible + lid + magnesium oxide = 9.12 g

d mass of magnesium (**b** − **a**) = **0.36 g**

mass of magnesium oxide (**c** − **a**) = **0.60 g**

mass of oxygen combined with magnesium

= 0.60 − 0.36 = 0.24 g

0.60 g of magnesium oxide is produced from heating 0.36 g of magnesium

Now try the similar calculation on practical data from the heating of hydrated magnesium sulfate crystals. Find the mass of water chemically combined in the sample of crystals.

i mass of empty crucible + lid = 10.20 g

ii mass of crucible + lid + crystals = 22.50 g

iii mass of crucible + lid + dehydrated crystals = 16.20 g

Use this data to find the mass of water chemically combined in the crystals of magnesium sulfate.

The results show that:

- The more magnesium used, the more oxygen combines with it from the air and the more magnesium oxide is produced. The graph is a straight line, showing that the ratio of magnesium to oxygen in magnesium oxide is fixed. A definite compound is formed by a chemical reaction.
- A particular compound always contains the same elements and these elements are always present in the same proportions by mass. It does not matter where the compound is found, or how it is made, these proportions cannot be changed.

In this example, magnesium oxide always contains 60% magnesium and 40% oxygen by mass. Experiments on other compounds show the same pattern. For example, ammonium nitrate always contains 35% nitrogen, 60% oxygen and 5% hydrogen by mass.

Similar experiments can be done to show that the **water of crystallisation** present in a particular hydrated salt, such as hydrated copper(II) sulfate ($CuSO_4 \cdot 5H_2O$), is always the same fraction of the total mass of the salt.

KEY WORDS

water of crystallisation: water included in the structure of certain salts as they crystallise, e.g. copper(II) sulfate pentahydrate ($CuSO_4 \cdot 5H_2O$) contains five molecules of water of crystallisation per molecule of copper(II) sulfate

Calculating reacting masses from equations

Relative molecular or formula masses can also be used to calculate the amounts of compounds reacted together or produced in reactions (Worked example 4.6).

Calculations of quantities like these are a very important part of chemistry. These calculations show how there is a great deal of information 'stored' in chemical formulae and equations. The equation for the reaction between magnesium and oxygen defines the proportions in which the two elements always react (Figure 4.21).

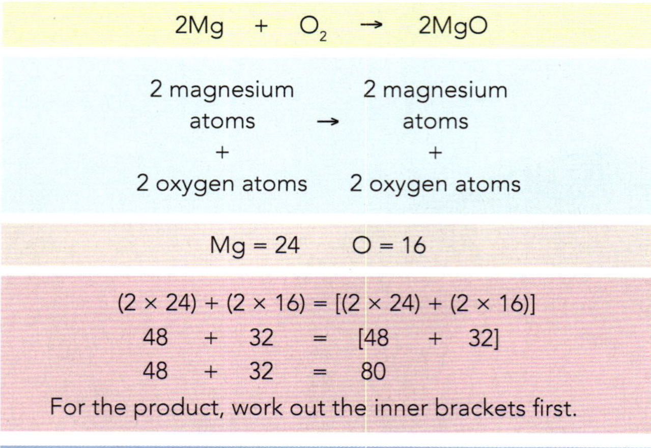

Figure 4.21: The proportions in which magnesium and oxygen react are defined by the chemical equation for the reaction.

WORKED EXAMPLE 4.6

Calculating the amounts of compounds reacted together or produced in reactions

If 0.24 g of magnesium react with 0.16 g of oxygen to produce 0.40 g of magnesium oxide (Figure 4.19), how much magnesium oxide (MgO) will be produced by burning 12 g of magnesium?

We have:

0.24 g Mg producing 0.40 g MgO

so 1 g Mg produces $\frac{0.40}{0.24}$ g MgO

= 1.67 g MgO

so 12 g Mg produces 12 × 1.67 g MgO

= 20 g MgO

Using this method based on simple proportions, calculate the mass of calcium oxide formed if 8.0 g of calcium carbonate are decomposed completely by heating. A previous experiment had shown that 20.0 g of calcium carbonate produces 11.2 g of calcium oxide when heated.

Questions

7 Figure 4.23 represents the structures of six different compounds A–F.

- **a** What type of bonding is present in compounds A, C, D, E and F?
- **b** What type of bonding is present in compound B?
- **c** State the simplest formula for each compound A–F.

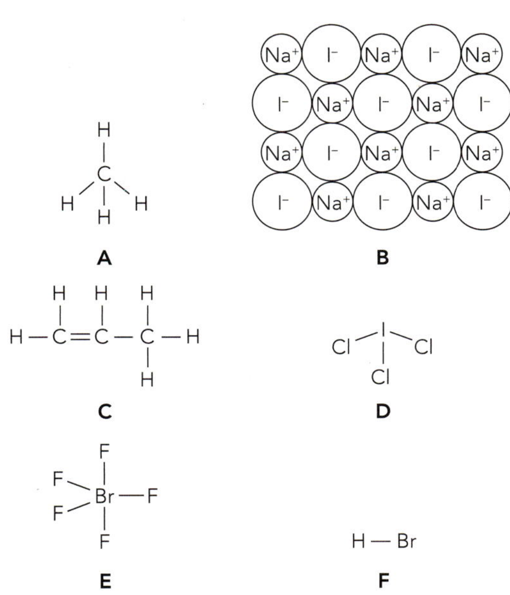

Figure 4.22: Diagrams representing the structures of several different compounds.

8 Calculate the relative molecular or formula masses (M_r) of the following substances:

a oxygen, O_2

b ammonia, NH_3

c sulfuric acid, H_2SO_4

d potassium bromide, KBr

e copper nitrate, $Cu(NO_3)_2$

(A_r: H = 1, N = 14, O = 16, S = 32, K = 39, Cu = 64, Br = 80)

9 Calcium carbonate has the formula $CaCO_3$.

a What is its relative formula mass (M_r)?

(A_r: Ca = 40, C = 12, O = 16)

b If 100 g of calcium carbonate contain 40 g of calcium, how much calcium is present in 15 g of calcium carbonate?

REFLECTION

Do you have a successful strategy to help you balance equations? What aspects do you find easy or difficult?

Could you explain how to calculate relative molecular or formula masses to someone else?

SUMMARY

Elements and compounds can be represented by chemical formulae using the symbols of the elements present and showing the numbers of atoms present in the molecule.

The formula of a simple covalent compound can be deduced from a model or diagram of the structure.

The formula of a more complex compound can be deduced from models or diagrams of the structure.

The formula of an ionic compound can be determined from the charges on the ions present or from diagrams representing the structure.

The empirical formula of a compound is the simplest whole number ratio of the different atoms or ions present.

Chemical reactions can be represented by word equations and balanced symbol equations, and state symbols can be used to show the physical states of the reactants and products.

Ionic equations can be constructed for appropriate chemical reactions.

The symbol equation for a chemical reaction can be deduced from a description of the reaction involved.

Values for relative atomic mass (A_r) can be used to calculate the relative molecular (formula) mass (M_r) of a compound, and this information can then be used to calculate the masses of substances involved in a chemical reaction.

PROJECT

Chemical formula bingo

Bingo is a game where players try to match the numbers that someone calls out with the details on a card that they have been given (Figure 4.23).

Figure 4.23: Bingo players mark their cards as the numbers are called out randomly.

What to do:

1. On a piece of paper write down the numbers 1–20.
2. Alongside each number write down one positively charged cation and one negatively charged anion from Table 4.3.
3. Using these ions, deduce an ionic formula, e.g. sodium carbonate is Na_2CO_3 and ammonium chloride is NH_4Cl.
4. When you have written down your list of 20 ionic compounds, produce a set of three or four different 3×3 grids.
5. In each, add any nine formulae from your list (Figure 4.24). These must be selected in a completely random order.
6. You are now ready to play.

CONTINUED

FORMULA BINGO!		
KBr	Na_2CO_3	Al_2O_3
$MgSO_4$	NaCl	NH_4Cl
$Ca(NO_3)_2$	$AlCl_3$	CuO

Figure 4.24: A possible 3×3 grid for chemical formula bingo.

Rules:

- each player in the group has their own game card
- fold up the pieces of paper with the numbers 1–20 on them
- then place these numbered pieces of paper in a small box
- pull a number out at random and from your list of formulae read out the two ions present in the compound
- the players need to check their card for this formula and if they have it, tick it off
- repeat by drawing further numbers, continuing until someone shouts bingo (or house!)
- bingo is won when someone ticks off all the formulae on their card.

To play several rounds, you could either laminate the bingo cards or play by placing a counter over any formulae that have already been drawn.

The game can be extended by adding a set of 20 different covalent compounds, giving the valence and elements present and drawing up a further set of 3×3 grids.

EXAM-STYLE QUESTIONS

1 Every atom has its own atomic mass, A_r. These are added together to find the relative formula mass, M_r, of a substance. The value for calcium carbonate, $CaCO_3$, is 100.

What mass of carbon is present in 100 g of calcium carbonate?

A 12 g
B 60 g
C 48 g
D 36 g **[1]**

CONTINUED

2 Aqueous phosphoric acid reacts with magnesium carbonate to form magnesium phosphate, water and carbon dioxide gas.

a Phosphoric acid has the structure shown in the figure.

O
||
HO — P — OH
|
OH

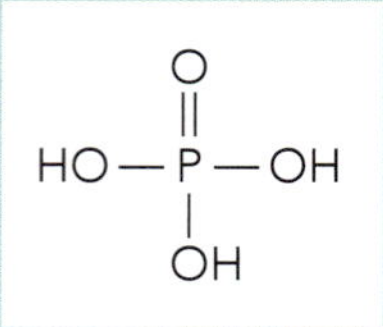

What is its molecular formula? [1]

b Write the word equation for the reaction between magnesium carbonate and phosphoric acid, including state symbols. [2]

c Solid magnesium carbonate has the formula $MgCO_3$. **Calculate** its relative formula mass. [2]

d **Define** the term *relative formula mass*. [1]

e 6 g of magnesium reacts to form 21 g of magnesium carbonate. Calculate the mass of magnesium carbonate formed from 1.2 g of magnesium. [2]

[Total: 8]

3 Butenedioic acid is a relatively complex acid molecule and has the structure shown in the figure.

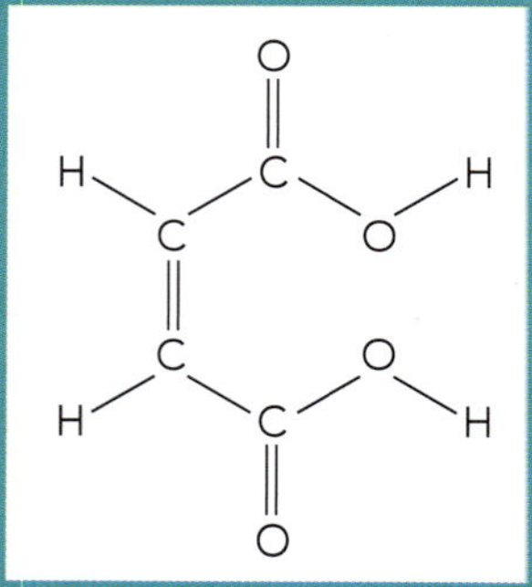

What is the empirical formula of butenedioic acid?

A $C_3H_2O_3$

B $C_2H_2O_2$

C CHO

D C_4HO_2 [1]

COMMAND WORDS

calculate: work out from given facts, figures or information

define: give precise meaning

CONTINUED

4 The structure of butanoic acid is shown in the figure.

H H H O
H—C—C—C—C
H H H O—H

a What is the empirical formula of butanoic acid? [1]

b Calculate the molecular mass of butanoic acid. [1]

c Butanoic acid dissolves in water forming butanoate and hydrogen ions.

$C_3H_7COOH \rightarrow C_3H_7COO^- + H^+$

Magnesium reacts with butanoic acid solution to form a salt, magnesium butanoate and hydrogen. Write a symbol equation for this reaction, including state symbols. [4]

d Butanoic acid has a *molecular mass* but magnesium butanoate has a *formula mass*. Both of these terms have the same definition. What is this definition? [2]

e Explain why different terms are used for the two quantities. [1]

[Total: 9]

SELF-EVALUATION CHECKLIST

After studying this chapter, think about how confident you are with the different topics. This will help you see any gaps in your knowledge and help you to learn more effectively.

I can	See Topic...	Needs more work	Almost there	Confident to move on
write the chemical formulae of elements and compounds	4.1			
deduce the formula of a simple covalent molecule from a model or diagram	4.1			
deduce the formula of a complex covalent molecule from a model or diagram	4.1			
understand that the empirical formula is the simplest ratio of the different atoms or ions in the compound	4.1			
write word equations and balanced symbol equations for chemical reactions, including the use of state symbols	4.2			
understand how to construct ionic equations for appropriate reactions	4.2			
deduce symbol equations for reactions form a relevant description of the reaction involved	4.2			
use relative atomic masses to calculate the relative molecular (formula) mass for a compound and use this information to calculate the masses involved in reactions	4.3			

Chapter 5

Chemical calculations

IN THIS CHAPTER YOU WILL:

- establish how the mole is the unit of amount of substance in chemistry
- calculate empirical and molecular formulae using appropriate data and the concept of the mole
- calculate the relationship between the number of moles of a substance and the mass or number of particles present
- calculate reacting masses, limiting reagents and amount of product for a stated reaction
- calculate the percentage composition by mass of a compound, and the percentage yield and purity of a product of a given reaction
- learn that the molar gas volume for any gas is $24\,dm^3$ at r.t.p. and use this value for calculations on reactions involving gases
- understand the different units used to express the concentration of a solution and use them when calculating the concentration of a solution from titration experiments.

GETTING STARTED

Working in groups of three or four, review the scenario described here and discuss the questions asked. To answer the questions, you will need to recall information covered in Chapters 1–3.

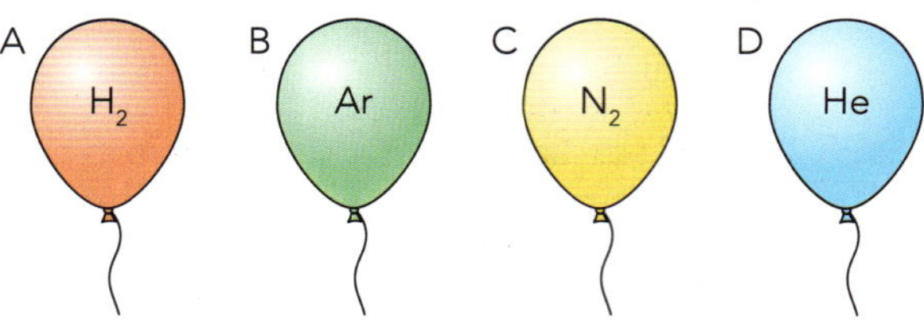

Figure 5.1: These balloons all contain the same volume of gas at the same pressure and temperature.

You have four balloons filled to the same volume and pressure. These balloons contain different gases as indicated in Figure 5.1.

- From what you know of the nature of gases, what can you say about the number of gas molecules in each of the balloons?
- Which of the four balloons will sink to the ground fastest?

Discuss these questions in your group and the reasons behind your answers.

CHEMICAL ACCOUNTANCY AND ATOM ECONOMY

The rather unusual word **stoichiometry** is derived from two Greek words – *stoicheion* (meaning element), and *metron* (meaning measure). It describes the relative measures, or amounts, of a reactant and a product in a chemical reaction. Chemists talk of stoichiometry when balancing symbol equations (Chapter 4), and it is how they work out what quantities of different substances will react to form particular amounts of product. You can think of stoichiometry as the link between what happens at an atomic level and what can be measured practically in a reaction. Although the pairing of elements may change in a chemical reaction, the amount of matter remains the same, so stoichiometry is a type of chemical accountancy that takes place at the atomic level. The standard amount of substance that contains a known number of particles at an atomic level is known as the **mole**.

The concept of the mole is important to new developments in industrial chemistry where the twelve principles of *Green Chemistry* have been proposed to guide more sustainable chemical engineering. Green Chemistry emphasises that industrial processes should reduce the use, or production, of hazardous or waste substances (Figure 5.2). By reducing the involvement of hazardous substances, the environmental safety of any new process is more effective.

Green Chemistry also stresses the *atom economy* of a reaction as a measure of percentage of substance present in the reactants that finishes up in the desired product. The concept of the mole is important here, as it is used to measure the amount of substance present in the reactants and the useful products. A high level of atom economy is important for sustainable development and the economic efficiency of an industrial process. Although atoms are neither created nor destroyed in a chemical reaction, not all the atoms in the reactants will necessarily become part of the desired product; some may end up forming by-products, or waste products. If all the atoms in the reactants end up in the desired product – that is, if there is no by-product from the reaction – this

Figure 5.2: Modern chemical plants should be designed for safety and atom economy.

CONTINUED

represents the maximum value of atom economy: 100%. These considerations show how important chemical formulae and equations are; they not only tell us what happens, but also put 'numbers' to it. Consideration of the quantities involved in reactions is vital to modern chemistry. Note that the ideas behind Green Chemistry and atom economy are important new developments, but are not required knowledge currently.

Discussion questions

1 What costs, other than those of the raw materials, should be considered when considering the economy of an industrial process?

2 In what way could the economic problem of by-products be reduced by positive and imaginative marketing?

KEY WORDS

stoichiometry: the ratio of the reactants and products in a balanced symbol equation

mole: the measure of amount of substance in chemistry; 1 mole of a substance has a mass equal to its relative formula mass in grams – that amount of substance contains 6.02×10^{23} (the Avogadro constant) atoms, molecules or formula units depending on the substance considered

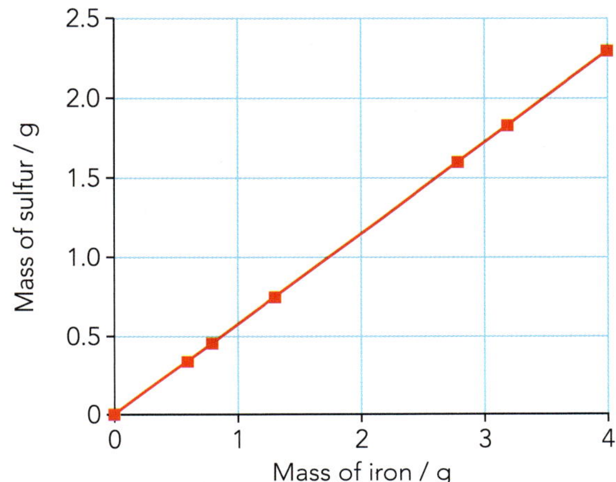

Figure 5.3: Experiments on heating iron with sulfur show that the two elements react in a fixed ratio by mass to produce iron sulfide.

5.1 The mole and Avogadro's constant

If you take any compound, e.g. iron sulfide (FeS), it will always contain the same elements, in this case iron (Fe) and sulfur (S). The two elements in iron sulfide are always present in the same ratio by mass. The relationship can be determined experimentally. When increasing amounts of iron are heated with sulfur, the mass of sulfur that combines also increases. When the results are plotted on a graph you see a straight-line relationship (Figure 5.3) indicating that the elements combine in a fixed ratio by mass. How can we make the link between the mass ratio and the chemical formula of a compound? To do this we need to use the concept of the mole.

The mole – the chemical counting unit

When carrying out an experiment, a chemist cannot weigh out a single atom or molecule and then react it with another atom or molecule. Atoms and molecules are simply too small. A 'counting unit' must be found that is useful in practical chemistry. This idea is not unusual when dealing with large numbers of small objects. For example, banks weigh coins rather than count them – they know that a fixed number of a particular coin will always have the same mass (Figure 5.4).

Figure 5.4: Coins weighed using a balance.

Chemists have applied this idea of counting large numbers of objects to the problem of relating a number of sub-microscopic particles to the amount of substance in a sample. Chemists count atoms and molecules by weighing them. The standard 'unit' of the 'amount' of a substance is taken as the relative molecular (or formula) mass of the substance in grams. This 'unit' is called 1 mole of the substance. The unit 'moles' is used to measure amounts of elements and compounds (mol is the symbol or shortened form of mole or moles). The mass of 1 mole of a substance is called the **molar mass** of that substance. To find the molar mass (mass of 1 mole) of any substance, you write down the formula of the substance; for example, ethanol is C_2H_5OH. Then work out its relative formula mass (for ethanol this is 46). For example, ethanol contains two carbon atoms ($A_r = 12$), six hydrogen atoms ($A_r = 1$) and one oxygen atom ($A_r = 16$).

So, for ethanol $M_r = (2 \times 12) + (6 \times 1) + 16 = 46$.

This value is expressed in grams per mole (1 mole of ethanol = 46 g).

Table 5.1 shows some further examples of how this idea is applied to other substances.

One mole of each of these different substances contains the same number of atoms, molecules or formula units (Table 5.1). That number per mole has been worked out by several different experimental methods. It is named after the 19th century Italian chemist Amedeo Avogadro and is 6.02×10^{23} particles per mole (this is called the **Avogadro constant** and it is given the symbol L) (Figure 5.5). Knowing that 46 g of ethanol is 1 mole of that compound, we now also know that there are 6×10^{23} ethanol molecules in that amount of ethanol.

KEY WORDS

molar mass: the mass, in grams, of 1 mole of a substance

Avogadro constant: the number (6×10^{23}) of characteristic particles in 1 mole of a substance

Substance	Formula	Relative formula mass, M_r	Mass of 1 mole (molar mass)	This mass (1 mole) contains
carbon	C	12	12 g	6.02×10^{23} carbon atoms
iron	Fe	56	56 g	6.02×10^{23} iron atoms
hydrogen	H_2	$2 \times 1 = 2$	2 g	6.02×10^{23} H_2 molecules
oxygen	O_2	$2 \times 16 = 32$	32 g	6.02×10^{23} O_2 molecules
water	H_2O	$(2 \times 1) + 16 = 18$	18 g	6.02×10^{23} H_2O molecules
magnesium oxide	MgO	$24 + 16 = 40$	40 g	6.02×10^{23} MgO 'formula units'
calcium carbonate	$CaCO_3$	$40 + 12 + (3 \times 16) = 100$	100 g	6.02×10^{23} $CaCO_3$ 'formula units'
silicon(IV) oxide	SiO_2	$28 + (2 \times 16) = 60$	60 g	6.02×10^{23} SiO_2 'formula units'

Table 5.1: Calculating the mass of 1 mole of various substances.

6.02×10^{23} atoms	6.02×10^{23} molecules	6.02×10^{23} ions	6.02×10^{23} formula units	6.02×10^{23} electrons

Figure 5.5: One mole always contains the same number of particles (Avogadro's constant).

One mole of a substance:

- contains 6.02×10^{23} (the Avogadro constant) atoms, molecules or formula units, depending on the substance considered
- has a mass equal to its relative molecular (or formula) mass (M_r) in grams.

Figure 5.6 shows how to convert between these important values.

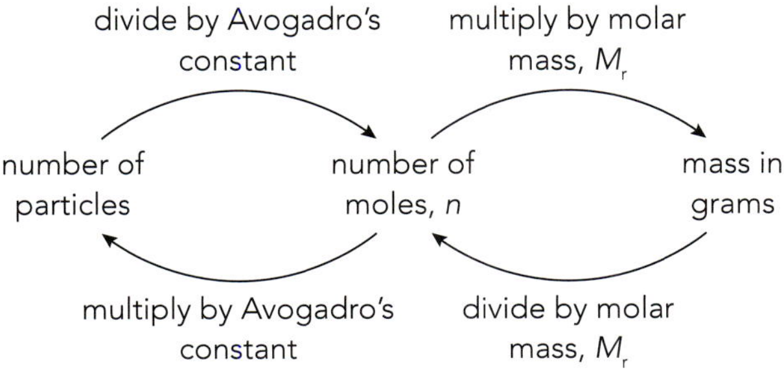

Figure 5.6: The number of moles of substances is related to the mass of substance in a sample and the number of particles present.

Calculations involving the mole

For any given mass of a substance, you can find the number of moles of atoms, molecules or formula units present using the following mathematical equation, where the mass is in grams and the molar mass is in grams per mole:

$$\text{number of moles} = \frac{\text{mass}}{\text{molar mass}}$$

This mathematical equation can be rearranged so that any one of the values can be calculated, provided the other two values are known. The 'calculation triangle' in Figure 5.7 is a useful check to make sure that you have rearranged the equation correctly: cover the quantity to be found and you are left with how to work it out.

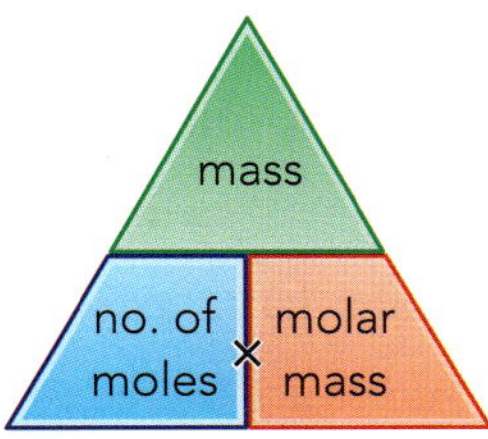

Figure 5.7: Calculation triangle for relating number of moles of substance to mass.

This shows that if we need to calculate the mass of one mole of some substance, the straightforward way is to work out the relative formula mass of the substance and write the word 'grams' after it. Using this mathematical equation, it is possible to convert any mass of a particular substance into moles, or vice versa. We shall look at two examples.

WORKED EXAMPLE 5.1

Using calculation triangles

1 How many moles are there in 60 g of sodium hydroxide?

The relative formula mass of sodium hydroxide is:

$M_r(NaOH) = 23 + 16 + 1 = 40$

molar mass of NaOH = 40 g/mol

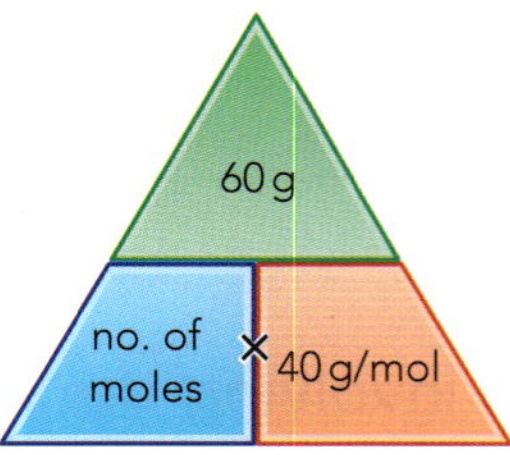

Figure 5.8: Calculation triangle for working out the number of moles in 60 g of sodium hydroxide.

$$\text{number of moles} = \frac{\text{mass}}{\text{molar mass}} \text{ (Figure 5.8)}$$

$$= \frac{60\text{ g}}{40\text{ g/mol}}$$

number of moles = 1.5

2 What is the mass of 0.5 mol of hydrated copper(II) sulfate crystals?

The relative formula mass of hydrated copper(II) sulfate is:

$M_r(CuSO_4{\cdot}5H_2O) = 64 + 32 + (4 \times 16) + (5 \times 18) = 250$

molar mass of $CuSO_4{\cdot}5H_2O$ = 250 g/mol

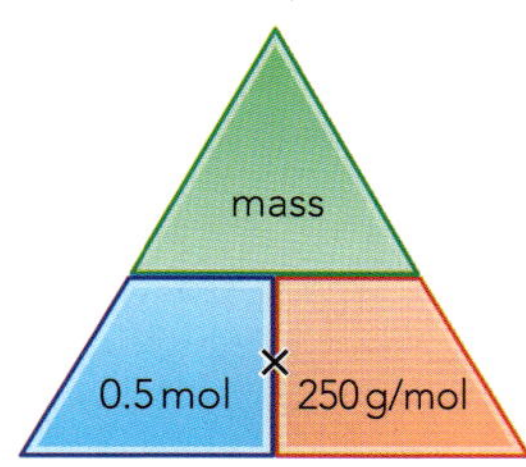

Figure 5.9: Calculating the mass of 0.5 mol of copper(II) sulfate crystals.

$$\text{number of moles} = \frac{\text{mass}}{\text{molar mass}}$$

Therefore,

$$0.5\text{ mol} = \frac{\text{mass}}{250\text{ g/mol}}$$

mass = 0.5 × 250 = 125 g

Note that here we have put the values into the mathematical equation before rearranging it. You can do this or rearrange the equation first, whichever you feel most confident about. The calculation triangle (Figure 5.9) confirms the calculation you should do in this case.

Now try these two examples for yourself, being careful to rearrange the mathematical equation correctly.

1 0.25 mol of water has a mass of 4.5 g. What is the molar mass of water?

2 How many moles are there in 12.75 g of aluminium oxide?

(A_r: H = 1; O = 16; Al = 27)

Working out chemical formulae experimentally

The concept of the mole means that we can now work out chemical formulae from experimental data on combining masses. The concept provides the link between the mass of an element in a compound and the number of its atoms present.

In experiments to make magnesium oxide (see description in Chapter 4), a constant ratio was found between the reacting amounts of magnesium and oxygen. If 0.24 g of magnesium is burnt, then 0.40 g of magnesium oxide is formed. This means that 0.24 g of magnesium combines with 0.16 g of oxygen (0.40 − 0.24 = 0.16 g). We can now use these results to find the formula of magnesium oxide (Figure 5.10).

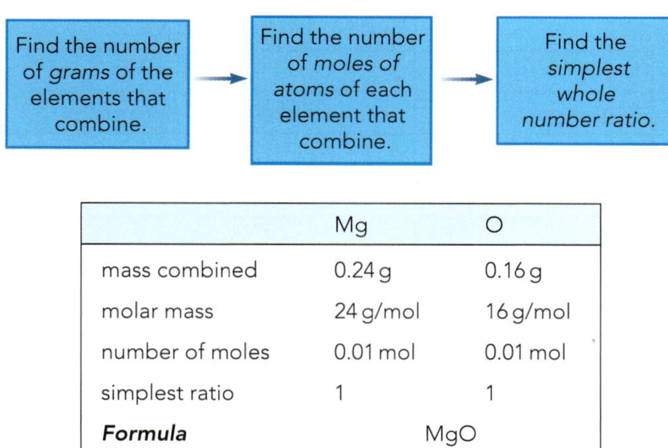

	Mg	O
mass combined	0.24 g	0.16 g
molar mass	24 g/mol	16 g/mol
number of moles	0.01 mol	0.01 mol
simplest ratio	1	1
Formula	MgO	

Figure 5.10: Calculating the formula of magnesium oxide from experimental data on the masses of magnesium and oxygen that react together.

The formula of magnesium oxide tells us that 1 mole of magnesium atoms combine with 1 mole of oxygen atoms. The atoms react in a 1 : 1 ratio to form a giant ionic lattice of Mg^{2+} and O^{2-} ions. For giant structures, the formula of the compound is the simplest whole number formula – in this example, MgO. A formula found by this method is also known as an empirical formula (Chapter 4).

A sample of silicon(IV) oxide is found to contain 47% by mass of silicon. How can we find its empirical formula? This is done in Figure 5.11. The empirical formula of silicon(IV) oxide is SiO_2. It consists of a giant molecular lattice of covalently bonded silicon and oxygen atoms in a ratio 1 : 2. Since it is a giant covalent structure, the formula we use for this compound is SiO_2.

Empirical and molecular formulae for simple covalent molecules

Not all compounds are giant structures – some are made up of simple molecules. Here we sometimes have to make a distinction between the empirical formula and the actual formula of the molecule, the molecular formula (Chapter 4).

Phosphorus burns in air to produce white clouds of phosphorus oxide. From experiments it is found that the oxide contains 44% phosphorus. The empirical formula of phosphorus oxide is P_2O_5 (Table 5.2).

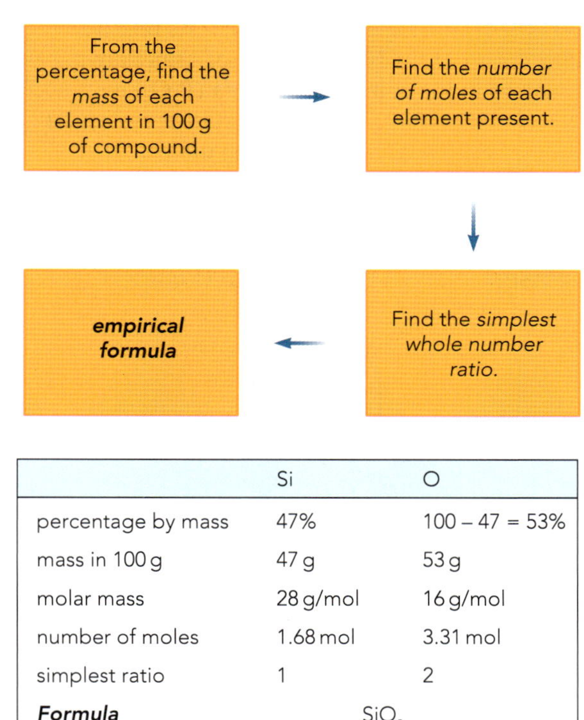

	Si	O
percentage by mass	47%	100 – 47 = 53%
mass in 100 g	47 g	53 g
molar mass	28 g/mol	16 g/mol
number of moles	1.68 mol	3.31 mol
simplest ratio	1	2
Formula	SiO_2	

Figure 5.11: Finding the empirical formula of silicon(IV) oxide from percentage mass data.

	P	O
percentage by mass	44%	100 – 44 = 56%
mass in 100 g	44 g	56 g
molar mass	31 g/mol	16 g/mol
number of moles	1.4 mol	3.5 mol
simplest ratio	1	2.5
or	2	5
Formula	P_2O_5	

Table 5.2: Calculating the empirical formula of phosphorus oxide.

However, it is found experimentally that its relative molecular mass (M_r) is 284. The sum of the relative atomic masses in the empirical formula (P_2O_5) is $(2 \times 31) + (5 \times 16) = 142$. The actual relative molecular mass is twice this value. Therefore, the molecular formula of phosphorus oxide is $(P_2O_5)_2$ or P_4O_{10}. The empirical formula is *not* the actual molecular formula of phosphorus oxide. A molecule of phosphorus oxide contains four P atoms and 10 O atoms (Figure 5.12).

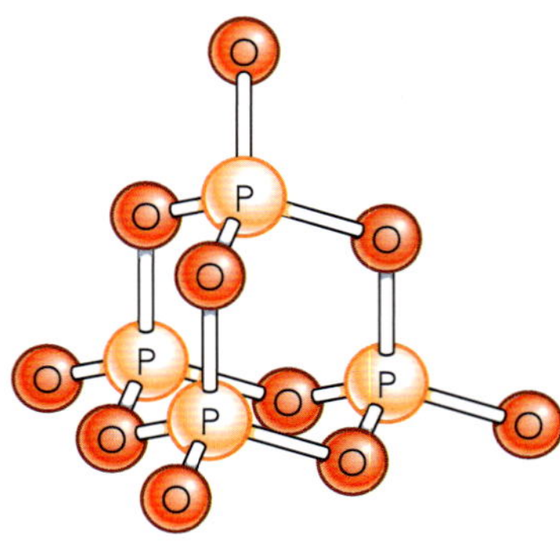

Figure 5.12: Phosphorus oxide, P_4O_{10}.

- The empirical formula of a compound is the simplest whole number formula.
- For simple molecular compounds, the empirical formula may not be the actual molecular formula. The molecular formula must be calculated using the relative molecular mass (M_r) of the compound as found by experiment.

Finding the formula of a hydrated salt

The mass of water present in crystals of hydrated salts is always a fixed proportion of the total mass. The formula of such a salt can be worked out by a method similar to that used to calculate the empirical formula of a compound.

If 5.0 g of hydrated copper(II) sulfate crystals are heated to drive off the water of crystallisation, the remaining solid has a mass of 3.2 g. The ratio of the salt and water in the crystal can be calculated. This gives the formula of the crystals (Table 5.3).

	$CuSO_4$	H_2O
Mass	3.2 g	5.0 – 3.2 = 1.8 g
Molar mass	160 g/mol	18 g/mol
Number of moles	0.02 mol	0.10 mol
Simplest ratio	1	5
Formula	$CuSO_4{\cdot}5H_2O$	

Table 5.3: Calculating the formula of hydrated copper(II) sulfate.

Questions

1 a Calculate the number of moles of sodium hydroxide in 16.0 g of sodium hydroxide (NaOH).

  b Calculate how many formula units of sodium hydroxide are present in 16.0 g of NaOH. From your answer, deduce how many sodium ions (Na^+) and hydroxide ions (OH^-) are present in this mass of sodium hydroxide.

2 One of the ores of copper is the mineral chalcopyrite. A laboratory analysis of a sample showed that 15.15 g of chalcopyrite had the following composition by mass: copper 5.27 g and iron 4.61 g. Sulfur is the only other element present. Use these figures to calculate the empirical formula of chalcopyrite.

  (A_r: S = 32, Fe = 56, Cu = 64)

3 A sample of antifreeze has the composition by mass: 38.7% carbon, 9.7% hydrogen, 51.6% oxygen.

  (A_r: H = 1, C = 12, O = 16)

  a Calculate its empirical formula.

  b The relative molecular mass of the compound is 62. What is its molecular formula?

  c This compound is a diol. The molecule contains two alcohol (–OH) groups attached to different carbon atoms. What is its displayed formula?

5.2 The mole and chemical equations

Calculating reacting amounts

We can now see that the chemical equation for a reaction is more than simply a record of what is produced. In addition to telling us *what* the reactants and products are, it tells us *how much* product we can expect from particular amounts of reactants.

The equation for the thermal decomposition of calcium carbonate (limestone) is:

calcium carbonate → calcium oxide + carbon dioxide

$$CaCO_3 \rightarrow CaO + CO_2$$

We can see that 1 mole of calcium carbonate gives 1 mole each of calcium oxide and carbon dioxide.

1 mol	1 mol	1 mol
$40 + 12 + (3 \times 16)$	$40 + 16$	$12 + (2 \times 16)$
= 100 g	= 56 g	= 44 g

The mass of the product is equal to the total mass of the reactants. This is the law of conservation of mass, which we met in Chapter 4 when discussing the balancing of chemical equations. Although the atoms have rearranged themselves, their total mass remains the same. A chemical equation must be balanced. In practice, we may not want to react such large amounts. We could scale down the quantities (use smaller amounts). However, the mass of calcium carbonate, calcium oxide and carbon dioxide will always be in the ratio 100 : 56 : 44.

We could use just 10 g of calcium carbonate, which would mean that we could not produce more than 5.6 g of calcium oxide (lime).

$$CaCO_3 \rightarrow CaO + CO_2$$

$$10\,g \rightarrow 5.6\,g + 4.4\,g$$

In industry, the reacting amounts given by an equation can also be scaled up (that is, use larger amounts). In industry, tonnes of chemical reactants may be used, but the ratios given by the equation still apply.

The industrial manufacture of lime is important for the cement industry and agriculture. Lime is made by heating limestone in lime kilns.

$CaCO_3$	→	CaO	+	CO_2
1 mol	→	1 mol		1 mol
100 g	→	56 g		44 g

This relationship between the reacting masses is fixed and can be scaled up to work in industrial proportions (tonnes):

100 tonnes   56 tonnes   44 tonnes

Similarly, if 10 tonnes of calcium carbonate were heated, we should expect to produce 5.6 tonnes of lime (calcium oxide).

Calculating reacting amounts – a chemical 'footbridge'

We can use the concept of the mole to find reactant or product masses from the equation for a reaction. There are various ways of doing these calculations. The balanced equation itself can be used as a numerical 'footbridge' between the two sides of the reaction (Figure 5.13).

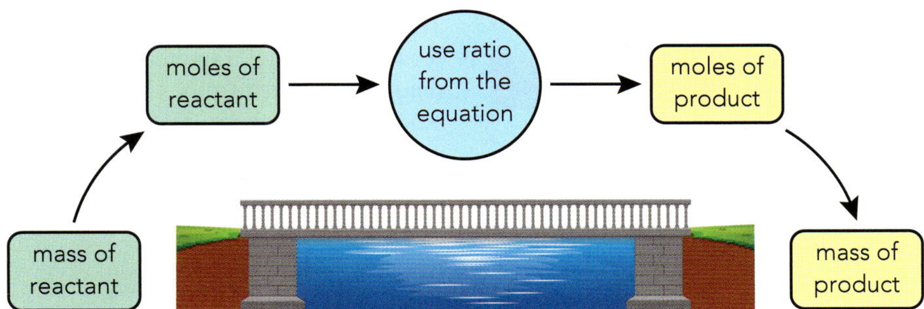

Figure 5.13: A chemical 'footbridge'. Following the sequence 'up–across–down' helps to relate the mass of product made to the mass of reactant used. The 'bridge' can also be used in the reverse direction.

WORKED EXAMPLE 5.2

Calculating the mass of aluminium oxide produced when aluminium reacts with oxygen

What mass of aluminium oxide is produced when 9.2 g of aluminium metal reacts completely with oxygen gas?

To answer this question, we first work out the balanced equation:

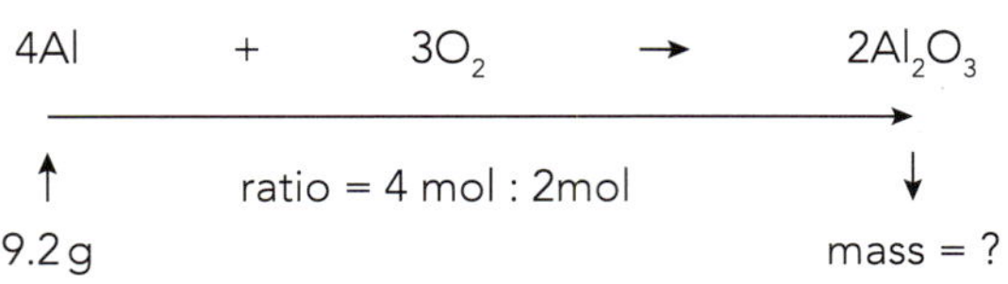

Figure 5.14: Using the chemical equation to calculate the amount of product in a reaction.

Then we work through the steps of the 'footbridge'.

Step 1. (the 'up' stage): Convert 9.2 g of Al into moles:

$$\text{number of moles} = \frac{9.2\text{ g}}{27\text{ g/mol}} = 0.34\text{ mol}$$

Step 2. (the 'across' stage): Use the ratio from the equation to work out how many moles of Al_2O_3 are produced:

4 mol of Al produce 2 mol of Al_2O_3

so

0.34 mol of Al produce 0.17 mol of Al_2O_3

Step 3. (the 'down' stage): Work out the mass of this amount of aluminium oxide (the relative formula mass of Al_2O_3 is 102):

$$0.17\text{ mol} = \frac{\text{mass}}{102\text{ g/mol}}$$

So, the mass of Al_2O_3 produced = 0.17×102 g = 17.34 g

Mass of Al_2O_3 produced = 17.3g to three significant figures

The following is a further example for you to try:

1 What mass of oxygen combines with 32 g of methane when that gas is burnt completely in air? (A_r: C = 12; H = 1; O = 16)

$CH_4(g) + 2O_2(g) \rightarrow CO_2(g) + 2H_2O(l)$

Remember to read questions on reacting masses carefully. If you set out the calculation carefully using the equation, as we have done here, you will be able to see which substances are relevant to your calculation. Remember to also take the balancing numbers into account in doing your calculation (this is called the stoichiometry of the equation).

In carrying out a reaction, one of the reactants may be present in excess. Some of this reactant will be left over at the end of the reaction. The **limiting reactant** is the one that is not in excess – there will be a smaller number of moles of this reactant present, taking into account the reacting ratio from the equation.

KEY WORDS

limiting reactant: the reactant that is not in excess

Limiting reactants

When we carry out a reaction, we do not always use the amounts of reactant defined by the equation. Sometimes, we deliberately use too much (an *excess*) of one reactant. Typically, the relative amounts available of reactants available to react together will dictate the amount of resulting product. One reactant, known as the limiting reactant, will determine this quantity. If the other reactants are not completely used in the process, they are said to be *in excess* (Figure 5.15).

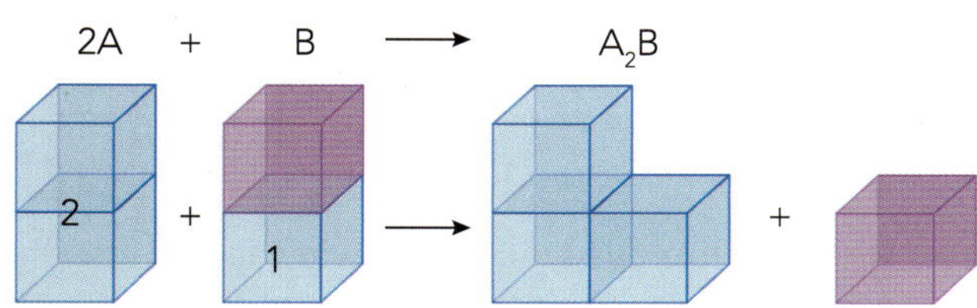

Figure 5.15: Illustrating how excess reactant B (in purple) remains unused after the reaction has taken place.

In the reaction between aluminium and oxygen (Worked example 5.2) we needed to make sure that the metal reacted completely with oxygen. We do this by using an excess of oxygen, some of which will not be used up. The amount of aluminium oxide produced is decided by the mass of aluminium we start with; that is, the limiting reactant in this case.

A reaction stops when the limiting reactant is used up. You can find which reactant is limiting by doing the following simple calculation:

- work out the number of moles of each reactant involved
- then divide each by its balancing number (coefficient) in the balanced symbol equation
- the smallest number indicates the limiting reactant.

WORKED EXAMPLE 5.3

Limiting reactants and reactants in excess

6.0 g of cobalt(II) carbonate was reacted with 0.08 mol of hydrochloric acid to produce cobalt(II) chloride. Show that the cobalt(II) carbonate was in excess.

$CoCO_3(s) + 2HCl(aq) \rightarrow CoCl_2(aq) + H_2O(l) + CO_2(g)$

(A_r: Co = 59; C = 12; O = 16)

M_r of cobalt carbonate = 59 + 12 + (3 × 16) = 119 g/mol

number of moles of cobalt carbonate = 6.0 / 119 = 0.05 mol

From the balanced equation:

- 1 mol cobalt(II) carbonate reacts with 2 mol hydrochloric acid
- 0.05 mol cobalt(II) carbonate would react with 0.10 mol hydrochloric acid
- the cobalt(II) carbonate is in excess.

The following is a further calculation of this type for you to try.

1 4 g of magnesium is reacted with a hydrochloric acid solution that contains 5.48 g of the acid.

$Mg(s) + 2HCl(aq) \rightarrow MgCl_2(aq) + H_2(g)$

Which of the reactants is the limiting reagent?

(A_r: Mg = 24; Cl = 35.5; H = 1)

ACTIVITY 5.1

Getting to grips with the mole

Which is heavier, 10 kg of feathers or 10 kg of bricks? This is a trick question as they have the same mass, they are both 10 kg. But what happens when we change the question and ask which is heavier – a dozen feathers or a dozen bricks? Now there is a definite answer, 12 bricks will have a much higher mass than 12 feathers. In chemistry, we could ask which is heavier, 1 g of hydrogen or 1 g of carbon? The answer is that both have a mass of 1 g. However, what if we were to ask the question which is heavier 1 mole of hydrogen or 1 mole of carbon? The answer is 1 mole of carbon.

The mole can be a tricky concept, but it is essential to chemical success as it links the number of particles present to the relative mass of that chemical. In simple terms, the mole is just the chemist's way of explaining the amount of substance present.

You are going to prepare a five-minute lesson to explain the concept of the mole to your classmates. To do this, you are going to use the equation for the decomposition of aluminium oxide by electrolysis:

$$2Al_2O_3 \rightarrow 4Al + 3O_2$$

Work in a pair to write down an explanation of moles and how they link to mass via M_r. Your explanation should include:

- what the mole is (you might want to include a reference to Avogadro's constant)
- how the balanced equation shows the mole ratios
- how these ratios remain constant, e.g. what would happen if you had 400 moles of Al_2O_3 in terms of the moles of Al produced or if you only had 0.6 moles of Al_2O_3
- how an understanding of the number of moles allows masses to be deduced (by use of the equation: number of moles = mass (g) / molar mass).

You might want to explain why the number of moles in the reactants and products does not have to be the same. In this case, for example, there are two moles of reactants but seven moles of products.

Having worked through this example, can you think of another balanced equation and produce a couple of questions to test your classmates understanding of moles?

Peer assessment

Having constructed a brief lesson on the mole, discuss how successful you thought it was with other members of your group. Use the following questions to help you:

- Were you able to link the ideas suggested in the activity convincingly?
- Did your group think your presentation was clear and understandable?
- What would you change if you were to repeat the activity?
- Could you extend your ideas to include reactions between gases, or in solution?

Percentage yield and percentage purity

A reaction may not always yield the total amount of product predicted by the equation. The loss may be due to several factors:

- the reaction may not be totally complete
- errors may be made in weighing the reactants or the products
- material may be lost in carrying out the reaction, or in transferring and separating the product.

The equation gives us an ideal figure for the yield of a reaction; in reality a lower yield is often produced. This can be expressed as the **percentage yield** for a particular experiment.

In other more complex reactions, a particular product may be contaminated by other products or unreacted material. The 'crude' product may prove to contain less than 100% of the required substance.

The **percentage purity** of a chemical product can be calculated in a similar way to the percentage yield.

KEY WORDS

percentage yield: a measure of the actual yield of a reaction when carried out experimentally compared to the theoretical yield calculated from the equation:

$$\text{percentage yield} = \frac{\text{actual yield}}{\text{predicted yield}} \times 100$$

percentage purity: a measure of the purity of the product from a reaction carried out experimentally:

$$\text{percentage purity} = \frac{\text{mass of pure product}}{\text{mass of impure product}} \times 100$$

WORKED EXAMPLE 5.4

Calculating percentage yield

Heating 12.4 g of copper(II) carbonate in a crucible produced only 7.0 g of copper(II) oxide. What was the percentage yield of copper(II) oxide?

$CuCO_3$	$\rightarrow$	CuO	+	CO_2
1 mol	$\rightarrow$	1 mol	+	1 mol
64 + 12 + 48		64 + 16		
= 124 g		= 80 g		

Therefore, heating 12.4 g of copper(II) carbonate should have produced 8.0 g of copper(II) oxide. So:

expected yield = 8.0 g

actual yield = 7.0 g

and percentage yield = $\frac{7.0}{8.0} \times 100 = \mathbf{87.5\%}$

Here is an example for you to try:

1 A student reacts 4.5 g of aluminium powder with chlorine gas. 17.8 g of aluminium chloride are produced. Calculate the percentage yield of this reaction.

$$2Al(s) + 3Cl_2(g) \rightarrow 2AlCl_3(s)$$

(A_r: Al = 27; Cl = 35.5)

The copper used for electrical circuits has to be exceptionally pure (99.99%). The following calculation uses the example of copper to show how percentage purity is calculated. An initial crude sample of copper is prepared industrially and then tested for purity. A sample of 10.15 g of the crude copper is analysed by various methods and shown to contain 9.95 g of copper, with the remaining mass being made up of other metals.

Generally,

$$\% \text{ purity} = \frac{\text{mass of pure product}}{\text{mass of impure product}} \times 100$$

Therefore:

% purity of the copper sample

$$= \frac{\text{mass of copper in sample}}{\text{mass of impure copper}} \times 100$$

$$= \frac{9.95}{10.15} \times 100$$

= 98.03% (answer to four **significant figures**)

This result shows that this batch of copper would need to be purified before it could be used for electrical circuits such as those inside TVs, smartphones and computers.

Percentage composition by mass

The importance of the relative molecular (or formula) mass of a compound (M_r) is central to calculations based on the concept of the mole. However, even the M_r value itself provides useful information about a compound. The M_r value enables us to calculate the **percentage composition** of a compound. Calculations of percentage composition are useful, for instance, in estimating the efficiency of one fertiliser compared with another (Chapter 9). Ammonium nitrate is a commonly used fertiliser. It is an important source of nitrogen.

KEY WORDS

significant figures: the number of digits in a number, not including any zeros at the beginning; for example the number of significant figures in 0.0682 is three

percentage composition: the percentage by mass of each element in a compound

WORKED EXAMPLE 5.5

What percentage of the mass of the compound is nitrogen?

The formula of ammonium nitrate is NH_4NO_3 (it contains the ions NH_4^+ and NO_3^-). Using the A_r values for N, H and O we get:

$$M_r = (2 \times 14) + (4 \times 1) + (3 \times 16) = 28 + 4 + 48 = 80$$

Then:

mass of nitrogen in the formula = 28

mass of nitrogen as a fraction of the total $= \frac{28}{80}$

mass of nitrogen as percentage of total mass $\frac{28}{80} \times 100 = 35\%$

Carry out a similar calculation for yourself to work out the percentage by mass of nitrogen in another fertiliser, ammonium sulfate, $(NH_4)_2SO_4$.

Similar calculations can be used to work out the percentage by mass of water of crystallisation in crystals of a hydrated salt (Worked example 5.6).

WORKED EXAMPLE 5.6

What is the percentage mass of water in crystals of hydrated magnesium sulfate?

The formula of hydrated magnesium sulfate is $MgSO_4{\cdot}7H_2O$. Using the A_r values for Mg, S, O and H we get:

$M_r = 24 + 32 + (4 \times 16) + (7 \times 18) = 246$

Then:

mass of water in formula = 126

mass of water as a fraction of the total $= \frac{126}{246}$

percentage mass of water in the crystals $= \frac{126}{246} \times 100 = 51.2\%$

Now carry out a similar calculation to work out the percentage mass of water in crystals of hydrated sodium carbonate, $Na_2CO_3{\cdot}10H_2O$.

Questions

4 Calculate the percentage by mass of nitrogen in the following fertilisers and nitrogen-containing compounds:

- **a** ammonium phosphate, $(NH_4)_3PO_4$
- **b** glycine, $CH_2(NH_2)COOH$ (an amino acid)

(A_r: H = 1, C = 12, N = 14, O = 16, P = 31, S = 32)

5 Define *the mole*.

6 Explain the difference between the percentage yield of a product from a reaction and its percentage purity.

Calculations involving gases

Molar volume of a gas

Many reactions, including some of those we have just considered, involve gases. Weighing solids or liquids is relatively straightforward. In contrast, weighing a gas is quite difficult. It is much easier to measure the volume of a gas. But how does gas volume relate to the number of atoms or molecules present?

Substance	Molar mass, g/mol	Molar gas volume, dm^3/mol	Number of particles
hydrogen (H_2)	2	24	6.02×10^{23} hydrogen molecules
oxygen (O_2)	32	24	6.02×10^{23} oxygen molecules
carbon dioxide (CO_2)	44	24	6.02×10^{23} carbon dioxide molecules
ethane (C_2H_6)	30	24	6.02×10^{23} ethane molecules

Table 5.4: Molar mass and molar volume of various gases.

In a gas, the particles are relatively far apart. Indeed, any gas can be regarded as largely empty space. Equal volumes of gases are found to contain the same number of particles (Table 5.4); this is **Avogadro's law**. This leads to a simple rule about the volume of one mole of a gas.

- one mole of any gas occupies a volume of approximately 24 dm^3 (24 litres) at room temperature and pressure (**r.t.p.**)
- the **molar gas volume** of any gas therefore has the value 24 dm^3/mol at r.t.p.
- remember that 1 dm^3 (1 litre) = 1000 cm^3.

This rule applies to all gases. This makes it easy to convert the volume of any gas into moles, or moles into volume:

$$\text{number of moles} = \frac{\text{volume}}{\text{molar volume}}$$

In this equation, the volume is in cubic decimetres (dm^3) and the molar volume is 24 dm^3/mol. The calculation triangle shown in Figure 5.16 will help you remember how to rearrange this mathematical equation to work out the different values.

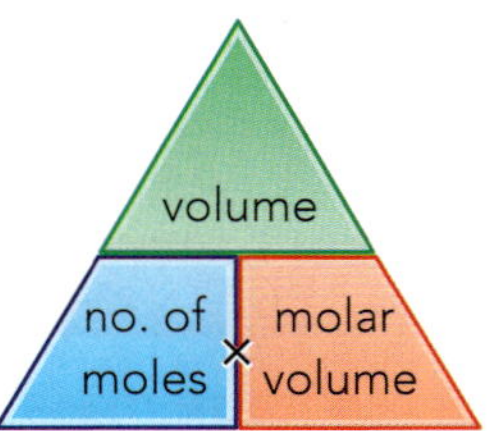

Figure 5.16: A calculation triangle for calculating the moles of gases.

KEY WORDS

Avogadro's law: equal volumes of any gas, under the same conditions of temperature and pressure, contain the same number of particles

r.t.p.: room temperature and pressure; the standard values are 25 °C/298K and 101.3 kPa/1 atmosphere pressure

molar gas volume: 1 mole of any gas has the same volume under the same conditions of temperature and pressure (24 dm^3 at r.t.p.)

WORKED EXAMPLE 5.7

If 8 g of sulfur are burnt, what volume of SO_2 is produced?

First, consider the reaction of sulfur burning in oxygen.

sulfur	+	oxygen	→	sulfur dioxide
S(s)	+	O_2(g)	→	SO_2(g)
1 mol		1 mol	→	1 mol
32 g		24 dm^3		24 dm^3

We have:

$$\text{number of moles of sulfur burnt} = \frac{8\text{ g}}{32\text{ g/mol}}$$

= 0.25 mol

From the equation:

1 mol of sulfur → 1 mol of SO_2

Therefore:

0.25 mol of sulfur → 0.25 mol of SO_2

So, from the above rule:

$$\text{number of moles} = \frac{\text{volume}}{\text{molar volume}}$$

$$0.25\text{ mol} = \frac{\text{volume}}{24\text{ dm}^3\text{/mol}}$$

volume of sulfur dioxide = 0.25 × 24 dm^3

= 6 dm^3 at r.t.p.

Now, using the relationship between the volume of a gas and the number of moles it contains, work out the following:

a the volume of gas present in 22g of carbon dioxide at r.t.p.

b the mass of a sample of nitrogen gas with a volume of 36 dm^3 at r.t.p.

Reactions involving gases

For reactions in which gases are produced, the calculation of product volume is similar to those we have seen already. The approach used is an adaptation of the 'footbridge' method used earlier for calculations involving solids. It is shown in Figure 5.17.

Some important reactions involve only gases. For such reactions, the calculations of expected yield are simplified by the fact that the value for molar volume applies to any gas.

For example:

hydrogen	+	chlorine	→	hydrogen chloride
$H_2(g)$	+	$Cl_2(g)$	→	$2HCl(g)$
1 mol		1 mol		2 mol
$24\,dm^3$		$24\,dm^3$		$48\,dm^3$

The volumes of the gases involved are in the same ratio as the number of moles given by the equation:

$H_2(g)$	+	$Cl_2(g)$	→	$2HCl(g)$
1 volume		1 volume		2 volumes

So, if we react $20\,cm^3$ of hydrogen with sufficient chlorine, it will produce $40\,cm^3$ of hydrogen chloride gas.

5.3 Moles and solution chemistry

We have seen how calculations based on the mole can provide very useful information about chemical reactions involving solids and gases. However, there are many other reactions of importance. These reactions all take place in solution. The usual solvent is water. When setting up such reactions, we normally measure out the solutions by volume. To know how much of the reactants we are actually mixing, we need to know the concentrations of the solutions.

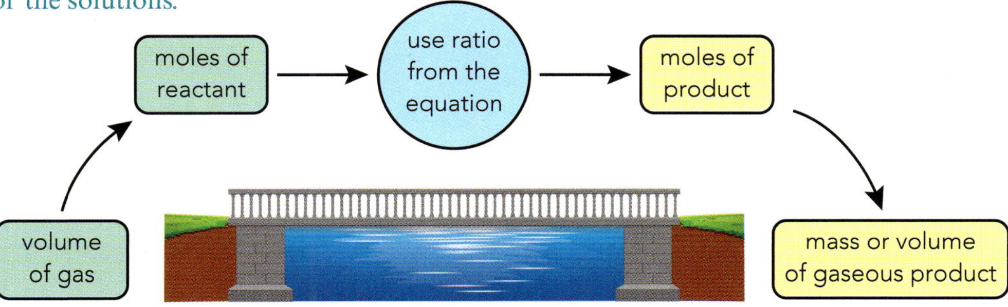

Figure 5.17: An outline of the 'footbridge' method for calculations involving gases.

Concentration of solutions

When a chemical substance (the solute) is dissolved in a volume of solvent, we can measure the 'quantity' of solute in two ways. We can measure either its mass (in grams) or its amount (in moles). The final volume of the solution is normally measured in cubic decimetres, dm^3 ($1\,dm^3 = 1$ litre or $1000\,cm^3$). When we measure the mass of the solute in grams, it is the **mass concentration** that we obtain, in grams per cubic decimetre of solution (g/dm^3). But it is more useful to measure the amount in moles, in which case we get the **molar concentration** in moles per cubic decimetre of solution (mol/dm^3):

$$\text{concentration} = \frac{\text{amount of solute}}{\text{volume of solution}}$$

- the mass concentration of a solution is measured in grams per cubic decimetre (g/dm^3)
- the molar concentration of a solution is measured in moles per cubic decimetre (mol/dm^3)
- when 1 mol of a substance is dissolved in water and the solution is made up to $1\,dm^3$ ($1000\,cm^3$), a solution with a concentration of $1\,mol/dm^3$ is produced.

KEY WORDS

mass concentration: the measure of the concentration of a solution in terms of the mass of the solute, in grams, dissolved per cubic decimetre of solution (g/dm^3)

molar concentration: the measure of the concentration of a solution in terms of the number of moles of the solute dissolved per cubic decimetre of solution (mol/dm^3)

For example, a 1 mol/dm^3 solution of sodium chloride contains 58.5 g of NaCl (1 mole) dissolved in water and made up to a final volume of 1000 cm^3. Figure 5.18 shows how the units are expressed for solutions of differing concentrations. It also shows how solutions of the same final concentration can be made up in different ways.

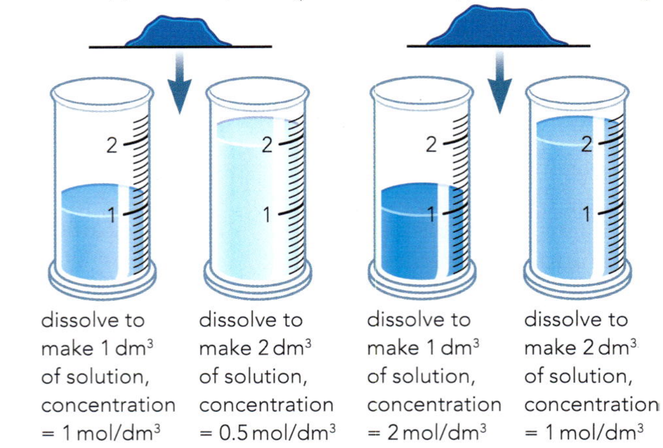

Figure 5.18: Making copper(II) sulfate solutions of different concentrations.

Calculations using solution concentrations

The following equation is useful when working out the number of moles present in a particular solution:

number of moles in solution = molar concentration × volume of solution (in dm^3)

This mathematical equation can be rearranged to work out an unknown value provided the other two values are known and can be represented by a calculation triangle (Figure 5.19).

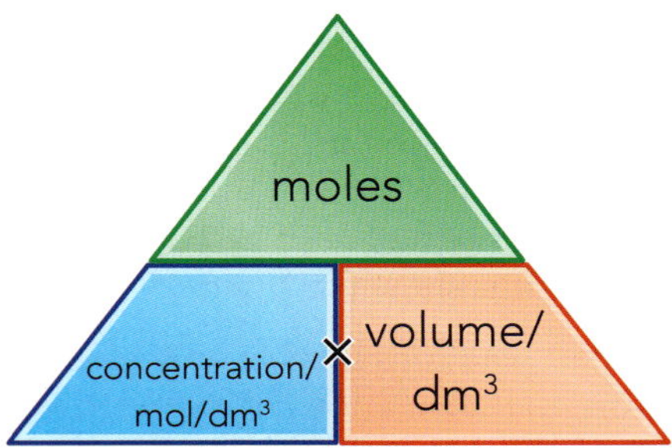

Figure 5.19: A calculation triangle for working with solution concentrations in dm^3.

In practice, however, we are usually dealing with solution volumes in cubic centimetres (cm^3). Therefore, the equation is usefully adapted to:

number of moles in solution =

$$\frac{\text{concentration}}{1000} \times \text{volume of solution (in cm}^3\text{)}$$

In this equation, concentration is in moles per cubic decimetre, but volume is in cubic centimetres (Figure 5.20).

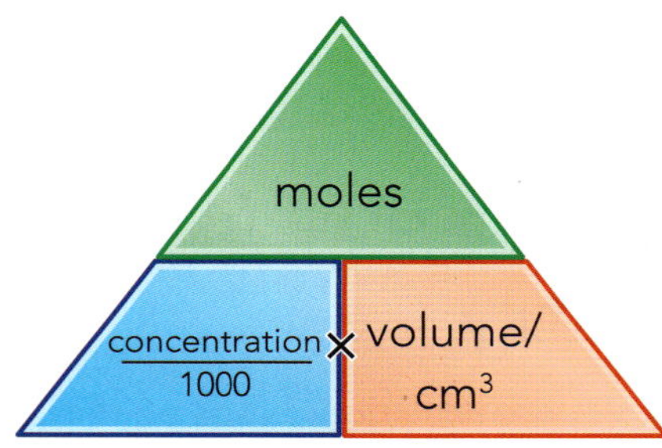

Figure 5.20: A calculation triangle for working with solution concentrations in cm^3.

For example, how many moles of sugar are there in 500 cm^3 of a 3.0 mol/dm^3 sugar solution?

We get:

$$\text{number of moles} = \frac{3.0}{1000} \times 500 = 1.5\,\text{mol}$$

In practice, a chemist still has to weigh out a substance in grams. So, questions and experiments may also involve converting between moles and grams.

WORKED EXAMPLE 5.8

Calculate the concentration of a solution of sodium hydroxide, NaOH, that contains 10 g of NaOH in a final volume of 250 cm³

Step 1. Find out how many moles of NaOH are present:

relative formula mass of NaOH = 23 + 16 + 1 = 40

$$\text{number of moles of NaOH} = \frac{10}{40} = 0.25\,\text{mol}$$

Step 2. Find the concentration:

$$\text{number of moles} = \frac{\text{concentration}}{1000} \times \text{volume (in cm}^3\text{)}$$

$$0.25 = \frac{\text{concentration}}{1000} \times 250$$

$$\text{concentration} = 0.25 \times \frac{1000}{250}$$

$= 1\,\text{mol/dm}^3$

Now find the molar concentration of a solution of 14.3 g of hydrated sodium carbonate, $Na_2CO_3{\cdot}10H_2O$ in 500 cm³ of distilled water.

Acid–base titration calculations

The concentration of an unknown acid solution can be found if it is reacted with a standard solution of an alkali. A **standard solution** is one that has been carefully made up so that its concentration is known precisely. The reaction is carried out in a carefully controlled way. The volumes are measured accurately using a pipette and a burette. Just sufficient acid is added to the alkali to neutralise the alkali. This end-point is found using an indicator. The method is known as **titration** and can be adapted to prepare a soluble salt. It is summarised in Figure 5.21.

The practical method of titration will be described later as a method of preparing a salt (Chapter 12) and as an analytical technique (Chapter 22). In this chapter, we are dealing with the application of the concept of the mole to the type of calculation that can be carried out.

KEY WORDS

standard solution: a solution whose concentration is known precisely – this solution is then used to find the concentration of another solution by titration

titration: a method of quantitative analysis using solutions: one solution is slowly added to a known volume of another solution using a burette until an end-point is reached

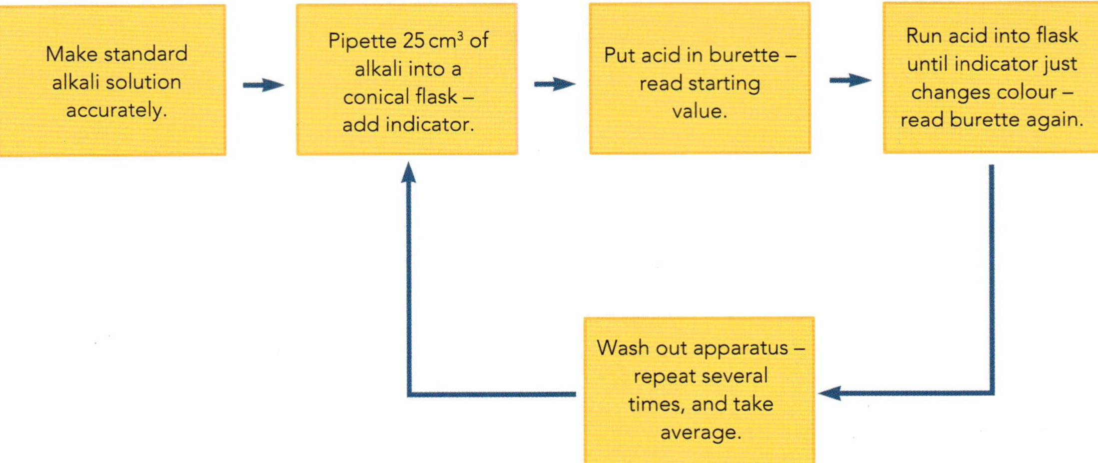

Figure 5.21: Summary of the titration method.

WORKED EXAMPLE 5.9

Calculating the concentration of a hydrochloric acid solution

A solution of hydrochloric acid is titrated against a standard sodium hydroxide solution. It is found that $20.0\,cm^3$ of acid neutralise $25.0\,cm^3$ of $0.10\,mol/dm^3$ NaOH solution. What is the concentration of the hydrochloric acid solution?

Step 1. Use information about the standard solution. How many moles of alkali are in the flask?

number of moles of NaOH

$$= \frac{\text{concentration}}{1000} \times \text{volume (in cm}^3)$$

$= 2.5 \times 10^{-3}\,mol$

Step 2. Use the chemical equation. How many moles of acid are used?

The equation is:

$HCl + NaOH \rightarrow NaCl + H_2O$

1 mol   1 mol

1 mol of NaOH neutralises 1 mol of HCl and so:

$2.5 \times 10^{-3}\,mol$ of NaOH neutralises $2.5 \times 10^{-3}\,mol$ of HCl

Step 3. Use the titration value. What is the concentration of the acid?

The acid solution contains $2.5 \times 10^{-3}\,mol$ in $20.0\,cm^3$.

$$\text{number of moles} = \frac{\text{concentration}}{1000} \times \text{volume (in cm}^3)$$

$$2.5 \times 10^{-3} = \frac{\text{concentration}}{1000} \times 20.0$$

concentration of acid $= 2.5 \times 10^{-3} \times 1000 / 20$

$=$ **0.125** mol/dm^3

Try working through the following question for practice:

Sulfuric acid can be neutralised using sodium hydroxide solution.

$H_2SO_4(aq) + 2NaOH(aq) \rightarrow Na_2SO_4(aq) + 2H_2O(l)$

$25.0\,cm^3$ of a sulfuric acid solution of concentration $0.2\,mol/dm^3$ reacted with $10.0\,cm^3$ of sodium hydroxide. Calculate:

- **a** the number of moles of sulfuric acid used
- **b** the number of moles of NaOH reacted
- **c** the concentration of the NaOH solution in mol/dm^3.

The calculation method uses a further variation of the 'footbridge' approach to link the reactants and products (Figure 5.22).

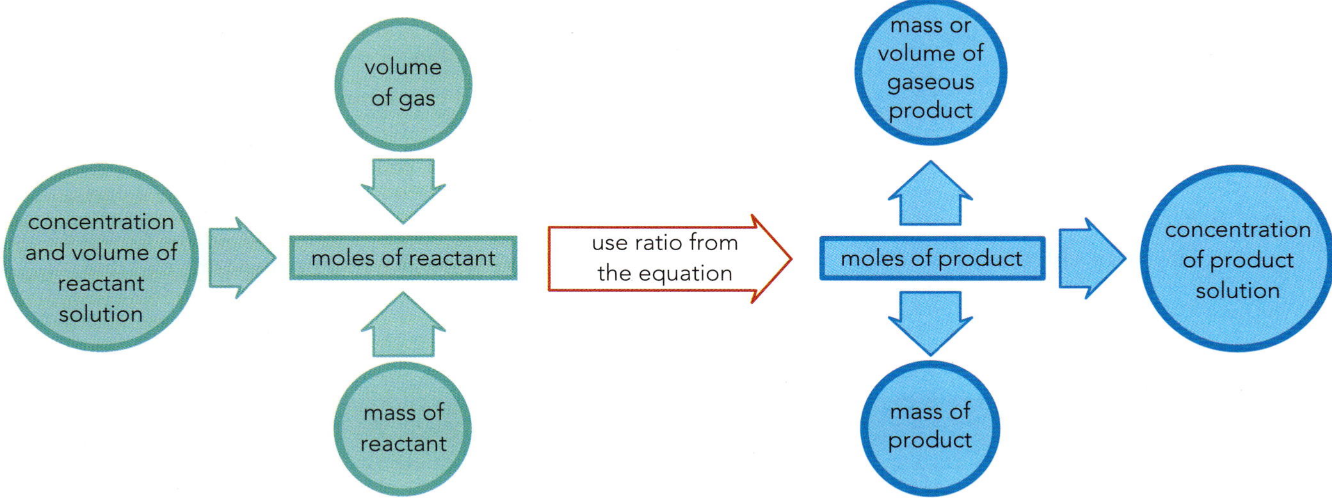

Figure 5.22: A summary of the different ways in which a balanced equation acts as a 'footbridge' in calculations.

Figure 5.23 is a visual reminder of the key relationships involved in discussing the meaning and usefulness of the mole as a measure of amount of substance.

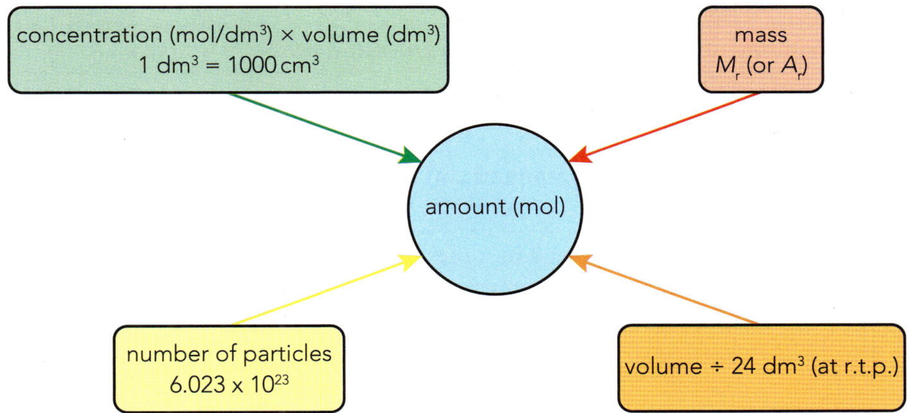

Figure 5.23: The mole as a measure of amount of substance in different situations.

REFLECTION

Think about the following questions:

- Did you face any problems when learning about the concept of the mole in this chapter?
- Are there any maths skills that you feel it would be helpful to improve?

Questions

7 Calculate the number of moles of gas there are in:
   a 480 cm^3 of argon
   b 48 dm^3 of carbon dioxide
   c 1689 cm^3 of oxygen.

8 Calculate the volume (in cm^3) of the following at r.t.p.
   a 1.5 moles of nitrogen
   b 0.06 moles of ammonia
   c 0.5 moles of chlorine.

9 Calculate the concentration (in mol/dm^3) of the following solutions.
   a 1.0 mol of sodium hydroxide is dissolved in distilled water to make 500 cm^3 of solution.
   b 0.2 mol of sodium chloride is dissolved in distilled water to make 1000 cm^3 of solution.
   c 0.1 mol of sodium nitrate is dissolved in distilled water to make 100 cm^3 of solution.
   d 0.8 g of solid sodium hydroxide is dissolved in distilled water to a final volume of 1 dm^3.

   (A_r: H = 1, O = 16, Na = 23)

SUMMARY

The mole is the unit that contains Avogadro's number of constituent particles (atoms, ions or molecules) of a substance and is used to express the amount of a substance taking part in a reaction.
The mole is equal to the relative molecular or formula mass of a substance in grams.
The empirical formula of a compound can be calculated using the concept of the mole.
The balanced chemical equation for a reaction can be used to calculate the reacting masses of substances involved and the amount of product formed.
1 mole of any gas has a volume of 24 dm^3 at room temperature and pressure (r.t.p.).
The percentage yield and the percentage purity of a product from a chemical reaction can be calculated from the theoretical yield from an equation.
The percentage by mass of an element in a compound can be found from the relative molecular or formula mass.
The concentration of a solution can be expressed in moles per cubic decimetre (mol/dm^3) and these values are useful in calculating the results of titration experiments.

PROJECT

The mole and a green approach to industrial chemistry

The concept of the mole and its use in calculating the yield and purity of the product of a chemical reaction is very important in discussing the conditions used in key industrial processes such as the Haber and Contact processes. These ideas have become part of the Green Chemistry approach to industrial processes (Figure 5.24). The ideas of Green Chemistry are proposed in a set of 12 principles (Figure 5.25).

Figure 5.24: The cover of the science journal published by the Royal Society of Chemistry concerned with the development of ideas of Green Chemistry.

These principles cover such ideas as:

- maximising the amount of the raw materials that ends up in the desired product; measured as atom economy
- using renewable raw materials and energy sources
- using the least environmentally harmful substances and solvents
- using energy efficient processes
- avoiding the production of waste.

In groups, discuss the idea of atom economy and the other principles of Green Chemistry. Focus on these questions in your discussion:

Why might a green approach to chemistry be important for industrial processes?

Which of the principles do you think are the most important? Why?

How do these ideas link to what you have learnt in this chapter?

You could also link your ideas to the industrial method for producing ammonia (the Haber process; Chapter 9 Topic 3) and think about the following:

- the economy of the supply of the raw materials
- the use of a catalyst and the effect of that on the conditions used
- the recycling of the major reactants.

Following your discussion, work in groups to prepare a podcast for the class to listen to explaining the importance of the green approach to industrial chemistry.

In your podcast, include ideas on the following:

- percentage yield, how to calculate it, and why it is important to industrial chemistry
- why it is important to consider limiting reactants and those in excess in an industrial reaction.

1. Prevent waste
2. Maximise atom economy
3. Less hazardous chemicals
4. Safer chemicals and products
5. Safer solvents and reaction conditions
6. Increase energy efficiency
7. Use renewable feedstocks
8. Reduce derivatives
9. Use catalysts
10. Design degradable products
11. Analyse for pollution control
12. Accident prevention

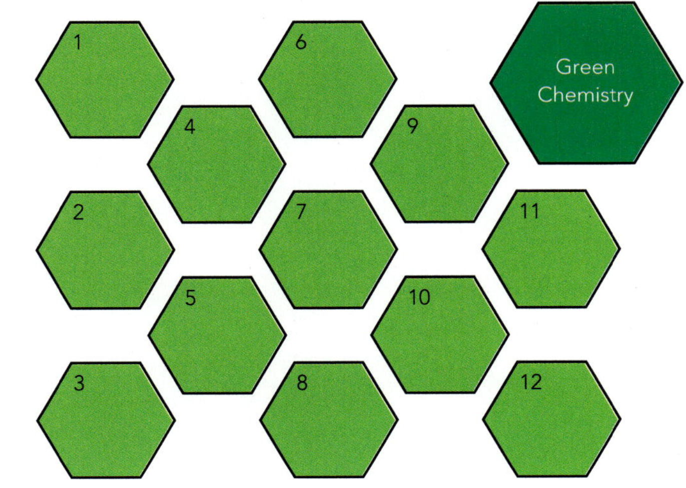

Figure 5.25: The 12 principles of Green Chemistry.

EXAM-STYLE QUESTIONS

1 Hydrogen peroxide solution is used as a bleach and to clean oil paintings. A solution of hydrogen peroxide contains 10 g of hydrogen peroxide in every 100 cm^3 of solution. What is the concentration of the solution in g/dm^3?

A 10 g/dm^3

B 20 g/dm^3

C 50 g/dm^3

D 100 g/dm^3 **[1]**

2 An acid has the following composition by mass.

carbon 40% hydrogen 6.7% oxygen 53.3%

a **Calculate** its empirical formula. **[3]**

b Its relative molecular mass is 60. Calculate its molecular formula. **[1]**

c 12 g of this acid is dissolved in 250 cm^3 of water. What is the concentration of the solution formed? **[3]**

[Total: 7]

3 25 cm^3 of a 0.25 molar solution of potassium carbonate (K_2CO_3) is mixed with 50 cm^3 of a 0.1 molar solution of magnesium sulfate ($MgSO_4$) to form insoluble magnesium carbonate.

$K_2CO_3(aq) + MgSO_4(aq) \rightarrow K_2SO_4(aq) + MgCO_3(s)$

a Which solution is in excess? **[1]**

b The magnesium carbonate is separated by filtration, washed and dried. What mass of magnesium carbonate should be obtained?

moles of magnesium sulfate in 50 cm^3 = ____________ mol

moles of magnesium carbonate produced in residue = ____________ mol

mass of magnesium carbonate produced = ____________ g **[3]**

c The actual mass obtained was 0.35 g. What was the percentage yield of the reaction? **[2]**

[Total: 6]

4 When propane, C_3H_8, burns completely in air, carbon dioxide and water are formed.

a Write a symbol equation for the reaction. **[2]**

b What volume of carbon dioxide, measured at r.t.p, will be formed when 100 cm^3 of propane is burned? **[2]**

c 50 cm^3 of methane is burned in 150 cm^3 of oxygen;

$CH_4 + 2O_2 \rightarrow CO_2 + 2H_2O$

What is the total volume of the mixture after reaction measured at r.t.p? **[3]**

[Total: 7]

> **COMMAND WORD**
>
> **calculate:** work out from given facts, figures or information

CONTINUED

5 An excess of hydrochloric acid was added to 1.23 g of impure barium carbonate. The volume of carbon dioxide collected at r.t.p. was 0.120 dm^3. The impurities did not react with the acid. Calculate the percentage purity of the barium carbonate.

$BaCO_3 + 2HCl \rightarrow BaCl_2 + CO_2 + H_2O$

Molar volume of a gas at r.t.p. = 24 dm^3.

a The number of moles of CO_2 collected = __________ **[1]**

b The number of moles of $BaCO_3$ reacted = __________ **[1]**

c Mass of one mole of $BaCO_3$ = __________ g **[1]**

d Mass of barium carbonate = __________ g **[1]**

e Percentage purity of the barium carbonate = __________ **[1]**

[Total: 5]

SELF-EVALUATION CHECKLIST

After studying this chapter, think about how confident you are with the different topics. This will help you see any gaps in your knowledge and help you to learn more effectively.

I can	See Topic...	Needs more work	Almost there	Confident to move on
understand that the mole is a standard number (Avogadro's constant) of characteristic particles (atoms, ions or molecules) of a substance	5.1			
calculate the empirical formula (and molecular formula) of a compound using appropriate data	5.1			
calculate the relationship between the number of moles of a substance and the mass or number of particles present in a sample of a substance	5.1			
use the information given in an equation to calculate reacting masses, limiting reagents and amount of product for a stated reaction	5.2			
calculate the percentage composition by mass of a compound, and the percentage yield and purity of a product of a given reaction	5.2			
understand that the molar gas volume for any gas is 24 dm^3 at room temperature and pressure, and be able to use this value for calculations on reactions involving gases	5.2			
understand the different units for used to express the concentration of a solution and use them when calculating the concentration of a solution from titration experiments	5.3			

Chapter 6 Electrochemistry

IN THIS CHAPTER YOU WILL:

- describe metals as electrical conductors and non-metallic materials as non-conducting insulators
- define electrolysis and identify the components of an electrolytic cell
- describe and predict the electrolysis products of binary compounds in the molten state
- describe the electrolysis of concentrated sodium chloride solution and dilute sulfuric acid using inert electrodes
- describe how to electroplate a metal object
- state that a hydrogen–oxygen fuel cell generates electricity

> predict the products of electrolysis of a halide compound in dilute and concentrated solutions

> describe how charge is transferred in electrolysis and learn how to construct ionic half-equations

> identify the products of the electrolysis of copper(II) sulfate solution using graphite or copper electrodes

> describe the advantages and disadvantages of hydrogen–oxygen fuel cells.

GETTING STARTED

Arrange yourselves into groups within the class and discuss your background understanding of electrical conductivity. Discuss the following questions:

1 From your everyday experience, which metals have you seen most often used in electrical wiring, both domestically and in computer circuit boards?

2 Why is wiring usually covered with a plastic coating?

3 Do liquid metals conduct electricity? What other kinds of liquid are able to conduct electricity?

Summarise the answers from your group and report back to the whole class.

THE HYDROGEN ECONOMY

The chemical reaction between hydrogen and oxygen is a simple reaction. The reacting substances are gaseous elements and are easy to mix. There is a single, simple non-polluting product: water. The reaction gives out a great amount of energy. This is what makes the prospect of using hydrogen as a fuel for transport very attractive. The hydrogen–oxygen fuel cell uses an electrochemical process to convert the chemical energy of the reaction into electricity and seems to be one of the better options to reduce the dependence of our transport systems on fossil fuels (Figure 6.1).

A future 'hydrogen economy' has been talked about, with hydrogen being used as an energy source in a variety of situations, including bulk transport, trains and public transport, shipping and aeroplanes. However, there are problems of storage and transport of hydrogen because of its low density, although companies are beginning to invest in fuelling stations to improve its distribution as a transport fuel. Hydrogen is not cheap. The main method of obtaining hydrogen currently is by the steam-reforming of natural gas (see Chapter 9 for how hydrogen is produced for ammonia production in the Haber process). This means that this hydrogen (sometimes called 'grey hydrogen') is not independent of fossil fuel production.

The hope is that hydrogen can be produced on a large scale by the electrolysis of water. However, this is currently not very economical. It is possible that cheap surplus electricity from nuclear, wind or solar power may make the production of hydrogen ('green hydrogen') by electrolysis more economical. Hydrogen could also be produced from waste plastics and paper that would otherwise go into landfill. The development of these technologies form significant steps in generating the hydrogen necessary to act as a pillar of a decarbonised economy.

Discussion questions

1 What is the word equation for the overall reaction in a hydrogen–oxygen fuel cell? Is the reaction exothermic or endothermic, and what type of energy is involved?

2 The use of hydrogen fuel cells is regarded as non-polluting for the environment, but what factors need to be considered for it to be regarded as carbon-neutral?

Figure 6.1: A hydrogen fuel cell bus operating in Perth, Australia.

6.1 Types of electrical conductivity

Conductivity in solids

Electricity has had a great effect on our way of living. Large urban areas (Figure 6.2) could not function without an electricity supply. The results of the large-scale supply of electricity can be seen in the pylons and power lines that mark our landscape. But electricity is also important on the very small scale. The silicon chip enables a vast range of products to work, and many people now have access to products containing advanced electronic circuits – from smartphones to washing machines.

Conductors and insulators

The ability to conduct electricity is the major simple difference between elements that are metals and elements that are non-metals. All metals conduct electricity, but carbon in the form of graphite is the only non-metallic element that conducts electricity. A simple circuit can be used to test if a solid conducts electricity (Figure 6.3). The circuit is made up of a battery (a source of direct current), some connecting copper wires fitted with clips, and a light bulb to show when a current is flowing. The material to be tested is clipped into the circuit. If the bulb lights up, then the material is an **electrical conductor**.

KEY WORDS

electrical conductor: a substance that conducts electricity but is not chemically changed in the process

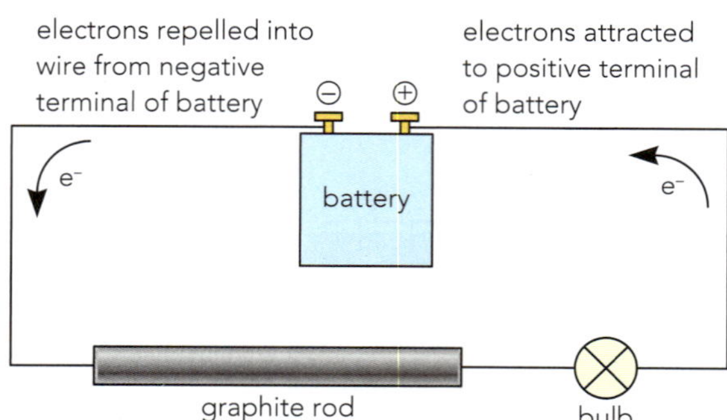

Figure 6.3: Testing a solid material (a graphite rod) to see if the material conducts electricity. The bulb will light if the material is an electrical conductor.

Figure 6.2: Akihabara Electric Town, Tokyo.

For a solid to conduct, it must have a structure that contains 'free' electrons (e^-) that are able to flow through the structure. There is a flow of electrons in the completed circuit. The battery acts as an 'electron pump'. Electrons are repelled (pushed) into the circuit from the negative terminal of the battery. They are attracted to the positive terminal. The battery 'pumps' all the free electrons in one direction. Metals (and graphite) conduct electricity because they have mobile free electrons in their structure (see Chapter 3 on metallic bonding and the structure of graphite). Metallic alloys are held together by the same type of bonding as the metal elements, so they can also conduct electricity.

Solid covalent non-metals do not conduct electricity. Whether they are giant molecular or simple molecular structures, there are no electrons that are not involved in bonding – there are no free electrons. Such substances are called non-conductors or **insulators** (Table 6.1).

There is no chemical change when an electric current is passed through a metal or graphite. The copper wire is still copper when the current is switched off!

Conductors	Insulators (non-conductors)	
	Giant molecular structures	Simple molecular structures
copper	diamond	sulfur
silver	poly(ethene)	iodine
aluminium	poly(chloroethene), commonly known as polyvinyl chloride, PVC	
steel	poly(tetrafluoroethene), PTFE	
brass		
graphite		

Table 6.1: Solid electrical conductors and insulators.

Conductivity in liquids

Some liquids conduct electricity. The process in which an electric current flows through a liquid compound or solution is called **electrolysis**. Unlike electrical conductivity in metals, electrolysis in liquid compounds or solutions produces a chemical reaction. Electrolysis has many uses in industry.

> **KEY WORDS**
>
> **insulator:** a substance that does not conduct electricity
>
> **electrolysis:** the breakdown of an ionic compound, molten or in aqueous solution, by the use of electricity

Electrolysis and the movement of ions

The conductivity of liquids can be tested in a similar way to solids, but the simple testing circuit is changed (Figure 6.4). Instead of clipping the solid material to be tested into the circuit, graphite rods are dipped into the test liquid. Liquid compounds, solutions and molten materials can all be tested in this way. Molten metals, and mercury, which is liquid at room temperature, conduct electricity. Electrons are still able to move through the liquid metal to carry the charge. As in solid metals, no chemical change takes place when liquid metals conduct electricity.

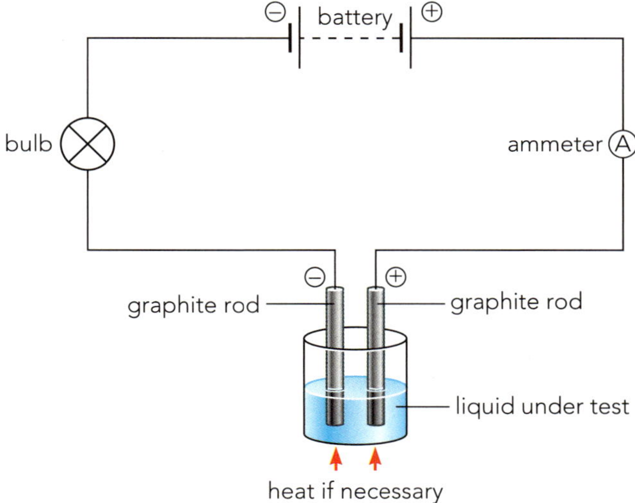

Figure 6.4: Apparatus for testing the conductivity of liquids (an electrolytic cell).

If liquid compounds or solutions are tested using the apparatus in Figure 6.4, then the result will depend on the type of bonding in the compound (Chapter 3). If the compound is bonded covalently, then the compound will *not* conduct electricity as a liquid or as a solution. Examples of such liquids are ethanol, **petrol (gasoline)**, pure water and sugar solution. Ionic compounds *will* conduct electricity if they are either molten or dissolved in water. Examples of these ionic compounds are molten lead bromide, sodium chloride solution and copper(II) sulfate solution.

When these ionically bonded liquids conduct electricity, they do so in a different way from metals. In this case, they conduct electricity because the ions present can move through the liquid; when metals conduct electricity, electrons move through the metal.

Ionic compounds will not conduct electricity when they are solid because their ions are fixed in position and cannot move. Remember that in Chapter 1 we discussed the fact that in solids the particles present can only vibrate about a fixed position. Liquids that conduct electricity by the movement of ions are called **electrolytes** and liquids that do not conduct in this way are called **non-electrolytes** (Table 6.2).

Electrolytes	Non-electrolytes
sulfuric acid	distilled water
molten lead bromide	ethanol
sodium chloride solution	petrol
hydrochloric acid	paraffin
copper(II) chloride solution	molten sulfur
sodium hydroxide solution	sugar solution

Table 6.2: Some electrolytes and non-electrolytes.

When electrolytes conduct electricity during electrolysis, a chemical change takes place and the ionic compound is broken down (it is decomposed). For example, lead(II) bromide is changed to lead and bromine:

lead(II) bromide → lead + bromine

$PbBr_2(l) \rightarrow Pb(l) + Br_2(g)$

The conductivity of ionic compounds is explained by the fact that ions move in a particular direction in an electric field. This can be shown in experiments using coloured salts.

For example, copper(II) chromate(VI) ($CuCrO_4$) dissolves in water to give a green solution. This solution is placed in the lower part of a U-tube. A colourless solution of dilute hydrochloric acid is then layered on top of the salt solution in each arm of the tube and graphite rods are fitted (Figure 6.5). These rods carry the current into and out of the solution. They are known as **electrodes**. In electrolysis, the negative electrode is called the **cathode** and the positive electrode is called the **anode**.

KEY WORDS

petrol (or **gasoline**): a clear flammable liquid derived from petroleum used primarily as a fuel in most combustion engines; it is obtained by the fractional distillation of petroleum

electrolyte: an ionic compound that will conduct electricity when it is molten or dissolved in water; electrolytes will not conduct electricity when solid

non-electrolytes: liquids or solutions that do not take part in electrolysis: they do not contain ions

electrodes: the points where the electric current enters or leaves a battery or electrolytic cell

cathode: the electrode in any type of cell at which reduction (the gain of electrons) takes place; in electrolysis it is the negative electrode

anode: the electrode in any type of cell at which oxidation (the loss of electrons) takes place – in electrolysis it is the positive electrode

After passing the current for a short time, the solution around the cathode becomes blue. Around the anode the solution becomes yellow. These colours are produced by the movement (migration) of the ions in the salt. The positive copper ions (Cu^{2+}) are blue in solution. They are attracted to the cathode (the negative electrode). The negative chromate ions (CrO_4^{2-}) are yellow in solution. They are attracted to the anode (the positive electrode). The use of coloured ions in solution has shown the direction that positive and negative ions move in an electric field.

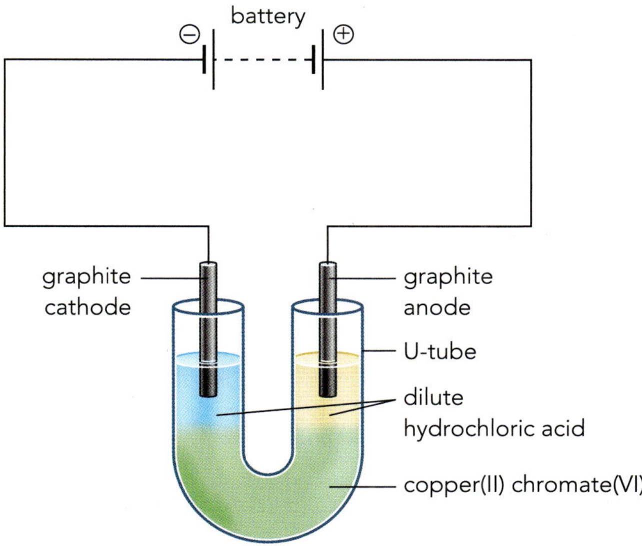

Figure 6.5: An experiment to show ionic movement by using a salt solution containing coloured ions. The acid solution was colourless at the start of the experiment.

Electrolytic cells

The apparatus in which electrolysis is carried out is known as an **electrolytic cell** (Figure 6.4). The direct current is supplied by a battery or power pack. Graphite electrodes carry the current into and out of the electrolyte. Graphite is chosen because it is quite unreactive (inert). It will not react with the electrolyte or with the products of electrolysis. Platinum electrodes are sometimes used in some cells instead of graphite. In the external circuit, electrons flow from the negative terminal of the battery to the cathode, and then from the anode back to the positive terminal. In the electrolyte it is the ions that move to carry the current.

During electrolysis:

- positive ions (metal ions or H^+ ions) move towards the cathode; they are known as **cations**
- negative ions (non-metal ions) move towards the anode; they are known as **anions**.

Difference between conductivity in solids and electrolytes

It is important to remember that it is the electrons that move through the wire when a metal conducts electricity. However, when a salt solution conducts electricity, it is the ions in the solution that move to the electrodes where they lose their charge (they are discharged). A solid ionic compound will not conduct electricity; because the ions are in fixed positions in a solid, they cannot move. The electrolyte must be melted or dissolved in water for it to conduct.

The two distinct types of electrical conductivity are called metallic and electrolytic conductivity. They differ from each other in important ways.

Conductivity in metals:

- electrons flow
- a property of elements (metals, and carbon as graphite) and alloys
- takes place in solids and liquids
- no chemical change takes place.

Electrolytic conductivity:

- ions flow
- a property of ionic compounds
- takes place in liquids and solutions (not solids)
- chemical **decomposition** takes place.

KEY WORDS

electrolytic cell: a cell consisting of an electrolyte and two electrodes (anode and cathode) connected to an external DC power source where positive and negative ions in the electrolyte are separated and discharged

cation: a positive ion that would be attracted to the cathode in electrolysis

anion: a negative ion that would be attracted to the anode in electrolysis

decomposition: a type of chemical reaction where a compound breaks down into simpler substances

Questions

1 a Which of the following will conduct electricity?
   i a strip of copper metal
   ii a solution of sugar in water
   iii aqueous sodium chloride solution
   iv mercury
   v dilute hydrochloric acid
   vi melted sugar
  b Which of i–vi are electrolytes?

2 a Why is solid potassium bromide a non-conductor of electricity?
  b Suggest two things that could be done to potassium bromide that would make it conduct electricity.

3 Explain the major difference between electrical conductivity in metals and solutions in terms of the particles that are carrying the current.

ACTIVITY 6.1

Key words in electrical conductivity and electrolysis

The topics of electrical conductivity and electrolysis include a number of key words that cover the nature of different substances and the features of an electrolytic cell. Work in pairs to create a list of the key words for this topic. You can include the key words identified in this chapter to help you.

Once you have completed your list, follow the steps:

1 Label yourselves Partner A and Partner B.

2 Partner A chooses a key word and describes it to Partner B. In your description, you are not allowed to include the key word.

3 Partner B then tries to guess the word that Partner A is explaining.

4 Swap over and take it in turns until you have discussed all of your key words.

After you have finished, create a word cloud of the terms you think are the most important to learn for this topic; using different sizes and colours to indicate the most straightforward and most difficult terms.

6.2 Products of electrolysis

Electrolysis of molten ionic compounds

An electrolytic cell can be used to electrolyse molten compounds. Heat must be supplied to keep the salt molten. Figure 6.6 shows the electrolysis of molten lead(II) bromide.

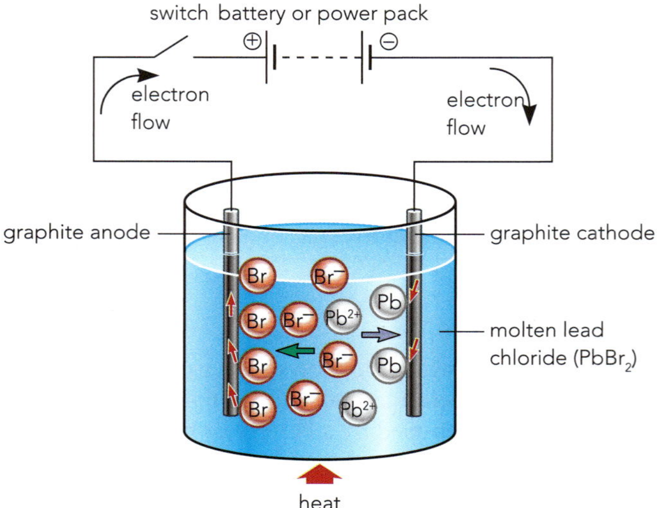

Figure 6.6: Movement of ions in the electrolysis of a molten salt, lead(II) bromide.

When the switch is closed and the circuit is complete, the current flows and bromine vapour (which is red–brown) begins to bubble off at the anode. After a short time, a bead of liquid lead collects at the cathode. The electrical energy from the cell has caused a chemical change (decomposition). The cell decomposes the molten lead(II) bromide because the Pb^{2+} and Br^{-} ions present move to opposite electrodes where they are discharged. Figure 6.6 shows this movement. This movement of the ions means that when a molten ionic compound is electrolysed:

- the metal is always formed at the cathode (the negative electrode)
- the non-metal is always formed at the anode (the positive electrode).

This means we can easily predict the products of electrolysis of a molten, binary ionic compound. Table 6.3 shows some further examples of this type of electrolysis. The electrolysis of molten salts is easier if the melting point of the salt is not too high. The electrolysis of molten ionic salts is important for the industrial extraction of reactive metals such as sodium and magnesium. The extraction of aluminium by the electrolysis of molten aluminium oxide is a very important process and is covered in detail in Chapter 16.

Electrolyte (compound electrolysed)	Product at cathode (negative electrode)	Product at anode (positive electrode)
lead bromide, $PbBr_2$	lead (Pb)	bromine (Br_2)
sodium chloride, NaCl	sodium (Na)	chlorine (Cl_2)
potassium iodide, KI	potassium (K)	iodine (I_2)
copper(II) bromide, $CuBr_2$	copper (Cu)	bromine (Br_2)
zinc chloride, $ZnCl_2$	zinc (Zn)	chlorine (Cl_2)
aluminium oxide (Al_2O_3)	aluminium (Al)	oxygen (O_2)

Table 6.3: Some examples of the electrolysis of molten salts.

The movement of ions to the different electrodes during electrolysis results in the decomposition of the ionic compound. Different reactions take place at the two electrodes. Look at the electrolysis of molten lead(II) bromide in more detail (Figure 6.7). During electrolysis, the bromide ions (Br^-) move to the anode. Each bromide ion gives up (donates) one electron to become a bromine atom:

At the anode $Br^- \rightarrow Br + e^-$

Then two bromine atoms bond together to make a bromine molecule:

$Br + Br \rightarrow Br_2$

The lead ions (Pb^{2+}) move to the cathode. There, each lead ion picks up (accepts) two electrons and becomes a lead atom:

At the cathode $Pb^{2+} + 2e^- \rightarrow Pb$

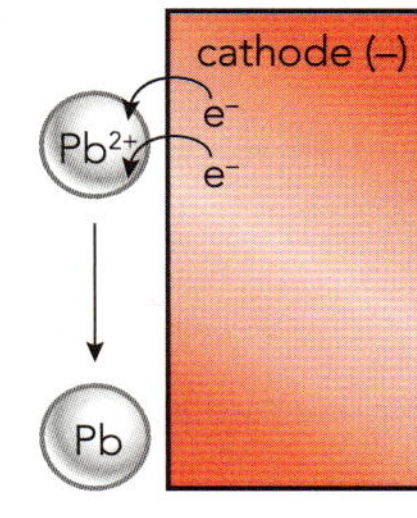

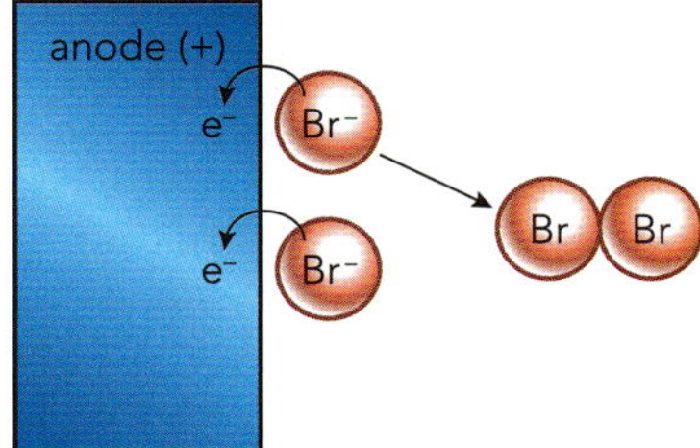

Figure 6.7: At the cathode, metal ions gain electrons. At the anode, non-metal ions lose electrons.

During electrolysis, the flow of electrons continues through the circuit. For every two electrons taken from the cathode by a lead ion, two electrons are set free at the anode by two bromide ions. So, overall, the electrons released at the anode flow through the circuit towards the cathode. During the electrolysis of molten salts, the metal ions, which are always positive (cations), move to the cathode and are discharged. Non-metal ions (except hydrogen), however, are always negative (anions) and move to the anode to be discharged. Table 6.4 shows the cathode and anode reactions (**half-equations**) for some examples of electrolysis of molten binary compounds. These electrode reactions are represented by half-equations; if the two half-equations are added together, they give the overall equation for the reaction taking place in the electrolytic cell.

KEY WORD

half-equations: ionic equations showing the separate oxidation and reduction steps in redox reactions, including the reactions at the anode (oxidation) and cathode (reduction) in an electrolytic cell

Electrolyte	Cathode reaction	Anode reaction
lead bromide, $PbBr_2$	$Pb^{2+} + 2e^- \rightarrow Pb$	$2Br^- \rightarrow Br_2 + 2e^-$
sodium chloride, NaCl	$Na^+ + e^- \rightarrow Na$	$2Cl^- \rightarrow Cl_2 + 2e^-$
aluminium oxide (Al_2O_3)	$Al^{3+} + 3e^- \rightarrow Al$	$2O^{2-} \rightarrow O_2 + 4e^-$
copper(II) bromide, $CuBr_2$	$Cu^{2+} + 2e^- \rightarrow Cu$	$2Br^- \rightarrow Br_2 + 2e^-$

Table 6.4: Electrode half-equations for some examples of the electrolysis of molten salts.

The anode reactions shown in Table 6.4 are the sum of the two stages written in the text: the loss of electrons from the ions followed by the joining together of two atoms to make a molecule of the element The loss of an electron from a negative ion like Cl^- can also be written $2Cl^- - 2e^- \rightarrow Cl_2$.

Electrolysis of solutions

Many ionic compounds dissolve in water. In such solutions the ions are free to move and therefore these solutions will conduct electricity. The electrolysis of ionic solutions also produces chemical change. However, the products from electrolysis of a solution of a salt may be different from those obtained by electrolysis of the molten salt. This is because water itself produces ions.

Although water is a simple molecular substance, a very small fraction of its molecules split into hydrogen ions (H^+) and hydroxide ions (OH^-):

$$H_2O \rightleftharpoons H^+ + OH^-$$

most molecules intact ⇌ only a very few molecules split into ions

Not enough ions are produced to make pure water a good conductor of electricity. Water is a weak electrolyte. During electrolysis, however, these hydrogen and hydroxide ions are also able to move to the electrodes. The hydrogen and hydroxide ions compete with the ions from the acid or salt to be discharged at the electrodes.

Electrolysis of dilute sulfuric acid

Although water is a very poor conductor of electricity, it can be made to decompose if the water is acidified with a few drops of sulfuric acid. An electrolytic cell such as the one shown in Figure 6.8, sometimes called a Hofmann voltameter, can be for the electrolysis of the dilute sulfuric acid as it keeps the gases produced separate.

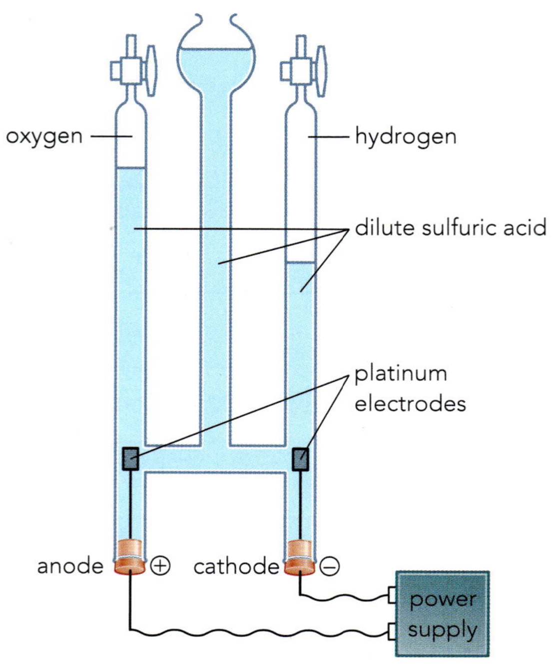

Figure 6.8: A Hofmann voltameter for the electrolysis of dilute sulfuric acid.

After a short time, the volume of gas in each arm of the apparatus can be measured and tested. The gas collected above the cathode is hydrogen. Oxygen collects above the anode. The ratio of the volumes is approximately 2 : 1. The ions present in the solution are H^+, OH^- and SO_4^{2-}. But at each electrode just *one* type of ion gets discharged. Only hydrogen ions move to the cathode so it is clear why hydrogen gas is produced there. However, both OH^- and SO_4^{2-} move to the anode, but only the hydroxide ions are discharged to give oxygen gas. Effectively, this experiment is the electrolysis of water.

$$2H_2O(l) \rightarrow \underset{\text{at the cathode}}{2H_2(g)} + \underset{\text{at the anode}}{O_2(g)}$$

This electrolysis begins to show that some ions are preferentially discharged ahead of others during electrolysis.

It is possible to write half-equations for the reactions that have taken place at the two electrodes.

At the cathode $2H^+ + 2e^- \rightarrow H_2$

At the anode $4OH^- \rightarrow 2H_2O + O_2 + 4e^-$

Electrolysis of salt solutions

The electrolysis of salt solutions is more complicated than for acids. Positive metal ions are also present in these solutions. These metal ions will compete with the hydrogen ions from water to be discharged at the cathode.

A concentrated solution of sodium chloride can be electrolysed in the laboratory (Figure 6.9). There are four different ions present in the solution. The positive ions (cations), Na^+ and H^+, flow to the cathode, attracted by its negative charge. The negative ions (anions), Cl^- and OH^-, travel to the anode.

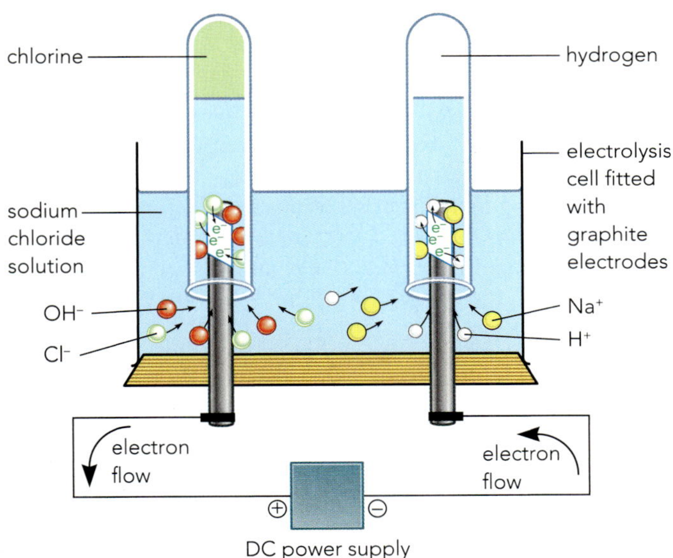

Figure 6.9: Movement and discharge of ions in the electrolysis of concentrated sodium chloride solution.

When this experiment is carried out, hydrogen is produced at the cathode, not sodium. Chlorine, a green gas, is collected at the anode. This chlorine is produced by the discharge of chloride ions at the anode. In this case, the hydroxide ions are not discharged and so no oxygen gas is produced.

It is important to understand that different products are formed when concentrated sodium chloride solution is electrolysed rather than molten sodium chloride (Table 6.5).

Electrolyte	Product at cathode	Product at anode
Molten sodium chloride, NaCl(l)	Sodium (Na)	Chlorine (Cl_2)
Concentrated sodium chloride solution, NaCl(aq)	Hydrogen (H_2)	Chlorine (Cl_2)

Table 6.5: Products of electrolysis of sodium chloride under different conditions.

A consideration of all the examples of electrolysis discussed so far gives us the following rules regarding the products obtained at the electrodes:

- a metal or hydrogen is always formed at the cathode
- a non-metal (other than hydrogen) is formed at the anode.

We can write half-equations for the electrode reactions taking place during the electrolysis of a concentrated sodium chloride solution.

At the cathode, it is the H^+ ions that accept electrons (Figure 6.10), as sodium is more reactive than hydrogen:

$$H^+ + e^- \rightarrow H$$

Then two hydrogen atoms combine to form a hydrogen molecule:

$$H + H \rightarrow H_2$$

So, overall, hydrogen gas bubbles off at the cathode:

$$2H^+ + 2e^- \rightarrow H_2$$

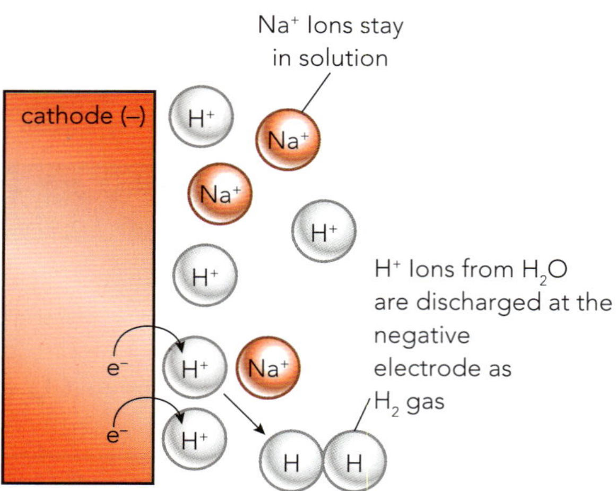

Figure 6.10: Hydrogen ions, rather than sodium ions, are discharged at the cathode. The atoms combine to form hydrogen molecules (H_2).

At the anode, the Cl^- ions are discharged more readily than the OH^- ions:

$$Cl^- \rightarrow Cl + e^-$$

Then two chlorine atoms combine to make a chlorine molecule:

$$Cl + Cl \rightarrow Cl_2$$

So, overall, pale-green chlorine gas bubbles off at the anode:

$$2Cl^- \rightarrow Cl_2 + 2e^-$$

The Na^+ and OH^- ions are left behind in solution; this is sodium hydroxide solution. The solution therefore becomes alkaline. This can be shown by adding indicator to the solution. Universal indicator will turn purple as the electrolysis takes place. The overall reaction taking place in the electrolytic cell is:

$$2NaCl(aq) + 2H_2O(l) \rightarrow 2NaOH(aq) + H_2(g) + Cl_2(g)$$

The products of the reaction (hydrogen, chlorine and sodium hydroxide) are very important industrially as the basis for the chlor-alkali industry. So, the electrolysis of concentrated brine (salt water) is a very important manufacturing process. Similar results are obtained if concentrated solutions of other metal halide compounds are electrolysed. Hydrogen gas is produced at the cathode and the halogen at the anode.

However, the effect of concentration is important. Electrolysis of a dilute solution of sodium chloride results in the formation of oxygen at the anode rather than chlorine. When the concentration of chloride ions is low, it is the hydroxide (OH^-) ions that are discharged instead.

$$4OH^- \rightarrow 2H_2O + O_2 + 4e^-$$

Hydrogen is still produced at the cathode. These results indicate that the preferential discharge of Cl^- ions ahead of OH^- ions only applies if the concentration of Cl^- is sufficiently high. Table 6.6 summarises these differences in the products formed for concentrated and dilute solutions of metal halide compounds.

Electrolyte solution	Product at the cathode	Product at the anode
concentrated NaCl(aq)	hydrogen	chlorine
concentrated KBr(aq)	hydrogen	bromine
dilute NaCl(aq)	hydrogen	oxygen
dilute KBr(aq)	hydrogen	oxygen

Table 6.6: Electrolysis of dilute and concentrated solutions of metal halide compounds.

Further information on the nature of electrode reactions can be found from experiments using different salt solutions. The electrolysis of a blue copper(II) sulfate solution using inert graphite electrodes (Figure 6.11) produces a deposit of red–brown copper metal on the cathode. Copper is less reactive than hydrogen and therefore it is the copper ions (Cu^{2+}) that are discharged at the cathode.

At the cathode $Cu^{2+}(aq) + 2e^- \rightarrow Cu(s)$

Figure 6.11: Copper is quite unreactive so it can be seen deposited on the cathode when copper(II) sulfate solution is electrolysed.

Oxygen gas is produced at the anode as the hydroxide ions from water are discharged rather than the sulfate ions.

At the anode $4OH^- \rightarrow 2H_2O + O_2 + 4e^-$

As the electrolysis proceeds, the blue colour of the solution will fade as the copper ions causing the colour are discharged. The electrolyte solution will also become more acidic as OH^- ions are discharged.

Consideration of the results of electrolysis experiments on salt solutions leads to some rules for ion discharge at the electrodes:

At the cathode:

- The more reactive a metal, the more it tends to stay as ions and not be discharged. The H^+ ions will accept electrons instead. Hydrogen molecules will be formed, leaving the ions of the reactive metal (e.g. Na^+ ions) in solution.
- In contrast, the ions of less reactive metals (e.g. Cu^{2+} ions) will accept electrons readily and form metal atoms. In this case, the metal will be discharged, leaving the H^+ ions in solution (Figure 6.11).
- For positive ions:
  Na^+ Mg^{2+} Al^{3+} Zn^{2+} $\mathbf{H^+}$ Cu^{2+} Ag
  → more likely to be discharged

At the anode:

- If the ions of a halogen (Cl^-, Br^- or I^-) are present in a high enough concentration, they will give up electrons more readily than OH^- ions will. Molecules of chlorine, bromine or iodine are formed. The OH^- ions remain in solution.
- If no halogen ions are present, or the halide solution is too dilute, the OH^- ions will give up electrons. When OH^- ions are discharged, oxygen is formed. Sulfate and nitrate ions are not discharged in preference to OH^- ions.
- For negative ions:
  SO_4^{2-} NO_3^- $\mathbf{OH^-}$ Cl^- Br^- I^-
  → more likely to be discharged

Electrode half-equations in electrolysis

The reactions that take place at the electrodes during electrolysis involve the loss and gain of electrons. These electrode reactions can be written as half-equations. Negative ions always travel to the anode, where they lose electrons. In contrast, positive ions always flow to the cathode, where they gain electrons. As we will discuss in more detail in Chapter 10, oxidation can be defined as the loss of electrons and reduction can be defined as the gain of electrons. Therefore, electrolysis can be seen as a process in which oxidation and reduction are physically separated. During electrolysis:

- the *oxidation* of non-metal ions always takes place at the anode
- the *reduction* of metal or hydrogen ions always takes place at the cathode.

The equation for the overall reaction taking place during a particular electrolysis can be found by adding the two half-equations together. Using the example of the electrolysis of copper(II) sulfate solution with graphite electrodes, we have:

At the cathode
$Cu^{2+}(aq) + 2e^- \rightarrow Cu(s)$
multiply by 2 to balance the electrons

At the anode
$4OH^-(aq) \rightarrow 2H_2O(l) + O_2(g) + 4e^-$

Overall reaction
$2Cu^{2+}(aq) + 4OH^-(aq) \rightarrow 2Cu(s) + 2H_2O(l) + O_2(g)$

The exchange of electrons at the anode and cathode is important in the transfer of charge around the electrolysis circuit as it is the bridge that continues the flow of electrons through the system. The transfer of charge during electrolysis consists of three distinct stages, which are illustrated in Figure 6.12:

- electrons move through the wires of the external circuit under the influence of the battery or DC power source (see Figure 6.9)
- the cations (positive ions) in the electrolyte move to the cathode where they gain electrons from the external circuit, and
- the anions (negative ions) move to the anode where they lose electrons, and these electrons complete the flow of charge by moving through the external wiring to the positive pole of the battery.

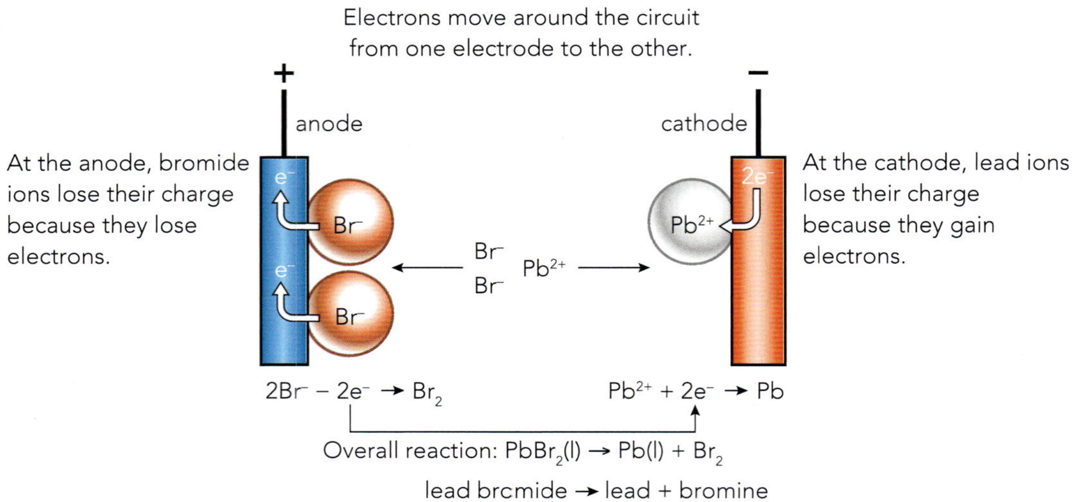

Figure 6.12: Movement of charge during the electrolysis of molten lead bromide.

Electrolysis of copper(II) sulfate with copper electrodes

Our discussion so far has been about the influence of the electrolyte on the products formed. However, it is possible to use electrodes other than graphite or platinum; electrodes that are not inert. A cell can be set up to electrolyse copper(II) sulfate solution using copper electrodes (not graphite electrodes). As electrolysis takes place, the cathode gains mass as copper is deposited on the electrode (Figure 6.13):

At the cathode $Cu^{2+}(aq) + 2e^- \rightarrow Cu(s)$

The anode, however, loses mass as copper dissolves from the electrode:

At the anode $Cu(s) \rightarrow Cu^{2+}(aq) + 2e^-$

So, overall, there is a transfer of copper from the anode to the cathode. The colour of the copper(II) sulfate solution does not change because the concentration of the Cu^{2+} ions in the solution remains the same. The idea of dissolving anodes is useful in the process of **electroplating** and in the purification of copper.

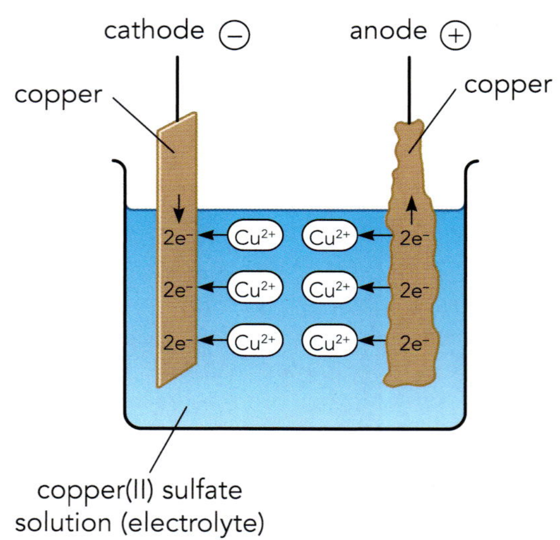

Figure 6.13: The electrolysis of copper(II) sulfate using copper electrodes. The movement of ions effectively transfers copper from one electrode to the other.

KEY WORD

electroplating: a process of electrolysis in which a metal object is coated (plated) with a layer of another metal

REFLECTION

An understanding of electrolysis involves considering several different physical situations: molten electrolytes, solutions of dilute acids or salt solutions. How have you organised the information that deals with these different situations? Do you have a clear strategy for predicting the different products formed in these varied examples? How would you explain the key features of electrolysis to someone else?

EXPERIMENTAL SKILLS 6.1

Electrolysis of copper(II) sulfate solution

This experiment is designed to demonstrate the different products obtained when the electrolysis of copper(II) sulfate solution is carried out first with inert graphite electrodes and then with copper electrodes. The use of copper electrodes illustrates how copper is refined industrially.

You will need:

- copper(II) sulfate solution (0.5 mol/dm^3)
- 200 cm^3 beaker
- graphite rods
- stand and clamp(s) for electrodes
- DC power supply
- small light bulb
- connecting leads and crocodile clips
- copper strips (cleaned with sandpaper).

Safety

Wear eye protection throughout. Be careful with chemicals. Never ingest them and always wash your hands after handling them.

Getting started

Consider what you have learnt about electrolytic cells and the process of electrolysis.

- Predict at which electrode you would expect a metal to be deposited.
- Predict at which electrode you would expect a non-metallic gas to be given off.

Finally, check that you can assemble the equipment for the cell for a test run.

Method

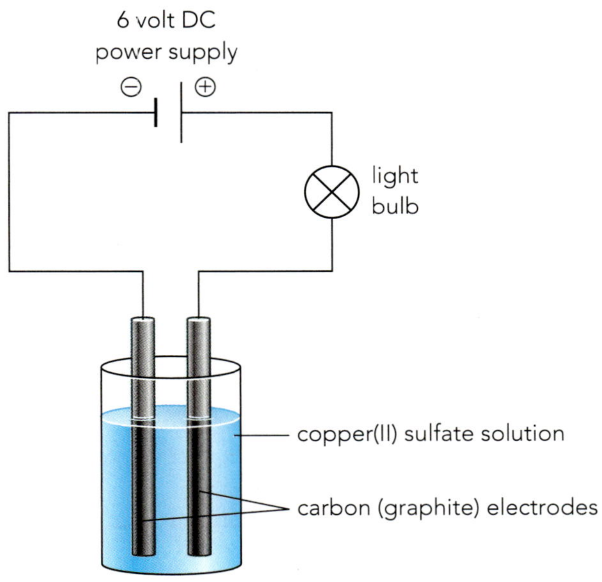

Figure 6.14: Electrolysis of copper(II) sulfate solution with graphite electrodes.

1 The electrolytic cell should be set up as shown in Figure 6.14 using graphite rods as electrodes (see also Figure 6.11).

A deposit of copper forms on the cathode; this will often be powdery and uneven. Bubbles of gas (oxygen) are formed at the anode.

cathode reaction:
$Cu^{2+}(aq) + 2e^- \rightarrow Cu(s)$

anode reaction:
$4OH^-(aq) \rightarrow 2H_2O(l) + O_2(g) + 4e^-$

CONTINUED

2 Repeat the experiment, replacing the graphite rods with clean copper strips as electrodes.

This produces a different anode reaction. No oxygen is produced, rather the copper anode dissolves. Copper can be seen disappearing from the surface of the copper-coated anode:

Anode reaction: $Cu(s) \rightarrow Cu^{2+}(aq) + 2e^{-}$

The reaction is the reverse of the cathode reaction.

During this electrolysis, the mass gained at the cathode is equal to the mass lost at the anode.

Questions

1 Were there ways in which the apparatus you had could be improved?

2 Comment on the steps you would need to take to make the electrolysis with copper electrodes **quantitative**.

Self-assessment

Setting up the apparatus for a functioning electrolytic cell requires organisation and practical skill. Would you be able to agree with the following statements as to how you carried out the experiment?

- I was confident when setting up the apparatus for this experiment and I set up everything correctly.
- My electrolytic cell worked and I could see the formation of copper on the cathode.
- I had time to repeat the experiment with copper electrodes.

KEY WORD

quantitative: the ability to put numerical values to the properties being studied

Electroplating

The fact that an unreactive metal can be coated on to the surface of the cathode by electrolysis (Figure 6.11) means that useful metal objects can be 'plated' with a chosen metal. Electroplating can be used to coat one metal with another (Figure 6.15a).

For electroplating, the electrolytic cell is adapted from the type usually used. The cathode is the object to be plated and the anode is made from the metal being used to plate the object. The electrolyte is a salt of the same metal as used for the anode. As the process proceeds, the anode dissolves away into the solution, replacing the metal plated on to the object. As a result of metal ions moving into solution at the anode the concentration, and any colour, of the solution remains the same throughout the electrolysis.

a

silver anode (+)

metal spoon cathode (–)

silver nitrate solution (electrolyte)

Figure 6.15 a: Cell for electroplating with silver. **b:** Industrial electroplating of metal objects.

The most commonly used metals for electroplating are copper, chromium, silver and tin. One purpose of electroplating is to give a protective coating to the metal underneath; an example is the tin plating of steel cans to prevent them rusting (Chapter 16). This is also the idea behind the chromium plating of articles such as motorcycle parts, door knobs, bath taps, etc. Chromium does not corrode; it is a hard metal that resists scratching and wear, and it can also be polished to give an attractive finish.

The attractive appearance of silver can be achieved by electroplating silver on to an article made from a cheaper metal such as nickel silver (Figure 6.15b). The 'EPNS' seen on cutlery and other objects stands for 'electroplated nickel silver'. 'Nickel silver' is an alloy of copper, zinc and nickel – it contains no silver at all! It is often used as the base metal for silver-plated articles.

The basic rules for electroplating an object with a metal (Figure 6.15a) are:

- The object to be plated must be made the cathode.
- The electrolyte must be a solution of a salt of metal to be plated on the object.
- The anode is made of a strip of metal to be plated on the object.

Questions

4 What products will you get when you pass an electric current through:
   a molten potassium chloride
   b a concentrated solution of potassium chloride?

5 Give a general rule for predicting the electrode products when a molten binary ionic compound is electrolysed.

6 A metal object is to be copper plated.
   a Which electrode should the object be made?
   b Name a solution that could be used as the electrolyte.

7 In the electrolysis of molten lead(II) bromide, the reaction occurring at the negative electrode was:
   $Pb^{2+} + 2e^- \rightarrow Pb$
   a Write the equation for the reaction taking place at the positive electrode.
   b Why is the reaction taking place at the negative electrode viewed as a reduction reaction?

6.3 Hydrogen as a fuel

Hydrogen–oxygen fuel cells

In electrolysis, we put in electrical energy to make a chemical reaction happen. The reaction that takes place is always a decomposition reaction. The change is endothermic; electrical energy is put in to bring about the change (endothermic and exothermic processes were introduced in Chapter 1, and will be discussed further in Chapter 7).

It is possible to reverse this process and use chemical reactions to produce electrical energy. This is what takes place in the electrochemical cells and batteries that we use so often in everyday situations (Figure 6.16). Simple electrochemical cells lose their power and need to be recharged from time to time. Many of the substances found in electrochemical cells are harmful and difficult to dispose of.

Figure 6.16: A collection of some different types and sizes of battery.

A much more efficient way of changing chemical energy into electrical energy is by using a **fuel cell**. Such a cell operates continuously, with no need for recharging. The cell supplies energy as long as the reactants are fed into the electrodes. A hydrogen–oxygen fuel cell (Figure 6.17) uses the reaction between hydrogen and oxygen:

$$2H_2(g) + O_2(g) \rightarrow 2H_2O(l)$$

This reaction releases a large amount of energy. Water is the only product. Hydrogen can be regarded as a non-polluting **fuel**.

KEY WORDS

fuel cell: a device for continuously converting chemical energy into electrical energy using a combustion reaction; a hydrogen fuel cell uses the reaction between hydrogen and oxygen

fuel: a substance that can be used as a source of energy, usually by burning (combustion)

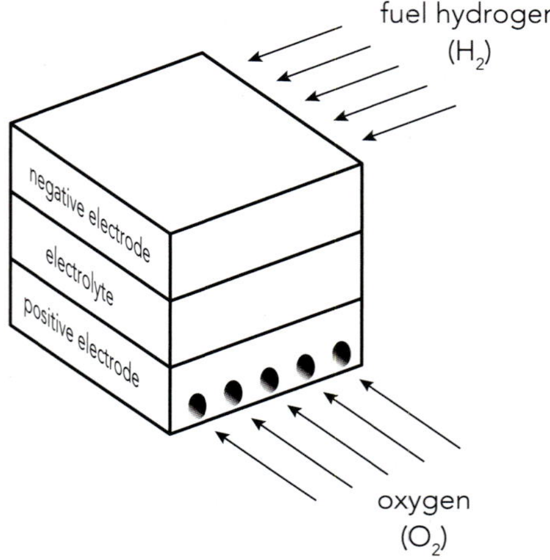

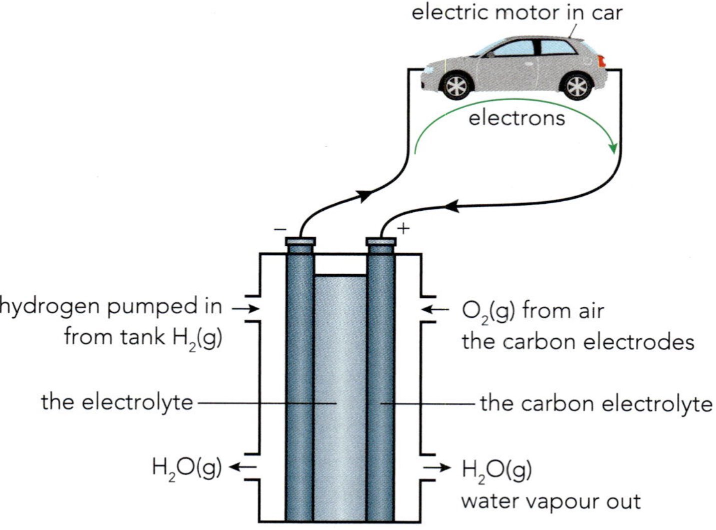

Figure 6.17: Sketches of a hydrogen–oxygen fuel cell. **a:** Hydrogen entering the cell at the negative electrode and oxygen at the positive electrode. **b:** Electrons flowing to a car.

Hydrogen-powered vehicles

Hydrogen gas has attractions as a fuel. All it produces on burning is water. When hydrogen burns, it produces more energy per gram than any other fuel (Figure 6.18).

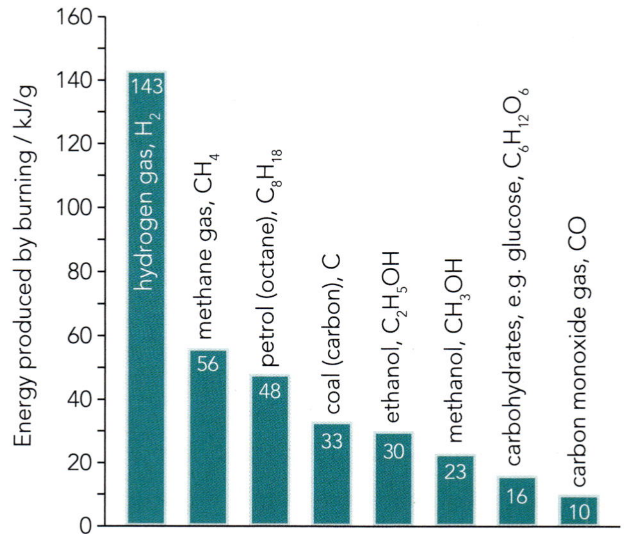

Figure 6.18: Energy produced by burning various fuels. Hydrogen produces more energy per gram than any other fuel.

Despite difficulties of production, safe storage and transport, several manufacturers have developed prototype hydrogen-powered cars. Some prototypes burn the hydrogen in a modified combustion engine. However, the majority use the hydrogen in a fuel cell. Electricity from this cell then powers an electric motor (Figure 6.17). Using a fuel cell operating an electric motor, hydrogen has an efficiency of 60% compared with 35% for a petrol engine. The advantages and disadvantages are summarised in Table 6.7.

A hydrogen fuel cell can be used to power a car and in recent years there have been prototypes developed by several motor manufacturers worldwide. The fuel cell operates continuously, with no need for recharging. The cell supplies energy so long as the reactants are fed to the electrodes.

A move to use hydrogen-powered vehicles more extensively would require a greater distribution of hydrogen filling stations. Indeed, fuel cells may be more suited for heavier vehicles and longer distances, suggesting they may be more use for public service vehicles and freight transport vehicles.

Advantages	Disadvantages
• renewable if produced using solar or wind energy • lower flammability than petrol • virtually emission-free • zero emissions of CO_2 • non-toxic	• non-renewable if generated using nuclear energy or energy from fossil fuels • large fuel tank required • as yet there are very few 'filling stations', where a car could be topped up with hydrogen • engine redesign needed or a fuel cell system • currently expensive

Table 6.7: Advantages and disadvantages of hydrogen as a fuel for motor vehicles.

Questions

8 What is currently the main source of hydrogen for hydrogen-powered vehicles?

9 For a future carbon-neutral approach, what is the preferred method of generating hydrogen? What sources of energy could be used in this approach?

10 In one type of hydrogen–oxygen fuel cell the electrode reactions are:

$$\_\_\ H_2(g) \rightarrow \_\_\ H^+(aq) + \_\_\ e^-$$

$$O_2(g) + 4H^+(aq) + 4e^- \rightarrow 2H_2O(l)$$

Balance the first of these half-equations so that together they add up to the overall equation:

$$2H_2(g) + O_2(g) \rightarrow 2H_2O(l)$$

SUMMARY

Metals can act as electrical conductors and non-metallic materials (e.g. plastics and ceramics) as non-conducting insulators.
Electrolysis is the decomposition of an ionic compound by the passage of an electric current when molten or dissolved is aqueous solution.
A simple electrolytic cell has different components, such as the cathode, anode, liquid electrolyte and a power supply.
The electrode products of molten lead(II) bromide during electrolysis using inert graphite electrodes are lead and bromine, and it is possible to predict the products of electrolysis of other molten binary ionic compounds.
The products of the electrolysis of concentrated sodium chloride solution are hydrogen and chlorine, and of dilute sulfuric acid solution are hydrogen and oxygen.
During electrolysis, metals or hydrogen are formed at the negative electrode (cathode), and non-metals are formed at the positive electrode (anode).
Ionic half-equations can be constructed to show the reactions taking place at the anode and cathode during different examples of electrolysis.
To the electrolysis products of dilute and concentrated solutions of metal halide compounds can be predicted provided the concentration of the solution is known.
The transfer of charge around the circuit during electrolysis consists of three distinct stages, including the movement of electrons in the external wires, the transfer of electrons at the electrodes and the movement of ions in the electrolyte.
Different electrolysis products are formed when copper(II) sulfate solution is electrolysed using graphite or copper electrodes, and these differences are shown in the electrolyte solution as well as the electrode reactions.
Electrolysis can be used to electroplate metal objects with another metal to improve their appearance and protect against corrosion.
A hydrogen–oxygen fuel cell can be used to generate electricity from the reaction between hydrogen and oxygen, with water as the only chemical product.
The hydrogen fuel cell has advantages and disadvantages as an alternative to petrol engines in vehicles.

PROJECT

Generating 'green' hydrogen for modern transport systems

You work for a scientific consultancy and have been asked to explain the process of electrolysis to members of the government to help inform their decisions to establish hydrogen-generating plants in several regions of your country.

Work in groups to prepare a presentation on the process of electrolysis and its environmental impacts.

In your presentation, include the following:

- An explanation of the process of electrolysing water to produce hydrogen. Use the structure provided by the explanatory triangle shown in Figure 6.19. This helps to explain the meaning of the word electrolysis and the process involved. Use the example of the electrolysis of dilute sulfuric acid in your explanation.

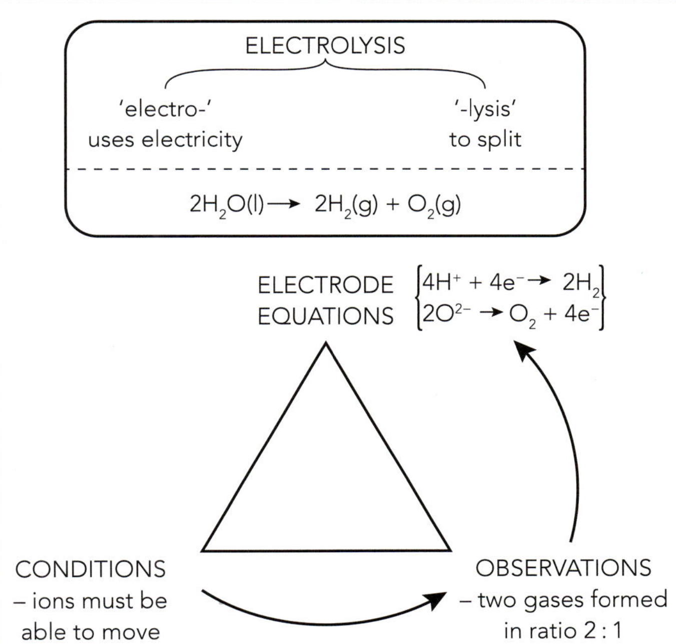

Figure 6.19: The electrolysis of water.

- Identify the sources of clean energy needed to carry out the electrolysis.
- Suggest how the hydrogen is used to power different types of transport system, and their environmental advantage.

Share your presentations with the whole class.

EXAM-STYLE QUESTIONS

1 There are certain requirements for electrolysis to take place. One of these is the nature of the substance between the electrodes. Which of the diagrams A–D in the figure shows a beaker in which electrolysis takes place?

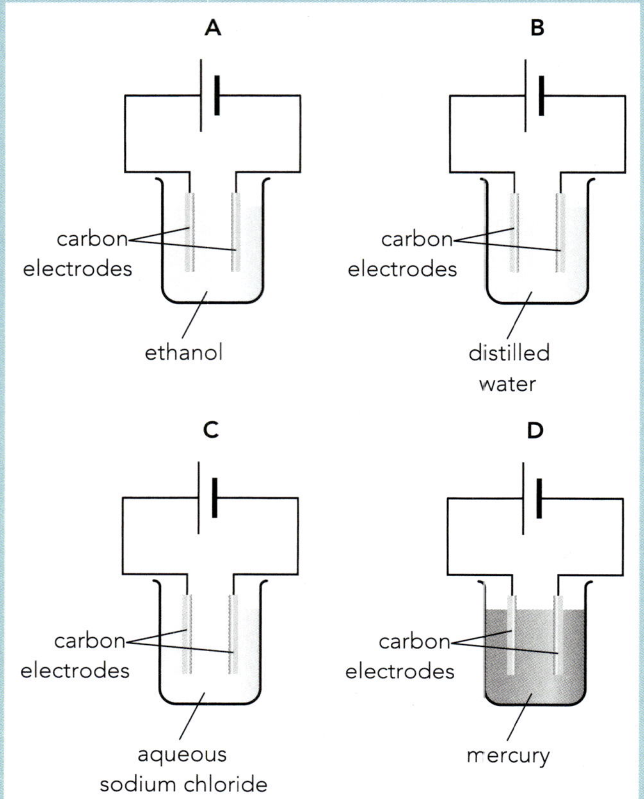

[1]

CONTINUED

2 Electroplating is a common industrial process. In which set of apparatus A–D in the figure would the metal key be electroplated with copper?

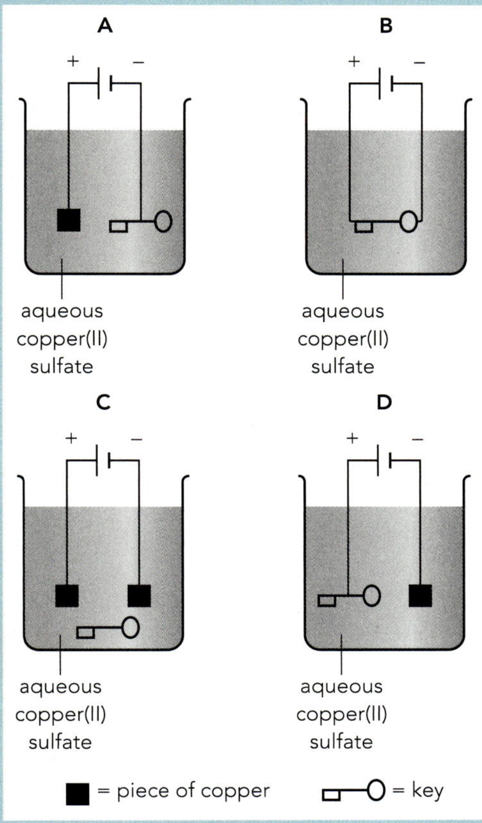

[1]

3 Eight substances A–G that all conduct electricity are listed here.

A aluminium
B chromium
C copper
D dilute sulfuric acid
E platinum
F molten sodium bromide
G sodium chloride solution

Name the substance in A–G that:

a is used to plate steel cutlery [1]
b is used for inert electrodes in some electrolytic cells [1]
c produces a metal at the cathode when electrolysed [1]
d produces oxygen at the anode when electrolysed [1]
e is used in household electrical wiring [1]

[Total: 5]

CONTINUED

4 The figure shows an electrolytic cell.

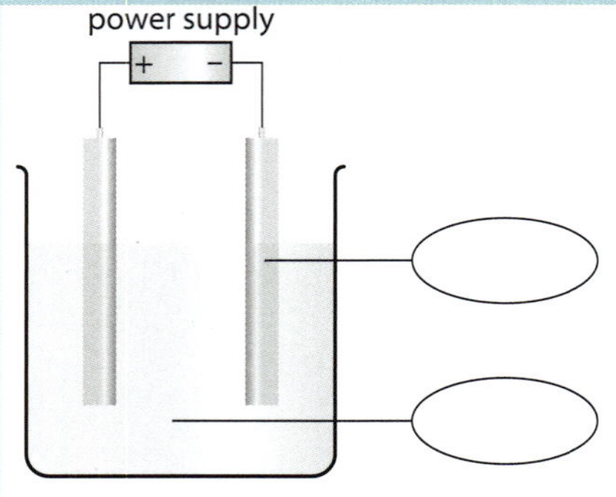

a Complete the labels for the following:
   i the electrode indicated [1]
   ii the contents of the beaker. [1]

b Draw arrows on the connecting wires to show the direction of the movement of electrons in the circuit. [1]

c Which of these solutions would give a colourless gas at both electrodes during electrolysis?
   dilute sulfuric acid
   copper sulfate solution
   sodium chloride solution [1]

d The electrodes shown are inert. **Suggest** a suitable substance that they could be made from. [1]

[Total: 5]

5 The figure shows a cell for the electrolysis of copper(II) sulfate solution using inert electrodes.

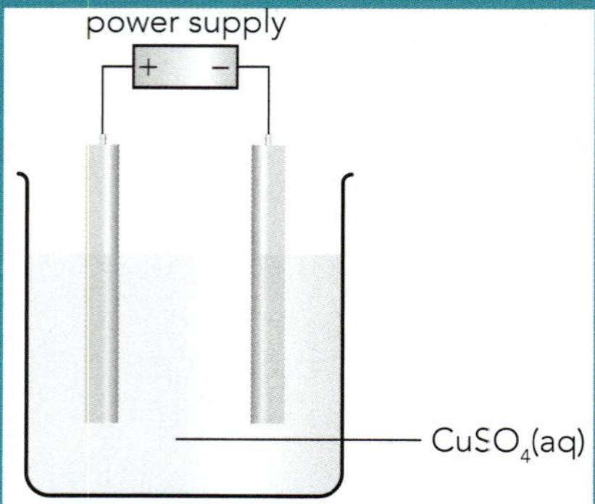

a What change in mass would take place at:
   i the cathode?
   ii the anode? [2]

b What colour change would take place in the solution over time? [1]

COMMAND WORD

suggest: apply knowledge and understanding to situations where there is a range of valid responses in order to make proposals / put forward considerations

CONTINUED

c Write an ionic half-equation for the reaction at the positive electrode. [2]

The electrolysis is repeated using copper electrodes.

d What difference would be observed in:

i the changes in the masses of the electrodes? [4]

ii the change in colour of the solution? [2]

Explain your answer in each case.

e **Describe** how the charge is transferred between the electrodes. [2]

[Total: 13]

COMMAND WORDS

explain: set out purposes or reasons/ make the relationships between things evident/ provide why and/or how and support with relevant evidence

describe: state the points of a topic / give characteristics and main features

SELF-EVALUATION CHECKLIST

After studying this chapter, think about how confident you are with the different topics. This will help you see any gaps in your knowledge and help you to learn more effectively.

I can	See Topic...	Needs more work	Almost there	Confident to move on
describe metals as electrical conductors and non-metallic materials as non-conducting insulators	6.1			
define electrolysis and identify the components of a simple electrolytic cell	6.1			
describe the electrolysis products of molten lead(II) bromide using graphite electrodes and predict the electrolysis products of other molten binary compounds	6.1			
describe the electrolysis of concentrated sodium chloride solution and dilute sulfuric acid using inert electrodes	6.2			
predict the products of electrolysis of dilute and concentrated halide solutions	6.2			
identify the products of the electrolysis of copper(II) sulfate solution using carbon or copper electrodes	6.2			
describe how charge is transferred around the circuit involved in electrolysis and construct ionic half-equations for the ionic reactions taking place at the anode and cathode	6.2			
describe how electrolysis can be used to electroplate a metal object	6.2			
consider how a hydrogen–oxygen fuel cell generates electricity	6.3			
describe the advantages and disadvantages of hydrogen–oxygen fuel cells as a means of powering road vehicles	6.3			

Chapter 7
Chemical energetics

IN THIS CHAPTER YOU WILL:

- identify and understand the differences between physical and chemical changes
- see how some chemical reactions and physical changes are exothermic while others are endothermic
- define exothermic and endothermic reactions in terms of thermal energy transfer
- interpret reaction pathway diagrams for exothermic and endothermic reactions

- understand that for an exothermic reaction the enthalpy change (ΔH) is negative and for an endothermic reaction ΔH is positive
- state that bond breaking is endothermic and bond making is exothermic, and use these ideas to calculate ΔH for a reaction
- define activation energy (E_a)
- learn how to draw reaction pathway diagrams, involving ΔH and E_a, for exothermic and endothermic reactions.

GETTING STARTED

Think back to the work that you have done on the different states of matter. How easy is it for matter to change from one state to another? Changing temperature has an obvious effect, but skiers and skaters will also know that pressure may turn solid ice into a film of liquid (Figure 7.1). Think about which of these changes of state require energy, and the reason for that energy being needed. Will the reverse changes give out energy?

Can you recall any other physical or chemical changes that give out heat? Briefly make a list of the most obvious ones and share ideas with the whole class.

Figure 7.1: Skiing in powder snow.

KEEPING COOL!

It is difficult to overestimate the importance of the invention of the modern refrigerator. The refrigerated transport and storage of food has led to important benefits in the home. Not only that, reliable refrigeration led to the globalisation of markets and the availability of important commodities across, and between, continents.

A refrigerator (Figure 7.2) takes advantage of the heat energy changes linked to evaporation and condensation. The liquid coolant in the system is volatile and has a low boiling point. When it evaporates it absorbs heat from the surroundings (the inside of the fridge). The inside of the fridge and its contents are cooled down.

The coolant is pumped round the system. The vapour is compressed by the pump and condenses back to a liquid again. This gives out heat to the air behind the fridge. In this way, heat is transferred from the inside of the fridge out into the room and the contents are kept cold.

A similar evaporation–condensation cycle is used in modern air conditioners. The technology is now being applied in reverse to heat exchangers for domestic central heating and hot water systems. The use of such technology is aimed at reducing the energy demands and costs of modern homes to make them more environmentally sustainable.

Discussion questions

1 Chlorofluorocarbons (CFCs) were used as the coolant in refrigerators and air conditioners, but their use has gradually been phased out. What environmental problems did they cause?

2 What were the benefits that refrigeration using coolants brought?

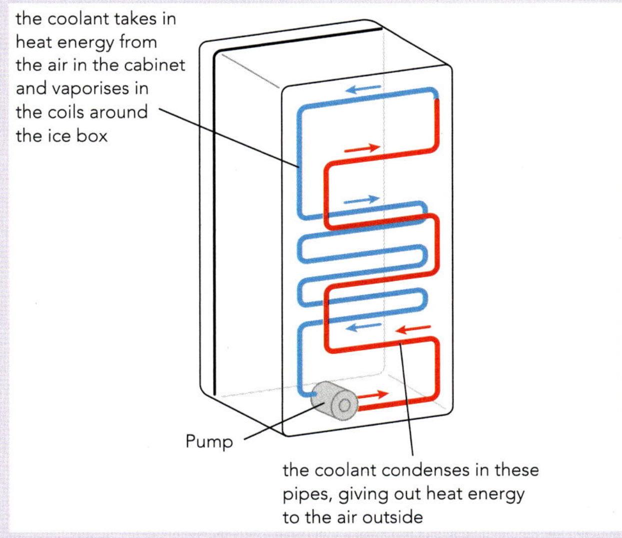

Figure 7.2: The evaporation–condensation cycle that works a refrigerator.

7.1 Physical and chemical changes

The Chinese characters for 'chemistry' literally mean 'change study' (Figure 7.3). Chemistry deals with how substances react with each other. Chemical reactions range from the very simple through to the interconnecting reactions that keep our bodies alive.

But what is a chemical reaction? How does it differ from a simple physical change?

Figure 7.3: Chinese characters for Chemistry.

Physical changes

Ice, snow and water may look different, but they are all made of water molecules (H_2O). They are different physical forms of the same substance – water – existing under different conditions of temperature and pressure. One form can change into another if those conditions change (Chapter 1). In such physical changes, no new chemical substances are formed. Dissolving sugar in ethanol or water is another example of a physical change. It produces a solution, but the substances can easily be separated again by distillation (Chapter 22).

This is what we know about **physical changes**:

- In a physical change, the substances present remain chemically the same: no new substances are formed.
- Physical changes are often easy to reverse. Any mixtures produced are usually easy to separate.

Physical changes can involve heat energy. We saw in Chapter 1 that changing physical state involves heat being taken in or given out. Melting a solid takes in heat; it is an endothermic change. The heat taken in overcomes the forces holding the structure together. Evaporation is also endothermic. The reverse changes of condensation and freezing are exothermic changes, as shown by the stages in a cooling curve.

Chemical changes

Chemical changes are the result of chemical reactions. When magnesium burns in oxygen (Figure 7.4), the white ash produced is a new substance – the compound, magnesium oxide. Burning magnesium produces a brilliant white flame. Energy is given out in the form of heat and light. The reaction is an exothermic change. The combination of the two elements, magnesium and oxygen, to form the new compound is difficult to reverse.

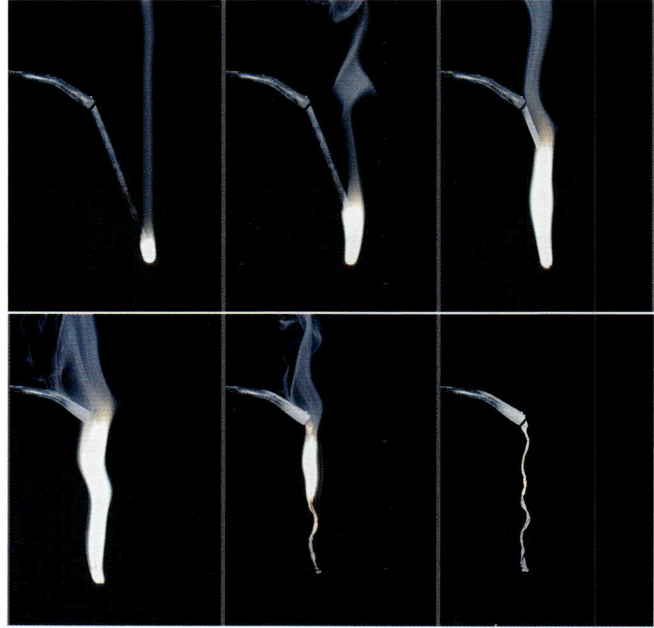

Figure 7.4: Magnesium burns strongly in air or oxygen producing brilliant white light. As a result of this reaction, white magnesium oxide is formed.

KEY WORDS

physical change: a change in the physical state of a substance or the physical nature of a situation that does not involve a change in the chemical substance(s) present

chemical reaction (change): a change in which a new substance is formed

Some other chemical reactions, such as those in fluorescent 'glow bracelets' (Figure 7.5), produce chemiluminescence. They give out energy in the form of light.

The reaction between nitrogen and oxygen to make nitrogen monoxide is an example of another type of reaction. During this reaction, heat energy is taken in from the surroundings. The reaction between these gases in the atmosphere takes place during lightning strikes. The reaction is an endothermic change. Endothermic reactions are much less common than exothermic reactions.

Figure 7.5: Glow-in-the-dark bracelets produce chemiluminescence.

This is what we know about chemical changes:

- the major feature of a chemical change, or reaction, is that new substance(s) are made during the reaction
- many reactions, but not all of them, are difficult to reverse
- during a chemical reaction, energy can be given out (exothermic change) or taken in (endothermic changes)
- there are many more exothermic reactions than endothermic reactions.

Questions

1 State whether the following changes are physical or chemical:
   a melting of ice
   b burning of magnesium
   c evaporation of ethanol
   d dissolving of sugar in water.

2 State whether the following changes are exothermic or endothermic:
   a condensation of steam to water
   b burning of magnesium
   c addition of concentrated sulfuric acid to water
   d evaporation of a volatile liquid.

3 What is the most important thing that shows us that a chemical reaction has taken place?

7.2 Exothermic and endothermic reactions

Energy comes in many forms such as heat, light, sound, electricity and, most importantly in this context, chemical energy. Chemical energy is the energy released or absorbed during chemical reactions.

Some chemical reactions are capable of releasing vast amounts of energy. Forest fires can rage impressively, producing overpowering and devastating waves of heat (Chapter 17). Yet we use similar reactions, under control, to provide heat for the home and for industry. Natural gas, which is mainly methane, is burnt under controlled conditions to produce heat in homes and industry (Figure 7.6).

Figure 7.6: Methane burning on a domestic cooker.

In some reactions, energy is absorbed. In one of the most important chemical reactions on the planet, plants use the energy in sunlight to convert carbon dioxide and water into carbohydrates by the process of photosynthesis (Chapter 17) (Figure 7.7).

Figure 7.7: Photosynthesis takes place in the green leaves of plants.

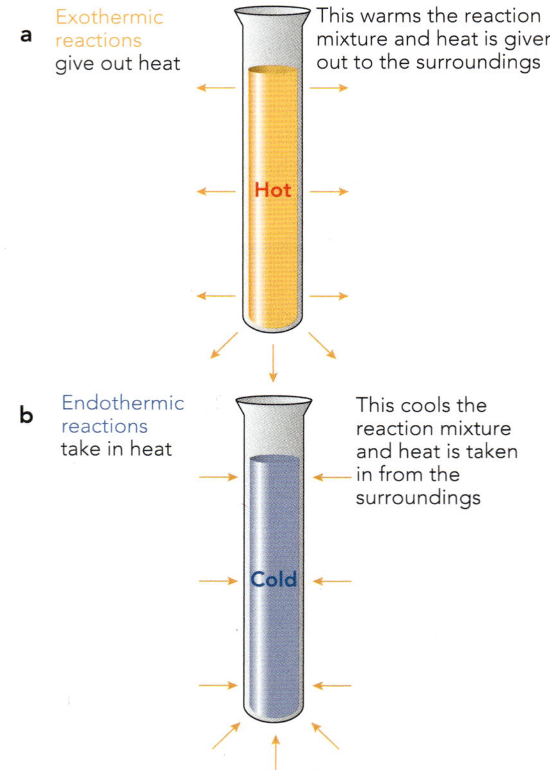

Figure 7.8 a: An exothermic reaction gives out heat energy to the surroundings. **b:** An endothermic reaction takes in thermal energy from the surroundings.

Processes that release heat energy (thermal energy) to the surroundings are called exothermic processes (Figure 7.8a). Dissolving magnesium chloride in water is an exothermic physical process. Adding zinc to copper(II) sulfate solution is an exothermic chemical reaction (see Experimental Skills 15.1). The thermal energy released in these exothermic reactions leads to an increase in the temperature of the surroundings. The surroundings include:

- the reaction mixture in the test-tube
- the air around the test-tube
- the test-tube itself
- the thermometer or anything dipping into the test-tube.

This final point is important: it is the thermal energy (heat) transferred to the thermometer that causes the rise in temperature shown on the thermometer

Processes that take in heat energy from the surroundings are endothermic processes (Figure 7.8b). Endothermic reactions absorb heat. The heat energy is taken in by the reacting substances (the system) and so leads to a decrease in the temperature of the surroundings. When you try to remember these particular terms, concentrate on the first letters of the words involved:

- **EX**othermic means that heat **EX**its the reaction.
- **EN**dothermic means that heat **EN**ters the reaction.

Reaction pathway diagrams

Exothermic and endothermic changes for reactions can be shown using **reaction pathway diagrams** (energy level diagrams). Such diagrams show:

- the energy of the reactants and products on the vertical axis (y-axis)
- the 'reaction pathway' (progress of reaction) on the horizontal axis (x-axis), with reactants on the left and products on the right
- the energy change indicated by an arrow (upward or downward) between the reactants and products.

KEY WORDS

reaction pathway diagram (energy level diagram): a diagram that shows the energy levels of the reactants and products in a chemical reaction and shows whether the reaction is exothermic or endothermic

Methane is the simplest hydrocarbon molecule and the main component of natural gas. When it burns, it reacts with oxygen. The products are carbon dioxide and water vapour:

methane + oxygen → carbon dioxide + water

$$CH_4(g) + 2O_2(g) \rightarrow CO_2(g) + 2H_2O(g)$$

Figure 7.9 shows a reaction pathway diagram for the reaction of methane and oxygen.

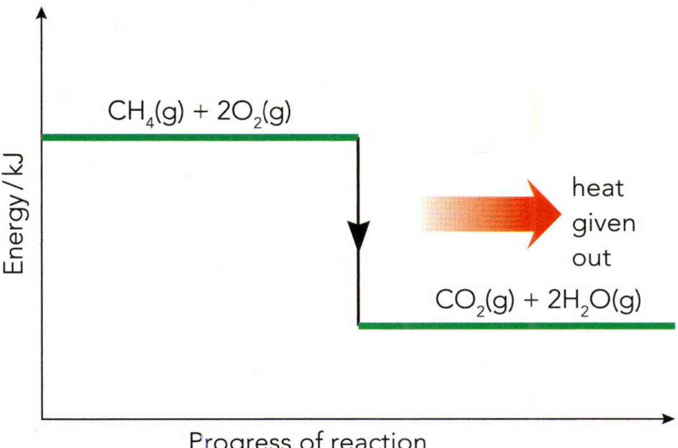

Figure 7.9: Reaction pathway diagram for the exothermic reaction between methane and oxygen.

For an exothermic reaction you can see

- that the energy of the reactants is higher than the energy of the products
- the black arrow points downwards to show that energy is given out (released).

The reaction between nitrogen and oxygen is endothermic. It is one of the reactions that take place in the hot engine when fuel is burnt in car engines. The equation for this reaction is:

nitrogen + oxygen → nitrogen monoxide

$$N_2(g) + O_2(g) \rightarrow 2NO(g)$$

Figure 7.10 shows a reaction pathway diagram for this reaction.

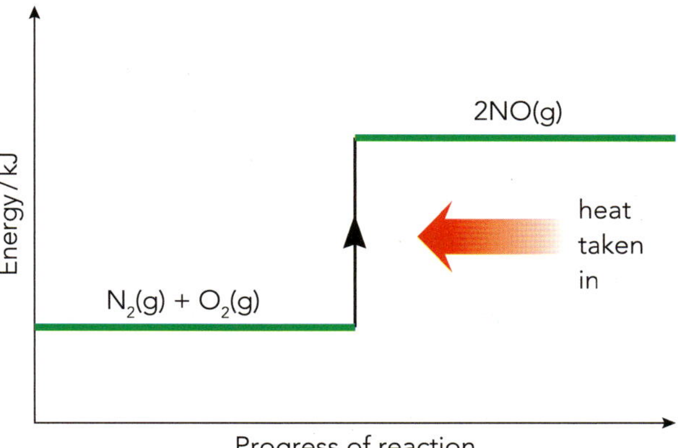

Figure 7.10: A reaction pathway diagram for the reaction between nitrogen and oxygen.

For an endothermic reaction you can see

- that the energy of the reactants is lower than the energy of the products
- the black arrow points upwards to show that energy is absorbed (taken in).

Questions

4 Which type of reaction takes in heat from its surroundings?

5 Draw a reaction pathway diagram for the following reaction, which is exothermic.

$Zn(s) + CuSO_4(aq) \rightarrow ZnSO_4(aq) + Cu(s)$

6 What key features of a reaction pathway diagram would indicate an endothermic reaction?

Enthalpy and enthalpy change (ΔH)

The thermal energy (heat) content of a system is called the **enthalpy (H)** of the system. The transfer of thermal energy during a reaction is called the **enthalpy change (ΔH)** (the symbol Δ (delta) means 'change in') of the reaction.

- when heat is released from the system to the surroundings in an exothermic reaction, the enthalpy of the system decreases and ΔH is negative
- when heat is taken in (absorbed) by the system from the surroundings in an endothermic reaction, the enthalpy of the system increases and ΔH is positive.

These ideas fit with the direction of the black arrows shown in the energy diagrams (Figures 7.9 and 7.10). The energy given out or taken in is measured in kilojoules (kJ); 1 kilojoule (1 kJ) = 1000 joules (1000 J). It is usually calculated per mole of a specific reactant or product (kJ/mol). The enthalpy change (ΔH) for the burning of methane is high (Figure 7.11). This high value explains why methane is such a useful fuel.

$CH_4(g) + 2O_2(g) \rightarrow CO_2(g) + 2H_2O(g)$
$\Delta H = -728$ kJ/mol

KEY WORDS

enthalpy (H): the thermal (heat) content of a system

enthalpy change (ΔH): the heat change during the course of a reaction (also known as heat of reaction); can be either exothermic (a negative value) or endothermic (a positive value)

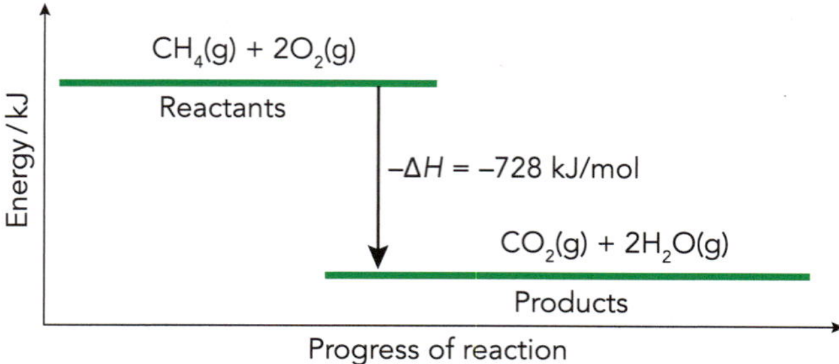

Figure 7.11: Detailed reaction pathway diagram for the exothermic reaction between methane and oxygen, showing the enthalpy change (ΔH).

Bond breaking and bond making

Most substances, apart from the noble gases (Group VIII), are involved in chemical bonding of some type (Chapter 3). During the reaction between methane and oxygen, as with all other reactions, bonds are first broken and then new bonds are made (Figure 7.12). In methane molecules, carbon atoms are covalently bonded to hydrogen atoms. In oxygen gas, the atoms are held together in diatomic molecules. During the reaction, all these bonds must be broken. Chemical bonds are forces of attraction between atoms or ions. Breaking these bonds requires energy; energy must be taken in to pull the atoms apart. Breaking chemical bonds takes in energy from the surroundings. Bond breaking is an endothermic process.

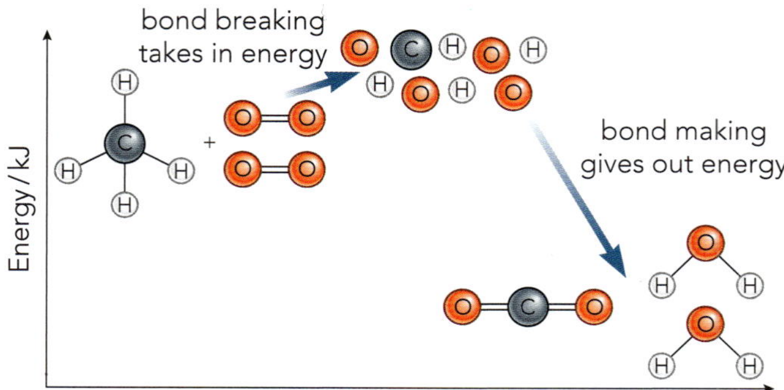

Figure 7.12: Burning of methane first involves the breaking of bonds in the reactants. This is followed by the formation of the new bonds of the products.

New bonds are then formed: between carbon and oxygen to make carbon dioxide (CO_2), and between hydrogen and oxygen to form water (H_2O). Making these chemical bonds gives out energy to the surroundings. Bond making is an exothermic process. Some bonds are stronger than others. They require more energy to break them, but they give out more energy when they are formed. Here the bonds in the products are stronger than those in the reactants. So, overall, this reaction gives out thermal energy (heat) – it is an exothermic reaction.

Experiments have been carried out to find out how much energy is needed to break various covalent bonds in compounds. The average value obtained for a particular bond is known as the **bond energy** (Table 7.1). It is a measure of the strength of the bond.

Bond	Bond energy / kJ/mol	Comment
H–H	436	in hydrogen
C–H	435	average of four bonds in methane
O–H	464	in water
C–C	347	average of many compounds
O=O	498	in oxygen
C=O	803	in carbon dioxide
N≡N	945	in nitrogen

Table 7.1: Bond energies for some covalent bonds.

The relationship between making and breaking bonds and the energy involved is summarised in the memory aid 'MEXOBENDO': **M**aking bonds = **EXO**thermic and **B**reaking bonds = **ENDO**thermic (Figure 7.13).

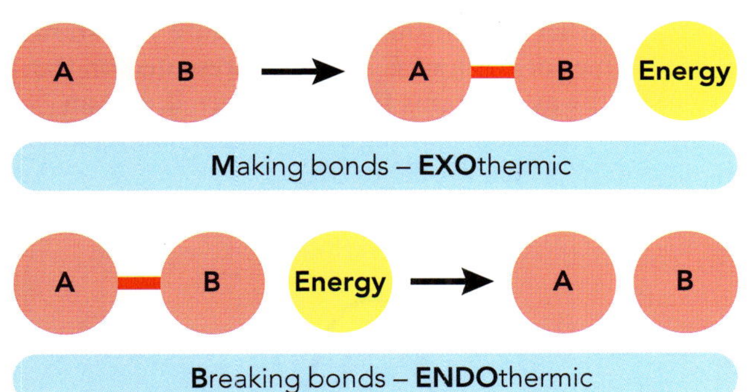

Figure 7.13: Memory aid for calculating the enthalpy change of reaction: MEXOBENDO.

We can use bond energy values to find the enthalpy change, ΔH, for the burning of methane. The equation is:

$$CH_4(g) + 2O_2(g) \rightarrow CO_2(g) + 2H_2O(g)$$

The left-hand side involves bond breaking and needs energy:

four C–H bonds $4 \times 435 = 1740$ kJ/mol

two O=O bonds $2 \times 498 = 996$ kJ/mol

total energy needed $= 2736$ kJ/mol

The right-hand side involves bond making and gives out energy:

two C=O bonds $2 \times 803 = 1606$ kJ/mol

four O–H bonds $4 \times 464 = 1856$ kJ/mol

total energy given out $= 3462$ kJ/mol

The enthalpy of reaction, ΔH, is the energy change on going from reactants to products. So, for the burning of methane:

enthalpy of reaction = energy difference

ΔH = (total energy needed) – (total energy given out)

$\Delta H = 2736 - 3462$

$\mathbf{\Delta H = -726}$ **kJ/mol**

It is useful to remember that combustion reactions are always exothermic; so, you have an indication that you have calculated the value correctly.

The important equation to remember when completing these calculations of the enthalpy of reaction (ΔH) is:

ΔH = (energy needed to break bonds) – (energy given out when bonds form)

KEY WORDS

bond energy: the energy required to break a particular type of covalent bond

REFLECTION

How helpful do you find the memory aids such as 'EXothermic means that heat EXits the reaction; ENdothermic means that heat ENters the reaction' and 'MEXOBENDO'? Do you find them helpful? Do you have a specific way to remember how to draw a reaction pathway diagram?

What strategies could you use to help you learn the ideas about chemical energetics covered in this chapter?

Activation energy

Although the vast majority of reactions are exothermic, only a few are totally spontaneous and begin without help at normal temperatures (e.g. sodium or potassium reacting with water). More usually, additional energy is required to start the reaction. When fuels are burnt, for example, energy is needed to ignite them (Figure 7.14). This energy may come from a spark, a match or sunlight. It is called the **activation energy (E_a)**. It is required because initially some bonds must be broken before any reaction can take place. Sufficient atoms or fragments of molecules must be freed for the new bonds to begin forming. Once started, the energy released as new bonds are formed causes the reaction to continue.

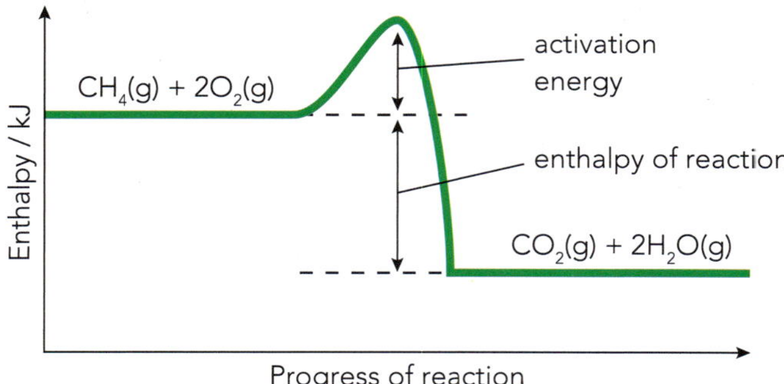

Figure 7.14: A reaction pathway diagram for the burning of methane, showing the need for activation energy to start the reaction.

KEY WORDS

activation energy (E_a): the minimum energy required to start a chemical reaction – for a reaction to take place the colliding particles must possess at least this amount of energy

For a chemical reaction to happen, some bonds in the reactants must first break before any new bonds can be formed. That is why all reactions require some activation energy. For the reaction of sodium or potassium with water, the activation energy is low, and there is enough energy available from the surroundings at room temperature for the reaction to begin spontaneously. Other exothermic reactions have a higher activation energy (e.g. the burning of magnesium can be started with heat from a Bunsen burner). Reactions can be thought of as the result of collisions between atoms, molecules or ions. In many of these collisions, the colliding particles do not have enough energy to react and they just bounce apart. A chemical reaction will only happen if the total energy of the colliding particles is greater than the required activation energy of the reaction (Chapter 8).

A consideration of activation energy (E_a) and the enthalpy of reaction (ΔH) means that we can develop the idea of a reaction pathway diagram further than we saw in the earlier part of this chapter. Figure 7.15 shows how the values of ΔH and E_a can be represented on the reaction pathway diagrams for an exothermic and an endothermic reaction. The data for a particular reaction will determine which profile to use.

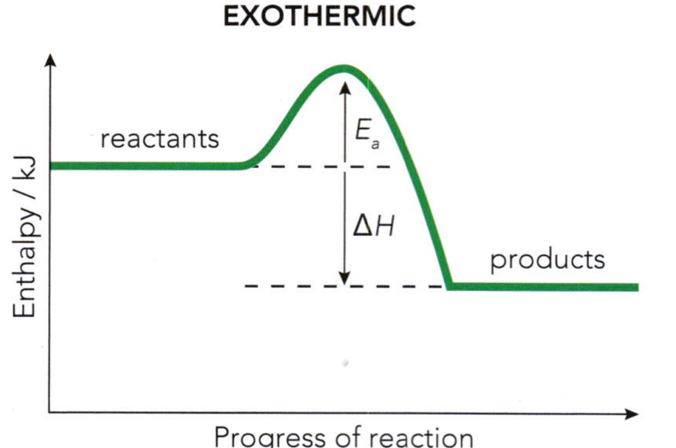

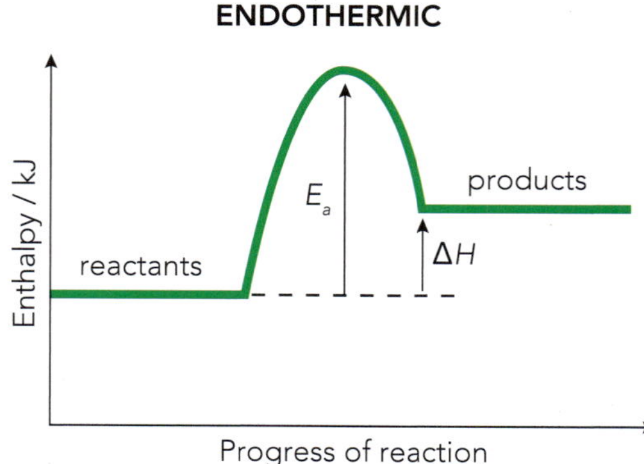

Figure 7.15: Reaction pathway diagrams for exothermic and endothermic reactions showing the activation energies (E_a) and enthalpies of reaction (ΔH) in each case.

Questions

7 Is bond breaking an endothermic or an exothermic process?

8 Hydrogen peroxide decomposes to produce water and oxygen. The equation is:

$2H_2O_2(g) \rightarrow 2H_2O(g) + O_2(g)$

Using the following values, calculate the heat change for the reaction and say whether it is exothermic or endothermic. Bond energies:

O–O = 144 kJ/mol　　O–H = 464 kJ/mol

O=O = 498 kJ/mol

9 Explain why every chemical reaction must have an activation energy, even though for some reactions it may have a low value.

ACTIVITY 7.1

The chemical roller coaster

Have you noticed that reaction pathway diagrams (reaction profiles) look a little bit like a roller-coaster ride? They go steeply up (taking in energy from the surroundings) and then there is peak before they come racing down (releasing energy back into the surroundings as they do).

Using the exothermic reaction for the reaction between hydrogen and oxygen create a cartoon strip that explains the main stages in the reaction pathway diagram (Figure 7.16). Think of the hydrogen and oxygen molecules being split into atoms before re-combining to form water molecules.

Your cartoon strip should include:

- the key scientific words: activation energy, enthalpy change, exothermic, bond breaking, bond making
- at least five diagrams – start with just the reactants, then show the process of bond breaking, the activation energy (peak), the process of bond making and end with the products

You might want to see the reaction from the view of one of the atoms involved, giving them the stage to explain what is happening.

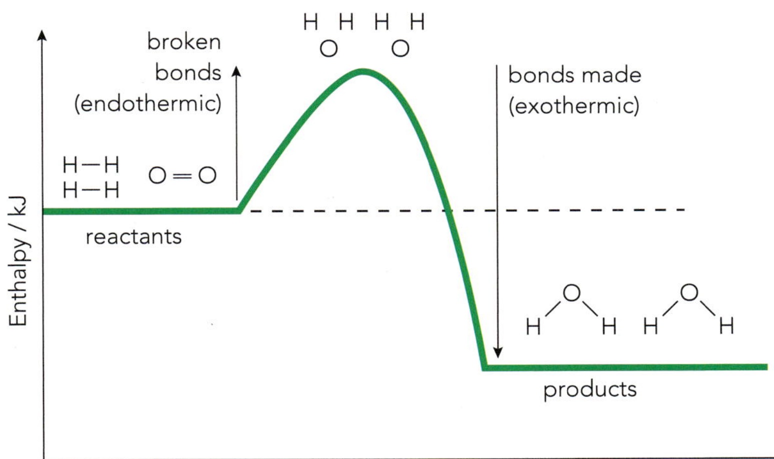

Figure 7.16: Reaction pathway diagram for the reaction between hydrogen and oxygen.

CONTINUED

Peer assessment

With a partner, or in a small group, discuss your cartoon strips. Do they clearly show the progress of the reaction and the different processes involved? Assess the ideas in each other's cartoon strips to determine which individual diagrams best illustrate the different stages.

SUMMARY

Physical changes are changes in the physical properties, state or situation of a substance that do not involve a change in the chemical substance(s) present; chemical changes are changes in which a new substance is formed.
Some chemical and physical changes are exothermic, while others are endothermic.
An exothermic reaction transfers thermal energy to the surroundings leading to an increase in the temperature of the surroundings.
An endothermic reaction takes in thermal energy from the surroundings leading to a decrease in the temperature of the surroundings.
Reaction pathway diagrams can be used to show the energy changes for both exothermic and endothermic reactions.
The transfer of thermal energy during a reaction is called the enthalpy change (ΔH) for the reaction.
ΔH for an exothermic reaction is negative (thermal energy is given out) and ΔH for an endothermic reaction is positive (thermal energy is taken in).
Bond breaking is endothermic and bond making is exothermic.
To calculate the ΔH for a reaction using bond energies, you use the equation: ΔH = (energy needed to break bonds) – (energy given out when bonds form).
Activation energy (E_a) is defined as the minimum energy required for a reaction to take place.
Reaction pathway diagrams showing ΔH and E_a can be drawn for both exothermic and endothermic reactions.

PROJECT

Mastermind

Writing and answering questions is useful to help embed deeper levels of understanding of a given topic. You are therefore going to write a quiz for other members of your class. Can you write four rounds of different types of question?

Round 1: The warm-up. Write four simple questions worth one point. These should be either true/false, multiple choice or fill in the gap.

Round 2: What is that word? Look back through this chapter and identify four key words that you need to know the definition for. Look carefully at each definition to see whether the answer should be worth one or two points.

Round 3: Picture this. Provide a diagram with missing labels and ask for the labels to be completed, one point should be given for each label.

Round 4: Summing up. Set one or two maths-based questions, think carefully about how to award the points.

All questions need to be based on materials covered in this chapter, but you must not repeat any questions in this coursebook. Plan your questions carefully, ideally there should be an increasing level of difficulty. You will need to write out the answers to accompany your questions on a separate page.

In a group of up to four people, one person needs to be the quiz master and set their test for the other group members. You then need to award points. Once completed, you can swap the quiz master role and allow a different student to test people on the questions they wrote.

If you have time, why not, within your team, look at the sets of questions you have written and produce a master quiz with the questions you think test understanding of this chapter in the best way? You could even set this quiz for your teacher to complete!

EXAM-STYLE QUESTIONS

1 Which of the following statements best describes an endothermic reaction?

A The surroundings of the reaction get warmer.

B The reaction absorbs thermal energy.

C The reaction releases heat energy.

D The temperature of the reaction mixture increases. **[1]**

2 Both physical and chemical changes involve changes of energy.
Complete the information in the table. **[5]**

Process	Temperature change	Chemical or physical
water vapour changing to water	**a**	physical
salt solution to salt and water	increase	**b**
magnesium plus hydrochloric acid	**c**	chemical
burning hydrogen to form water	increase	**d**
iron rusting	increase	**e**

3 The equation shows the reaction between sodium hydrogen carbonate and hydrochloric acid.

$NaHCO_3 + HCl \rightarrow NaCl + CO_2 + H_2O$

a What observation is made as the reaction proceeds? **[1]**

At the beginning of the reaction the temperature of the acid is 21 °C, when the reaction is complete the temperature is 16 °C.

b How do you know when the reaction has finished? **[1]**

c Is the reaction exothermic or endothermic? **Explain** your answer. **[2]**

A different carbonate is reacted with hydrochloric acid. The reaction pathway diagram for this second reaction is shown in the figure.

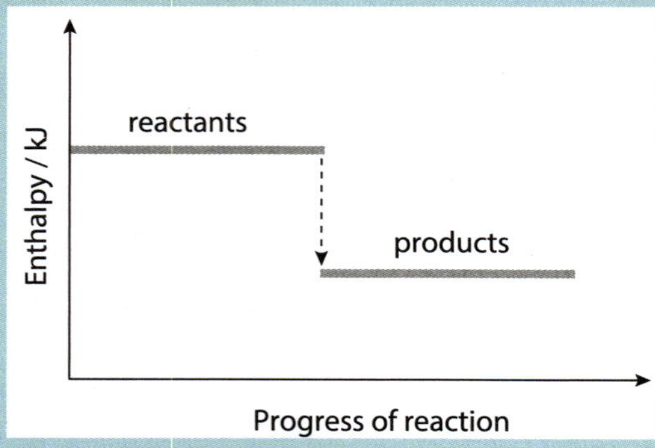

d Is this reaction endothermic or exothermic? Explain your answer using ideas about energy. **[2]**

[Total: 6]

COMMAND WORD

explain: set out purposes or reasons/make the relationships between things evident/provide why and/or how and support with relevant evidence

CONTINUED

4 Nitrogen reacts with hydrogen to form ammonia.

$N_2 + 3H_2 \rightarrow 2NH_3$

The bond energies involved:

H–H 436 kJ/mol

N–N 945 kJ/mol

N–H 391 kJ/mol

a **Calculate** the enthalpy change, ΔH, for the reaction. **[4]**

This reaction can be reversed.

$2NH_3 \rightarrow N_2 + 3H_2$

b What would be the enthalpy change, ΔH, for this reaction? **[2]**

c The figure shows the energy pathway diagrams for the reaction between hydrogen and nitrogen. On the energy profile in part **a**, label 'overall enthalpy change ΔH' and 'activation energy E_a'. **[2]**

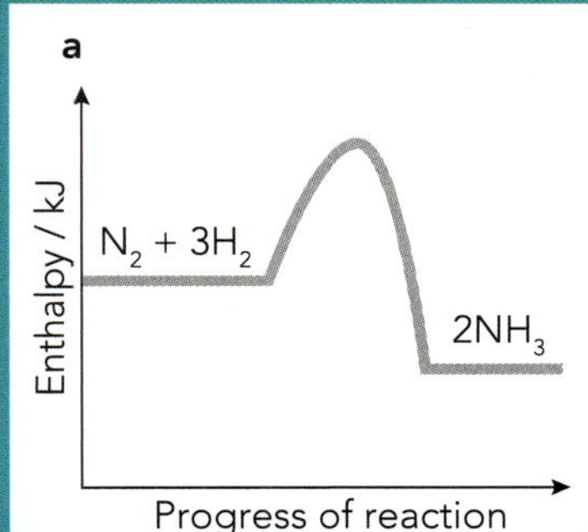

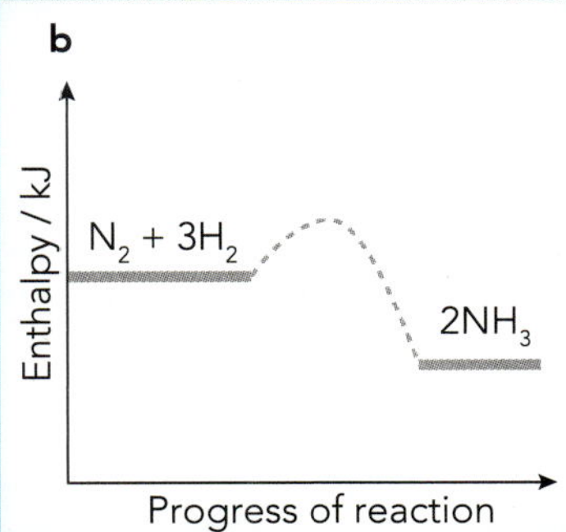

Part **b** of the figure shows the reaction pathway for the same reaction when a catalyst is used.

d What difference does the catalyst make? **[1]**

[Total: 9]

COMMAND WORD

calculate: work out from given facts, figures or information

SELF-EVALUATION CHECKLIST

After studying this chapter, think about how confident you are with the different topics. This will help you see any gaps in your knowledge and help you to learn more effectively.

I can	See Topic...	Needs more work	Almost there	Confident to move on
identify physical and chemical changes and understand the differences between them	7.1			
explain how some chemical reactions and physical changes are exothermic, while others are endothermic	7.2			
state that an exothermic reaction transfers thermal energy to the surroundings leading to an increase in the temperature of the surroundings	7.2			
state that an endothermic reaction takes in thermal energy from the surroundings leading to a decrease in the temperature of the surroundings	7.2			
interpret reaction pathway diagrams for exothermic and endothermic reactions	7.2			
understand that the enthalpy change (ΔH) for an exothermic reaction is negative as heat is given out to the surroundings and that ΔH for an endothermic reaction is positive as heat is taken in from the surroundings	7.2			
state that bond breaking is endothermic and bond making is exothermic, and be able to use data for bond energies to calculate ΔH for a reaction	7.2			
define activation energy (E_a) as the minimum energy required for a reaction to take place	7.2			
draw and label detailed reaction pathway diagrams for reactions involving ΔH and E_a for both exothermic and endothermic reactions	7.2			

Chapter 8

Rates of reaction

IN THIS CHAPTER YOU WILL:

- describe the effects of various factors on the rate of a chemical reaction
- learn how a catalyst increases the rate of a reaction and is unchanged at the end of that reaction
- learn how to investigate the rates of various different reactions

- evaluate different practical methods for investigating the rate of a reaction
- describe collision theory
- explain the effects of various factors on reaction rate using collision theory
- explain how an increase in temperature produces an increase in reaction rate
- discuss the role of a catalyst in increasing reaction rate.

GETTING STARTED

Some chemical reactions are very fast indeed and other reactions are rather slow (Figure 8.1).

Figure 8.1: Rusting is a slow chemical reaction.

Working in groups of two or three, discuss the following questions:

1. What reactions can you think of that are particularly slow?
2. What reactions can you think of that are particularly fast?
3. Think about reactions in the body, for instance, and occurrences such as the browning of sliced fruit open to the air. What is the most obvious thing that speeds up, or slows down, reactions?

Having made notes, join with the whole class to compare ideas.

STUDYING INCREDIBLY FAST REACTIONS

Ahmed Zewail, Linus Pauling Professor of Chemistry at the California Institute of Technology (Caltech), was the first Egyptian to win a Nobel Prize for science. He was awarded the 1999 Nobel Prize in Chemistry for his pioneering laser techniques that allow scientists to observe atoms while chemical reactions are taking place.

Figure 8.2: Professor Ahmed Zewail, in 1986, sometimes known as 'the father of femtochemistry'.

Zewail worked at Caltech for almost 40 years. During this time, he developed a way of observing chemical reactions despite the speed at which occur – over a time scale of a millionth of a billionth of a second (a femtosecond = 10^{-12} seconds) – using ultrashort laser flashes. This field of reaction kinetics came to be known as femtochemistry. It allows scientists to observe and describe the changes taking place in reactions, and to analyse the short-lived transition states that occur as bonds in the reactants are broken and new bonds made. Zewail's research paved the way for greater control over the result of chemical reactions, for new approaches to analysing chemical and biological reactions, as well as for faster electronics and ultra-precise machinery. Professor Zewail died in 2016, aged 70.

Discussion questions

1. Photosynthesis is a very fast series of biochemical reactions and has been described as the most significant reaction on Earth. Why do you think it is described as such? Can you suggest other possible reactions that could be described in this way?
2. Why do you think achieving the best possible rate of reaction is important in industry? What factors should be thought of when considering this?

8.1 Factors affecting the rate of reaction

The possibility of a dust explosion is a major industrial hazard where fine dusts are present. An accidental spark from machinery or another source of ignition can be disastrous. Any solid material that can burn in air will do so with a violence and speed that increases with a high degree of subdivision of the material. Under carefully controlled conditions such explosions are often used to create the spectacular special effects we see in action movies.

Coal dust in the atmosphere of mines has always been a great concern. Figure 8.3 shows a demonstration of the extent of the fireball that can be produced by a coal dust explosion.

Figure 8.3: Demonstration of a coal dust explosion showing the size of the fireball that can be generated.

This type of explosion can also occur with fine powders in flour mills (Figure 8.4a), sugar factories and wood mills. Dust particles have a large surface area relative to their volume in contact with the air. A simple spark can set off an explosive reaction; for example, powdered *Lycopodium* moss piled in a dish does not burn easily – but if the powder is sprayed across a Bunsen flame, it produces a spectacular reaction. Even metal powders can produce quite spectacular effects (Figure 8.4b). A massive explosion at the Kunshan car parts factory in China in 2014 is thought to have been caused by flames igniting dust from metal polishing. The explosion killed 146 workers and injured 114 others.

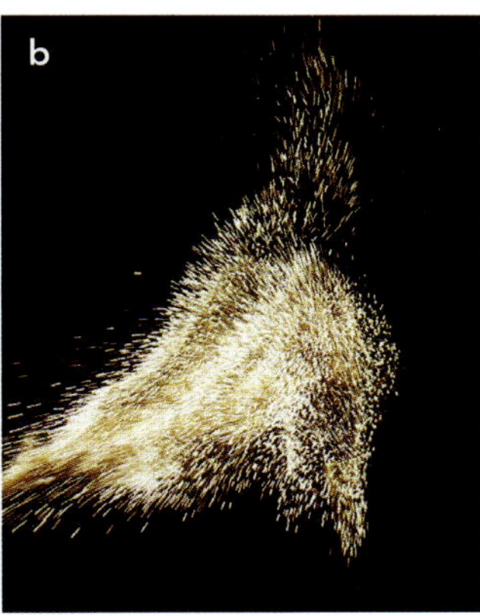

Figure 8.4 a: Fireball produced by dropping powdered flour into a flame. **b:** Iron dust ignited in a Bunsen flame.

Some reactions, such as neutralisation and precipitation reactions (Chapters 11 and 12), are very fast. Other reactions, such as the enzymatic browning of fruits and the fermentation of sugars to form ethanol (Chapter 19), are slow.

What factors influence the speed of a reaction? Experiments have been carried out to study a wide range of reactions, and here we will consider the following major influences on the speed of a reaction (**reaction rate**):

- concentration effects that bring more of the reacting molecules closer together, including:
  - surface area of any solid reactants
  - concentration of the reactant solutions
  - pressure of reacting gases
  - temperature at which the reaction is carried out
- use of a **catalyst**.

KEY WORDS

reaction rate: a measure of how fast a reaction takes place

catalyst: a substance that increases the rate of a chemical reaction but itself remains unchanged at the end of the reaction

Effect of reactant concentration

Surface area of reacting solids

Where one or more of the reactants is a solid, the more finely powdered (or finely divided) the solid(s) are, the greater the rate of reaction. This is because reactions involving solids take place on the surface of the solids. A solid has a much larger surface area when it is powdered than when it is in larger pieces (Figure 8.5). Effectively, we have increased the 'concentration' of the solid. For reactions involving two solids, grinding the reactants means that they can be better mixed. The mixed powders are then in greater contact with each other and are more likely to react.

If a solid is being reacted with a liquid (or a solution), the greater the surface area and the more the solid is exposed to the liquid. A good demonstration of this is the reaction between limestone or marble chips (two forms of calcium carbonate) and dilute hydrochloric acid:

calcium carbonate + hydrochloric acid → calcium chloride + water + carbon dioxide

$$CaCO_3(s) + 2HCl(aq) \rightarrow CaCl_2(aq) + H_2O(l) + CO_2(g)$$

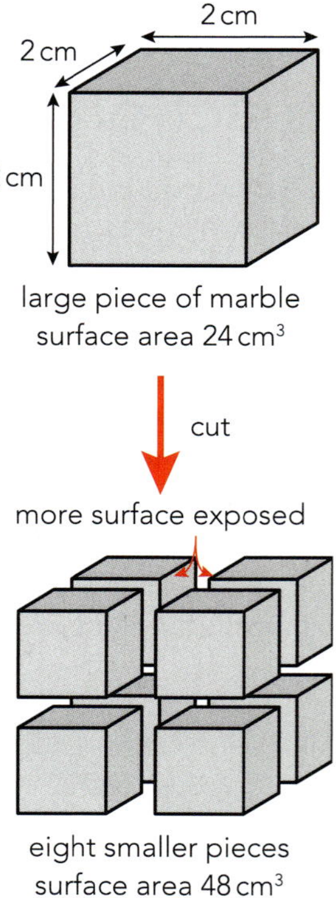

Figure 8.5: Breaking a cube into smaller pieces increases the surface area.

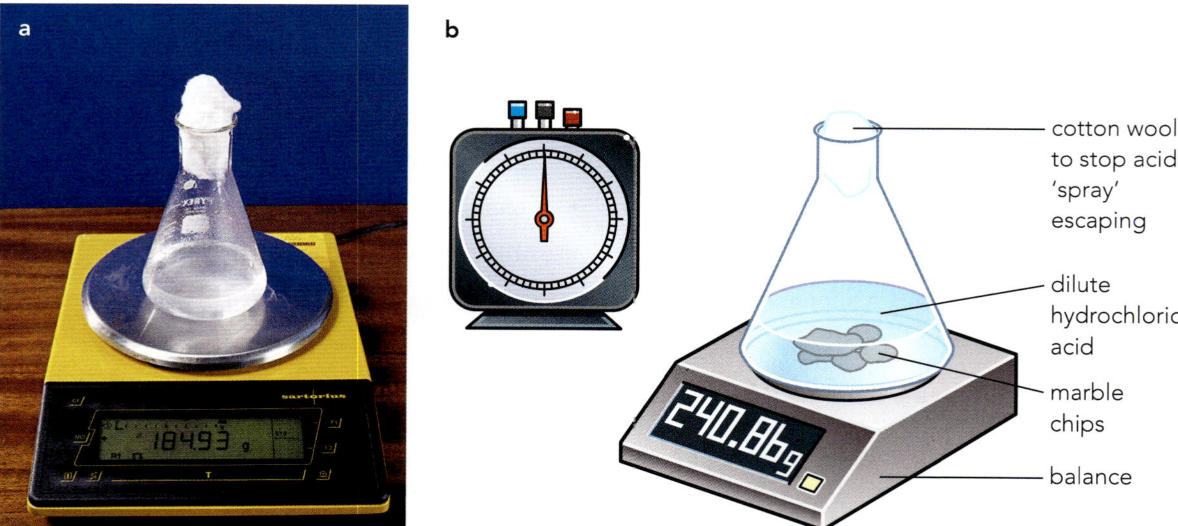

Figure 8.6: Apparatus for experiments **A** and **B**: the reaction of marble chips with dilute hydrochloric acid. The loss of carbon dioxide from the flask produces a loss in mass. This is detected by the balance.

The experiment can be done as shown in Figure 8.6. Using this arrangement, we can compare two samples of marble chips, one sample (**B**) being in smaller pieces than the other sample (**A**). The experiment is carried out twice, once with sample **A** and once with sample **B**. In each experiment the mass of sample used is the same. The same volume and concentration of hydrochloric acid is used in both experiments. The flask is placed on the balance during the reaction. A loose cotton wool plug prevents liquid spraying out of the flask but allows the carbon dioxide gas to escape into the air. This means that the flask will lose mass during the reaction. The balance is set to zero (tared) as the stopclock is started. Mass readings are taken at regular time intervals and the loss in mass can be calculated. When the loss in mass is plotted against time, curves such as those in Figure 8.7 are obtained.

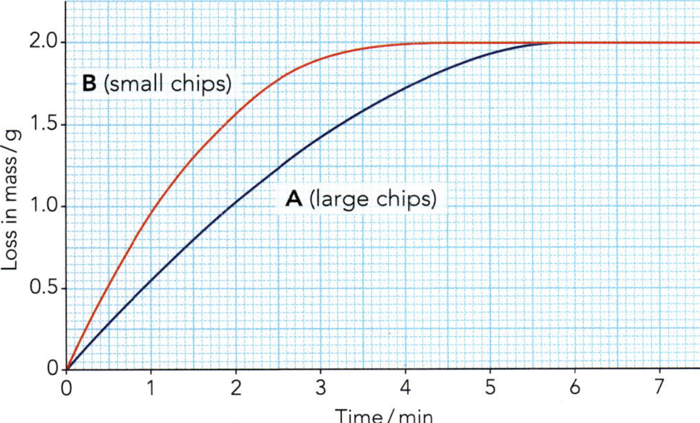

Figure 8.7: The graph shows the loss in mass against time for experiments **A** and **B**. The reaction is faster if the marble chips are broken into smaller pieces (curve **B**).

There are several important points about the graph.

1. The reaction is fastest at the start. This is shown by the steepness of the curves over the first few minutes. Curve **B** is steeper than curve **A**. This means that gas (CO_2) is being produced faster with sample **B**. The sample with smaller chips, with a greater surface area, reacts faster. Beyond this part of the graph, both reactions slow down as the reactants are used up (Figure 8.8).
2. The total volume of gas released is the same in both experiments. The mass of $CaCO_3$ and the amount of acid are the same in both cases. Both curves flatten out at the same final volume. Sample **B** reaches the horizontal part of the curve (the plateau) first.
3. A numerical value for the rate of reaction at any given time can be found by drawing a tangent to the curve at that time. The slope of that tangent gives a value for the rate of reaction at that point. A tangent drawn at the start of the curve will give the initial rate of reaction.

These results show that the rate (speed) of a reaction increases when the surface area of a solid reactant is increased.

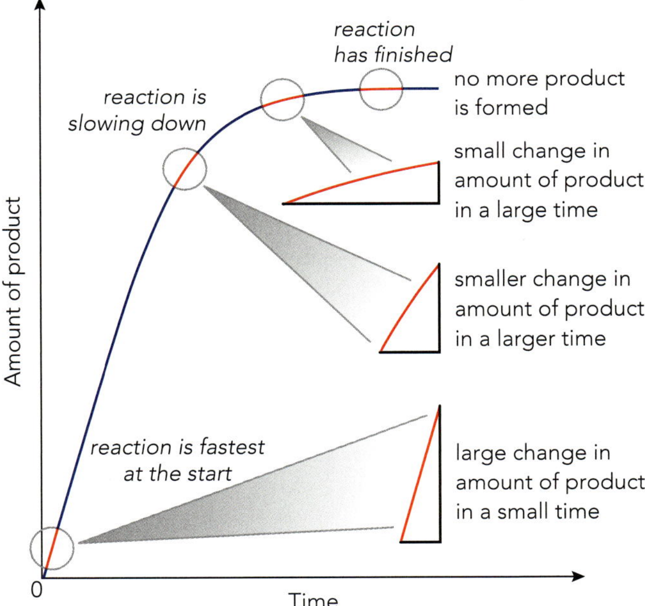

Figure 8.8: A chemical reaction is fastest at the start. It slows down as the reactants are used up.

Concentration of reacting solutions

Reactions that produce gases are also very useful in studying the effect of solution concentration on the reaction rate. The reaction between marble chips and acid can be adapted for this (Figure 8.9).

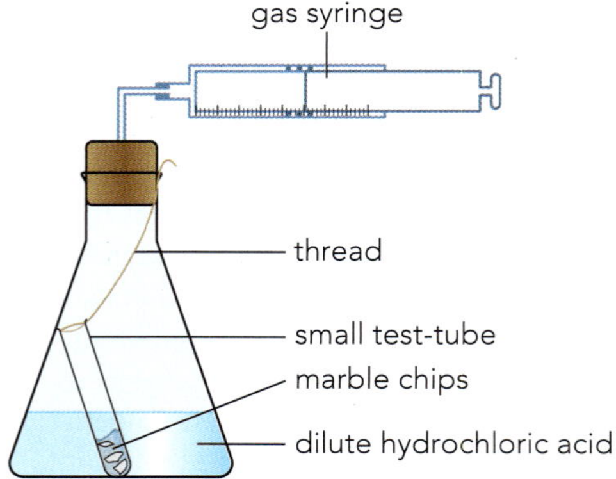

Figure 8.9: Recording the production of a gas using a gas syringe.

As in the previous experiment, we will compare two different situations, which we will call **C** and **D**. Apart from changing the concentration of the acid, everything else must stay the same, making the experiments a 'fair test'. So, the volume of acid, the temperature and the mass of marble chips used must be the same in both experiments. The two reactants are kept separate while the apparatus is set up. To time the start as accurately as possible the stopper is released just long enough to loosen the thread and let the tube drop into the acid. The stopper is then replaced and the flask gently shaken.

The acid in experiment **C** is twice as concentrated (1 mol/dm^3) as in experiment **D** (0.5 mol/dm^3). The carbon dioxide gas produced is collected in a gas syringe. The volume of gas produced is measured at frequent time intervals. We can then plot a graph of volume of gas collected against time, like that in Figure 8.10.

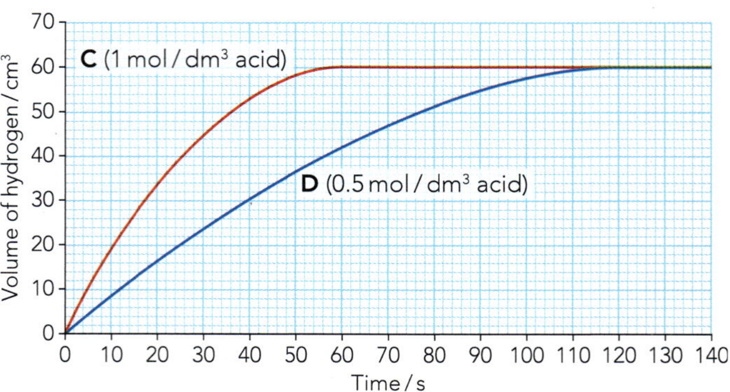

Figure 8.10: Graph showing the volume of carbon dioxide collected in experiments **C** and **D**.

Again, the graph shows some important points.

1. The curve for experiment **C** is steeper than for **D**. This shows clearly that reaction **C**, using more concentrated acid, is faster than reaction **D**.
2. The curve for experiment **C** starts off twice as steeply as for **D**. This means that the reaction in **C** is twice as fast as in experiment **D** initially. So, doubling the concentration of the acid doubles the rate of reaction.
3. The total volume of carbon dioxide produced is the same in both experiments. Both reactions produce the same volume of gas, although experiment **C** produces it faster.

These results show that the rate of a reaction increases when the concentration of a reactant in solution is increased.

Note that here we have stated the concentration of the acid in mol/dm^3. These units are one of the two ways to give the concentration of a solution; the other units are g/dm^3.

This apparatus could be used to study the rates of other reactions that produce a gas. The reaction between magnesium and excess dilute hydrochloric acid to give hydrogen would be one example.

magnesium + hydrochloric acid → magnesium chloride + hydrogen

$$Mg(s) + 2HCl(aq) \rightarrow MgCl_2(aq) + H_2(g)$$

Pressure of reacting gases

Another type of reaction that shows a concentration effect on reaction rate is reactions involving gases. In a reaction involving gases, increasing the pressure has a similar effect to increasing the concentration in a liquid. Increasing the pressure pushes the gas particles closer together so that they collide and react more readily, so that the rate of reaction is increased. Changing the pressure on a reaction that involves only solids or liquids does not affect the reaction rate. One reaction that shows this effect of increasing the pressure is the combustion of fuel in a car engine. Such a reaction explosively generates energy. The reaction is that of the hydrocarbons in petrol (gasoline) burning with the oxygen in the air as they are compressed by the movement of the piston (Figure 8.11).

The gases are ignited by the spark plug. The resulting explosion drives the piston down and that movement is transmitted to the wheels of the car. The exhaust gases are ejected through the exhaust valve.

The effect of pressure on reactions taking place between gases is discussed in more detail in Chapter 9 in connection with two important industrial processes. The **Haber process** is important for the production of ammonia, while the **Contact process** is used for the making of sulfuric acid.

Figure 8.11: Computer graphic of the explosion within the cylinder of a petrol engine as the piston compresses the gases.

KEY WORDS

Haber process: the industrial manufacture of ammonia by the reaction of nitrogen with hydrogen in the presence of an iron catalyst

Contact process: the industrial manufacture of sulfuric acid using the raw materials sulfur and air

REFLECTION

The effects of solution concentration, solid surface area and gas pressure on reaction rate have been grouped together here because they are all 'concentration' related. Does emphasising relationships between ideas help you understand different concepts in chemistry? Could you explain the effect of concentration on the rate of reaction to someone else using these relationships? Why or why not?

Effect of temperature

A reaction can be made to go faster or slower by changing the temperature of the reactants. Some food is stored in a refrigerator because the food will stay fresher for longer. The rate of decay and oxidation by the air is slower at lower temperatures.

The previously described experiments could be altered to study the effect of temperature on the rate of production of gas. An alternative approach is to use the reaction between sodium thiosulfate and hydrochloric acid. In this case, the formation of a precipitate is used to measure the rate of reaction (experiment **E**).

sodium thiosulfate + hydrochloric acid → sodium chloride + sulfur + sulfur dioxide + water

$Na_2S_2O_3(aq) + 2HCl(aq) \rightarrow 2NaCl(aq) + S(s) + SO_2(g) + H_2O(l)$

The experiment is shown in Figure 8.12. A cross is marked on a piece of paper. A flask containing sodium thiosulfate solution is placed on top of the paper. Hydrochloric acid is added quickly. The yellow precipitate of sulfur produced is very fine and stays suspended in the liquid. With time, as more and more sulfur is formed, the liquid becomes cloudier and more difficult to see through. The time taken for the cross to 'disappear' is measured. The faster the reaction, the shorter the length of time during which the cross is visible.

Experiment **E** is carried out several times with solutions pre-warmed to different temperatures. The solutions and conditions of the experiment must remain the same; only the temperature is altered. A graph can then be plotted of the time taken for the cross to disappear against temperature, like that shown in Figure 8.13.

The graph shows two important points.

1. The cross disappears more quickly at higher temperatures. The shorter the time needed for the cross to disappear, the faster the reaction.
2. The curve is not a straight line.

These results show that the rate of a reaction increases when the temperature of the reaction mixture is increased.

To be more precise, the rate of the reaction is faster when the time taken for the reaction to finish is shorter:

rate of reaction = 1/time

A graph of 1/time against temperature would show how the rate increases with a rise in temperature.

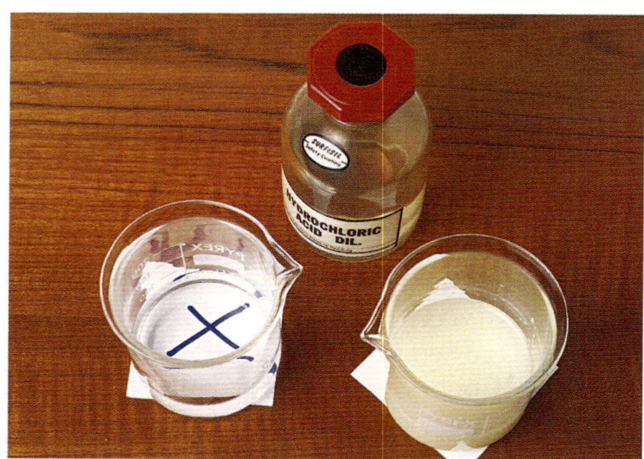

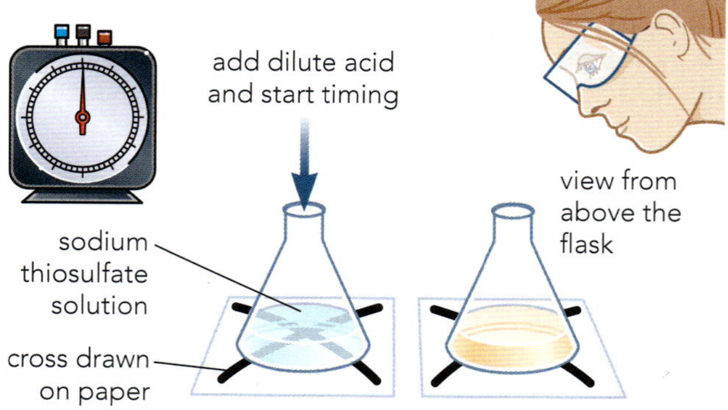

Figure 8.12: Apparatus for the reaction between hydrochloric acid and sodium thiosulfate (experiment **E**).

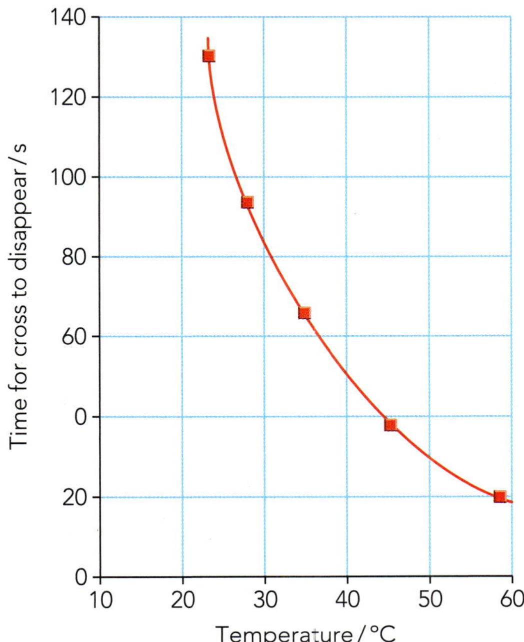

Figure 8.13: Graph for experiment **E** on the reaction between hydrochloric acid and sodium thiosulfate.

Questions

1. What will we observe happening to the rate of a chemical reaction in response to the following?
   - **a** an increase in temperature
   - **b** an increase in the surface area of a solid reactant
   - **c** an increased concentration of a reacting solution.
2. Why is perishable food kept in a refrigerator?
3. When is a chemical reaction at its fastest?

Presence of a catalyst

Hydrogen peroxide is a colourless liquid with the formula H_2O_2. It is a very reactive oxidising agent. Hydrogen peroxide decomposes to form water and oxygen:

hydrogen peroxide → water + oxygen

$$2H_2O_2(l) \rightarrow 2H_2O(l) + O_2(g)$$

We can follow the rate of this reaction by collecting the oxygen in a gas syringe (Figure 8.14a). The formation of oxygen is very slow at room temperature. However, the addition of 0.5 g of powdered manganese(IV) oxide (MnO_2) makes the reaction go much faster (experiment **F**).

The black powder does not disappear during the reaction. Indeed, if the solid is filtered and dried at the end of the reaction, the same mass of powder remains (Figure 8.14b).

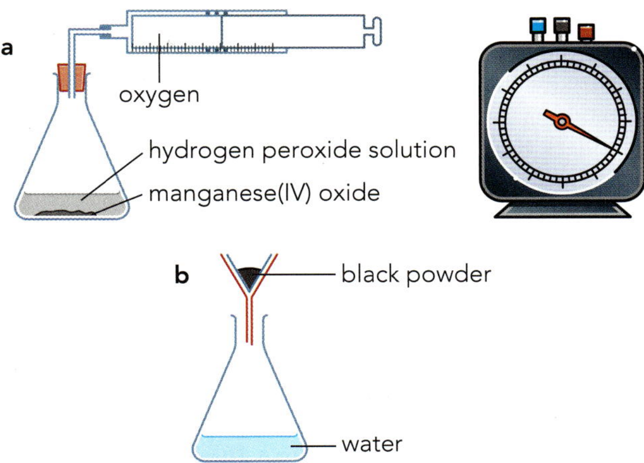

Figure 8.14 a: Apparatus for the experiments on the decomposition of hydrogen peroxide to water and oxygen. **b:** The black powder does not disappear during the reaction.

If the amount of MnO_2 powder added is doubled (experiment **G**), the rate of reaction increases (Figure 8.15). If the powder is more finely divided (powdered), the reaction also speeds up. Both these results suggest that it is the surface of the manganese(IV) oxide powder that is important here. By increasing the surface area, the rate of reaction is increased. We say that manganese(IV) oxide is a catalyst for this reaction.

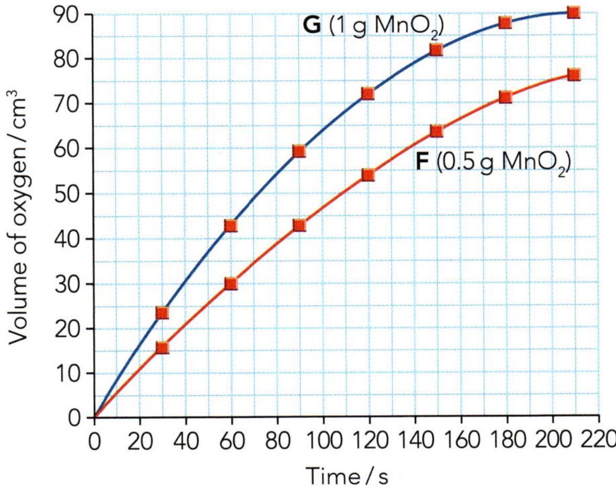

Figure 8.15: Increasing the amount of catalyst increases the rate of reaction.

A catalyst is a substance that increases the rate of a chemical reaction. The catalyst remains chemically unchanged at the end of the reaction. Catalysts have been found for a wide range of reactions. They are useful because a small amount of catalyst can produce a large change in the rate of a reaction. Also, since they are unchanged at the end of a reaction, they can be re-used. Industrially, they are very important. Industrial chemists use catalysts to make everything from polythene and painkillers, to fertilisers and fabrics. If catalysts did not exist, many chemical processes would go very slowly and some reactions would need much higher temperatures and pressures to proceed at a reasonable rate. All these factors would make these processes more expensive, so that they may be uneconomic. Transition elements (see Chapter 13) or their compounds make particularly good catalysts and are used in catalytic converters (Chapter 17).

Living cells also produce catalysts. They are protein molecules called **enzymes** (Figure 8.16). Many thousands of reactions happen in every kind of organism. Enzymes speed up these reactions. Each enzyme works only for a particular reaction, or a reaction type. We say that the enzyme is specific for that reaction.

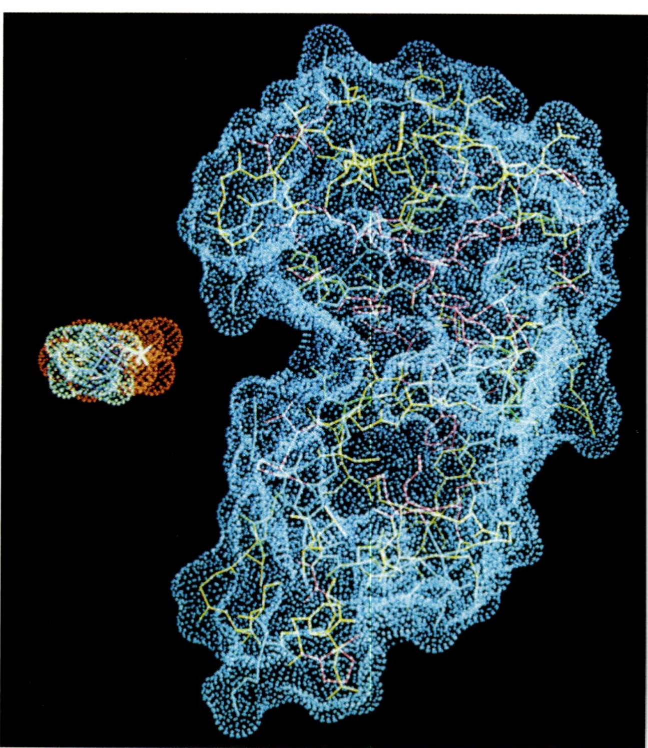

Figure 8.16: A computer image of an enzyme and the smaller molecule it is about to carry out a reaction with (the substrate).

> **KEY WORD**
>
> **enzymes:** protein molecules that act as biological catalysts

The general features of enzymes:

- enzymes are proteins
- they are very specific: each enzyme controls one reaction or reaction type
- they are generally sensitive to temperature: enzymes are inactivated (denatured) by heat (most stop working above 45 °C)
- they are sensitive to pH: most enzymes work best in neutral conditions around pH 7.

Enzymes are being used increasingly as catalysts in industry. Biological washing powders use enzymes to remove biological stains such as sweat, blood and food. The enzymes in these powders are those that break down proteins and fats. Because the enzymes are temperature-sensitive, these powders are used at a wash temperature of around 30–40 °C.

Questions

4 What is a catalyst?

5 What is an enzyme?

6 Which solid catalyst will speed up the decomposition of hydrogen peroxide?

ACTIVITY 8.1

Planning and design of an experiment

Working in your practical group, plan and design an experiment to find which of three possible solid catalysts, **A**, **B** or **C**, works best to catalyse the formation of oxygen from a dilute solution of hydrogen peroxide. You may find that the information in Chapter 21 is useful for this activity. Also, think about other practicals you have performed.

Your plan should include a safety note, the equipment and chemicals required, the method to be followed, and how the results will be analysed (see experiments **F** and **G**).

Hydrogen peroxide decomposes to form water and oxygen.

hydrogen peroxide → water + oxygen

$$2H_2O_2 \rightarrow 2H_2O + O_2$$

This reaction is very slow at room temperature but can be speeded up by using a catalyst. You have access to the three possible catalysts, a dilute solution of hydrogen peroxide, and normal laboratory apparatus. The comparison must be a 'fair test' (only one variable should be changed, the others kept constant) and consider the relevant factors that influence the rate of a reaction.

When you have planned your method, come back together as a class and compare notes, discussing the merits of the different approaches taken.

Peer assessment

Swap your plan with another group. Have they ...

- included a safety note?
- included the equipment and chemicals required?
- included a clear method that is easy to follow?
- said how the results will be analysed?

Write down one thing the group did really well and one thing they could improve on next time they design an experiment.

Evaluating different practical methods

Practical methods of studying the effects of various factors on reaction rate depend on reliably measuring something that changes with time (Chapter 21). The rate can be measured in terms of either:

- how quickly a reactant is used up
- how quickly a product is formed.

The examples we have used in this chapter have all used the second of these approaches. It is possible to use the first approach but the methods are usually more complex, with samples being removed for testing at set time intervals. For experiments not involving the formation of a gas, it may be possible to follow the formation or disappearance of a substance in a reacting mixture by one of the following:

- a change in colour
- a change in the electrical conductivity
- a change in the pH.

Reactions involving the formation of a gas can be followed by either measuring the gas volume produced with time (e.g. using a gas syringe) or by tracking the loss of mass (e.g. using a balance). However, care is needed with the use of this second method. Following the change in mass is not suitable for investigating a reaction where hydrogen is given off. Hydrogen has a very low density and the change in the balance reading will be too small to measure.

When choosing how to measure the volume of a gas formed in a reaction it is important to consider the ease of reading the volume. The use of a gas syringe offers advantages in terms of ease of setting up and of reliably obtaining measurements. Alternatives such as using an inverted measuring cylinder mean collecting the gas over water and are less easy to set up.

When evaluating an experiment, you need to think about:

- the accuracy of the measuring apparatus used, as the accuracy of the whole experiment depends on the least accurate measuring instrument
- the number of times an experiment can be repeated, to make sure results obtained are consistent
- how easy it is to control the variables, as difficulties doing this would invalidate the experiment as a 'fair test'; for example, trying to maintain the temperature of a water-bath with a Bunsen burner is difficult; an electric water-bath should be used.

8.2 Collision theory of reaction rate

One way to reduce the polluting effects of car exhaust fumes is to use a catalytic converter (Figure 8.17). Car exhaust fumes contain pollutant gases such as carbon monoxide, oxides of nitrogen and unburnt hydrocarbons. The catalytic converter converts these gases to less harmful products such as carbon dioxide, nitrogen and water.

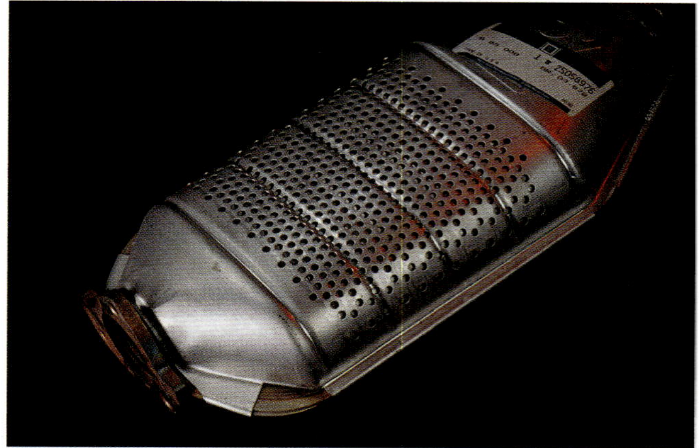

Figure 8.17: A catalytic converter can be fitted to a car exhaust system.

The catalytic converter therefore 'removes' polluting oxides and completes the oxidation of unburnt hydrocarbon fuel. It speeds up these reactions considerably by providing a 'honeycombed' surface on which the gases can react. The converter contains a thin coating of rhodium and platinum catalysts on the solid honeycomb surface. These catalysts have many tiny pores that provide a large surface area for the reactions.

Collision frequency

The importance of surface area in reactions involving solids helps us understand how reactions take place. In these cases, reactions can only occur when particles collide with the surface of a solid. If a solid is broken into smaller pieces, there is more surface exposed. This means there are more places where collisions can take place and so there is more chance of a reaction taking place. Iron reacts more readily with oxygen if it is powdered (Figures 8.4b and 8.18).

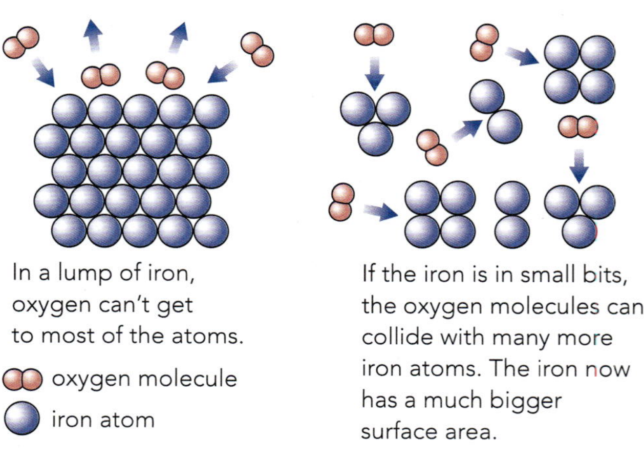

Figure 8.18: The reaction between iron and oxygen is speeded up if the iron is in small pieces.

We can see how these ideas (**collision theory**) apply in other situations. When solutions are more concentrated, the speed of a reaction is faster. A more concentrated solution means that there are more reactant particles in a given volume (a unit volume). Collisions will occur more

KEY WORDS

collision theory: a theory that states that a chemical reaction takes place when particles of the reactants collide with sufficient energy to initiate the reaction

often. The more often they collide, the more chances the particles have to react. This means that the rate of a chemical reaction will increase if the concentration of the reactants is increased. A more concentrated acid reacts more vigorously with a piece of magnesium ribbon than a dilute acid (Figure 8.19). The increased acid concentration means that the frequency of collision between the H^+ ions and the metal creates a greater chance of the reaction occurring.

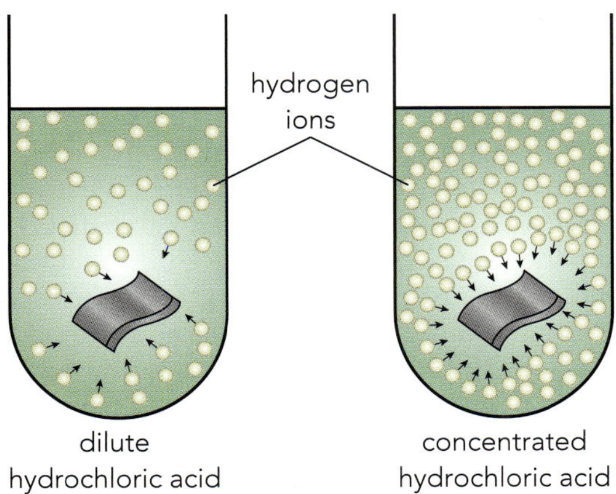

Figure 8.19: The more concentrated the acid solution, the faster the reaction with the magnesium.

For reactions involving gases, increasing the pressure has the same effect as increasing the concentration, so the rate of a reaction between gases increases with pressure (Figure 8.20).

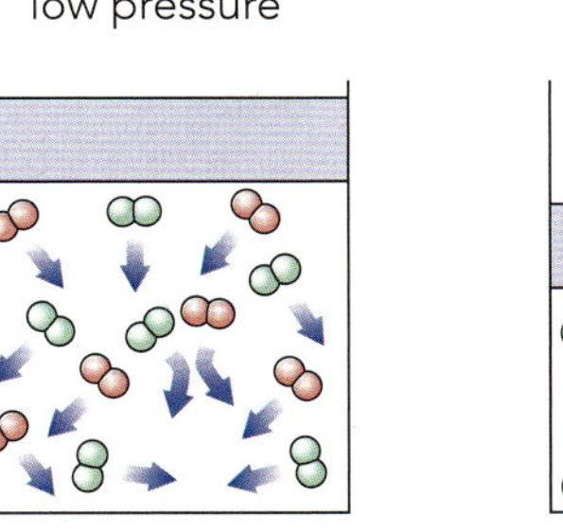

Collisions between different molecules do not happen very often.

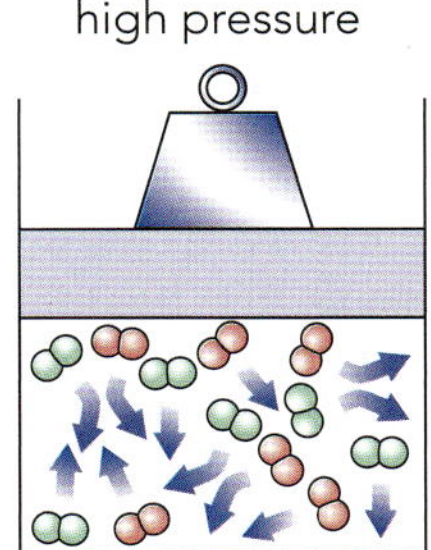

Collisions between different molecules are much more frequent.

Figure 8.20: Increasing the pressure of gases effectively increases the concentration of the gas.

When the temperature is raised, a reaction takes place faster. At higher temperatures, the particles have more kinetic energy and are moving faster. Again, this means that collisions will occur more often, giving more chance of reaction. Also, the particles have more energy at the higher temperature. This increases the chances that a collision will result in bonds in the reactants breaking and new bonds forming to make the products. If we look at the reaction between zinc and hydrochloric acid, we can see how the rate of reaction is affected by changes in collision frequency (Figure 8.21).

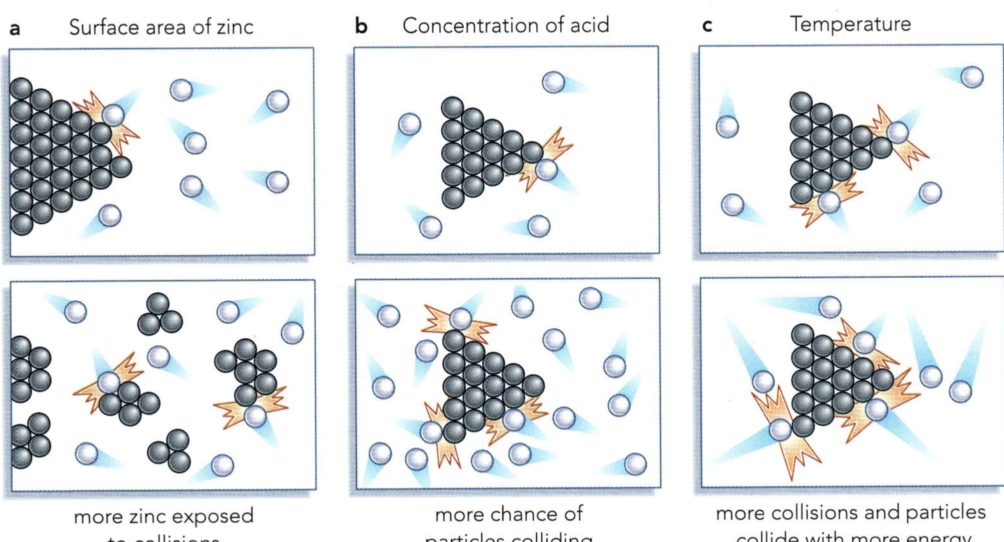

Figure 8.21: Collision theory can be used to explain how various factors affect the rate of reaction. Here we use the reaction between zinc and hydrochloric acid as an example.

When solutions are more concentrated, the speed of a reaction is faster. A more concentrated solution means that there are more reactant particles in a given volume. Collisions will occur more often. The more often they collide, the more chance the particles have of reacting. This means that the rate of a chemical reaction will increase if the concentration of the reactants is increased (Figure 8.21b). When the temperature is raised, a reaction takes place faster. At higher temperatures, the particles are moving faster. Again, this means that collisions will occur more often, giving more chance of reaction. Also, the particles have more energy at the higher temperature. This increases the chances that a collision will result in bonds in the reactants breaking and new bonds forming to make the products (Figure 8.21c).

Activation energy

Not every collision between particles in a reaction mixture produces a reaction. We have seen in Chapter 7 that a certain amount of energy is needed to begin to break bonds in the reactant molecules. This minimum amount of energy is known as the activation energy (E_a) of the reaction.

- Each reaction has its own different value of activation energy.
- When particles collide, they must have a combined energy greater than this activation energy, otherwise they will not react.
- Chemical reactions occur when the reactant particles collide with each other.

There are two factors that explain why an increase in temperature increases the rate of a reaction.

1 The increased temperature means that the molecules are moving faster and therefore they collide more often. The frequency of collision is increased and so the rate of reaction is faster.

2 At the higher temperature, the particles have higher kinetic energy. This means that when they collide more particles will have the minimum amount of energy needed to react. At the higher temperature more of the collision will produce a reaction; there will be more successful collisions. The number of effective collisions increases with increasing temperature.

This second factor is the more important of the two in explaining the level of increased rate produced by raising the temperature of a reaction.

Action of a catalyst

A catalyst increases the rate of reaction by reducing the amount of energy that is needed to break the bonds. This reduces the activation energy of the reaction and makes sure that more collisions are likely to produce products. The rate of the reaction is therefore increased. The catalyst remains chemically unchanged at the end of the reaction. Because it is unchanged, a catalyst can be re-used.

We can think of an 'analogy' for this. Suppose we are hiking in the mountains (Figure 8.22). We start on one side of a mountain and want to get to the other side. We could go right over the summit of the mountain. This would require us to be very energetic. What we might prefer to do would be to find an alternative route along a pass through the mountains. This would be less energetic. In our analogy, the starting point corresponds to the reactants and the finishing point to the products. The route over the top of the mountain would be the uncatalysed path. The easier route through the pass would be a catalysed path.

Different chemical reactions need different catalysts. One broad group of catalysts works by adsorbing molecules onto a solid surface. This process of **adsorption** brings the molecules of reactants closer together. The process of adsorption is also thought to weaken the bonds in the reactant molecules. This makes them more likely to react. Some of the most important examples of industrial catalysts work in this way, for example iron in the Haber process and vanadium(V) oxide in the Contact process (Chapter 9). The efficiency of the solid catalysts in the Haber and Contact processes is improved by increasing the surface area available to interact with the gaseous reactants.

KEY WORD

adsorption: the attachment of molecules to a solid surface

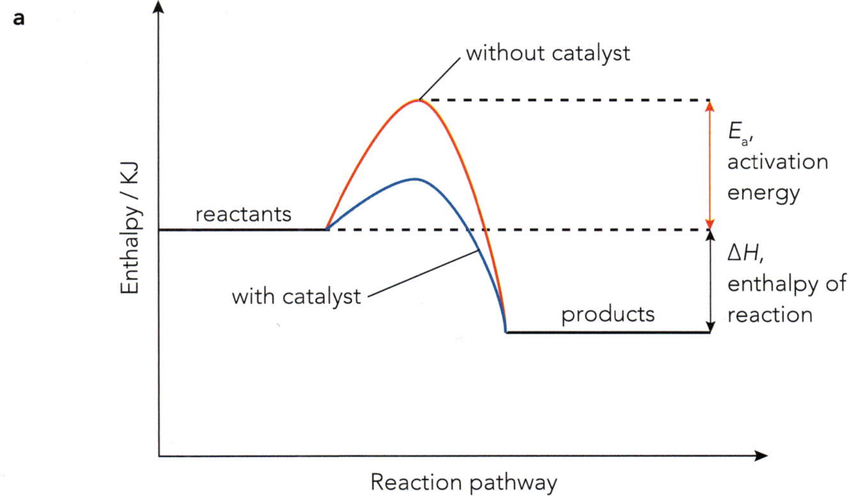

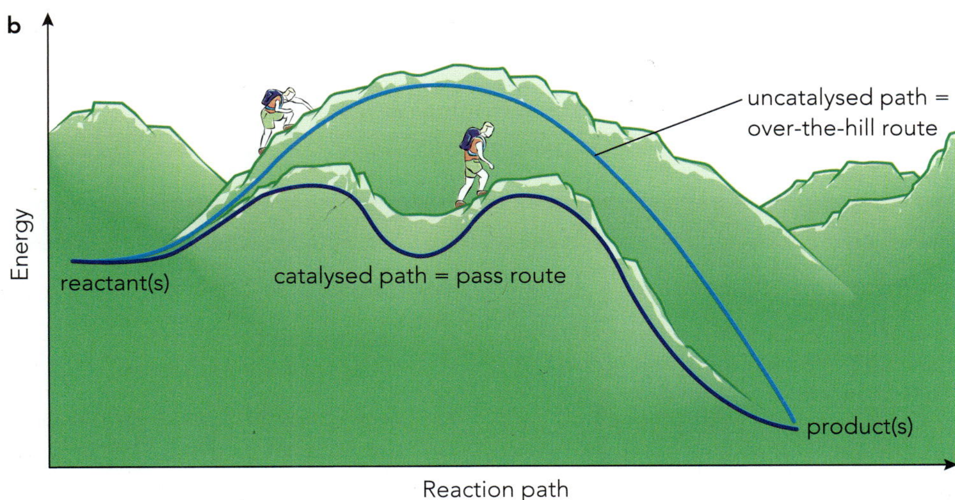

Figure 8.22 a: The reaction pathway diagram for an exothermic reaction showing how a catalyst lowers the activation energy of the reaction. **b:** The barrier between reactant(s) and product(s) may be so high that it defeats all but the most energetic; the catalysed route is an easy pass through the mountains.

Questions

7 What changes in physical conditions are enzymes particularly sensitive to?

8 Does the presence of a catalyst increase or decrease the activation energy for a reaction?

9 In terms of the collision theory, explain why the rate of a reaction increases with:

- **a** an increase in temperature
- **b** an increase in the surface area of a solid reactant
- **c** an increased concentration of a reacting solution.

SUMMARY

Factors such as reactant concentration, gas pressure, the surface area of solids, temperature and the presence of a catalyst will affect the rate of a reaction.
A catalyst increases the rate of a reaction and is unchanged at the end of that reaction.
The rates of various different reactions can be investigated by suitable practical methods and the results and graphs plotted after using show particular patterns and give information about rate of reaction.
Different practical methods of investigating rates of reaction, including reactions involving changes in mass and the formation of gases, can be evaluated by assessing the accuracy of the measuring apparatus, the number of times an experiment can be repeated and the ease with which the variables can be controlled to make the experiments a 'fair test'.
Collision theory states that a chemical reaction takes place when particles of the reactants collide with sufficient energy to initiate the reaction increasing the concentration of the reactants increases the number of particles per unit volume and therefore the collision frequency, producing an increased rate of reaction.
An increase in temperature produces an increase in reaction rate resulting from an increased collision frequency and an increase in the kinetic energy of the particles present, which means that more collisions exceed the required activation energy (E_a).
A catalyst increases reaction rate by lowering the activation energy of a reaction.

PROJECT

Modelling concentration and pressure

This chapter looks at the factors that affect the rates of reaction (Figure 8.23). Working in small groups, create a visual tool or model that would help to teach someone else how concentration and pressure can affect the rate of reaction. You could use a cardboard box or some type of carton (possibly transparent), and some polystyrene spheres (or other similar materials) to help you make this.

Figure 8.23: Marbles can be used to model how the concentration of the particles in a reacting mixture will affect the rate of reaction.

Think about the following:

- how to use one type of sphere to represent different concentrations
- how you would produce movement of the spheres
- how you would represent different reactant molecules and the process of reacting
- how you could use a piece of card to show the effect of changing pressure on gas molecules.

Present your visual tool/model to the class and explain your ideas.

EXAM-STYLE QUESTIONS

1 Various factors affect the rate of a chemical reaction. One of these is the surface area of a solid that is available to react when in contact with a liquid. A chemical reaction between pieces of a solid and an acid is very fast.

Which of the changes in the table would make the reaction slower?

	Acid	Pieces of solid
A	more concentrated	larger
B	less concentrated	larger
C	less concentrated	smaller
D	more concentrated	smaller

[1]

2 The rate of the reaction between lumps of calcium carbonate and dilute hydrochloric acid was investigated by collecting the gas produced as shown in the figure.

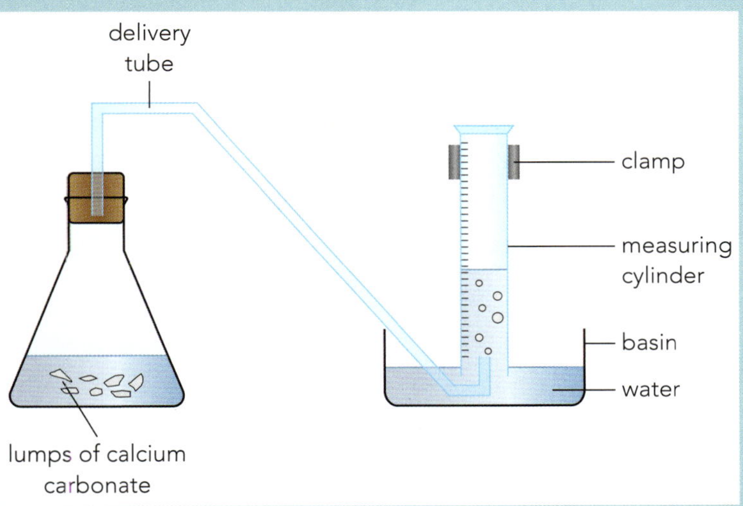

The results obtained are given in the table.

Time / s	0	30	60	90	120	150
Volume / cm^3	10	50	70	80	90	90

a Why are the volumes at 120 and 150 seconds the same? **[1]**

b The rate for the first 30 seconds is $40/30 = 1.33\,cm^3/s$. **Calculate** the rate for the second 30 seconds. **[2]**

c **Suggest** an alternative piece of apparatus that could be used to measure the volume of gas. **[1]**

d **Give** two changes that could be made to the conditions that would speed up the reaction. **[2]**

[Total: 6]

COMMAND WORDS

calculate: work out from given facts, figures or information

suggest: apply knowledge and understanding to situations where there are a range of valid responses in order to make proposals / put forward considerations

give: produce an answer from a given source or recall/ memory

CONTINUED

3 The reaction between sodium thiosulfate and hydrochloric acid produces a finely divided precipitate of sulfur. This precipitate remains in suspension and eventually masks a cross written on a piece of paper placed below the reaction (as shown in the figure).

$Na_2S_2O_3(aq) + 2HCl(aq) \rightarrow 2NaCl(aq) + S(s) + SO_2(g) + H_2O(l)$

The rate of this reaction was investigated using the following experiment. A beaker containing 50 cm³ of sodium thiosulfate solution was placed on a black cross. 5.0 cm³ of dilute hydrochloric acid was added and the clock was started. Initially the cross was clearly visible. When the solution became cloudy and the cross could no longer be seen, the clock was stopped and the time recorded.

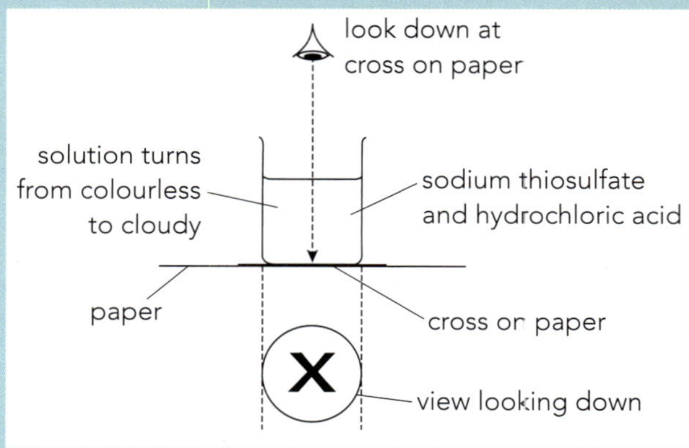

The experiment was repeated several times at different temperatures and the results are given in the table.

Temperature / °C	Time to obscure cross / s	1/time
20	47	0.021
30	23	0.043
40	12	0.083
50	6	0.167

a **Explain** why the value in the third column of the table is proportional to the reaction rate. **[1]**

b Use the information in the table to describe how the reaction rate varies with temperature. **[2]**

c Use ideas about reacting particles to explain why reaction rate changes with temperature. **[3]**

[Total: 6]

COMMAND WORD

explain: set out purposes or reasons/make the relationships between things evident/provide why and/or how and support with relevant evidence

CONTINUED

4 Zinc reacts with dilute sulfuric acid to form hydrogen.

$Zn + H_2SO_4 \rightarrow ZnSO_4 + H_2$

If lumps of zinc are used the reaction is quite slow, but small pieces of zinc react more rapidly.

a Explain why smaller pieces of zinc react more rapidly. Use ideas about particles and collisions. **[2]**

b If the acid is heated, the reaction takes place more rapidly. Explain why the reaction speeds up when the acid is heated. Use ideas about collisions and energy. **[2]**

c When small pieces of copper metal are added, the fizzing becomes much faster. Copper does not react with dilute sulfuric acid and is still there when effervescence stops. Explain these observations. **[2]**

d The rate of some reactions that produce gases can be followed by recording loss of mass on a chemical balance. Explain why this method is not suitable for the reaction between zinc and dilute sulfuric acid. **[2]**

[Total: 8]

SELF-EVALUATION CHECKLIST

After studying this chapter, think about how confident you are with the different topics. This will help you see any gaps in your knowledge and help you to learn more effectively.

I can	See Topic...	Needs more work	Almost there	Confident to move on
describe the effects of factors such as reactant concentration, gas pressure, the surface area of solids, temperature and the presence of a catalyst on the rate of a reaction	8.1			
understand that a catalyst (e.g. an enzyme) increases the rate of a reaction but remains unchanged at the end of the reaction	8.1			
investigate the rates of various different reactions and interpret the results and graphs obtained from such methods	8.1			
evaluate different practical methods of investigating rates of reaction including reactions involving changes in mass and formation of gases	8.1			
describe collision theory	8.2			
explain the effects of concentration on reaction rates in terms of particles per unit volume and collision frequency	8.2			
explain how increased temperature produces an increase in reaction rate in terms of collision frequency, kinetic energy and activation energy	8.2			
explain that a catalyst increases the reaction rate by lowering the activation energy of a reaction	8.2			

> Chapter 9

Reversible reactions and equilibrium

IN THIS CHAPTER YOU WILL:

- understand that some chemical reactions are reversible
- describe how changing conditions can alter the direction of a reversible reaction
- describe the use of reversible reactions as a chemical test for the presence of water
- state that ammonium salts and nitrates can be used as fertilisers
- describe the use of NPK fertilisers for improved plant growth

> state that in a closed system a reversible reaction can reach an equilibrium where the rate of the forward reaction is equal to the rate of the reverse reaction

> predict and explain how the position of an equilibrium is affected by various changes in the conditions

> state the symbol equations for the reversible reactions used to produce ammonia in the Haber process and sulfuric acid in the Contact process and give the sources of the reactants for these two processes

> outline the typical conditions used in the Haber and Contact processes.

GETTING STARTED

A major topic in this chapter centres on an industrial process that revolutionised agricultural food production worldwide. Work with a partner to draw a mind map that includes the following questions:

1 What is a crop?

2 What crops are grown in your local area?

3 What do crops need to grow?

4 What is a fertiliser?

5 What do the words 'artificial' and 'organic' mean, and how could these words be linked to fertiliser?

REVOLUTIONISING FOOD PRODUCTION

Figure 9.1: Fritz Haber (1868–1934) at work in his laboratory in Karlsruhe, Germany.

Ammonia is one of the major products of industrial chemistry. A key ingredient of synthetic fertilisers, approximately 175 million tonnes of ammonia are produced annually around the world. About 50% of the world's food production is dependent on ammonia-based fertilisers. The figure for worldwide ammonia production is predicted to rise to well over 250 million tonnes by the year 2050. Ammonia is produced globally by the Haber process, in which nitrogen is reacted directly with hydrogen in a reversible reaction that needs carefully controlled conditions.

$$N_2(g) + 3H_2(g) \rightleftharpoons 2NH_3(g)$$

Fritz Haber (Figure 9.1) was born in Breslau (now Wrocław, Poland) in 1868. Many important developments in chemistry were taking place in Germany at that time and Haber became heavily involved in this progress. During the 1890s, there were predictions that the world's population would soon overtake food production unless crop yields could be increased using nitrogen fertilisers. Atmospheric nitrogen is unavailable to most plants unless it is first converted into water-soluble compounds such as ammonia or nitrates. Haber used high pressure and temperature together with an osmium catalyst to achieve the reaction (Figure 9.8). The process was developed to an industrial scale by the chemical engineer Karl Bosch. The Haber–Bosch process became of major importance in the production of fertilisers and is perhaps the most recognised chemical process in the world. In 1918, Haber received the Nobel Prize in Chemistry for his part in the invention of the ammonia production process.

Although the Haber process represents a huge technological advancement, it has always been an energy-demanding one. The high carbon footprint of the process is added to by the fact that most of the hydrogen used is obtained from the 'steam reforming' of natural gas and generates carbon dioxide. It is estimated that every ammonia molecule produced releases one molecule of carbon dioxide. Consequently, current research and development is focused on making 'green ammonia' using hydrogen from water electrolysis and electricity generated by solar or wind power. Trial plants aimed at sustainable use of the Haber process are being developed in several countries, with some exploring the use of more efficient catalysts such as ruthenium.

Discussion questions

1 Haber originally used osmium as the catalyst in his experiments. However, when adapting the process for industrial production this was replaced by iron. Why was this advantageous?

2 What issues are involved in using rare metals such as osmium or ruthenium as catalysts for industrial scale processes?

9.1 Reversible reactions

The physical and biological world is the product of a complex set of chemical interactions and reactions. Some reactions can even be reversed if we change the conditions.

Our life depends on the reversible attachment of oxygen to a protein called haemoglobin found in our red blood cells. Oxygen is picked up as these cells pass through the blood vessels of the lungs. It is then carried to tissues in other parts of the body. As the conditions change in other regions of our body (e.g. the muscles and brain), the oxygen is detached and used by the cells of these organs.

Simpler reactions that can be reversed by changing the conditions include the re-formation of hydrated salts by adding back the water to the dehydrated powder. However, there are **reversible reactions** that are more complex. In these reactions the products can interact to reverse the reaction. No sooner are the products formed than some molecules of the product react to give back the original reactants. One industrially important example of this is the reaction between nitrogen and hydrogen to produce ammonia:

$$N_2(g) + 3H_2(g) \rightleftharpoons 2NH_3(g)$$

The double arrow symbol ($\rightleftharpoons$) indicates that the reaction is reversible; in the forward reaction, nitrogen and hydrogen combine to form ammonia. In the reverse reaction, ammonia decomposes back into nitrogen and hydrogen.

The German chemist, Fritz Haber (see the Science in Context feature), was the first to show how this reaction could be controlled to make useful amounts of ammonia by the Haber process (Topic 9.2).

Reversible hydration of salts

Thermal decomposition of salts such as hydrated copper(II) sulfate ($CuSO_4{\cdot}5H_2O$) results in the dehydration of the salt:

$$CuSO_4{\cdot}5H_2O(s) \xrightarrow{\text{heat}} CuSO_4(s) + 5H_2O(g)$$

$$\text{light blue crystals} \xrightarrow{\text{heat}} \text{white powder}$$

In this case, the reaction results in a colour change from blue to white. The physical structure of the crystals is also destroyed. The water driven off can be condensed separately (Figure 9.2).

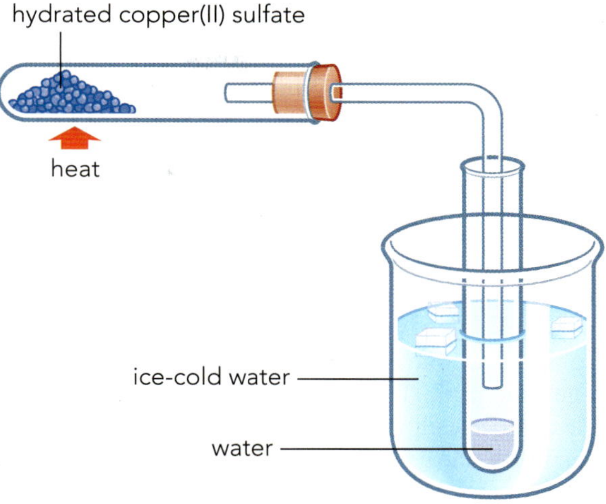

Figure 9.2: Apparatus for condensing the water vapour driven off from blue crystals of hydrated copper(II) sulfate by heating.

The white **anhydrous** copper(II) sulfate and the water are cooled down. Then the dehydration reaction can be reversed by slowly adding the water back to the white anhydrous powder (Figure 9.3). This reaction is strongly exothermic and the colour of the powder returns to blue.

Figure 9.3: Adding water back to dehydrated copper(II) sulfate.

KEY WORDS

reversible reaction: a chemical reaction that can go either forwards or backwards, depending on the conditions

anhydrous: an adjective to describe a substance without water combined with it

Other **hydrated salts** can be dehydrated in a similar way to copper(II) sulfate. Examples of such salts are cobalt(II) chloride and iron(II) sulfate. These salts crystallise with a defined amount of water chemically combined in their crystals (Chapter 12). Hydrated cobalt(II) chloride is a pink crystalline salt. When heated it is dehydrated to a form that is blue. The reaction is endothermic.

$$\underset{\text{pink}}{CoCl_2{\cdot}6H_2O} \xrightarrow{\text{heat}} \underset{\text{blue}}{CoCl_2 + 6H_2O}$$

Adding water to blue anhydrous cobalt(II) chloride turns the solid pink. The dehydration of cobalt(II) chloride can be easily reversed. The two reactions can be combined into one equation using the ⇌ sign.

$$\underset{\text{pink}}{CoCl_2{\cdot}6H_2O} \rightleftharpoons \underset{\text{blue}}{CoCl_2 + 6H_2O}$$

These dehydration reactions are useful as they involve a colour change and can be used as a test for the presence of water (Chapter 17).

Chemical test for the presence of water

Not all neutral colourless liquids are water. The presence of water can be detected using anhydrous copper(II) sulfate or cobalt(II) chloride. Water will turn anhydrous copper(II) sulfate from white to blue and anhydrous cobalt(II) chloride from blue to pink. Cobalt chloride paper contains blue anhydrous cobalt chloride. It turns pink if water is present (Figure 9.4).

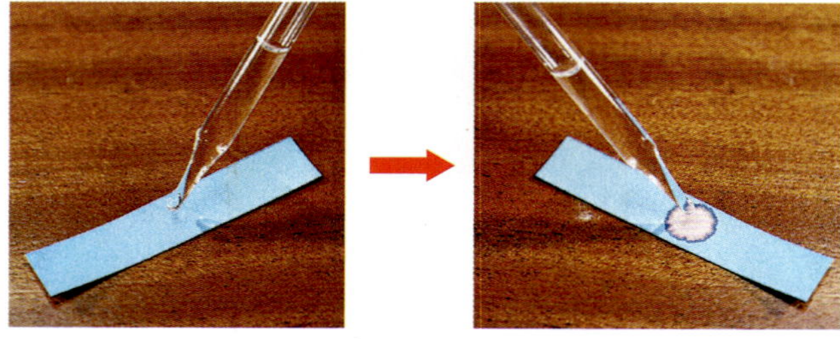

Figure 9.4: The test for the presence of water using cobalt chloride paper. The paper turns from blue to pink.

However, the cobalt chloride test will only tell you that water is present, not that it is pure water. To decide whether a liquid is pure water, you would need to test to show that its boiling point is exactly 100 °C (Chapter 1).

KEY WORDS

hydrated salts: salts whose crystals contain combined water (*water of crystallisation*) as part of the structure

Questions

1. a What colour change do we see when water is added to anhydrous copper(II) sulfate powder?
   b What can the colour change seen in part **a** be used as a test for?
   c Write a balanced symbol equation for the reaction in part **a**.
2. Which test would you carry out to show that a colourless liquid was pure water?
3. How do we know that the reaction to form pink hydrated cobalt(II) chloride from blue anhydrous cobalt(II) chloride is exothermic?

Chemical equilibria

When we change between blue copper(II) sulfate crystals and the white anhydrous powder the two reactions are separate. We don't heat and add water at the same time. The two reactions, although reversible, are clearly separate.

A simple reversible reaction where we can see both reactions taking place at the same time is the sublimation of ammonium chloride. The thermal decomposition of ammonium chloride is an example of a reversible change taking place in different parts of the same test-tube (Figure 9.5). When warmed in a test-tube, the white solid decomposes to ammonia and hydrogen chloride:

$$NH_4Cl(s) \rightarrow NH_3(g) + HCl(g)$$

However, on the cooler surface of the upper part of the tube, the white solid is re-formed:

$$NH_3(g) + HCl(g) \rightarrow NH_4Cl(s)$$

Figure 9.5: The reversible reaction involving ammonium chloride.

The ideas about reversible reactions can be extended to reactions taking place in a **closed system**. A closed system is one in which no reactants or products can escape from the reaction mixture. Figure 9.6 shows the difference between a closed system and an open system when calcium carbonate is heated at a high temperature. In the closed system, the carbon dioxide gas produced cannot escape. Eventually, an equilibrium is set up in the container.

It is not always necessary to study chemical equilibria in closed containers. If no gas is produced as a result of the reaction (i.e. it takes place entirely in solution), equilibrium may be established in an open container such as a flask.

At equilibrium, the reactants are continuously combining to make a product. At the same time and, importantly, at the same rate, the product is continuously breaking apart and turning back into its reactants. This is a **dynamic equilibrium**. At equilibrium, the rate of the forward reaction equals the rate of the reverse (backward) reaction. For reactions where the reaction mixture is all in the same phase the equilibrium can be approached from either direction. One example of such a reaction is heating a mixture of hydrogen and iodine in a sealed tube:

$$H_2(g) + I_2(g) \rightleftharpoons 2HI(g)$$

At a particular temperature, we get the same fixed concentration of hydrogen, iodine and hydrogen iodide whether we start from hydrogen iodide or the hydrogen/iodine mixture.

KEY WORDS

closed system: a system where none of the reactants or products can escape the reaction mixture or the container where the reaction is taking place

dynamic (chemical) equilibrium: two chemical reactions, one the reverse of the other, taking place at the same time, where the concentrations of the reactants and products remain constant because the rate at which the forward reaction occurs is the same as that of the reverse reaction

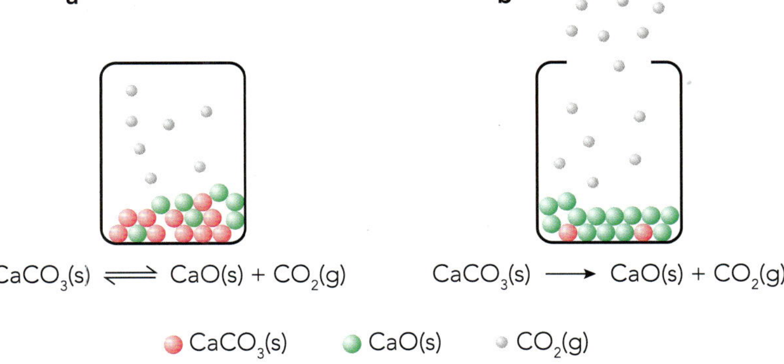

Figure 9.6: Comparison of a closed and open system for the decomposition of calcium carbonate.

The **position of equilibrium** tells us how far a reaction has gone in favour of reactants or products. If the concentration of products is greater, then we say the position of equilibrium is to the right – it favours the products. If the concentration of reactants is greater than the products, then the equilibrium position favours the reactants – the position of equilibrium is to the left.

An equilibrium reaction has four particular features under constant conditions:

- it is dynamic: reactants are continuously being changed to products and products are continuously being changed back to reactants
- the forward and reverse (backward) reactions occur at the same rate
- the concentrations of reactants and products (the position of equilibrium) remain constant
- it requires a closed system.

Analogies that have been used to illustrate this idea of a dynamic equilibrium are either the person running up a down escalator or a fish swimming upstream against the water current (Figure 9.7a and b). If the speed of the person, or the fish, in their direction is equal to the flow in the opposite direction, then they will appear to be stationary.

KEY WORDS

position of equilibrium: the mixture of reactants and products at which a reversible reaction is in equilibrium under a particular set of physical conditions of temperature and pressure

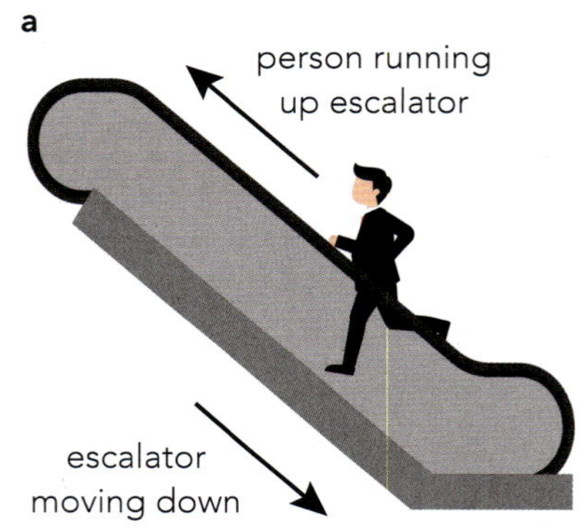

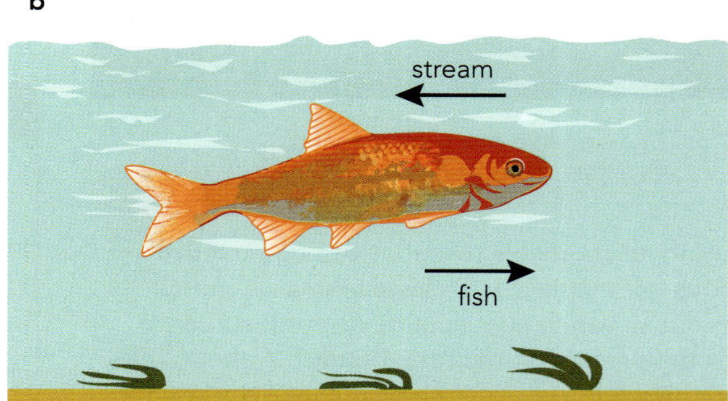

Figure 9.7: Dynamic equilibrium: **a:** the person and **b:** the fish appear to be stationary.

REFLECTION

The idea of an equilibrium set up by two competing reactions is difficult to imagine. Various different analogies have been suggested to help understand what is taking place in the reaction mixture (Figure 9.7). An analogy is a comparison of two things to show their similarities. Consider the following questions:

- How do you picture what is happening as the forward and reverse reactions compete but keep the concentrations of all the chemicals present constant?
- Do you find analogies helpful?
- Could you use an analogy to teach a difficult idea to someone else?

Chemical equilibria and reaction conditions

The position of equilibrium set up for a particular reaction depends on the conditions used. Changing the temperature alters the equilibrium position. For some reactions involving gases, changing the pressure can also alter the equilibrium position. The addition or removal of one of the reactants or products alters their concentration and this can change the position of equilibrium.

For all these possible effects can we predict what the change will be? There is one rule that helps us predict the results of a change in conditions (it is sometimes called Le Chatelier's principle, named after the French chemist Henry-Louis Le Chatelier). The rule is based on practical observation of many reversible reaction and states that:

- when a change is made to the conditions of a system in dynamic equilibrium, the system moves so as to oppose that change.

Let us consider some reversible reactions and the effect of changing conditions when they are at equilibrium.

Effect of temperature on the position of an equilibrium

The **Contact process** is used for the manufacture of sulfuric acid. The main reaction that converts sulfur dioxide (SO_2) to sulfur trioxide (SO_3) in the Contact process is reversible:

$$2SO_2(g) + O_2(g) \rightleftharpoons 2SO_3(g) \qquad \Delta H = -197\ \text{kJ/mol}$$

The reaction to produce sulfur trioxide is exothermic. So, if this reaction were at equilibrium and the temperature raised, the system would try to absorb the heat. The endothermic reaction would be favoured and the equilibrium position would move to the left. Less sulfur trioxide would be produced. For an exothermic reaction, when temperature increases the equilibrium shifts in favour of the reverse reaction. Raising the temperature favours the endothermic change where heat is taken in.

KEY WORDS

Contact process: the industrial manufacture of sulfuric acid using the raw materials sulfur and air

For an endothermic reaction the opposite is true. Increasing the temperature moves the reaction to the right, favouring the products. Raising the temperature in this case favours the exothermic change where heat is given out.

Effect of pressure on the position of an equilibrium

Changes of pressure can only affect reactions that involve a gas in the equation. Increasing the pressure moves the equilibrium to the side which occupies the smaller volume (the side with less gas).

$2SO_2(g) + O_2(g)$	$\rightleftharpoons$	$2SO_3(g)$
3 moles of gas		2 moles of gas
= 3 volumes		= 2 volumes

There are fewer gas molecules on the right of the equation. Therefore, increasing the pressure would favour the production of sulfur trioxide. This happens because increased pressure compresses the gas, pushing the gas molecules into a smaller space (increasing the concentration). The system responds to counteract that and shifts to the side that involves fewer gas molecules. A decrease in pressure would produce the opposite shift in equilibrium position, favouring the decomposition of sulfur trioxide.

A gas phase reaction such as that between hydrogen and iodine vapour would not be affected by a change in pressure as there are equal volumes of gas on both sides of the equation.

$H_2(g) + I_2(g)$	$\rightleftharpoons$	$2HI(g)$
2 moles of gas	$\rightleftharpoons$	2 moles of gas

Effect of concentration on the position of an equilibrium

An organic carboxylic acid reacts with an alcohol to produce an ester (Chapter 19).

CH_3COOH	+	C_2H_5OH	$\rightleftharpoons$	$CH_3COOC_2H_5$	$+ H_2O$
ethanoic acid		ethanol	$\rightleftharpoons$	ethyl ethanoate	water

The reaction is catalysed by the presence of H^+ ions provided by the addition of a few drops of concentrated sulfuric acid. Concentrated sulfuric acid is also a dehydrating agent and removes some of the water from the reaction mixture. The removal of water shifts the equilibrium position to the right as the reaction tries to restore the concentration of water.

Predicting how the position of equilibrium is affected by changing conditions

The addition or removal of a reactant or product will affect the position of equilibrium in any reversible reaction. The overall effects of these different changes in the conditions on the equilibrium set up by a reversible reaction are summarised in Table 9.1. We will see the effect of these factors as we consider the industrial production of ammonia in the Haber process. Remember that, when the equilibrium conditions are changed, the reaction always tends to oppose the change and act in the opposite direction.

Condition	Effect on equilibrium position
temperature	Increasing the temperature makes the reaction move in the direction that takes in heat (the endothermic direction). Decreasing the temperature makes the reaction move in the direction that gives out heat (the exothermic direction).
pressure	This only affects reactions involving gases. Increasing the pressure shifts the equilibrium in the direction that produces fewer gas molecules. Decreasing the pressure shifts the equilibrium in the direction that produces more gas molecules.
concentration	Increasing the concentration of one substance in the mixture makes the equilibrium move in the direction that produces less of that substance. Decreasing the concentration of one substance in the mixture makes the equilibrium move in the direction that produces more of that substance.
catalyst	Using a catalyst does not affect the position of equilibrium, but the reaction reaches equilibrium faster.

Table 9.1: Predicting the effect of changing conditions on a chemical equilibrium.

9.2 Haber process and Contact process

Haber process

The Haber process for the synthesis of ammonia was one of the most significant new ideas of the 20th century. Ammonia was first manufactured using the Haber process on an industrial scale in 1913 following Fritz Haber's earlier laboratory experiments (Figure 9.8), and it allowed industrial chemists to make ammonia cheaply and in large quantities. Now, over 150 million tonnes of ammonia are produced each year by this process.

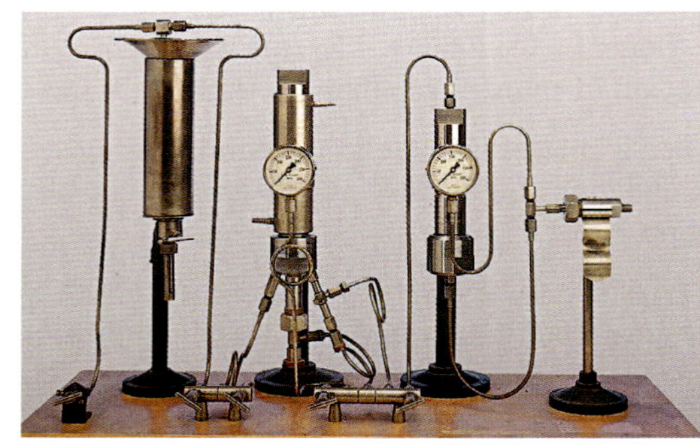

Figure 9.8: Haber's original experimental apparatus, designed for adjusting the pressure of the reacting mixture.

Ammonia is a colourless gas with a distinctive smell, is less dense than air, and is very soluble in water to give an alkaline solution. As a raw material for both fertilisers and explosives, ammonia has played a large part in human history. It helped to feed a growing population in peacetime and it was used to manufacture explosives in wartime (see the Science in Context feature).

Nitrogen is an unreactive gas and changing it into compounds useful for plant growth (nitrogen fixation) is important for agriculture. Most plants cannot directly use (or fix) nitrogen from the air. The main purpose of the industrial manufacture of ammonia is to make agricultural **fertilisers**.

KEY WORD

fertiliser: a substance added to the soil to replace essential elements lost when crops are harvested, which enables crops to grow faster and increases the yield

Conditions for the Haber process

The reaction to produce ammonia from nitrogen and hydrogen is a reversible reaction.

$$N_2(g) + 3H_2(g) \rightleftharpoons 2NH_3(g)$$

Because of its importance, the Haber process for making ammonia has been studied under a wide range of conditions of temperature and pressure (Figure 9.9). The percentage amount of ammonia in the equilibrium mixture depends on both the temperature and the pressure. Under the conditions Haber first used, only 8% of the equilibrium mixture was ammonia.

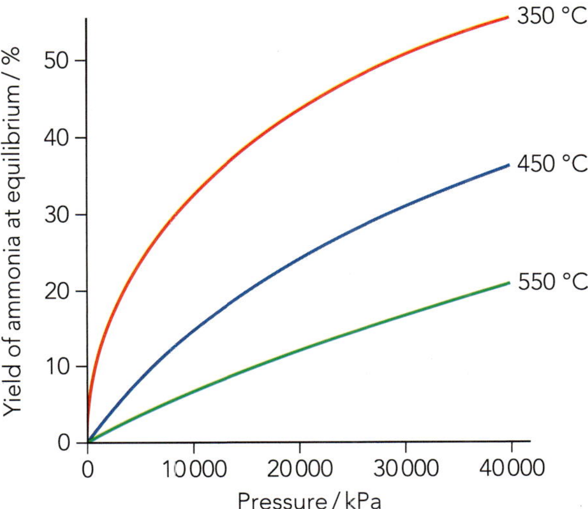

Figure 9.9: A wide range of conditions of temperature and pressure have been tried for the Haber process. The curves show the yields that would be obtained for some of the conditions.

How could conditions be changed to improve this yield? Le Chatelier's principle gives clues as to how this can be done. When a change is made to the conditions of a system in dynamic equilibrium, the system moves to oppose that change. So how can this reaction system be changed to produce more ammonia at equilibrium – to shift the equilibrium to the right (Figure 9.10)?

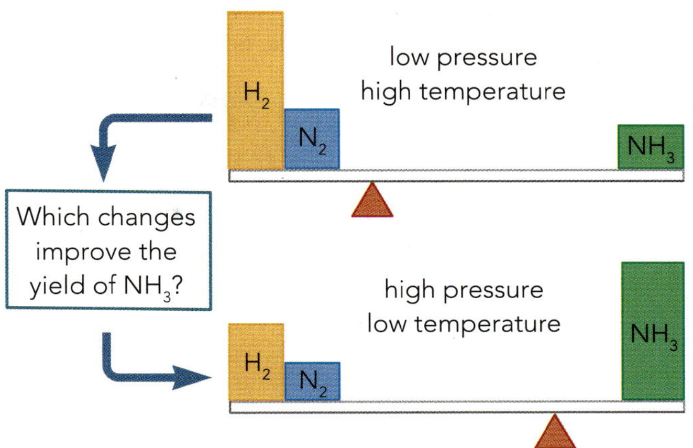

Figure 9.10: For the Haber process, increasing the pressure and lowering the temperature both move the equilibrium to the right to give more ammonia in the mixture.

Changing the pressure: How will increasing the pressure affect the amount of ammonia produced? The pressure of a gas is caused by collisions of the gas particles with the walls of the container – the fewer molecules present, the lower the pressure. If we apply more pressure to the equilibrium, the system will shift to favour the side of the equation that has fewer molecules:

$$N_2(g) + 3H_2(g) \quad \rightleftharpoons \quad 2NH_3(g)$$

$$\text{4 moles} \quad \rightleftharpoons \quad \text{2 moles}$$

With increased pressure, there will be a shift to the right. More ammonia will form to reduce the number of molecules in the mixture. High pressures will increase the yield of ammonia (Figures 9.9 and 9.10). Modern industrial plants use a pressure of 20000 kPa.

The use of a high pressure in the process also increases the rate of the reaction (Chapter 6).

The high pressure used means that there are more gas molecules in a unit volume; the concentration of the reactants has been increased, increasing the rate at which the equilibrium position is reached. Higher pressures could be used, but high-pressure reaction vessels are expensive to build and there is a greater risk of explosion.

Changing the temperature: The forward reaction producing ammonia is exothermic, and the reverse reaction is therefore endothermic:

$$N_2(g) + 3H_2(g) \rightarrow 2NH_3(g)$$
exothermic – heat given out

$$2NH_3(g) \rightarrow N_2(g) + 3H_2(g)$$
endothermic – heat taken in

If we raise the temperature of the system, more ammonia will break down to take in the heat supplied. Less ammonia will be produced at high temperatures. Lowering the temperature will favour ammonia production (Figures 9.9 and 9.10). However, the rate at which the ammonia is produced will be so slow as to be uneconomical. In practice, a **compromise (or optimum) temperature** is used to produce enough ammonia at an acceptable rate. Modern plants use temperatures of about 450 °C.

Reducing the concentration of ammonia: If the system was at equilibrium and then some of the ammonia was removed, more ammonia would be produced to replace what had been taken away. Industrially, it is easy to remove ammonia. Ammonia has a much higher boiling point than nitrogen or hydrogen (Table 9.2) and condenses easily, leaving the nitrogen and hydrogen still as gases. In modern plants, the gas mixture is removed from the reaction chamber when the percentage of ammonia is about 15%. The ammonia is condensed by cooling, and the remaining nitrogen and hydrogen are recycled.

Compound	Boiling point / °C
nitrogen (N_2)	−196
hydrogen (H_2)	−253
ammonia (NH_3)	−33

Table 9.2: Boiling points of nitrogen, hydrogen and ammonia.

Use of a catalyst: This reaction is a difficult reaction to get to work at a reasonable rate. A catalyst can be added. Chemists have tried more than 2500 different combinations of metals and metal oxides as catalysts for this reaction. Finely divided iron has been used as the chosen catalyst industrially. However, as we have seen, the presence of a catalyst does not alter the equilibrium concentrations of the reactants and product (N_2, H_2 and NH_3). The catalyst shortens the time taken to reach equilibrium by increasing the rates of both the forward and reverse reactions.

KEY WORDS

compromise temperature: a temperature that gives sufficient product and a reasonable and economic rate of reaction

Industrial plant for the Haber process

In the Haber process (Figure 9.11), nitrogen and hydrogen are directly combined to form ammonia:

nitrogen + hydrogen ⇌ ammonia

$$N_2(g) + 3H_2(g) \rightleftharpoons 2NH_3(g)$$

Nitrogen is obtained from air by fractional distillation or by reacting the oxygen with hydrogen. Hydrogen is obtained from the 'steam-reforming' of natural gas, or from the breakdown (catalytic cracking) of hydrocarbons such as ethane.

The reaction of natural gas with steam in the presence of a nickel catalyst forms hydrogen and carbon monoxide.

$$CH_4(g) + H_2O(g) \rightarrow CO(g) + 3H_2(g)$$

methane + steam → carbon monoxide + hydrogen

The carbon monoxide formed could poison the catalyst and is therefore removed by reacting with more steam.

$$CO(g) + H_2O(g) \rightarrow CO_2(g) + H_2(g)$$

Alternatively, the hydrogen can be obtained from the cracking of ethane using a high temperature and a catalyst (Chapter 20).

$$C_2H_6(g) \rightarrow C_2H_4(g) + H_2(g)$$

ethane → ethene + hydrogen

The reactant gases are cleaned of contaminants ('scrubbed') before entering the reactor. The two gases, nitrogen and hydrogen, are mixed in a 1 : 3 ratio and compressed to 20 000 kPa. The gases are then passed over a series of catalyst beds containing finely divided iron (Figure 9.11). The temperature of the converter is about 450 °C. The reaction is reversible and does not go to completion. The proportion of ammonia in the mixture is about 15%. Ammonia has a much higher boiling point than nitrogen or hydrogen, so it condenses easily on cooling. The unchanged nitrogen and hydrogen gases are re-circulated over the catalyst. By re-circulating in this way, an eventual yield of 98% can be achieved. The ammonia produced is stored as a liquid under pressure.

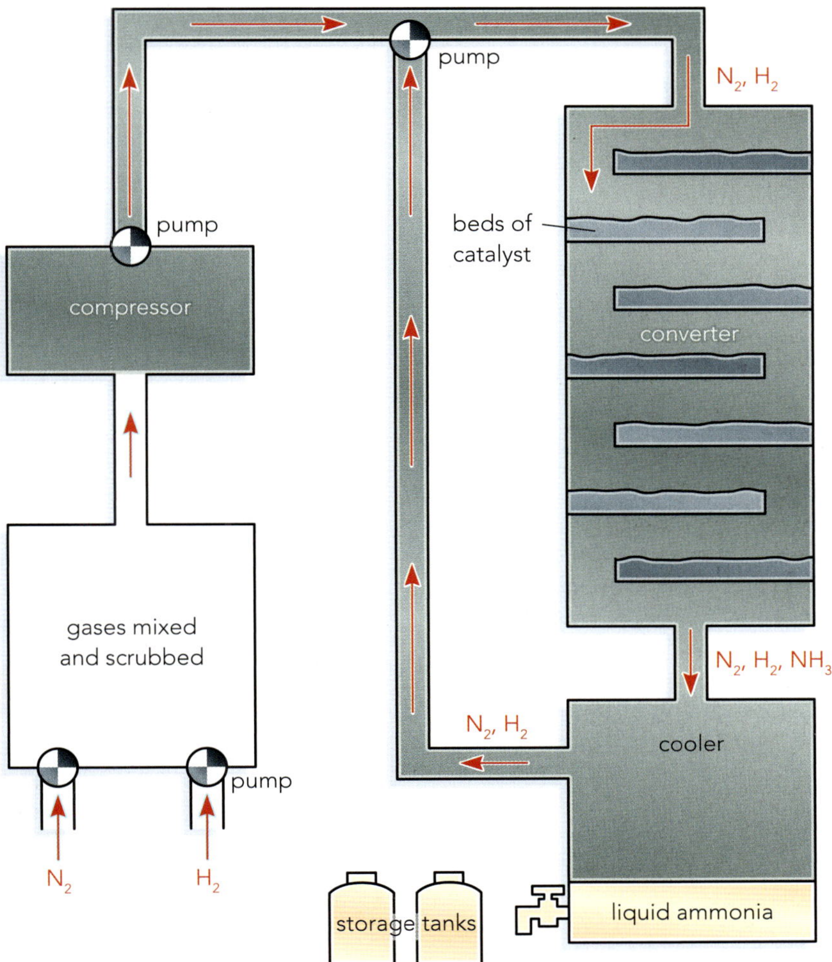

Figure 9.11: A schematic drawing of the different stages of the Haber process.

The typical conditions used in the Haber process are:

- N_2 and H_2 are mixed in a ratio of 1 : 3
- an optimum (or compromise) temperature of 450 °C
- a pressure of 20 000 kPa (200 atmospheres)
- a catalyst of finely divided iron.

The ammonia is condensed out of the reaction mixture and the remaining N_2 and H_2 recycled.

Most of the ammonia produced is used to manufacture fertilisers. Liquid ammonia itself can be used directly as a fertiliser, but it is an unpleasant liquid to handle and transport. The majority of the ammonia is converted into a variety of solid fertilisers. A substantial amount of ammonia is converted into nitric acid by oxidation (Figure 9.12). Much of the nitric acid produced is also used in the fertiliser industry.

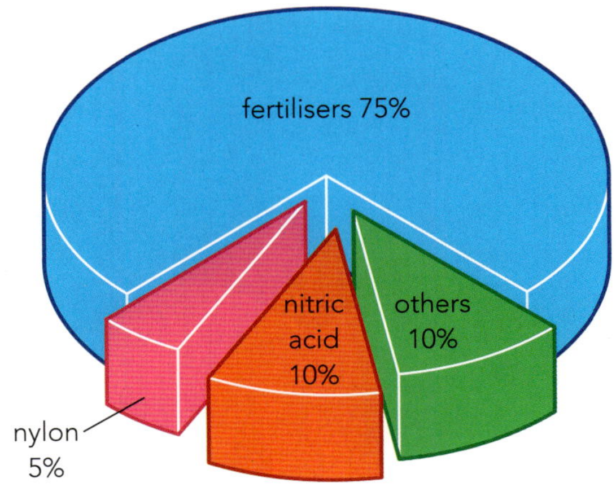

Figure 9.12: Uses of ammonia produced by the Haber process.

Contact process

Sulfuric acid is a major product of the chemical industry. Industrially, sulfuric acid is made from sulfur dioxide in the Contact process. The main reaction in the Contact process is the reversible reaction in which sulfur dioxide and oxygen combine to form sulfur trioxide.

Conditions for the Contact process

In the manufacture of sulfuric acid, the main reaction that converts sulfur dioxide (SO_2) to sulfur trioxide (SO_3) is reversible:

$$2SO_2(g) + O_2(g) \rightleftharpoons 2SO_3(g)$$
$$\text{exothermic } (\Delta H = -197 \text{ kJ/mol})$$

The ideas of Le Chatelier can also be applied to this equilibrium. The reaction to produce sulfur trioxide is exothermic. This means that sulfur trioxide production would be favoured by low temperatures. The reaction would be too slow to be economic if the temperature were too low. A compromise temperature of 450 °C is used. This gives sufficient sulfur trioxide at an economical rate. A catalyst of vanadium(V) oxide is also used to increase the rate and ensure that the reaction is fast enough for a dynamic equilibrium to be established. There are fewer gas molecules on the right of the equation. Therefore, increasing the pressure would favour the production of sulfur trioxide. In fact, the process is run at about twice atmospheric pressure (200 kPa) because the conversion of sulfur dioxide to sulfur trioxide is about 98% complete under these conditions. The minor increase in yield that could be obtained by increasing the pressure further is not considered to be economical.

The typical conditions used in the Contact process are:

- a compromise (optimum) temperature of about 450 °C
- a catalyst of vanadium(V) oxide (V_2O_5)
- an operating pressure of 200 kPa.

Industrial plant for the Contact process

The raw materials needed for the Contact process are sulfur, air and water. Sulfur can be mined in various parts of the world, but large quantities are now obtained by removing the impurities from fossil fuels (see **desulfurisation** in Chapter 17) (Figure 9.13).

KEY WORD

desulfurisation: an industrial process for removing contaminating sulfur from fossil fuels such as petrol (gasoline) or diesel

Figure 9.13: A pile of sulfur from desulfurising fossil fuels can be seen in Vancouver, Canada.

The sulfur dioxide required for the Contact process is obtained industrially by burning sulfur or roasting sulfide ores such as zinc blende (ZnS).

$$S(s) + O_2(g) \rightarrow SO_2(g)$$

$$\text{metal sulfide} + \text{oxygen} \xrightarrow{\text{heat}} \text{metal oxide} + \text{sulfur dioxide}$$

$$2ZnS(s) + 3O_2(g) \xrightarrow{\text{heat}} 2ZnO(s) + 2SO_2(g)$$

The sulfur dioxide is then mixed with air and cleaned before passing into the reactor tower containing the trays of catalyst pellets (vanadium(V) oxide, V_2O_5) (Figure 9.14).

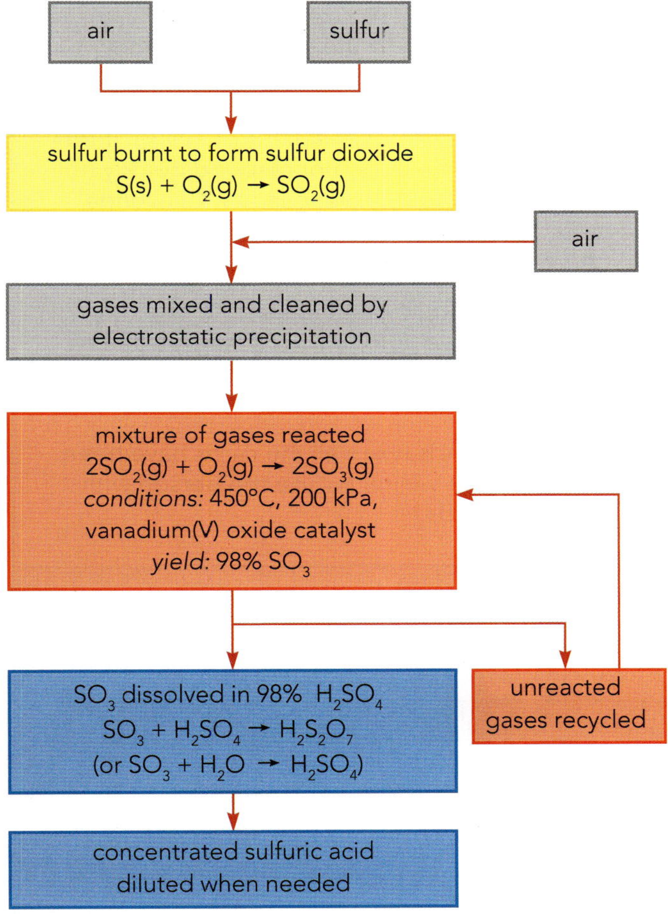

Figure 9.14: Flow chart for making sulfuric acid by the Contact process.

The sulfur trioxide produced is dissolved in 98% sulfuric acid, and not water, to prevent environmental problems of an acid mist that is formed if sulfur trioxide is reacted directly with water. The reaction between sulfur trioxide and water is extremely exothermic. The solution formed means that the acid can be transported in concentrated form (98.5% acid, sometimes known as oleum) and then diluted on site.

Sulfuric acid is important for the fertiliser industry because it is needed to make ammonium sulfate and phosphoric acid.

Questions

4 Oxygen will react with both nitrogen (N_2) and nitrogen(II) oxide (NO) in reversible reactions.

$$N_2(g) + O_2(g) \rightleftharpoons 2NO(g)$$

$$2NO(g) + O_2(g) \rightleftharpoons 2NO_2(g)$$

a Describe how an increase in pressure will affect the position of equilibrium of

i the reaction between nitrogen and oxygen

ii the reaction between oxygen and nitrogen(II) oxide.

b The reaction between nitrogen and oxygen is endothermic. Describe how an increase in temperature will affect the reaction. Explain your answer.

5 The availability of raw materials is essential for an industrial process.

a How is hydrogen obtained for use in the Haber process?

b Describe two sources of sulfur dioxide for the Contact process.

c What is the common source of the nitrogen and oxygen required for the Haber and Contact processes?

6 a What are the conditions used for the Haber process?

b Will increasing the pressure in the Haber process produce more or less ammonia?

c What would be the effect of increasing the temperature in the Haber process on the level of ammonia produced?

d Why are the unreacted gases re-circulated?

ACTIVITY 9.1

A game of 'Higher or Lower?'

The Haber process produces ammonia for use in fertilisers. It is without a doubt one of, if not the, most important industrial scale reactions. Changes in pressure, concentration and temperature are all important in determining the product yield.

Reaction:
$N_2(g) + 3H_2(g) \rightleftharpoons 2NH_3(g) \quad \Delta H = -92$ kJ/mol

Working in a group of three to five people:

1 Write the equation nice and big on a large piece of cardboard. This is the game board and needs to be in the centre of the players.

2 Cut out eight pieces of card, all to the same size, and write on them 'pressure' (×2), 'concentration of N_2' (×2), 'concentration of H_2', concentration of 'NH_3' , and 'temperature' (×2). Shuffle these cards and place them in a pile (to the left of the board).

3 Cut out a further eight pieces of card, all the same size, and write on them 'increase' (×4) and 'decrease' (×4). Shuffle these cards and place them in a second pile (to the right of the board).

4 You are now ready to play the game!

Select a chairperson. This person will not play but needs to:

- check that turns move fairly between the rest of the players
- make decisions on whether challenges are fair.

To play the game:

1 The person on the left of the chairperson takes one card from the first pile and one from the second pile. They then need to explain what will happen to the yield of ammonia under these conditions without hesitation or error. The answer must refer to the position of equilibrium and Le Chatelier's principle. If they make a mistake or do not respond quickly (they hesitate), another player can say 'Challenge' and take the cards from them and be offered the chance to explain.

2 The game should continue around the other players (always going back to the person on the left of the player who started to give the previous explanation).

3 The game ends when all the cards have been used and the winner will be the person with the most cards.

Self-assessment

How confident were you in explaining what would happen to the yield of ammonia under these conditions without hesitation or error? Complete the table in your notebooks.

Green very confident

Amber quite confident

Red not very confident

Conditions	Colour
Changes in pressure	
Changes in concentration of N_2	
Changes in concentration of H_2	
Changes in concentration of NH_3	
Changes in temperature	

Look at your colours in the table you have created. Is there anything you could learn more about to improve your understanding of reversible reactions and equilibrium?

9.3 Fertilisers

We rely on many crop plants for our food. Plants make their own food by photosynthesis from carbon dioxide and water, but they also need other chemical nutrients for producing healthy leaves, roots, flowers and fruit. They get these chemicals from minerals in the soil. When many crops are grown on the same piece of land, these minerals get used up and have to be replaced by artificial fertilisers. Fertilisers enable plants to grow faster and provide larger yields to feed the world's growing population.

Ammonium salts (which contain the NH_4^+ ion, e.g. ammonium phosphate, $(NH_4)_3PO_4$) and nitrates (which contain the NO_3^- ion, e.g. potassium nitrate, KNO_3) can be used as fertilisers.

Ammonium nitrate is the most important of the nitrogenous fertilisers, and provides both NH_4^+ and NO_3^- ions. Ammonium nitrate contains 35% by mass of nitrogen and is produced when ammonia solution reacts with nitric acid:

ammonia + nitric acid → ammonium nitrate

$$NH_3(aq) + HNO_3(aq) \rightarrow NH_4NO_3(aq)$$

The ammonium nitrate can be crystallised into pellet form suitable for spreading on the land.

Ammonium nitrate is soluble in water, as are all other ammonium salts, for example ammonium sulfate, $(NH_4)_2SO_4$. This solubility is important because plants need soluble nitrogen compounds that they can take up through their roots.

Ammonium salts tend to make the soil slightly acidic. To overcome this, they can be mixed with chalk (calcium carbonate), which will neutralise this effect. Calcium ammonium nitrate (CAN), also known as nitrochalk, is an example of a **compound fertiliser**.

Nitrogen (N) is important for producing the proteins needed for plant growth and also for the development of healthy leaves, but other additional elements that plants need are:

- phosphorus (P), especially important for healthy roots
- potassium (K), which is important for the production of flowers and fruit.

Different plants need different combinations of these elements, which is why **NPK fertilisers** are produced. The NPK value (Figure 9.15) informs the farmer how much of each element is present. Fruits such as apples and tomatoes need a lot of potassium, whereas leafy vegetables such as cabbage need a lot of nitrogen and root crops such as carrots need a lot of phosphorus.

Figure 9.15: Bags of NPK (21:8:11) and potassium phosphate fertiliser.

A modern fertiliser factory will produce two main types of product:

- Straight N fertilisers are solid nitrogen-containing fertilisers sold in pellet form, for example, ammonium nitrate (NH_4NO_3), ammonium sulfate ($(NH_4)_2SO_4$) and urea ($CO(NH_2)_2$)
- NPK compound fertilisers (Figure 9.15) are mixtures that supply the three most essential elements lost from the soil by extensive use: nitrogen (N), phosphorus (P) and potassium (K). They are usually a mixture of ammonium nitrate, ammonium phosphate and potassium chloride, in different proportions to suit different conditions. The numbers on a bag of NPK fertiliser correspond to the percentage of that nutrient in the make-up of the fertiliser, e.g. 21:8:11 indicates 21% nitrogen, 8% phosphorus and 11% nitrogen, with the remaining 60% being filler ingredients that help disperse the chemicals. Nitrogen is sometimes omitted from these fertilisers because it washes into streams and rivers causing algal growth (Chapter 17).

KEY WORDS

compound fertiliser: a fertiliser such as an NPK fertiliser or nitrochalk that contains more than one compound to provide elements to the soil

NPK fertiliser: fertilisers to provide the elements nitrogen, phosphorus and potassium for improved plant growth

Questions

7 a Why do farmers use fertilisers?
  b Why do many fertilisers contain the elements N, P and K?

8 State the names of an acid and alkali that could be used to make the following fertilisers.
  a ammonium nitrate
  b ammonium phosphate
  c ammonium sulfate

9 Fertilisers are used by farmers for beneficial effects, but how can excessive or inappropriate use of fertilisers cause pollution (Chapter 17)?

SUMMARY

Some chemical reactions are reversible and this is indicated by the use of the symbol $\rightleftharpoons$.

Changing conditions can alter the direction of a reversible reaction as shown by the effect of heat on hydrated salts and the addition of water to the dehydrated product, e.g. copper(II) sulfate or cobalt(II) chloride.

These reactions can be used as a chemical test for the presence of water, and the melting point and boiling point can be used as tests for pure water.

In a closed system, a reversible reaction can reach an equilibrium where the rate of the forward reaction is equal to the rate of the reverse reaction, and the concentrations of reactants and products are no longer changing.

The position and achievement of an equilibrium is affected by various changes in the conditions, such as temperature, pressure, concentration and the addition of a catalyst, and these factors enable you to predict and explain what is happening.

The symbol equations for the reversible reactions in the Haber process and the Contact process are $N_2(g) + 3H_2(g) \rightleftharpoons 2NH_3(g)$ and $2SO_2(g) + O_2(g) \rightleftharpoons 2SO_3(g)$, respectively.

The raw materials for the Haber process are hydrogen and nitrogen and for the Contact process they are sulfur dioxide and oxygen.

Typical conditions for the Haber process (450 °C, 20 000 kPa pressure and an iron catalyst) and the Contact process (450 °C, 200 kPa and a vanadium(V) oxide catalyst) are chosen to maintain the rate of reaction and equilibrium position to safely give an optimal yield at an economic rate.

Ammonium salts and nitrates can be used as fertilisers.

NPK fertilisers are used to provide the elements nitrogen, phosphorus and potassium for improved plant growth.

PROJECT

Maximising yield

Increasing product yield obviously means producing more product from a given mass of starting material and more product means more profit! More product may also mean less waste or less energy needed to recycle unreacted materials.

Imagine you have been offered a significant loan to set up a new Chemical Consultancy business that will specialise in maximising profits, reducing costs and improving sustainability of your country's chemical industry. Your business is based on two key ideas: an understanding of reaction rates and an understanding of reaction yield.

The first challenge is to come up with a company name, logo and advertising slogan.

You then need to set about designing your company's online presence. The website pages can be produced on paper, on computer slides or even using a web design program.

You should have a main home page with links to two further pages:

- one page should explain the importance of reaction rate (concerns about reactions being too fast or too slow) and the science behind reaction rates
- one page should explain the importance of chemical equilibria (reversible reactions).

From each of these pages, you could add links to the different factors affecting them. For example, for equilibrium you would need to include concentration, pressure and temperature.

If you have time, you might also want to have a final page that explains how there may be a need to find compromise conditions.

EXAM-STYLE QUESTIONS

1 When hydrated iron(II) sulfate is heated the following reaction takes place.

$FeSO_4{\cdot}7H_2O \rightleftharpoons FeSO_4 + 7H_2O$

The colour changes from green to white.

a What is the meaning of the symbol $\rightleftharpoons$? **[1]**

b What two observations are made when water is added to anhydrous iron(II) sulfate? **[2]**

c **Describe** how cobalt chloride can be used to test for the presence of water. **[2]**

[Total: 5]

2 Which of these compounds contains all the elements needed for a balanced compound fertiliser?

A $(NH_4)_2SO_4$

B $K(NH_4)_2PO_4$

C K_2SO_4

D $(NH_2)_2CO$ **[1]**

3 This equation represents an exothermic reversible reaction:

$2A(g) + B(g) \rightleftharpoons 2C(g)$

Which change would move the equilibrium to the left?

A an increase in the concentration of substance A

B an increase in pressure

C an increase in the temperature

D removal of substance C from the reaction vessel **[1]**

4 Ethanol, C_2H_5OH, is manufactured from ethene, C_2H_4, and steam, H_2O.

$C_2H_4(g) + H_2O(g) \rightleftharpoons C_2H_5OH(g)$

This reaction is reversible and exothermic. It is usually carried out at 300 °C and 60 atmospheres pressure (6000 kPa).

a **Give** an advantage and a disadvantage of using a high temperature for this reaction. **[2]**

b Using a high pressure makes the process more costly. Use ideas about equilibrium to explain why a high pressure is used. **[2]**

c Why is a catalyst used for this reaction? **[1]**

d **Predict** the effect of a catalyst on the position of equilibrium. **[1]**

In the industrial process, the gases are passed continuously over the catalyst in a reaction vessel.

e Explain why the reaction never reaches equilibrium. **[2]**

[Total: 8]

COMMAND WORD

describe: state the points of a topic / give characteristics and main features

COMMAND WORDS

give: produce an answer from a given source or recall/ memory

predict: suggest what may happen based on available information

CONTINUED

5 Ammonia is an important chemical manufactured in the Haber process. Nitrogen and hydrogen are reacted in an exothermic reversible reaction.

$N_2(g) + 3H_2(g) \rightleftharpoons 2NH_3(g)$

The process uses an iron catalyst and is carried out at 450 °C and 20 000 kPa pressure.

a How are the nitrogen and hydrogen needed for the process obtained? [2]

b The temperature of 450 °C is a compromise temperature. What is meant by the expression 'compromise temperature'? [2]

c The 20 000 kPa pressure used is also a compromise. A higher pressure would give a higher yield. What is the disadvantage of using a higher pressure? [1]

d Catalysts are not consumed by a reaction. Why is the iron catalyst in the Haber process changed regularly? [1]

e The catalysts in the Haber process and the Contact process are used either in finely divided or pellet form. Explain the advantage of their use in these forms. [1]

[Total: 7]

SELF-EVALUATION CHECKLIST

After studying this chapter, think about how confident you are with the different topics. This will help you see any gaps in your knowledge and help you to learn more effectively.

I can	See Topic...	Needs more work	Almost there	Confident to move on
understand that some chemical reactions are reversible, and their equation includes the symbol $\rightleftharpoons$	9.1			
describe how changing conditions alter the direction of the reaction using the dehydration of hydrated salts as examples	9.1			
describe the use of these reactions as a chemical test for the presence of water	9.1			
state that in a closed system a reversible reaction can reach an equilibrium where the concentrations of reactants and products are no longer changing	9.1			
predict and explain how the position of an equilibrium is affected by various changes in the conditions	9.1			
state the sources for the reactants and the factors affecting the reversible reactions used to produce ammonia in the Haber process and sulfuric acid in the Contact process	9.2			
outline the typical conditions used in the Haber process (450 °C, 20 000 kPa pressure and an iron catalyst) and the Contact process (450 °C, 200 kPa and a vanadium(V) oxide catalyst)	9.2			
explain how the conditions used for the Haber process and the Contact process are used to safely obtain an optimal yield at an economic rate	9.2			
state that ammonium salts and nitrates can be used as fertilisers	9.3			
describe the use of NPK fertilisers to provide the elements nitrogen, phosphorus and potassium for improved plant growth	9.3			

Chapter 10
Redox reactions

IN THIS CHAPTER YOU WILL:

- learn that combustion reactions involve oxidation and reduction (redox) reactions
- define oxidation as the gain of oxygen and reduction as the loss of oxygen
- use Roman numerals to indicate oxidation number

- define oxidation and reduction in terms of loss and gain of electrons
- define the terms oxidising agent and reducing agent
- identify oxidation/oxidising agents and reduction/reducing agents in redox reactions
- understand oxidation numbers, and define oxidation and reduction in terms of an increase and decrease in oxidation number
- use oxidation numbers to identify oxidation/oxidising agents and reduction/reducing agents
- describe the use of colour to identify oxidising or reducing agents and redox reactions.

GETTING STARTED

Oxidation is one of the most important types of reaction in chemistry. You have most likely seen magnesium burn or a sample of hydrogen 'pop' when ignited. What gas in the air is involved in these reactions?

Discuss in groups any other reactions that you have studied that could be classified as oxidation. Write these down in your notes.

Reduction is the opposite of oxidation. How would you define reduction from the experiments you have seen so far in your studies?

FUELLING THE INTERNATIONAL SPACE PROGRAMME

The new generation of space missions is truly international. Many countries, including the 22 member states of the European Space Agency and Russia, have collaborated on the International Space Station. China, Japan and India have ongoing projects related to the Moon and Mars. NASA has re-focused on the potential of reaching Mars, collaborating with private companies such as SpaceX to achieve its aims (Figure 10.1), while countries in Africa and the Middle East are developing small-scale satellite programmes. But what is the chemistry that powers this exploration?

Space exploration is dependent on the energy and explosive thrust produced by a group of powerful oxidation–reduction reactions. The redox reaction between hydrogen and oxygen is highly exothermic and has been used to fuel rockets, most notably the now-retired Space Shuttle. Large tanks beneath the Shuttle contained liquid hydrogen and oxygen. In 1986, cracked rubber seals on the fuel tanks of the shuttle *Challenger* caused a catastrophic explosion and loss of life. Despite that accident, the hydrogen/oxygen combination continues to be used for many of the most powerful rockets.

Another fuel used extensively for powerful rocket launches is known as 'rocket propellant-1' (RP-1). This is a form of highly purified paraffin (kerosene) obtained from petroleum (Chapter 20). Petroleum rocket fuels are usually used in combination with liquid oxygen as the oxidiser.

On a smaller scale, the firing of rockets is needed to reposition satellites and landing modules. For this type of use the fuel used is often hydrazine (N_2H_4). Mixing hydrazine with oxidising agent dinitrogen tetroxide creates a mixture so explosive that no ignition is required. Alternatively, hydrazine can be used by itself in the presence of an iridium catalyst. *The Curiosity Rover*, which has been working on Mars since 2012, used hydrazine to land on the red planet. The basis of all the different areas of space exploration is dependent on the energy and explosive thrust produced by a group of powerful oxidation–reduction reactions.

Discussion questions

1 What type of chemical reaction is useful to drive an engine or launch a rocket? What else, besides energy, would be useful from the reaction to generate explosive power? What physical state do you think the products of an explosive reaction are likely to be in?

2 Hydrazine, N_2H_4, is often used as a rocket propellant. When it burns explosively in oxygen it produces nitrogen and water as the only products. Can you write a balanced symbol equation for this reaction, including state symbols?

Figure 10.1: The SpaceX *Falcon 9* rocket lifts off from Kennedy Space Center carrying two astronauts to the International Space Station.

10.1 Combustion, oxidation and reduction

Combustion reactions are of great importance and can be very useful or destructive. The combustion of natural gas is an important source of energy for homes and industry. Natural gas is mainly methane. The complete combustion of methane produces carbon dioxide and water vapour:

$$\text{methane} + \text{oxygen} \rightarrow \text{carbon dioxide} + \text{water}$$

$$CH_4(g) + 2O_2(g) \rightarrow CO_2(g) + 2H_2O(g)$$

Substances such as methane, which undergo combustion readily and give out a large amount of energy, are known as fuels.

Our bodies need energy to make the reactions that take place in our cells possible. These reactions allow us to carry out our everyday activities. We get this energy from food. During digestion, food is broken down into simpler substances. For example, the carbohydrates in rice, potatoes and bread are broken down to form glucose. The combustion of glucose with oxygen in the cells of our body provides energy:

$$\text{glucose} + \text{oxygen} \rightarrow \text{carbon dioxide} + \text{water}$$

$$C_6H_{12}O_6 + 6O_2 \rightarrow 6CO_2 + 6H_2O$$

This reaction is exothermic and is the overall reaction of the process known as **respiration**.

In combustion reactions, the substance involved is oxidised. Oxygen is added and oxides are formed. Not all reactions with oxygen produce a great amount of energy. For example, when air is passed over heated copper, the surface becomes coated with black copper(II) oxide. There is no flame, nor is the reaction very exothermic. But it is still an oxidation reaction (Figure 10.2a):

$$\text{copper} + \text{oxygen} \xrightarrow{\text{heat}} \text{copper(II) oxide}$$

$$2Cu + O_2 \xrightarrow{\text{heat}} 2CuO$$

This process can be reversed, and the copper surface regenerated, if hydrogen gas is passed over the heated material. The black coating on the surface turns pink as the reaction takes place (Figure 10.2b):

$$\text{copper(II) oxide} + \text{hydrogen} \xrightarrow{\text{heat}} \text{copper} + \text{water}$$

> **KEY WORDS**
>
> **combustion:** a chemical reaction in which a substance reacts with oxygen – the reaction is exothermic
>
> **respiration:** the chemical reaction (a combustion reaction) by which biological cells release the energy stored in glucose for use by the cell or the body; the reaction is exothermic and produces carbon dioxide and water as the chemical by-products

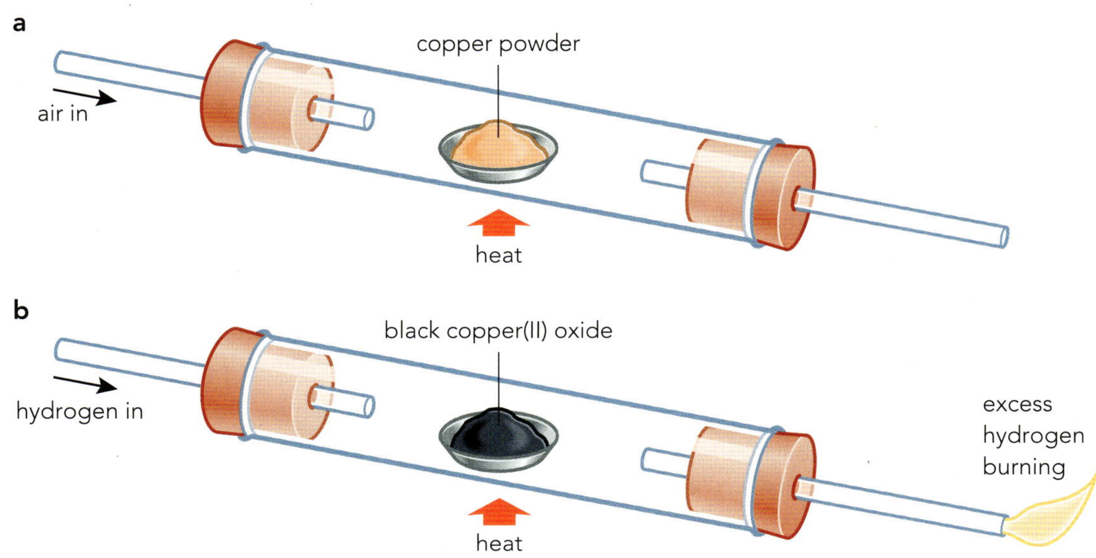

Figure 10.2 a: Oxidation of copper to copper(II) oxide. **b:** Reduction of copper(II) oxide back to copper using hydrogen.

During this reaction, the copper(II) oxide is losing oxygen. The copper(II) oxide is undergoing reduction – it is losing oxygen and being reduced (Figure 10.3). The hydrogen is gaining oxygen; hydrogen is being oxidised.

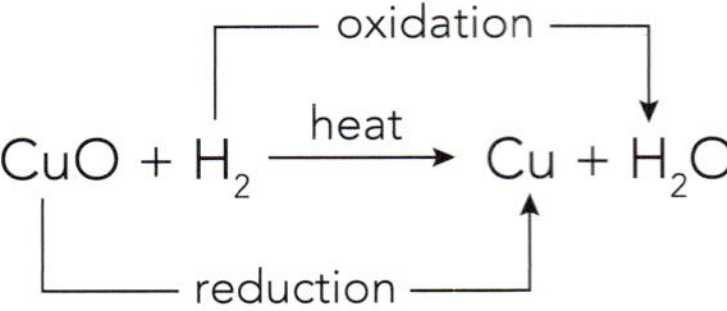

Figure 10.3: Reduction of copper(II) oxide with hydrogen.

It is important that you notice the words we use in describing these reactions:

- if a substance gains oxygen during a reaction, it is *oxidised*
- if a substance loses oxygen during a reaction, it is *reduced*.

Notice that the two processes of oxidation and reduction take place together during the same reaction. This is true for a whole range of similar reactions. Consider the following reaction (Figure 10.4):

zinc oxide + carbon → zinc + carbon monoxide

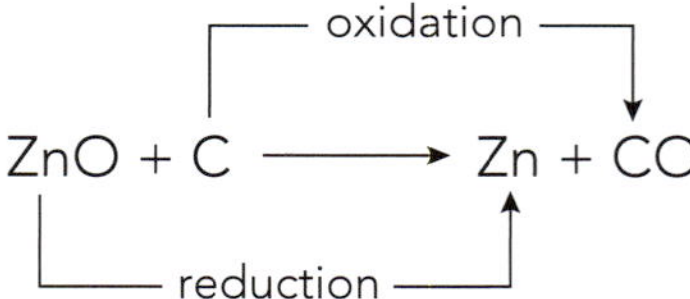

Figure 10.4: Reduction of zinc oxide by carbon.

Again, in this reaction, the two processes occur together. The zinc oxide has been reduced by the carbon. In carrying out this reaction the carbon has itself been oxidised to carbon monoxide.

You will notice from these reactions that oxidation never takes place without simultaneous reduction. It is better to call these reactions oxidation–reduction reactions or **redox reactions**. During redox reactions there is both gain and loss of oxygen. Even in combustion reactions, the oxygen can be thought of as being reduced as it is no longer present as the free element.

Reduction is very important in industry as it provides a way of extracting metals from the metal oxide ores that occur in the Earth's crust. A good example is the blast furnace for extracting iron from hematite (Fe_2O_3) (see Chapter 16). There are two reduction reactions taking place in the blast furnace extraction of iron. First, carbon dioxide formed in the furnace is reduced to carbon monoxide by reaction with carbon:

carbon dioxide + carbon → carbon monoxide

The carbon monoxide formed then reduces the iron(III) oxide to iron in the reaction that is central to the process.

iron(III) oxide + carbon monoxide
→ iron + carbon dioxide

Other moderately reactive metals, such as zinc, lead and copper, can be extracted from their ores by reduction with carbon.

There are two common examples of oxidation reaction that we might meet in our everyday lives.

- **Corrosion**. If a metal is reactive, its surface may be attacked by air, water or other substances. The effect is called corrosion. When iron or steel slowly corrodes in damp air, the product is a brown, flaky substance we call rust. Rust is a form of iron(III) oxide. Rusting weakens structures such as car bodies, iron railings, ships' hulls and bridges. The corrosion of iron and its prevention are discussed in detail in Chapter 16.
- **Rancid**. Oxidation also has damaging effects on food. When the fats and oils in butter and margarine are oxidised, they become rancid. Their taste and smell change and become very unpleasant. Manufacturers sometimes add antioxidants to fatty foods and oils to prevent oxidation. Keeping food in a refrigerator or an airtight containers can slow down the oxidation process.

KEY WORDS

redox reaction: a reaction involving both reduction and oxidation

corrosion: the name given to the process that takes place when metals and alloys are chemically attacked by oxygen, water or any other substances found in their immediate environment

rancid: a term used to describe oxidised organic material (food) – usually involving a bad smell

Questions

Some chemical reactions are listed below.

A hexane + oxygen → carbon dioxide + water

B magnesium + oxygen → magnesium oxide

C calcium carbonate → calcium oxide + carbon dioxide

D magnesium + copper(II) oxide → magnesium oxide + copper

E hydrochloric acid + sodium hydroxide → sodium chloride + water

1. Which of these reactions involve oxidation and reduction?
2. Which of these reactions usually involve burning?
3. What type of reaction has happened to the copper(II) oxide in equation **D**?

In the main reaction in the extraction of iron in the blast furnace (Chapter 15), carbon monoxide removes oxygen from iron(III) oxide. Carbon monoxide is an example of a **reducing agent**. The commonest reducing agents are hydrogen, carbon and carbon monoxide.

Some substances are capable of giving oxygen to others. These substances are known as **oxidising agents** (or sometimes as oxidants). The commonest oxidising agents are oxygen (or air), hydrogen peroxide, potassium manganate(VII) and potassium dichromate(VI).

Remember that, in the process of acting as a reducing agent, the substance will itself be oxidised. The reducing agent will gain the oxygen it is removing from the other compound. The reverse is true for an oxidising agent; the oxidising agent will itself be reduced during the reaction.

KEY WORDS

reducing agent: a substance that reduces another substance during a redox reaction and is itself oxidised during the reaction

oxidising agent: a substance that oxidises another substance during a redox reaction and is itself reduced during the reaction

10.2 Redox reactions

Electron loss and gain in redox reactions

Chemists' ideas about oxidation and reduction have expanded as a wider range of reactions have been studied. Look again at the reaction between copper and oxygen:

$$\text{copper} + \text{oxygen} \xrightarrow{\text{heat}} \text{copper(II) oxide}$$

$$2Cu + O_2 \xrightarrow{\text{heat}} 2CuO$$

It is clear that copper has been oxidised, but what has been reduced? We can apply the ideas behind ionic equations to analyse the changes taking place during this reaction. It then becomes clear that:

- the copper atoms in the metal have become copper ions (Cu^{2+}) in copper(II) oxide
- the oxygen molecules in the gas have split and become oxide ions (O^{2-}) in the black solid copper(II) oxide.

The copper atoms were clearly oxidised during the reaction as they have gained oxygen. However, in the process these copper atoms have lost electrons. Note that the oxygen atoms have gained electrons in the process.

A new, broader definition of oxidation and reduction can now be put forward.

- oxidation is the loss of electrons
- reduction is the gain of electrons.

We can remember this by using the memory aid 'OIL RIG' (Figure 10.5).

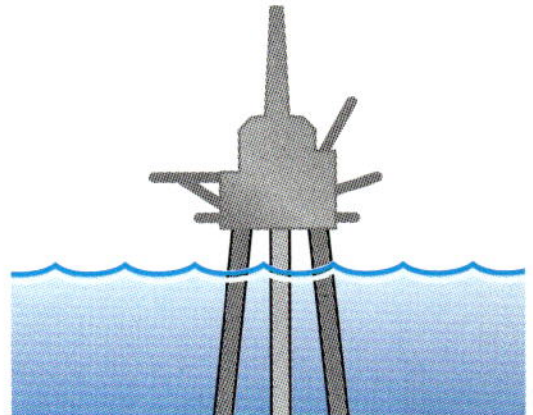

OIL RIG

Oxidation **I**s the **L**oss of electrons

Reduction **I**s the **G**ain of electrons

Figure 10.5: A memory aid for remembering one definition of oxidation and reduction.

This new definition of redox increases the number of reactions that can be called redox reactions. For instance, metal **displacement reactions** where there is no transfer of oxygen are now included (Figure 10.7a). This is best seen by looking at an ionic equation. For example:

$$Zn(s) + CuSO_4(aq) \rightarrow ZnSO_4(aq) + Cu(s)$$

As an ionic equation this becomes (Figure 10.6):

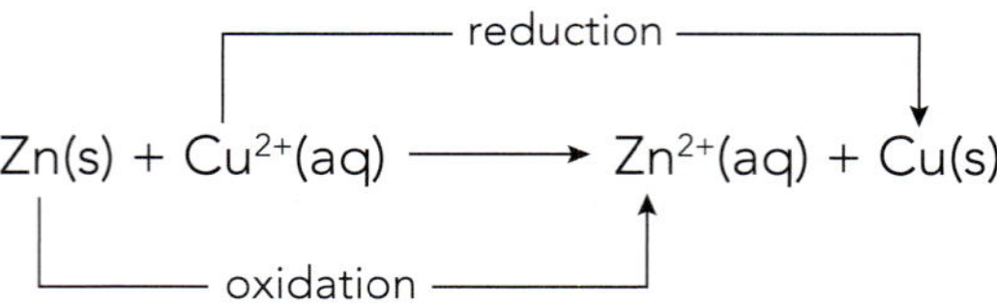

Figure 10.6: Ionic equation for the reaction between zinc and copper(II) sulfate solution is a redox reaction.

Zinc has lost two electrons and copper has gained two electrons. This reaction is a redox reaction as there has been both loss and gain of electrons by different elements during the reaction.

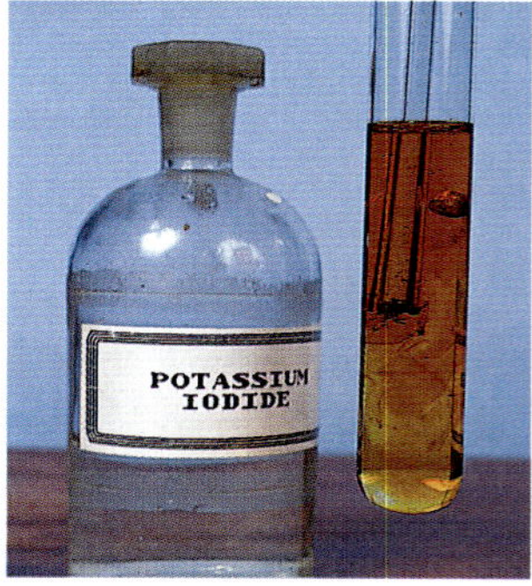

Figure 10.7: Displacement reactions. **a:** reaction between zinc and copper(II) sulfate solution. **b:** chlorine displaces iodine from potassium iodide solution.

It is on the basis of this definition that chlorine, for instance, is a good oxidising agent. It displaces iodine from potassium iodide solution (Figure 10.7b). Is this reaction a redox reaction?

$$Cl_2(aq) + 2I^-(aq) \rightarrow 2Cl^-(aq) + I_2(aq)$$

From the ionic equation we can see that the chlorine atoms of the molecule have gained electrons to become chloride ions; they have been reduced.

$$Cl_2(g) + 2e^- \rightarrow 2Cl^-(aq) \qquad \text{a reduction}$$

The iodide ions have lost electrons to form iodine; they have been oxidised.

$$2I^-(aq) \rightarrow I_2(aq) + 2e^- \qquad \text{an oxidation}$$

As can be seen from this example, equations for redox reactions can be separated into half-equations representing the gain and loss of electrons.

Oxidation and reduction during electrolysis and in fuel cells

The reactions that take place at the electrodes during electrolysis involve the loss and gain of electrons. Negative ions always travel to the anode, where they lose electrons. In contrast, positive ions always flow to the cathode, where they gain electrons. As we saw earlier, oxidation can be defined as the loss of electrons and reduction as the gain of electrons. Therefore, electrolysis (Chapter 6) can be seen as a process in which oxidation and reduction are physically separated.

KEY WORDS

displacement reaction: a reaction in which a more reactive element displaces a less reactive element from a solution of its salt

For example, in the electrolysis of concentrated sodium chloride solution, hydrogen is formed at the cathode (negative electrode) and chlorine is formed at the anode (positive electrode).

At the cathode: $2H^+(aq) + 2e^- \rightarrow H_2(g)$ a reduction

At the anode: $2Cl^-(aq) \rightarrow Cl_2(g) + 2e^-$ an oxidation

During electrolysis:

- the oxidation of non-metal ions always takes place at the anode
- the reduction of metal or hydrogen ions always takes place at the cathode.

This general rule for electrolysis can be remembered by extending the 'OIL RIG' memory aid to 'AN OIL RIG CAT'; reminding you of the electrode where each change takes place. Just as in redox reactions, where both oxidation and reduction occur at the same time, the two processes must take place together to produce electrolytic decomposition of an electrolyte.

A redox reaction also takes place in a hydrogen–oxygen fuel cell. This fuel cell generates energy by controlling the highly exothermic oxidation reaction of hydrogen to produce water (Figure 10.8).

$$2H_2(g) + O_2(g) \rightarrow 2H_2O(g)$$

Figure 10.8 a: Balloon filled with hydrogen. **b:** Hydrogen reacts explosively when ignited.

The hydrogen–oxygen fuel cell is discussed further in Chapter 6.

Questions

4 Define the following terms, giving one example of each.
   a an oxidising agent
   b a reducing agent

5 Complete the following statement:
   __________ is the gain of electrons; __________ is the loss of electrons. During a redox reaction the oxidising agent __________ electrons; the oxidising agent is itself __________ during the reaction.

6 Redox means *reduction and oxidation*. It can be defined by loss and gain of oxygen or by loss and gain of electrons.
   a Which definition is more useful?
   b Is it possible to have oxidation without reduction in a chemical reaction? Explain your answer.

Oxidation numbers

Throughout this book, you will have noticed that the names of several compounds are written with Roman numerals in them, most often after the names of **transition metals**. Examples of such compounds include copper(II) sulfate, iron(III) oxide and cobalt(II) chloride. The reason for these numbers is that transition metal elements show variable *valency*. Iron in iron(III) oxide is present as the Fe^{3+} ion, copper in copper(II) sulfate is present as the Cu^{2+} ion and so on. These metals can form other ions in different compounds. Iron can be found as Fe^{2+} ions; copper as Cu^{+} ions (see Table 10.1 for examples of compounds). This number is known as the **oxidation number** (or oxidation state) of the element in a compound.

KEY WORDS

transition metals (transition elements): elements from the central region of the Periodic Table – they are hard, strong, dense metals that form compounds that are often coloured

oxidation number: a number given to show whether an element has been oxidised or reduced; the oxidation number of a simple ion is simply the charge on the ion

Transition metal	Oxidation number of metal	Examples of compounds
copper	+1	copper(I) oxide, CuO_2
	+2	copper(II) oxide, CuO
iron	+2	iron(II) chloride, $FeCl_2$
	+3	iron(III) chloride, $FeCl_3$
manganese	+2	manganese(II) oxide, MnO
	+4	manganese(IV) oxide, MnO_2
	+7	manganese(VII) oxide, Mn_2O_7

Table 10.1: Some examples of the variable oxidation numbers of transition metals.

The oxidation number is a number given to each atom in a compound to show how oxidised or reduced an element is in a compound. As we have seen, some elements can have more than one oxidation number in their compounds (Table 10.1 and Figure 10.9).

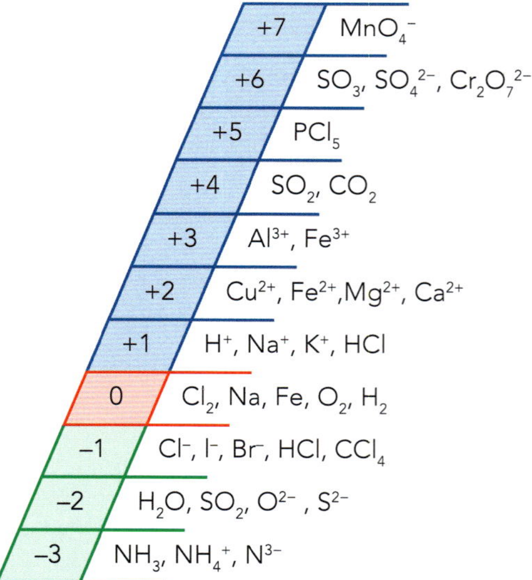

Figure 10.9: The range of oxidation numbers shown by various elements. Note how oxidation numbers are written, with the sign first followed by the number.

Figure 10.9 shows the range of oxidation numbers that various elements can have in their compounds, and also some simple rules for deciding the usual value for certain elements. Table 10.2 gives more detail of the rules we can use to find the value of the oxidation number if an element in a particular compound.

Rule	Example
The oxidation number of the uncombined element is zero (0)	In H_2 the oxidation number of H is 0 In Cl_2 the oxidation number of Cl is 0
The oxidation number of the element in a simple monoatomic ion is the charge in the ion	In Zn^{2+} the oxidation number of Zn is +2 In O^{2-} the oxidation number of O is −2
The oxidation number of hydrogen is usually +1	In HCl, the oxidation number of H is +1 and the oxidation number of Cl is −1
The oxidation number of oxygen is usually −2	In water, the oxidation number of H is +1 and the oxidation number of O is −2
The sum of the oxidation numbers in a compound is zero (0)	In water (H_2O), the oxidation number of H is +1 and the oxidation number of O is −2 $[2 \times (+1)] + [1 \times (-2)] = 0$
The sum of the oxidation numbers in an ion is equal to the charge on the ion	In the manganate(VII) ion (MnO_4^-), the oxidation number of Mn is +7 $[1 \times (+7)] + [4 \times (-2)] = -1$

Table 10.2: Rules for working out the oxidation number of an element.

Oxidation numbers and redox reactions

Oxidation numbers are not only important in naming certain compounds. They can also be very useful in deciding whether oxidation or reduction has taken place during a particular reaction. This is because if the oxidation number of an element increases during a reaction, then that element is oxidised during that reaction. However, if the oxidation number decreases then the element is reduced.

If we look closely at the reactions we have considered earlier we can see that a further definition of oxidation is possible. In the metal displacement reaction, each zinc atom has lost two electrons and the copper ions have gained two electrons. The loss or gain of electrons in the reaction means that changes in the oxidation number have taken place, as shown in Figure 10.10.

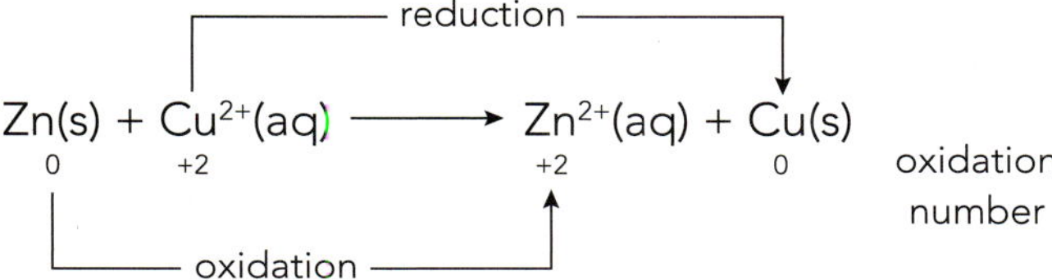

Figure 10.10: Changes in oxidation number during metal displacement reaction.

During the reaction, the oxidation number of zinc has increased by 2, from 0 to +2. Meanwhile the oxidation number of copper has decreased by 2, from +2 to 0. Zinc has acted as a reducing agent as it has decreased the oxidation number of the copper ions. Copper ions (Cu^{2+}) have oxidised the zinc atoms to Zn^{2+} ions.

In the halogen displacement reaction, chlorine displaces iodine from potassium iodide solution. Consider the changes in oxidation number during the reaction (Figure 10.11):

- The oxidation number of iodine changes from −1 to 0. It has increased. Iodide ions are *oxidised* to iodine.
- The oxidation number of chlorine changes from 0 to −1. It has decreased. Chlorine is *reduced* to chloride ions.

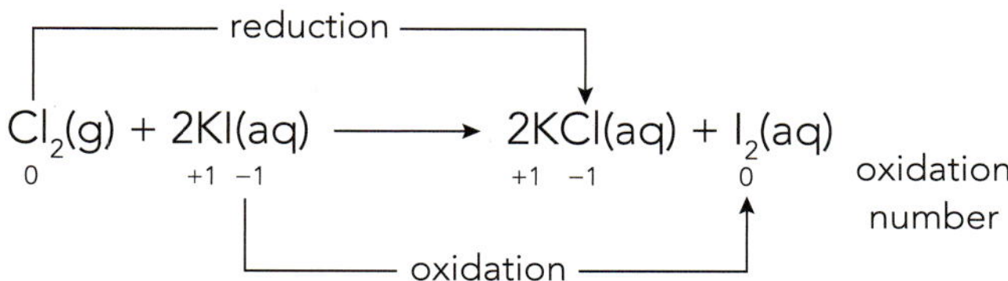

Figure 10.11: Changes in oxidation number during halogen displacement reaction.

This discussion leads to a further definition of oxidation and reduction:

- **oxidation** is an *increase* in oxidation number of an atom or ion during a reaction
- **reduction** is a *decrease* in oxidation number of an atom or ion during a reaction.

REFLECTION

Think about what you have found easy or difficult when learning about the different definitions of redox reactions. Are you clear about the progression from one definition to the next?

KEY WORDS

oxidation: there are three definitions of oxidation:

i a reaction in which oxygen is added to an element or compound
ii a reaction involving the loss of electrons from an atom, molecule or ion
iii a reaction in which the oxidation state of an element is increased

reduction: there are three definitions of reduction:

i a reaction in which oxygen is removed from a compound
ii a reaction involving the gain of electrons by an atom, molecule or ion
iii a reaction in which the oxidation state of an element is decreased

Colour tests for oxidising and reducing agents

Reactions involving potassium iodide can be very useful as a test for any oxidising agent, because a colour change is produced. The iodide ion (I^-) is oxidised to iodine (I_2). The colour of the solution changes from colourless to yellow–brown (see Figure 10.6b). If starch indicator is added, then a dark blue colour is produced because of the presence of the iodine.

Reactions involving acidified potassium manganate(VII) are useful for detecting a reducing agent. The manganese is in a very high oxidation state (oxidation number +7) in the manganate(VII) ion (MnO_4^-). A solution containing the manganate(VII) ion has a purple colour. When it is reduced, the manganate(VII) ion loses its purple colour and the solution appears colourless because of the formation of the pale pink Mn^{2+} ion (Figure 10.12). In the test the oxidation number of manganese is reduced from +7 to +2.

Figure 10.12: Testing for a reducing agent using acidified potassium manganate(VII) solution.

Questions

7 What is the oxidation number of the element underlined in the following elements, compounds and ions?

a $\underline{Al}^{3+}$

b $\underline{Cl}O_3^-$

c $\underline{O}_3$

d $\underline{P}Cl_3$

e $\underline{Cr}_2O_7^{2-}$

8 Which is the oxidising agent in each of the following reactions?

a $Mg(s) + ZnSO_4(aq) \rightarrow MgSO_4(aq) + Zn(s)$

b $Br_2(aq) + 2KI(aq) \rightarrow 2KBr(aq) + I_2(aq)$

c $5Fe^{2+}(aq) + MnO_4^-(aq) + 8H^+(aq) \rightarrow 5Fe^{3+}(aq) + Mn^{2+}(aq) + 4H_2O(l)$

9 Describe the colour change you would see in reactions **b** and **c** in Question 8.

ACTIVITY 10.1

Never a Cross Word

'Criss-cross' puzzles are a popular puzzle format where words are fitted together in a unique way to create a grid. A simple example is shown in Figure 10.13.

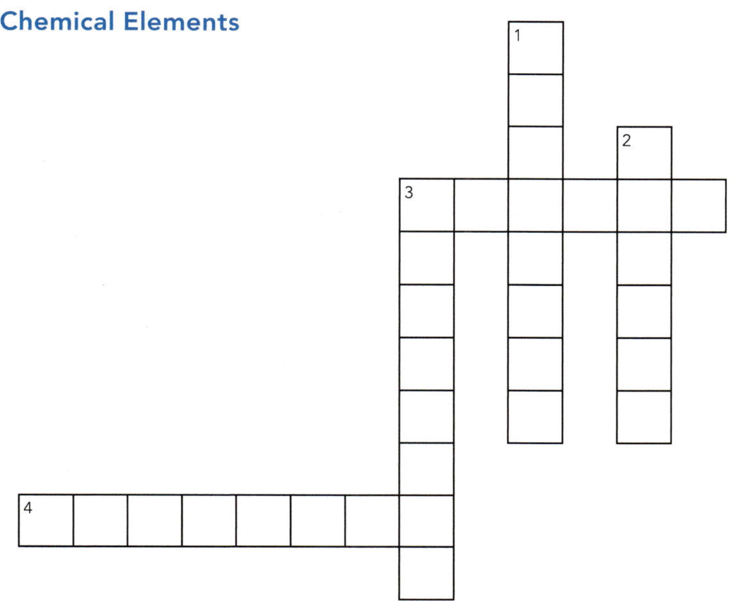

Across
3 This element has the symbol C (6 letters)
4 This is the first element on the periodic table (8 letters)

Down
1 This element makes up almost 88% of the atmosphere (8 letters)
2 Has the symbol Na (6 letters)
3 A toxic green gas found in Group VII (8 letters)

Figure 10.13: An example of a 'criss-cross' puzzle.

Create a 'criss-cross' puzzle based on the redox reactions.

1. Find a minimum of 10 important words or short phrases in this chapter. You should be creative, e.g. 'reducing agent' could be as a single line of 13 boxes with the clue stating '(8,5 letters)' or could be split across two areas of the crossword as '1 down (8 letters)' and '4 across (5 letters)'.
2. Write down the clues. These could be definitions or sentences that include a blank space for each word/phrase you have chosen.
3. Assemble a 'criss-cross' grid to fit your words/phrases (similar to the example shown in Figure 10.13). You could use various online puzzle maker apps to help you do this.

When you have completed your puzzle, challenge your partner to a race. The winner is the person to complete the other person's puzzle the quickest.

Peer assessment

When completing your partner's criss-cross puzzle, did you know all the definitions or clues? Which words did you find the easiest? Which words or clues did you find the most difficult?

SUMMARY

Combustion reactions such as the burning of natural gas involve oxidation and reduction reactions.
Oxidation can be defined as the gain of oxygen by and element or compound, and reduction the loss of oxygen, during a chemical reaction.
Redox reactions involve simultaneous oxidation and reduction, and can identify the gain and loss of oxygen during a reaction.
Oxidation can be defined as the loss of electrons during a reaction and reduction as the gain of electrons during a reaction.
An oxidising agent is a substance that oxidises another substance and is itself reduced during the redox reaction.
A reducing agent is a substance that reduces another and is itself oxidised during the reaction.
A reducing agent will gain the oxygen it is removing from the other compound, or lose electrons during the reaction.
An oxidising agent will lose the oxygen, or gain electrons, during the redox reaction.
Roman numerals are used to indicate the oxidation number of an element in the formula of a compound.
Oxidation numbers are given to atoms of elements in a compound or ion.
Oxidation can be defined as the increase in oxidation number of an element in a compound, and reduction a decrease in oxidation number, during a reaction.
Oxidation numbers can be used to identify oxidation and oxidising agents, and reduction and reducing agents, in the equations of redox reactions.
Colour tests involving potassium iodide or acidified potassium manganate(VII) solutions can be used to identify oxidising or reducing agents and redox reactions.

PROJECT

I'd like to teach the world to ...

An educational broadcaster wants to expand their selection of chemistry-based materials and needs your help to produce either a podcast or a simple video.

In this chapter, you have learnt about redox reactions. In particular, you have covered what is meant by oxidation in terms of the gain of oxygen, e.g. magnesium is oxidised to form magnesium oxide, or the loss of electrons, e.g. aluminium atoms are oxidised by the loss of three electrons to form aluminium ions (Al^{3+}). You have also learnt what is meant by reduction and seen how reduction and oxidation occur in parallel through redox reactions.

You can choose to produce:

1 A podcast. To do this you must initially produce an interview between a 'reporter' and a 'chemist'. The reporter needs to ask questions to the chemist on the topic of redox chemistry. The questions need to be designed so that the listener develops their understanding of redox reactions during the report. Having written the script, rehearse it and then record it in the form of a podcast.

2 A video. To do this, you will need to produce a set of slides that give an overview of what is meant by reduction, oxidation and redox. You should then record yourself presenting your slides and perhaps share it with your teacher.

EXAM-STYLE QUESTIONS

1 The equations for four different chemical reactions are listed here:

A $\underline{CaCO_3} \rightarrow CaO + CO_2$

B $\underline{Fe_2O_3} + 3CO \rightarrow 2Fe + 3CO_2$

C $\underline{CH_4} + 2O_2 \rightarrow CO_2 + 2H_2O$

D $\underline{Ca(OH)_2} + H_2SO_4 \rightarrow CaSO_4 + 2H_2O$

a **State** the reaction in which the underlined substance is oxidised. [1]

b In which reaction is the underlined substance reduced? [1]

[Total: 2]

2 The reaction shown occurs in catalytic converters.

$2NO_2(g) + 4CO(g) \rightarrow 4CO_2(g) + N_2(g)$

Which statement about this reaction is true?

A Carbon monoxide is an oxidising agent and is oxidised to carbon dioxide.

B Carbon monoxide is a reducing agent and is oxidised.

C Nitrogen dioxide is an oxidising agent and oxidises nitrogen.

D Nitrogen dioxide is a reducing agent and reduces carbon monoxide. [1]

3 The equations for six different chemical reactions are listed here:

A $2CaO + C \rightarrow Ca + CO_2$

B $2KBr + Cl_2 \rightarrow 2KCl + Br_2$

C $2CuI \rightarrow Cu + CuI_2$

D $2NaOH + H_2SO_4 \rightarrow Na_2SO_4 + 2H_2O$

E $MnO_2 + 4HCl \rightarrow MnCl_2 + 2H_2O$

F $2Mg + O_2 \rightarrow 2MgO$

Use these reactions to answer the questions that follow. Each alternative may be used once, more than once, or not at all.

Which reaction:

a shows a metal being oxidised? [1]

b shows a non-metal acting as a reducing agent? [1]

c shows a substance being both reduced and oxidised? [1]

d shows a non-metal acting as an oxidising agent? [1]

e shows a reaction that is not a redox reaction? [1]

f shows a substance changing its oxidation number from (+4) to (+2)? [1]

g shows a metal acting as a reducing agent? [1]

[Total: 7]

COMMAND WORD

state: express in clear terms

CONTINUED

4 In an experiment, a purple solution of potassium manganate(VII) was added to a pale-green mixture of iron(II) sulfate and sulfuric acid. The purple colour disappeared and the solution turned pale yellow.

- **a** Which substance was reduced? **[1]**
- **b** Which substance was oxidised? **[1]**
- **c** Write the formula of the ion of iron that was produced. **[1]**
- **d** In a second experiment, chlorine gas was bubbled through colourless potassium iodide solution. The solution turned yellow/orange. Explain this observation using ideas about oxidation and reduction. **[2]**

[Total: 5]

5 In each substance listed (a–f), **give** the oxidation number of the underlined element.

- **a** $\underline{C}O$
- **b** $H_2\underline{S}$
- **c** $\underline{Zn}$
- **d** $\underline{S}O_2$
- **e** $\underline{N}H_3$
- **f** $H_2\underline{S}O_4$ **[6]**

COMMAND WORD

give: produce an answer from a given source or recall/memory

SELF-EVALUATION CHECKLIST

After studying this chapter, think about how confident you are with the different topics. This will help you see any gaps in your knowledge and help you to learn more effectively.

I can	See Topic...	Needs more work	Almost there	Confident to move on
state that combustion reactions such as the burning of natural gas involve oxidation and reduction reactions	10.1			
define oxidation as the gain of oxygen by an element or compound and reduction as the loss if oxygen by a compound	10.1			
define redox reactions as involving simultaneous oxidation and reduction, and identify the gain or loss of oxygen during a reaction	10.1			
understand that oxidation is the loss of electrons in a reaction and reduction is the gain of electrons in a reaction	10.2			

CONTINUED

I can	See Topic...	Needs more work	Almost there	Confident to move on
understand that an oxidising agent is a substance that oxidises another substance and is itself reduced during the reaction	10.2			
define a reducing agent is a substance that reduces another substance and is itself oxidised during the reaction	10.2			
identify oxidation and oxidising agents, and reduction and reducing agents, in the equations of redox reactions	10.2			
understand how oxidation numbers are given to the atoms of elements in compounds and ions	10.2			
define oxidation as the increase in oxidation number of an element in a compound, and reduction as the decrease in oxidation number during a reaction	10.2			
use Roman numerals to indicate the oxidation number of an element in the formula of a compound	10.2			
use oxidation numbers to identify oxidation and oxidising agents, and reduction and reducing agents, in the equations for redox reactions	10.2			
describe the use of colour tests involving potassium iodide or acidified potassium manganate(VII) solutions to identify oxidising or reducing agents and redox reactions	10.2			

Chapter 11

Acids and bases

IN THIS CHAPTER YOU WILL:

- describe acids and alkalis in terms of their effect on indicators
- learn that bases are the oxides and hydroxides of metals; those bases that are soluble are referred to as alkalis
- learn that aqueous solutions of acids contain an excess of hydrogen ions, while alkaline solutions contain an excess of hydroxide ions
- compare the relative acidity or alkalinity, hydrogen ion concentration or pH of a solution using universal indicator
- describe how acids and alkalis react together in neutralisation reactions, and that bases displace ammonia from ammonium salts
- learn that all metal oxides and hydroxides can act as bases, while many oxides of non-metals can be classified as acidic oxides
- describe the characteristic reactions of acids

- learn that some metal oxides (amphoteric oxides) can react with both acids and alkalis
- define strong and weak acids in terms of ion dissociation
- define an acid as a proton donor and a base as a proton acceptor.

GETTING STARTED

On a visit to a local grocery store or supermarket you can find numerous examples of everyday items that involve acids or alkalis. Get together in groups and draw up a list of such items and divide them into the two categories: acids and alkalis.

Create a table of the two categories and discuss the following questions:

1 What are the different items in your table used for?

2 How safe do you think these items are?

THE SIGNIFICANCE OF LANGUAGE

Chemistry was a field of study in ancient Egypt and classical Greece. The name *kimia* came into use in these cultures and Arabic-speaking scientists modified this to *al-Kimya*. The first ideas on acidity were put forward by the Ancient Greeks and Egyptians when they defined sour-tasting substances as acids. The Greeks and the Egyptians had both discovered that one substance particularly was very sour. This sour substance was produced following the fermentation of fruits and the subsequent air oxidation of the product formed. This substance became known as *vinegar* (a solution of *ethanoic acid*). Consequently, there developed a new category of substances, termed *acids*, that included all things that were sour, or sharp tasting, including lemon juice (see Figure 11.2). The Latin word *acidus* means sour. The eighth century Islamic scholar, Jabir ibn-Hayyan, is often described as the 'father of chemistry' and he is credited with the discovery of the three strong acids: sulfuric, hydrochloric and nitric acids.

The Greeks developed things further by identifying three slippery substances left behind as residue after certain materials were burnt. Potash, from burning wood, and lime, from burning seashells, were among these substances. A new terminology grew up based on these physical observations about substances. The term *alkali* was derived from the Arabic word *al-qali*, meaning 'the ashes', and was used to talk about substances that felt slippery, or soapy, to the touch. A major industry of soap-making grew out of some of the substances made using alkalis.

Indeed soap making has a long pre-industrial history and the processes involved have been exploited in different parts of the world. Black soap is a handmade soap from natural raw materials that has been used for centuries throughout Western Africa (Figure 11.1).

Figure 11.1: Making black soap in a Nigerian village.

The soap making process uses unrefined shea butter, local coconut oil or palm oil. These oils and fats are heated with the ashes of native African plant materials (e.g. plantain leaves and cocoa pods). The ashes are alkaline and act as a source of potassium hydroxide. Acid–base theory has a long and multicultural history, and its vocabulary has been influenced by the languages used by the scientists involved.

Discussion questions

1 Can you think of other words in science and mathematics that have an Arabic origin? Was there a 'golden era' of Arabic and Islamic science when such words were introduced?

2 Making soap involves using strong alkali. Why is it important that excess alkali is removed before the soap is used? What pH are most soaps you can buy?

11.1 The nature of acids and bases

What is an acid?

Vinegar, lemon juice, grapefruit juice and spoilt milk are all sour tasting because of the presence of acids (Figure 11.2). These **acids** are present in animal and plant material and are known as organic acids (Table 11.1).

Carbonic acid from carbon dioxide dissolved in water is present in soft fizzy drinks. The acids present in these circumstances are weak and dilute. But taste is not a test that should be tried – some acids would be dangerous, even deadly, to taste!

A number of acids are also **corrosive**. They can eat their way through clothing, are dangerous on the skin, and some are able to attack stonework and metals. These powerful acids are often called mineral acids (Table 11.1).

Table 11.1 shows how commonly acids occur.

Figure 11.2: Citrus fruits have an 'acidic' sharp taste.

KEY WORDS

acid: a substance that dissolves in water, producing $H^+(aq)$ ions – a solution of an acid turns litmus red and has a pH below 7.
Acids act as proton donors.

corrosive: a corrosive substance (e.g. an acid) is one that can dissolve or 'eat away' at other materials (e.g. wood, metals or human skin)

Type	Name	Formula	Strong or weak?	Where found or used
Organic acids	ethanoic acid	CH_3COOH	weak	in vinegar
	methanoic acid	$HCOOH$	weak	in ant and nettle stings; used in kettle descaler
	lactic acid	$CH_3CH(OH)CO_2H$	weak	in sour milk
	citric acid	C_6H_8O7	weak	in lemons, oranges and other citrus fruits
Mineral acids	carbonic acid	H_2CO_3	weak	in fizzy soft drinks
	hydrochloric acid	HCl	strong	used in cleaning metal surfaces; found as the dilute acid in the stomach
	nitric acid	HNO_3	strong	used in making fertilisers and explosives
	sulfuric acid	H_2SO_4	strong	in car batteries; used in making fertilisers, paints and detergents
	phosphoric acid	H_3PO_4	strong	in anti-rust paint; used in making fertilisers

Table 11.1: Some common acids.

Indicators

The easiest way to detect whether a solution is acidic or not is to use an **indicator**. Indicators are substances that change colour if they are put into an acid or alkaline solution. Three commonly used indicators are **litmus**, **thymolphthalein** and **methyl orange**.

Litmus is extracted from lichens and is purple in neutral solution. When added to an acidic solution, it turns red. This colour change of litmus is the result of a chemical reaction. Substances with the opposite chemical effect to acids are needed to reverse the change, and these are called **bases**. Bases turn litmus solution blue.

You can also use litmus paper. This is paper that has been soaked in litmus solution. It is available in blue and red forms. The blue form of litmus paper changes colour to red when dipped into acid solutions. Red litmus paper turns blue in alkaline solutions. **Alkalis** are soluble bases. Note that litmus only gives a single colour change.

Figure 11.3 shows a simple visual memory aid to help you to remember the colour change that litmus shows for acids and bases.

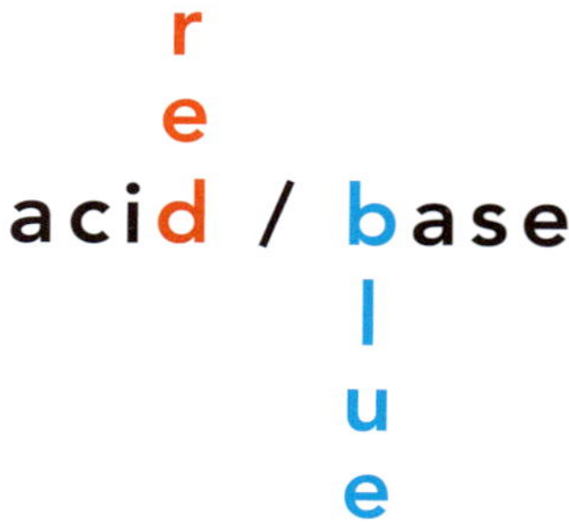

Figure 11.3: The colour change of litmus in acid (red) and base (blue).

The presence of water is very important in the action of acids and alkalis. One practical consequence of this is that, when we use litmus paper to test gases, it must always be damp. The gas needs to dissolve in the moisture to bring about the colour change. This is important in your practical work (see the tests for different gases in Chapter 22).

Other frequently used indicators include thymolphthalein and methyl orange. These substances give different colour changes from litmus (Table 11.2) and the colour changes are sometimes easier to detect than for litmus.

Indicator	Colour in acid	Neutral colour	Colour in alkali
litmus	red	purple	blue
thymolphthalein	colourless	colourless	blue
methyl orange	red	orange	yellow

Table 11.2: Some common indicator colour changes.

Universal indicator

Another commonly used indicator, **universal indicator** (or full-range indicator), is a mixture of indicator dyes. Such an indicator is useful because it gives a range of colours (a 'spectrum') depending on the relative strength of the acid or alkali added (Figure 11.4). When you use universal indicator paper, you see that solutions of different acids produce different colours depending on their relative acidity. Solutions of the same acid with different concentrations will also give different colours.

The more acidic solutions (e.g. battery acid) turn universal indicator bright red. A less acidic solution (e.g. vinegar) will only turn universal indicator orange–yellow. There are also colour differences produced with different alkali solutions. The most alkaline solutions give a violet colour.

KEY WORDS

indicator: a substance that changes colour when added to acidic or alkaline solutions, e.g. litmus or phenolphthalein

litmus: the most common indicator; turns red in acid and blue in alkali

thymolphthalein: an acid–base indicator that is colourless in acidic solutions and blue in alkaline solutions

methyl orange: an acid–base indicator that is red in acidic and yellow in alkaline solutions

base: a substance that neutralises an acid, producing a salt and water as the only products. Bases act as proton acceptors.

alkalis: soluble bases that produce $OH^-(aq)$ ions in water – a solution of an alkali turns litmus blue and has a pH above 7

universal indicator: a mixture of indicators that has different colours in solutions of different pH

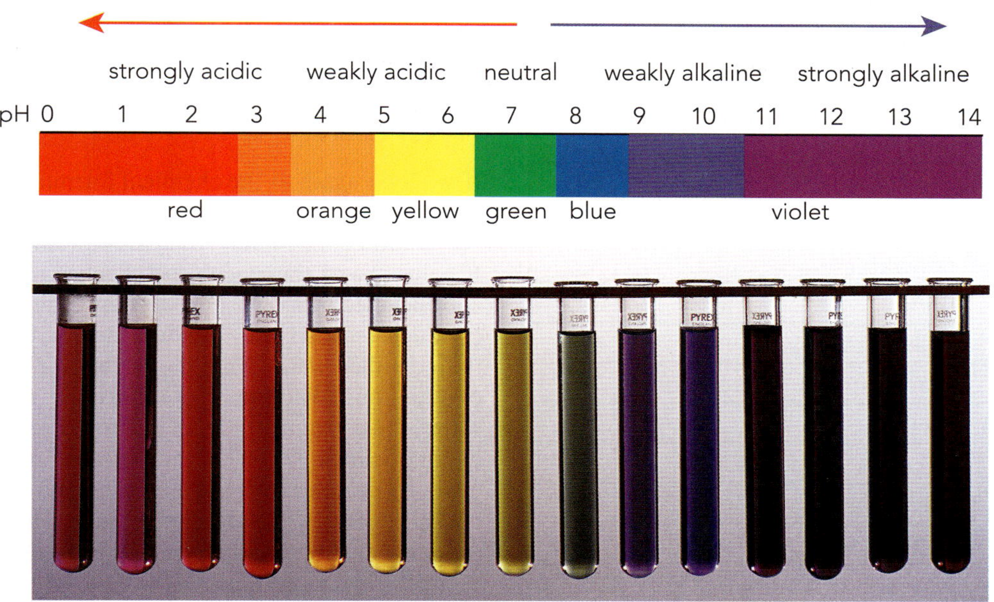

Figure 11.4: How the colour of universal indicator changes in solutions of different pH values.

The pH scale

The most useful measure of the relative strength of an acid or alkaline solution is the **pH scale**. The scale runs from 1 to 14 (Figure 11.4), and the following general rules apply:

- acids have a pH less than 7
- the more acidic a solution, the lower the pH
- neutral substances, such as pure water, have a pH of 7
- alkalis have a pH greater than 7.

The pH of a solution can be measured in several ways. Universal indicator papers that are sensitive over the full range of values can be used. Alternatively, if the approximate pH value is known, then we can use a more accurate test paper that is sensitive over a narrow range. The most accurate method is to use a pH meter (Figure 11.5), which uses an electrode to measure pH. The pH values of some common solutions are shown in Table 11.3.

> **KEY WORDS**
>
> **pH scale:** a scale running from below 0 to 14, used for expressing the acidity or alkalinity of a solution; a neutral solution has a pH of 7

Figure 11.5: pH meter for in use in the laboratory.

It is very important to remember that the 'reference point' when measuring pH is neutrality, pH 7 – the mid-point of the scale. Therefore:

- As we move down from pH 7, the solution is getting more acidic.
- Moving up from pH 7, the solution is getting more alkaline.

	Substance	pH
highly acidic	hydrochloric acid (HCl)	0.0
↑	gastric juices	1.0
	lemon juice	2.5
	vinegar	3.0
	acid rain	4.4
	rainwater	5.6
↓	urine	6.0
poorly acidic	milk	6.5
NEUTRAL	pure water, sugar solution	7.0
poorly alkaline	blood	7.4
↑	baking soda solution	8.5
	toothpaste	9.0
	limewater	11.0
↓	household ammonia	12.0
highly alkaline	sodium hydroxide (NaOH)	14.0

Table 11.3: The pH values of some common solutions.

It is important to realise that the pH scale is a logarithmic scale in which two adjacent values change by a factor of 10. Each pH unit means a 10-fold difference in H^+ ion concentration. An acid of pH 1.0 has 10 times the H^+ ion concentration of an acid of pH 2.0.

Questions

1 **a** What do you understand by the word corrosive?
 b Which acid is present in orange or lemon juice?
 c What acid is present in vinegar?

2 **a** Methyl orange is an indicator. What does this mean?
 b Is a solution acidic, alkaline or neutral if its pH is:
 i 11 **ii** 7 **iii** 8 **iv** 3?

3 Which solution is more acidic: an acid with a pH of 4 or an acid with a pH of 1?

Acid and alkali solutions: the importance of hydrogen and hydroxide ions

If we look again at the chemical formulae of some of the best-known acids (Table 11.1), we see that one element is common to them all. All acids contain hydrogen. If solutions of these acids are checked to see if they conduct electricity, we find that they are all electrical conductors. Also, they conduct electricity much better than distilled water. This shows that the solutions contain ions. Water itself contains very few ions. In pure water, the concentrations of hydrogen ions (H^+) and hydroxide ions (OH^-) are equal. All acids dissolve in water to produce H^+ ions. Therefore, all acid solutions contain more H^+ ions than OH^- ions. The pH scale is designed around the fact that acid solutions have this excess of hydrogen ions

Alkali solutions also conduct electricity better than distilled water. All alkalis dissolve in water to produce OH^- ions. Therefore, all alkali solutions contain an excess of OH^- ions. An indicator such as litmus, is affected by the presence of H^+ or OH^- ions (Figure 11.6):

- the hydrogen ions (H^+) in acid solutions turn litmus red
- the hydroxide ions (OH^-) in alkali solutions turn litmus blue.

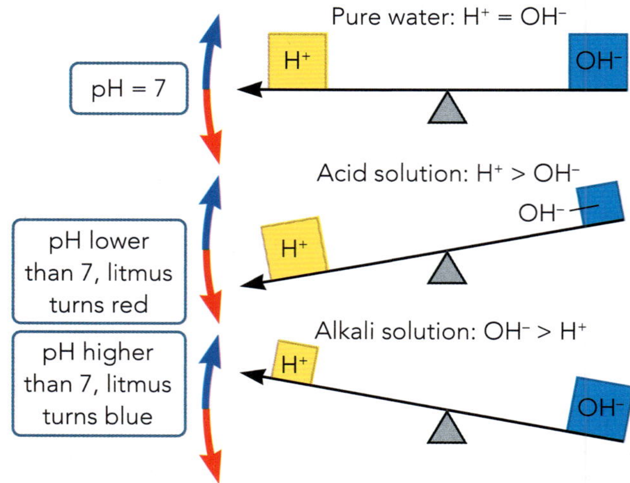

Figure 11.6: pH and the balance of hydrogen ions and hydroxide ions in solution.

The ions present in some important acid and alkali solutions are given in Table 11.4.

	Name	Ions present
Acids	hydrochloric acid	$H^+(aq)$ and $Cl^-(aq)$
	nitric acid	$H^+(aq)$ and $NO_3^-(aq)$
	sulfuric acid	$H^+(aq)$, $HSO_4^-(aq)$ and $SO_4^{2-}(aq)$
Alkalis	sodium hydroxide (caustic soda)	$Na^+(aq)$ and $OH^-(aq)$
	potassium hydroxide (caustic potash)	$K^+(aq)$ and $OH^-(aq)$
	calcium hydroxide (limewater)	$Ca^{2+}(aq)$ and $OH^-(aq)$
	ammonia solution (ammonium hydroxide)	$NH_4^+(aq)$ and $OH^-(aq)$

Table 11.4: Ions present in solutions of some acids and alkalis.

Alkalis and bases

What types of substance are alkalis and bases?

Alkalis are substances that dissolve in water to give solutions with a pH greater than 7 and turn litmus blue. The solutions contain an excess of hydroxide, OH^-, ions. If an alkali is added to an acid then the effect of the hydroxide ions will neutralise the acid's excess H^+ ions. This neutralisation reaction between an acid and an alkali can be represented by the equation:

$$H^+(aq) + OH^-(aq) \rightarrow H_2O(l)$$

When investigated further, it was found that all metal oxides and hydroxides would neutralise acids, whether they dissolve in water or not. Therefore, the soluble alkalis are just a small part of a group of substances – the oxides and hydroxides of metals – that neutralise acids. These substances are known as bases (Figure 11.7). These bases all react in the same way with acids. A base will neutralise an acid, and in the process a salt is formed. This type of reaction is known as a **neutralisation** reaction. It can be summed up in a general equation:

$$\text{acid} + \text{base} \rightarrow \text{salt} + \text{water}$$

Most bases are insoluble in water. This makes the few bases that do dissolve in water more significant. They are given a special name – alkalis. Alkalis are generally used in the laboratory as aqueous solutions. The common alkalis are shown in Table 11.4.

Figure 11.7: This Venn diagram shows the relationship between bases and alkalis. All alkalis are bases, but not all bases are alkalis.

Our stomachs contain dilute hydrochloric acid to help digest our food. However, excess acid causes indigestion, which can be painful and eventually give rise to ulcers. To ease this, we can take an antacid treatment. **Antacids** (or 'anti-acids') are a group of bases such as magnesium hydroxide, or carbonates, with no toxic effects on the body. They are used to neutralise the effects of acid indigestion.

KEY WORDS

neutralisation: a chemical reaction between an acid and a base to produce a salt and water only; summarised by the ionic equation $H^+(aq) + OH^-(aq) \rightarrow H_2O(l)$

antacids: compounds used medically to treat indigestion by neutralising excess stomach acid

Properties and uses of alkalis and bases

Alkalis feel soapy to the skin. Alkalis react with the oils in your skin into soap. They are used as degreasing agents because they convert oil and grease into soluble soaps, which can be washed away easily. The common uses of some alkalis and bases are shown in Table 11.5.

In addition to their neutralisation reactions with acids, bases will also react with ammonium salts to produce ammonia gas. The production of ammonia can be detected as the gas turns damp red litmus paper blue.

ammonium nitrate + sodium hydroxide → sodium nitrate + water + ammonia

$$NH_4NO_3(s) + NaOH(aq) \rightarrow NaNO_3(aq) + H_2O(l) + NH_3(g)$$

This reaction occurs because ammonia is a more volatile base than sodium hydroxide. Ammonia is therefore easily displaced from its salts by sodium hydroxide. The reaction can be used to test an unknown substance for ammonium ions.

A base such as calcium oxide, or calcium hydroxide, can be used to prepare ammonia in the laboratory by this type of reaction (Figure 11.8). The two solids are mixed together and heated.

ammonium nitrate + calcium oxide → calcium nitrate + water + ammonia

$$2NH_4NO_3(s) + CaO(s) \rightarrow Ca(NO_3)_2(s) + H_2O(l) + 2NH_3(g)$$

Type	Name	Formula	Strong or weak?	Where found or used
Alkalis	sodium hydroxide (caustic soda)	NaOH	strong	in oven cleaners (degreasing agent); in making soap and paper; other industrial uses
	potassium hydroxide (caustic potash)	KOH	strong	in making soft soaps and biodiesel
	calcium hydroxide (limewater)	$Ca(OH)_2$	strong	to neutralise soil acidity and acidic gases produced by power stations; has limited solubility
	ammonia solution (ammonium hydroxide)	$NH_3(aq)$ or NH_4OH	weak	in cleaning fluids in the home (degreasing agent); in making fertilisers
Bases	calcium oxide	CaO		for neutralising soil acidity and industrial waste; in making cement and concrete
	magnesium oxide	MgO		in antacid indigestion tablets

Table 11.5: Some common alkalis and bases.

Figure 11.8: Preparation of ammonia. The gas is dried by passing it through a tower containing quicklime (CaO). Note that ammonia turns red litmus blue.

The properties of acids, bases and alkalis can be summarised as follows:

Acids:

- solutions contain an excess of H^+ ions and have a pH lower than 7.0
- turn blue litmus red
- are neutralised by a base to give a salt and water only.

Bases:

- are the oxides and hydroxides of metals
- neutralise acids to give a salt and water only
- are mainly insoluble in water.

Alkalis:

- are bases that dissolve in water, and feel soapy to the skin
- give solutions that contain an excess of OH^- ions
- give solutions with a pH greater than 7.0 and turn red litmus blue.

Questions

4 What is the formula for:

- **a** sulfuric acid
- **b** hydrochloric acid?

5 **a** What statement can we make about the concentrations of hydrogen ions and hydroxide ions in water?

- **b** Which ion is in excess in the solution of an alkali?
- **c** Which ions are present in:
  - **i** nitric acid solution
  - **ii** calcium hydroxide solution
  - **iii** ammonia solution?

6 Which of the following compounds are alkalis?

zinc oxide; magnesium oxide; potassium hydroxide; aluminium hydroxide; ammonium hydroxide

Metal oxides and non-metal oxides

Acidic and basic oxides

Venus, our nearest neighbour in the solar system, is identical in size and density to the Earth. But Venus has yielded its secrets reluctantly because it is veiled in clouds and has an atmosphere that is very damaging to space probes. The *Venus Express* spacecraft (Figure 11.9) launched by the European Space Agency conducted the most comprehensive study of the thick atmosphere responsible for the intense greenhouse effect on the planet.

Figure 11.9: Image of the Venus Express probe orbiting above the clouds of Venus' atmosphere.

The sulfuric acid clouds of Venus are the product of great volcanic activity. This has thrown out huge amounts of water vapour and the oxides of sulfur into the planet's atmosphere. Similar acidic clouds can be made in a gas jar by lowering burning sulfur into oxygen (Figure 11.10):

$$S(s) + O_2(g) \rightarrow SO_2(g)$$

Other burning non-metals (e.g. carbon) react in the same way to produce acidic gases:

$$C(s) + O_2(g) \rightarrow CO_2(g)$$

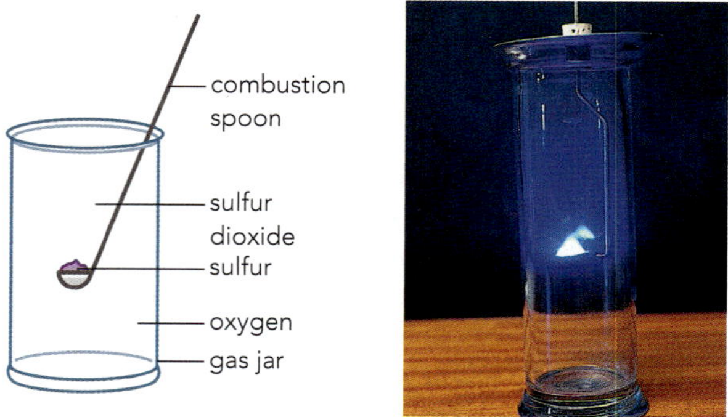

Figure 11.10: Burning sulfur in a gas jar of oxygen.

When water is added to the gas jars, it dissolves the gases and gives solutions that turn blue litmus paper red. Turning litmus paper red shows that these solutions contain acids. These solutions are the product of burning non-metals to produce **acidic oxides**.

Burning metals produces oxides that, if they dissolve, give solutions that turn litmus paper blue. The metal oxides produced in these reactions react with acids to neutralise them – they are said to be **basic oxides** (Table 11.6).

Element	How it reacts	Product	Effect of adding water and testing with litmus
Non-metals			
sulfur	burns with bright blue flame	colourless gas (sulfur dioxide, SO_2)	dissolves, turns litmus red
phosphorus	burns with yellow flame	white solid (phosphorus(V) oxide, P_2O_5)	dissolves, turns litmus red
carbon	glows red	colourless gas (carbon dioxide, CO_2)	dissolves slightly, slowly turns litmus red
Metals			
sodium	burns with yellow flame	white solid (sodium oxide, Na_2O)	dissolves, turns litmus blue
magnesium	burns with bright white flame	white solid (magnesium oxide, MgO)	dissolves slightly, turns litmus blue
calcium	burns with red flame	white solid (calcium oxide, CaO)	dissolves, turns litmus blue
iron	burns with yellow sparks	blue–black solid (iron oxide, FeO)	insoluble
copper	does not burn, turns black	black solid (copper oxide, CuO)	insoluble

Table 11.6: Reactions of certain elements with oxygen.

KEY WORDS

acidic oxides: oxides of non-metals that will react with bases and dissolve in water to produce acid solutions

basic oxide: oxide of a metal that will react with acids to neutralise the acid

Neutral and amphoteric oxides

Water can be thought of as hydrogen oxide. It has a pH of 7 and is therefore a neutral oxide. It is an exception to the broad 'rule' that the oxides of non-metals are acidic oxides. Neutral oxides do not react with either acids or alkalis. There are a few other exceptions to this 'rule' (Figure 11.11). The most important is the poisonous gas carbon monoxide (CO). The 'rule' that most non-metal oxides are acidic remains useful and important, however.

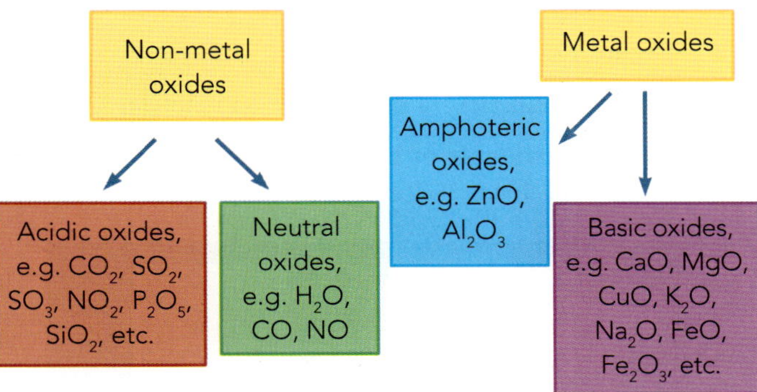

Figure 11.11: The classification of non-metal and metal oxides.

Of more importance is the unusual behaviour of some metal oxides). These metal oxides react and neutralise acids; they react as bases which would be expected. However, they can also neutralise alkalis, reacting as acids which is unusual. Metal oxides that can react as either an acid or a base to produce a salt and water are called **amphoteric compounds**.

The most important examples of metals that have amphoteric compounds are zinc and aluminium. Zinc oxide and aluminium oxide both react with acids to produce the expected salt and water. For example:

$$ZnO(s) + 2HCl(aq) \rightarrow ZnCl_2(aq) + H_2O(l)$$

However, they also react with an alkali to produce a salt and water.

acid + alkali → salt + water

$$ZnO(s) + 2NaOH(aq) \rightarrow Na_2ZnO_2(aq) + H_2O(l)$$

zinc oxide + sodium hydroxide → sodium zincate + water

$$Al_2O_3(s) + 2NaOH(aq) \rightarrow 2NaAlO_2(aq) + H_2O(l)$$

aluminium oxide + sodium hydroxide → sodium aluminate + water

The salts formed in the reactions with sodium hydroxide are soluble and contain compound ions with names ending *-ate* because they contain oxygen (this type of ion is discussed in Chapter 4).

The fact that zinc hydroxide and aluminium hydroxide are also amphoteric helps in the identification of salts of these metals using sodium hydroxide (Chapter 22).

KEY WORDS

amphoteric compound: a compound (hydroxide or metal oxide) that reacts with both an acid and an alkali to give a salt and water

Questions

7 a What colour is the flame when sulfur burns?
  b Write the symbol equation for the reaction when sulfur burns in oxygen.
  c Write the word equation for magnesium burning in air.

8 Define the term *amphoteric oxide*. Write the balanced symbol equations for the reaction of aluminium oxide with hydrochloric acid and sodium hydroxide.

9 Name one amphoteric metal hydroxide and write the word and symbol equations for its reaction with sodium hydroxide solution.

ACTIVITY 11.1

The environmental impact of acidic oxides

The image on page 230 shows the devastating effect of acid rain on trees in a national park in Poland. As we saw in Table 11.3, rain is normally slightly acidic, with a pH of 5.6, while acid rain generally has a pH between 4.2 and 4.4. Fossil fuels are contaminated with the non-metal, sulfur. When the fuels are burnt, the sulfur is converted into the gaseous acidic oxide, sulfur dioxide (SO_2). Fossil-fuel powered vehicles also produce acidic oxides of nitrogen (NO and NO_2, sometimes referred to as NO_x), which are released into the atmosphere.

CONTINUED

Once released, the pollutant gases react with water, oxygen and other chemicals in the atmosphere to form acidic compounds such as sulfuric and nitric acids. The problem of acid rain is not confined to the geographical location in which it forms, however. This pollution can be carried by prevailing winds through the atmosphere for hundreds of kilometres. Once acid rain falls, it sinks into the soil and also enters the water system through surface runoff. (For further discussion of acid rain see Chapter 17.)

In groups, create an educational flyer for distribution at a student environmental conference on the origins and consequences of acid rain. Your flyer should include text and artwork and could cover some of the following topics (though not necessarily all):

- the origins of acid rain (acid deposition) including an equation for the production of one of the acids mentioned
- the nature of the pH scale – its direction and the meaning of one division on the scale
- the effects of acid rain on trees, lakes (and the life in them), and on buildings (include an equation for the effect of, say, nitric acid on limestone)
- prevention and remedy of the effects; addition of lime to acidified lakes (a neutralisation reaction), for instance.

11.2 Characteristic reactions of acids

Reactions of acids

There are three major chemical reactions in which all acids will take part. These reactions are best seen using dilute acid solutions. In these reactions, the acid reacts with:

- a reactive metal (e.g. magnesium or zinc – Figure 11.12)
- a base (or alkali) – a neutralisation reaction
- a metal carbonate or hydrogen carbonate (Figure 11.13).

Figure 11.12 a: Magnesium ribbon. **b:** Zinc granules, reacting with hydrochloric acid and giving off hydrogen.

Figure 11.13: Some antacid tablets are designed to fizz – they contain sodium hydrogen carbonate and citric acid, which react together in water.

One type of product is common to all these reactions. They all produce a metal compound called a salt.

Normally, we use the word 'salt' to mean 'common salt', which is sodium chloride. This is the salt we put on our food, the main salt found in seawater, and the salt used over centuries to preserve food. However, in chemistry, the word has a more general meaning. A salt is a compound made from an acid when a metal takes the place of the hydrogen in the acid (Chapter 12). The acid from which the salt is made is often called the parent acid of the salt.

Reaction of acids with metals

Metals that are quite reactive can be used to displace the hydrogen from an acid safely. Hydrogen gas is given off. The salt made depends on the combination of metal and acid used:

metal + acid → salt + hydrogen

It is unsafe to try this reaction with very reactive metals such as sodium or calcium. The reaction is too violent. No reaction occurs with metals such as copper that are less reactive than lead. Even with lead, it is difficult to see any reaction in a short time. The reactivity of metals with acids is considered in more detail in Chapter 15.

The salt made depends on the acid reacted with the metal:

- hydrochloric acid always gives a chloride
- nitric acid always gives a nitrate
- sulfuric acid always gives a sulfate.

For example:

magnesium	+	nitric acid	→	magnesium nitrate	+	hydrogen
$Mg(s)$	+	$2HNO_3(aq)$	→	$Mg(NO_3)_2(aq)$	+	$H_2(g)$
zinc	+	hydrochloric acid	→	zinc chloride	+	hydrogen
$Zn(s)$	+	$2HCl(aq)$	→	$ZnCl_2(aq)$	+	$H_2(g)$

Reaction of acids with bases and alkalis

This is the neutralisation reaction that we saw earlier:

acid + base → salt + water

The salt produced by this reaction will depend on the combination of reactants used. To make a particular salt, you choose a suitable acid and base to give a solution of the salt you want. For example:

hydrochloric acid	+	sodium hydroxide	→	sodium chloride	+	water
$HCl(aq)$	+	$NaOH(aq)$	→	$NaCl(aq)$	+	$H_2O(l)$

In this reaction we can see that the hydrogen ions of the acid are matched by the hydroxide ions from the alkali to produce water:

$$H^+(aq) + OH^-(aq) \rightarrow H_2O(l)$$

If mixed in the correct amounts then the alkali exactly neutralises the acid.

A neutralisation reaction can also occur by reacting an insoluble base with the acid.

magnesium oxide + nitric acid → magnesium nitrate + water

$$MgO(s) + 2HNO_3(aq) \rightarrow Mg(NO_3)_2 + H_2O(l)$$

Other examples of salts made from different combinations of acid and base are shown in Table 11.7.

	Salt made with ...		
Base	**Hydrochloric acid (HCl)**	**Nitric acid (HNO_3)**	**Sulfuric acid (H_2SO_4)**
sodium hydroxide (NaOH)	sodium chloride, NaCl	sodium nitrate, $NaNO_3$	sodium sulfate, Na_2SO_4
potassium hydroxide (KOH)	potassium chloride, KCl	potassium nitrate, KNO3	potassium sulfate, K_2SO_4
magnesium oxide (MgO)	magnesium chloride, $MgCl_2$	magnesium nitrate, $Mg(NO_3)_2$	magnesium sulfate, $MgSO_4$
copper oxide (CuO)	copper chloride, $CuCl_2$	copper nitrate, $Cu(NO_3)_2$	copper sulfate, $CuSO_4$

Table 11.7: Some examples of making salts using alkalis or insoluble bases.

It is useful to realise the origins of a salt because it helps you predict which salt you get from a particular combination of acid and base (Chapter 12). The cubic crystals of sodium chloride come from the neutralisation of hydrochloric acid with sodium hydroxide solution (Figure 11.14).

Sodium chloride

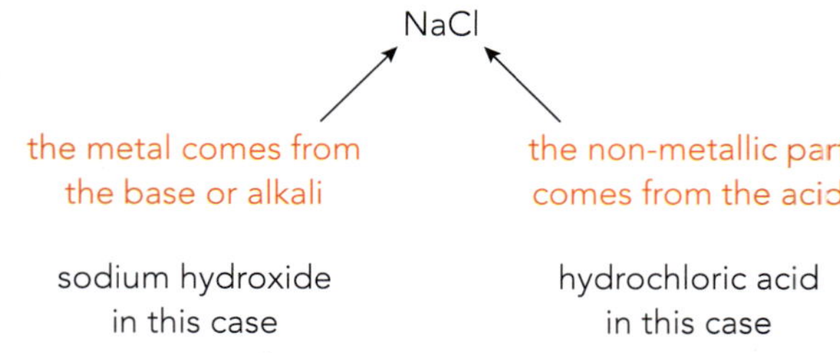

Figure 11.14: Sodium chloride is the product of a reaction between hydrochloric acid and sodium hydroxide.

Reaction of acids with carbonates

All carbonates give off carbon dioxide when they react with acids (Figure 11.15). We have seen that this reaction occurs with effervescent antacid tablets. The result is to neutralise the acid and produce a salt solution:

acid + metal carbonate → salt + water + carbon dioxide

Figure 11.15: Limestone (calcium carbonate) reacting with acid.

The normal method of preparing carbon dioxide in the laboratory is based on this reaction. Dilute hydrochloric acid is reacted with marble chips (calcium carbonate):

hydrochloric acid	+	calcium carbonate	→	calcium chloride	+	water	+	carbon dioxide
$2HCl(aq)$	+	$CaCO_3(s)$	→	$CaCl_2(aq)$	+	$H_2O(l)$	+	$CO_2(g)$

The test for a carbonate is also based on this type of reaction. Dilute hydrochloric acid is added to the unknown salt and if carbon dioxide is given off then the salt is a carbonate (Chapter 22).

zinc carbonate	+	hydrochloric acid	→	zinc chloride	+	water	+	carbon dioxide
$ZnCO_3(s)$	+	$2HCl(aq)$	→	$ZnCl_2(aq)$	+	$H_2O(l)$	+	$CO_2(g)$

EXPERIMENTAL SKILLS 11.1

Comparing the effectiveness of different antacid tablets

Antacid tablets relieve the symptoms of acid indigestion by neutralising the excess acid. Many of these indigestion tablets contain insoluble carbonates or bases such as calcium carbonate, magnesium carbonate or magnesium hydroxide. This activity involves titrating powdered samples of antacid tablets with dilute hydrochloric acid (0.5 mol/dm^3).

You will need:

- a selection of antacid tablets
- dilute hydrochloric acid (0.5 mol/dm^3)
- methyl orange indicator
- burette
- burette stand
- conical flask (100 cm^3)
- spatula
- small filter funnel
- balance
- mortar and pestle.

Safety

Wear eye protection throughout. Be careful with chemicals. Never ingest them and always wash your hands after handling them.

Getting started

Familiarise yourself with the apparatus you are about to use. While washing out the burette and filling with acid make sure that you can comfortably read the burette scale. Make sure you are clear about the colour change you will observe with methyl orange indicator. It can be useful for you to set up two small beakers showing the indicator in some acid and alkali and then keep these as a reminder to you as you carry out the experiment.

Method

1. Set up a burette and use a small filter funnel to fill the burette with the dilute hydrochloric acid solution.
2. Choose an antacid tablet and note its brand name. Weigh this tablet on the balance and record the mass of the tablet.
3. Draw up a results table in which to record your burette readings.
4. Grind up the tablet into a powder using the mortar and pestle. Transfer all the powder to the conical flask and add a little water.
5. Add about 10 drops of methyl orange indicator and swirl the flask.
6. In your results table, record the initial reading of the burette.

CONTINUED

7 Perform the titration by adding the acid from the burette and keep swirling the flask as you do so. Allow time for the acid to react with the powder. The colour of the solution will flash red and will get stronger as you add more acid. When you are close to the end-point, add the acid one drop at a time. When the solution stays permanently red after swirling, stop adding any more acid.

8 In your results table, record the final reading of the burette.

Questions

1 Which brand of tablet did you judge to be the most effective in neutralising acid indigestion? Comment on the criteria you used to make this judgement.

2 Examine the packaging of the tablets. Write word equations for the reaction between the active ingredient in each case and the hydrochloric acid. Write the balanced chemical equation for the reaction of at least one of the tablets.

3 Why were the tablets crushed before carrying out the reaction?

Peer assessment

Titrations require precise and accurate use of specific equipment and careful reading of the values obtained. Do you understand how to construct a results table to record your data? Discuss with your partner the reliability of the readings you took and your organisation of the work involved. What areas do you think you could improve?

Strong and weak acids and alkalis

Strong and weak acids

Not all acids are equally strong. The vinegar used in salad dressing and to pickle vegetables is significantly less acidic than a hydrochloric acid solution of the same concentration. If differences in concentration are not the reason for this, then what does cause the difference?

The difference lies in the ionic nature of acid solutions; more precisely, in the concentration of hydrogen ions (H^+ ions) in a solution. Earlier we stressed the importance of water as the necessary *solvent* for acid solutions. There is a relationship between H^+ ion concentration, acidity and pH: the higher the H^+ ion concentration, the higher the acidity and the lower the pH. The pH scale ranges from 1 to 14. Each pH unit means a ten-fold difference in H^+ ion concentration. An acid of pH 1.0 has 10 times the H^+ ion concentration of an acid of pH 2.0.

When hydrochloric acid is formed in water, the hydrogen chloride molecules separate (dissociate) completely into ions:

$$HCl(g) \xrightarrow{H_2O} H^+(aq) + Cl-(aq)$$

In a similar way, sulfuric acid and nitric acid molecules dissociate completely into ions when dissolved in water:

$$H_2SO_4(l) \xrightarrow{H_2O} H^+(aq) + HSO_4^-(aq)$$

$$HNO_3(l) \xrightarrow{H_2O} H^+(aq) + NO_3^-(aq)$$

Complete **dissociation** into ions (complete ionisation) produces the maximum possible concentration of H+ ions, and so the lowest possible pH for that solution.

When pure ethanoic acid (CH_3COOH) is dissolved in water, only a small fraction of the covalently bonded molecules is dissociated into hydrogen ions and ethanoate ions:

$$CH_3COOH(l) \xrightleftharpoons{H_2O} H^+(aq) + CH_3COO^-(aq)$$

most molecules intact

only a small number of molecules are dissociated into ions at any one time

Thus, an ethanoic acid solution will have far fewer hydrogen ions present in it than a hydrochloric acid solution of the same concentration. The ethanoic acid solution will have a higher pH. Carbonic acid (H_2CO_3) is an example of a weak mineral acid. The other organic acids, such as methanoic acid, citric acid and so on (Table 11.1), also only partially dissociate into ions when dissolved in water.

Because of their high concentration of ions, solutions of **strong acids** conduct electricity well. Compounds such as nitric acid and sulfuric acid are strong electrolytes in solution (Chapter 6). Ethanoic acid and other **weak acid** solutions are only weak electrolytes. The large difference in conductivity between weak and strong acids shows clearly the differences in ionisation in these solutions.

In solutions of weak acids, the process of dissociation is reversible: it can go in either direction. Once the weak acid or alkali has dissolved the solution reaches an equilibrium position (Chapter 9) hence the use of the reversible arrows in the equations above.

When comparing two acids and trying to decide whether they are weak or strong, it is very important to compare two solutions of the same concentration. If the concentrations are the same, then you can use the pH, the conductivity or the rate of a particular reaction (Figure 11.16) to help you make a judgement as to the extent of ionisation. Figure 11.16 shows the reaction of chalk (calcium carbonate, $CaCO_3$) with hydrochloric acid and ethanoic acid solutions of the same concentration (0.5 mol/dm^3).

KEY WORDS

dissociation: the separation of a covalent molecule into ions when dissolved in water

strong acid: an acid that is completely ionised when dissolved in water – this produces the highest possible concentration of H^+(aq) ions in solution, e.g. hydrochloric acid

weak acid: an acid that is only partially dissociated into ions in water – usually this produces a low concentration of H^+(aq) in the solution, e.g. ethanoic acid

Figure 11.16: Reaction of chalk with solutions of a strong acid (hydrochloric acid) on the left and a weak acid (ethanoic acid) of the same concentration on the right.

REFLECTION

On the pH scale, a change of 1 unit means a 10-fold increase or decrease. A solution of pH 4 is 10 times more acidic than a solution of pH 5. How confident are you in dealing with ratios and proportion? Can you think of a way to remember this?

What happens to the ions in neutralisation?

An acid can be neutralised by an alkali to produce a salt and water only, according to the general equation:

acid + alkali → salt + water

For example:

hydrochloric acid	+	sodium hydroxide	→	sodium chloride	+	water
HCl(aq)	+	NaOH(aq)	→	NaCl(aq)	+	H_2O(l)

All these compounds are completely ionised, except for the water produced. The hydrogen ions from the acid and the hydroxide ions from the alkali have combined to form water molecules. We can show this in the following equation:

H^+(aq)	+	OH^-(aq)	→	H_2O(l)
hydrogen ions		hydroxide ions		water

This is the ionic equation for this neutralisation reaction. The spectator ions (chloride and sodium ions) remain in solution, which becomes a solution of sodium chloride (Figure 11.17).

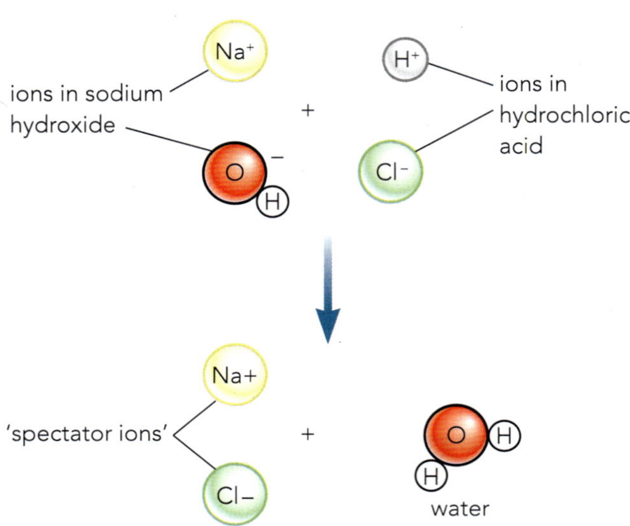

Figure 11.17: Reactions of ions when hydrochloric acid is mixed with sodium hydroxide.

By evaporating some of the water, the salt can be crystallised out. In fact, the same ionic equation can be used for any reaction between an acid and an alkali.

In these reactions, the acid is providing hydrogen ions to react with the hydroxide ions. In turn, the base is supplying hydroxide ions to accept the H^+ ions and form water. It is important to realise that a hydrogen ion (H^+) is simply a proton (Chapter 2). If the single electron of a hydrogen atom is removed to form the positive ion, all that remains is the proton of the nucleus (Figure 11.18).

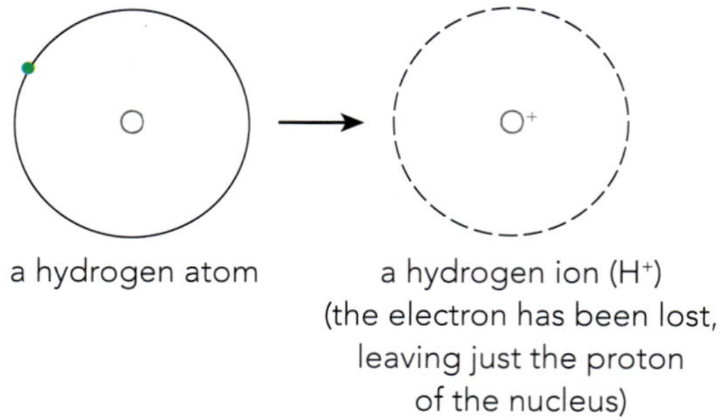

Figure 11.18: A hydrogen ion (H^+) is simply a proton.

This leads to a further definition of an acid and a base in terms of hydrogen ion (proton) transfer:

- an acid is a proton donor (it gives a proton, H^+ ion, to a base)
- a base is a proton acceptor (it accepts a proton, H^+ ion, from an acid).

Questions

10 Define an acid and a base using the ideas of proton (H^+ ion) transfer.

11 **a** Write the balanced symbol equations for the reaction between:

- **i** copper oxide and nitric acid
- **ii** zinc and hydrochloric acid

b The ionic equation for the reaction in **ii** is:

$Zn(s) + 2H^+(aq) \rightarrow Zn^{2+}(aq) + H_2(g)$

Explain why this can be regarded as a redox reaction. State which reactant has been oxidised.

12 **a** Write an equation to show what happens when hydrogen chloride dissolves in water.

b Explain why an ethanoic acid solution has a lower conductivity than hydrochloric acid of the same concentration.

SUMMARY

Acids turn litmus and thymolphthalein red, and leave thymolphthalein colourless.

Alkalis turn red litmus paper and thymolphthalein blue, and methyl orange yellow.

Bases are the oxides and hydroxides of metals, and that soluble bases are also referred to as alkalis.

Aqueous solutions of acids contain an excess of hydrogen ions (H_+), while alkaline solutions contain an excess of OH^- ions (hydroxide ions).

The relative acidity or alkalinity, hydrogen ion concentration or pH of a solution can be assessed using the colour observed using universal indicator.

Acids and bases react together in a neutralisation reaction to produce a salt and water only, and bases also displace ammonia from ammonium salts.

Most metal oxides and hydroxides are basic (e.g. CuO and CaO), while many oxides of non-metals can be classified as acidic oxides (e.g. CO_2 and SO_2). Acidic oxides, such as the oxides of nitrogen and sulfur dioxide, produced by human activity contribute to the pollution problem acid rain (see Chapter 17).

Some metal oxides, such as aluminium oxide and zinc oxide, can react with both acids and alkalis, and are therefore classified as amphoteric oxides.

Acids take part in certain characteristic reactions such as those with bases (neutralisation), with metals to produce hydrogen and with metal carbonates to form carbon dioxide.

Acids can be described as strong or weak depending on whether they are totally or only partially dissociated into ions in aqueous solution.

An acid can be defined as a proton donor and a base can be defined as a proton acceptor.

PROJECT

Ion balance and solution pH

You and a partner are consultants who have been asked to give a presentation to people working in the cosmetics industry on the meaning of the pH scale. As part of that presentation, you need to draw up a series of illustrations that show how the colour of Universal Indicator paper shows the pH of a solution and the balance of hydrogen and hydroxide ions in the solution tested.

Base your illustrations on those shown for litmus (Figure 11.6) adapted to show the range of Universal Indicator colours and pH number on the left of your illustrations. An example, for water, is shown in Figure 11.19.

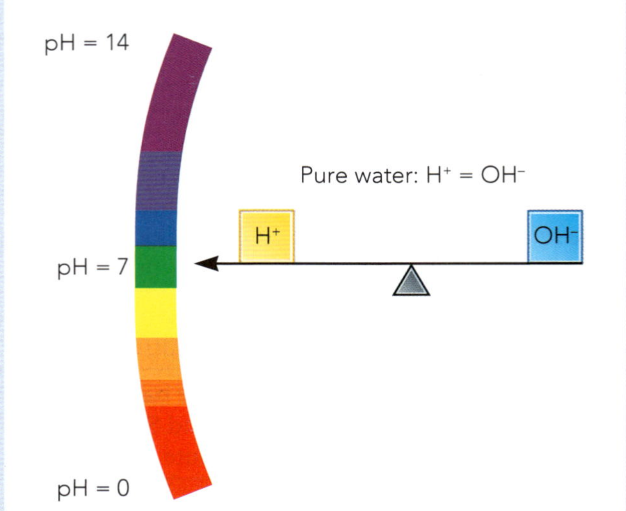

Figure 11.19: The balance of H^+ and OH^- ions and Universal Indicator colours.

Your series of illustrations should show the balance at several different pH values (and indicator colours); pHs of 2.0, 4.0, 6.0 and 12.0, for instance.

When discussing the illustrations with your partner, you should cover the following ideas:

- the range of values involved in the pH scale
- the meaning of the values in terms of acidity and alkalinity
- how the colours of the indicator show the pH values of different solutions.

EXAM-STYLE QUESTIONS

1 Which of the following elements reacts with dilute hydrochloric acid to produce hydrogen gas?

A copper

B zinc

C carbon

D chlorine **[1]**

2 Metals and non-metals generally produce different types of oxides when reacted with air or oxygen. Which row in the table correctly defines the nature of the oxides of the elements listed? **[1]**

	Forms an acidic oxide	Forms a basic oxide
A	phosphorus	sulfur
B	magnesium	sulfur
C	sulfur	phosphorus
D	sulfur	magnesium

3 A student tested 50 cm^3 of hydrochloric acid with methyl orange indicator. She then added the hydrochloric acid to powdered calcium carbonate to make carbon dioxide.

a What colour did the indicator turn? **[1]**

b Write a word equation for the reaction of hydrochloric acid with calcium carbonate, and include state symbols. **[2]**

c Carbon dioxide is a colourless gas which reacts with limewater. What observation is made when carbon dioxide is bubbled through limewater? **[1]**

d When carbon dioxide is tested with moist universal indicator paper, the paper turns orange. What does this show about carbon dioxide? **[1]**

e Limewater is a solution of calcium hydroxide. It changes thymolphthalein indicator blue.

Give the general rule about the characteristics of metal and non-metal oxides shown by these observations. **[2]**

f pH is a measure of how acidic or alkaline a substance is. Describe how you could use an indicator to find the pH of a solution of calcium hydroxide. **[3]**

[Total: 10]

COMMAND WORDS

give: produce an answer from a given source or recall/ memory

describe: state the points of a topic / give characteristics and main features

CONTINUED

4 Acids dissolve in water to produce hydrogen ions.

a Sulfuric acid, H_2SO_4, is a strong acid. Write an ionic equation, including state symbols, to show what happens when it is added to water. [2]

b Ethanoic acid, CH_3COOH, is a weak acid. Write an ionic equation, including state symbols, to show what happens when it dissolves in water. [2]

c Metal oxides react with water to form hydroxides.

$Na_2O(s) + 2H_2O(aq) \rightarrow 2NaOH(aq)$

Write an equation to show how aluminium oxide, Al_2O_3, reacts with hydrochloric acid. [2]

d Aluminium hydroxide reacts with sodium hydroxide as shown in the equation.

$Al(OH)_3(s) + NaOH(aq) \rightleftharpoons NaAl(OH)_4$

What type of compound is aluminium hydroxide? [1]

e When aqueous sodium hydroxide is added to a solution of aluminium chloride a white precipitate of insoluble aluminium hydroxide is formed. When excess sodium hydroxide is added the precipitate redissolves. Addition of hydrochloric acid causes the precipitate to reappear. **Explain** these observations. [3]

[Total: 10]

COMMAND WORD

explain: set out purposes or reasons/make the relationships between things evident/ provide why and/or how and support with relevant evidence

SELF-EVALUATION CHECKLIST

After studying this chapter, think about how confident you are with the different topics. This will help you see any gaps in your knowledge and help you to learn more effectively.

I can	See Topic...	Needs more work	Almost there	Confident to move on
describe acids and alkalis in terms of their effect on indicators such as litmus, thymolphthalein and methyl orange	11.1			
state that bases are the oxides and hydroxides of metals, and that soluble bases are referred to as alkalis	11.1			
describe how acidic solutions contain an excess of hydrogen ions, while alkaline solutions contain an excess of hydroxide ions	11.1			
compare the relative acidity or alkalinity, H^+ ion concentration or pH of solutions using the colour observed using universal indicator	11.1			
describe how acids and bases react together in a neutralisation reaction, and that bases displace ammonia from ammonium salts	11.1			
classify metal oxides and hydroxides as basic while many non-metal oxides are acidic oxides	11.2			
classify some metal oxides such as ZnO and Al_2O_3, as amphoteric as they can react with both acids and alkalis	11.2			
describe the characteristic reactions of acids with bases, and with metals to produce hydrogen, and metal carbonates to form carbon dioxide	11.2			
define a strong acid as being completely dissociated into ions in solution, while weak acids are only partly dissociated into ions	11.2			
define an acid as able to donate a proton to a base and a base as being able to accept a proton from an acid	11.2			

> Chapter 12

Preparation of salts

IN THIS CHAPTER YOU WILL:

- understand that salts are an important group of ionic compounds
- learn that some salts are soluble in water, while others are insoluble
- describe the general solubility rules for various different types of salt
- define hydrated and anhydrous substances in terms of water content
- describe the preparation, separation and purification of soluble salts by reaction of the parent acid with either excess metal, excess insoluble base or excess insoluble carbonate
- describe the preparation, separation and purification of a soluble salt by titration of an acid with an alkali

> define the term water of crystallisation

> describe the preparation and separation of insoluble salts by precipitation.

GETTING STARTED

You will have some previous knowledge of salts from earlier parts of your science studies. 'Salt' is a word used in everyday conversation, but usually to mean one particular salt, sodium chloride (NaCl). Get together in a small group and make a list of where, in everyday situations, you have come across this salt.

You could discuss the following questions when drawing up your list:

- Why is it sometimes called table salt? What use does this refer to?
- What do you know about salt in relation to the human body?
- Where do we find this salt in the environment?

THE SIGNIFICANCE OF SALT

Saltiness is one aspect of the basic human sense of taste and salt is essential to life. Salt has long been a crucial trading commodity. Thousands of years ago, people harvested salt by evaporating seawater (brine), or water from salt lakes, in the sun. There is evidence of this type of salt processing site in central Europe and China dating back some 8000 years. This method is still used in many parts of the world, especially in hot, dry areas located near seas or large salt lakes (Figure 12.1). Evaporated seawater also leaves exposed deposits of rock salt, which people began to quarry. They also developed methods of underground mining, to access salt deposits below ground.

Historically, salt was of such great importance that people were paid with salt. This is the origin of the word 'salary'. The importance of salt lay in its use in preserving food, which was particularly vital before there was access to refrigeration. Salt became an important article of trade. Tracking the movement of salt shows us the historical routes of commerce. One of the ancient roads in Italy, the Via Salaria, was used to carry salt from Rome to other areas of the country. Salt was shipped across the Mediterranean and transported by caravan route across the deserts of northern Africa.

Figure 12.1: Salt collection on the Sambhar salt lake. The lake is India's largest inland salt lake and is located southwest of the city of Jaipur in Rajasthan.

Discussion questions

1. What are the reasons for adding salt to food? What are the alternatives used in different cultures to replace salt in one of these functions?
2. Under certain conditions athletes can sweat so much that a fine layer of salt covers their skin. What does this tell you about salt? What physical effect can result from this loss of salt from the body and how can the salt be replaced?

12.1 The importance of salts – an introduction

In chemistry, a **salt** is defined as a compound formed from an acid by the replacement of the hydrogen in the acid by a metal or by ammonium ions. Salts are ionic compounds. There is a wide range of types of salt. Many of them play an important part in our everyday life (Table 12.1).

> **KEY WORD**
>
> **salts:** ionic compounds made by the neutralisation of an acid with a base (or alkali), e.g. copper(II) sulfate and potassium nitrate

Many important minerals are single salts. For example, fluorite (calcium fluoride, Figure 12.2) is used in a variety of chemical, metallurgical and ceramic processes, and gypsum (calcium sulfate) is used in the manufacture of wallboard, cement, plaster of Paris and soil treatment.

Figure 12.2: A crystal of fluorite. Pure crystals of fluorite are transparent, the colour is due to impurities.

Common salt (sodium chloride) is mined from underground deposits of 'rock salt' in many parts of the world (Figure 12.3). Salt can also be found where salt lakes dry up through evaporation. In other cases, the sea forms shallow inland pools where the water slowly evaporates and crystals of 'sea salt' are formed. 'Sea salt' is not just sodium chloride; it also contains magnesium chloride, calcium sulfate, potassium bromide and other salts.

Figure 12.3: View of a boat quay on the lake at the bottom of Turda salt mine, Transylvania, Romania; the mine has been known since ancient times and is a tourist site.

Sodium chloride is essential for life and is an important raw material for industries. Biologically, it has a number of functions: it is involved in muscle contraction; it enables the conduction of nerve impulses in the nervous system; it regulates osmosis (the passage of solvent molecules through membranes) and it is converted into the hydrochloric acid that aids digestion in the stomach. When we sweat, we lose both water and sodium chloride.

Salt	Parent acid	Colour and other characteristics	Uses
ammonium chloride	hydrochloric acid	white crystals	fertilisers; dry cells (batteries)
ammonium nitrate	nitric acid	white crystals	fertilisers; explosives
ammonium sulfate	sulfuric acid	white crystals	fertilisers
calcium carbonate (marble, limestone, chalk)	carbonic acid	white	decorative stonework; making lime and cement; extracting iron in the blast furnace
sodium carbonate (washing soda)	carbonic acid	white crystals or powder	in cleaning; water softening; making glass
magnesium sulfate (Epsom salts)	sulfuric acid	white crystals	health salts (laxatives)
copper(II) sulfate	sulfuric acid	blue crystals	fungicides used to kill parasitic fungi or their spores
calcium phosphate	phosphoric acid	white	making fertilisers

Table 12.1: Salts in common use.

Loss of too much salt during sport and exercise can give us muscle cramp. Isotonic drinks are designed to replace this loss of water and to restore energy and the balance of mineral ions in our body.

Solubility of salts

While a number of salts can be obtained by mining, others must be made by industry. Therefore, it is worth considering the methods available to make salts. Some of these can be investigated in the laboratory.

Two things are important in working out a method of preparation:

- Is the salt **soluble** or **insoluble** in water?
- Do crystals of the salt contain chemically combined water as part of the crystal structure?

The first point influences the preparation method chosen. The second point affects how the crystals are handled at the end of the experiment.

Soluble salts are made by neutralising an acid. Insoluble salts are made by other methods. Table 12.2 outlines the general patterns of solubility for the more usual salts.

Salts	Soluble	Insoluble
sodium salts	all are soluble	none
potassium salts	all are soluble	none
ammonium salts	all are soluble	none
nitrates	all are soluble	none
chlorides	most are soluble	silver chloride, lead(II) chloride
sulfates	most are soluble	barium sulfate, lead(II) sulfate, calcium sulfate
carbonates	sodium, potassium and ammonium carbonates	most are insoluble

Table 12.2: The patterns of solubility for various types of salts.

Table 12.2 is organised to help you to remember the following points:

- all the common sodium, potassium and ammonium salts are soluble
- all nitrates are soluble
- most chlorides and sulfates are soluble: the most important exceptions are silver chloride and barium sulfate, which are important precipitates in chemical analysis
- almost all carbonates are insoluble.

Figure 12.4 provides a pictorial image of the solubility of various types of salt and also the bases (oxides and hydroxides) that are essential for making salts in the laboratory. All the substances within the large central circle have a solubility greater than 1 g/dm^3, and so are considered to be 'soluble'. Calcium hydroxide and calcium sulfate have only limited solubility (about 1 g/dm^3). This is the reason why it is not possible to make an alkaline solution of limewater similar to dilute sodium hydroxide. The limited solubility of calcium sulfate is why the reaction between calcium carbonate and sulfuric acid stops after just a few seconds. The bold letters show those insoluble compounds (barium and silver) that are the basis of the analytical tests covered in Chapter 22.

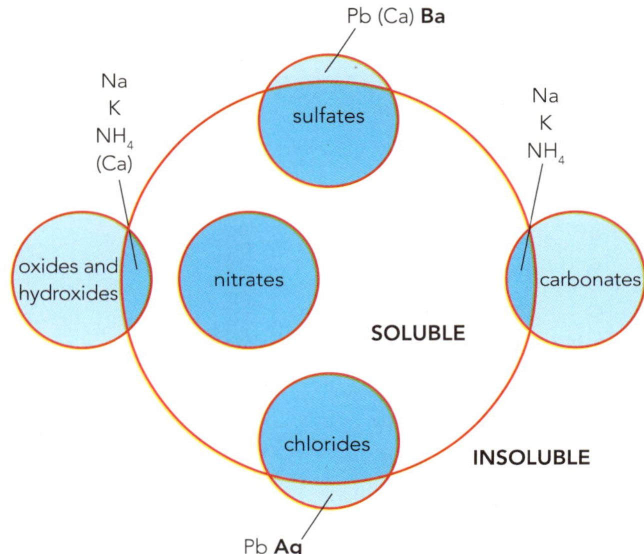

Figure 12.4: Venn diagram of the solubility of inorganic compounds.

We shall look at the preparation of both soluble salts and insoluble salts. But first we must consider the second point mentioned before. When crystallised, many salts contain water that is chemically combined in the crystal structure. Such salts are an important group of **hydrated substances**. They are known as **hydrated salts**. When preparing crystals of a hydrated salt, we must be careful not to heat them too strongly when drying them. If we do heat too strongly then the product is an **anhydrous** (dehydrated) powder, not crystals.

KEY WORDS

soluble: a solute that dissolves in a particular solvent

insoluble: a substance that does not dissolve in a particular solvent

hydrated substance: a substance that is chemically combined with water; **hydrated salts** are an important group of such substances

anhydrous: an adjective to describe a substance without water combined with it

Questions

1 What are the parent acids of the following salts?
 a potassium sulfate
 b silver nitrate
 c calcium carbonate

2 Which acid and alkali would react to form the following salts?
 a sodium chloride
 b calcium nitrate
 c ammonium sulfate

3 a The salts of which acid are all soluble in water?
 b Which two of the following salts are soluble in water?
 lead chloride; potassium sulfate; barium sulfate; silver chloride; ammonium nitrate

The water chemically combined in the hydrated crystals of some salts is known as **water of crystallisation**. This water of crystallisation is represented in the formulae of hydrated salts, e.g. $CoCl_2{\cdot}6H_2O$ (Table 12.3). This water gives the crystals their shape. Note that in some cases it also gives them their colour, for example copper sulfate crystals (Figure 12.5). These coloured salts are those of transition metals (Chapter 13).

Figure 12.5: Crystals of hydrated copper(II) sulfate are blue. They contain water of crystallisation in their structure ($CuSO_4{\cdot}5H_2O$).

Hydrated salt	Formula	Colour
copper(II) sulfate	$CuSO_4{\cdot}5H_2O$	blue
cobalt(II) chloride	$CoCl_2{\cdot}6H_2O$	pink
iron(II) sulfate	$FeSO_4{\cdot}6H_2O$	green
magnesium sulfate	$MgSO_4{\cdot}7H_2O$	white
sodium carbonate	$Na_2CO_3{\cdot}10H_2O$	white
calcium sulfate	$CaSO_4{\cdot}2H_2O$	white

Table 12.3: Some hydrated salts.

KEY WORDS

water of crystallisation: water included in the structure of certain salts as they crystallise, e.g. copper(II) sulfate pentahydrate ($CuSO_4{\cdot}5H_2O$) contains five molecules of water of crystallisation per molecule of copper(II) sulfate

When these hydrated salts are heated, their water of crystallisation is driven off as steam. The crystals lose their shape and become a powder. Copper(II) sulfate crystals are blue, but, when they are heated, they are dehydrated to form a white powder:

hydrated copper(II) sulfate crystals → anhydrous copper(II) sulfate + water vapour

$$CuSO_4{\cdot}5H_2O(s) \rightarrow CuSO_4(s) + 5H_2O(g)$$

Crystals that have lost their water of crystallisation are said to be anhydrous. If water is added back to the white anhydrous copper(II) sulfate powder, the powder turns blue again and heat is given out. This can be used as a test for the presence of water (see Chapters 9 and 22). Blue anhydrous cobalt(II) chloride can also be used for this test. The colour change with cobalt(II) chloride is from blue to pink.

ACTIVITY 12.1

Oceans of seawater

The Earth has been referred to as the blue planet because so much of its surface is covered by water. Only about 3% of this water is freshwater, with most of it being saltwater. Saltwater is the water that is present in the seas and oceans that cover some 71% of the surface of the Earth.

Working in small groups, create a colourful poster on the salt composition (salinity) of the water on the surface of the Earth. Use the following information to help you.

Figure 12.6 shows data on the overall concentration of various ions in seawater.

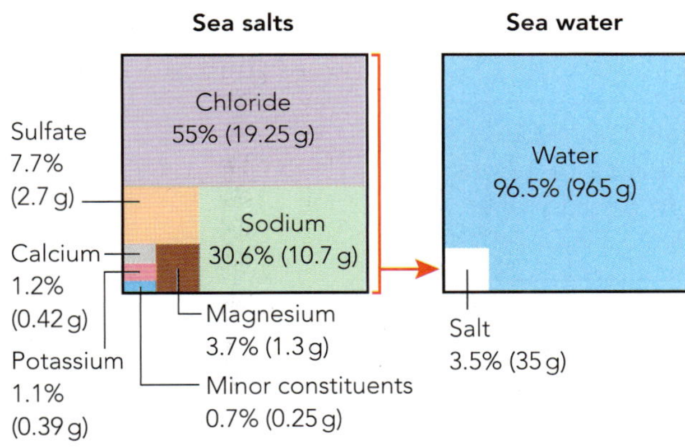

Figure 12.6: Data on the overall salt concentration of seawater.

Use the data in Figure 12.6 to suggest some of the major salts present in seawater. On your poster, make a table of these salts and suggest which acid and base or carbonate could have reacted to produce them. Relate your list to the solubilities of different types of salt (Figure 12.4).

The distribution of salts in the oceans is not uniform. For example, the Red Sea has a salinity of 40%, the Mediterranean 38%, while the Black Sea (18%) and the Baltic Sea (8%) are much lower.

- Research and comment on the factors that produce this variation in salt concentration in different regions on your poster. Does temperature influence the composition? Is the outflow from major rivers an influence?
- How would you prepare a sample of salt from a known volume of seawater to find a value for the total salt concentration in gram/dm^3?
- Include this, alongside a photograph or drawing of the apparatus, on your poster.

Prepare your poster and then present and discuss it with the whole class.

12.2 Preparation of salts

Preparing soluble salts

Soluble salts can be made from their parent acid using any of the three characteristic reactions of acids outlined in Chapter 11.

Choosing a method of salt preparation

The choice of method for preparing a soluble salt depends on two things:

1. Is the metal reactive enough to displace the hydrogen in the acid? If it is, is it too reactive and therefore unsafe?
2. Is the base or carbonate soluble or insoluble?

Figure 12.7 shows a flow chart summarising the choices to be made when preparing a soluble salt.

Method A – acid plus solid metal, base or carbonate

Method A is essentially the same whether you are starting with a solid metal, a solid base or a solid carbonate (Figure 12.8):

- Stage 1: An excess of the solid is added to the acid and allowed to react. Using an excess of the solid makes sure that all the acid is used up. If it is not used up at this stage, the acid would become more concentrated when the water is evaporated later (stage 3).
- Stage 2: The excess solid is filtered out.
- Stage 3: The filtrate is gently evaporated to concentrate the salt solution. This can be done on a heated water-bath (Figure 12.8) or sand tray (Figure 12.9).
- Stage 4: When crystals can be seen forming (crystallisation point), heating is stopped and the solution is left to crystallise.
- Stage 5: The concentrated solution is cooled to let the crystals form. The crystals are filtered off and washed with a little distilled water. Then the crystals are dried carefully between filter papers (Figure 12.10).

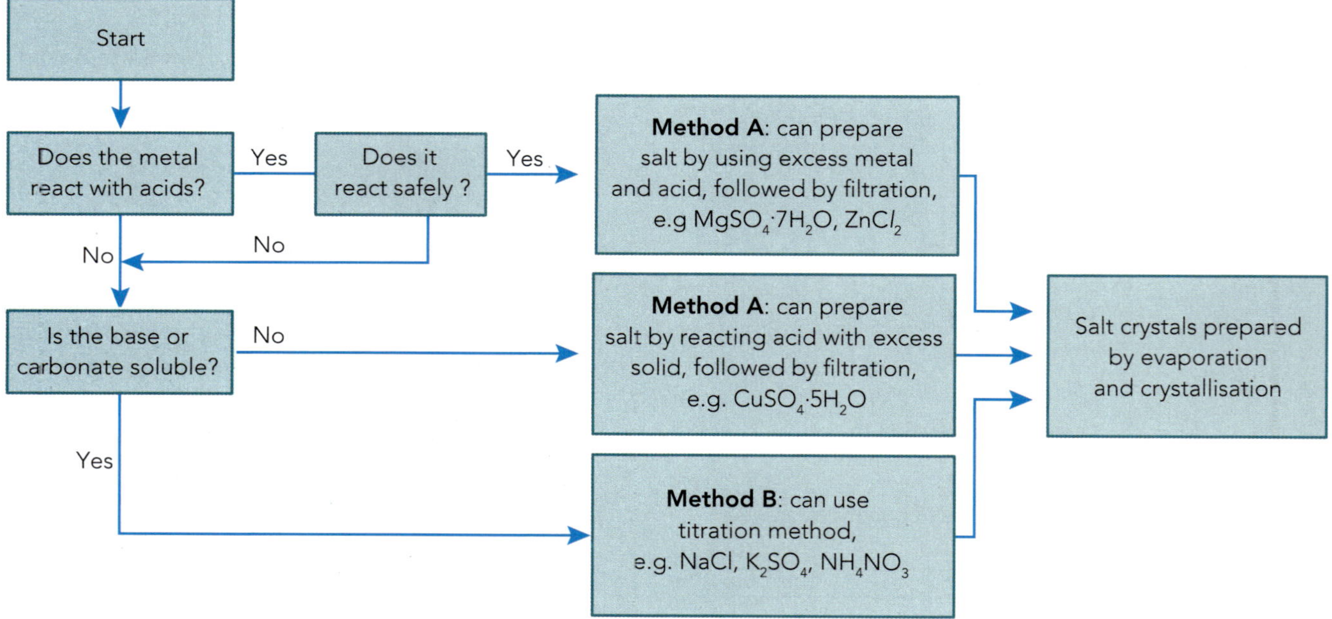

Figure 12.7: Flow chart showing which method to use for preparing soluble salts.

a

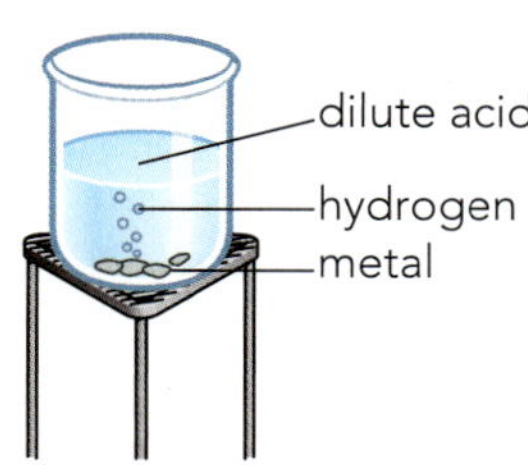

i
Warm the acid. Switch off the Bunsen burner. Add an excess of the metal to the acid. Wait until no more hydrogen is given off.

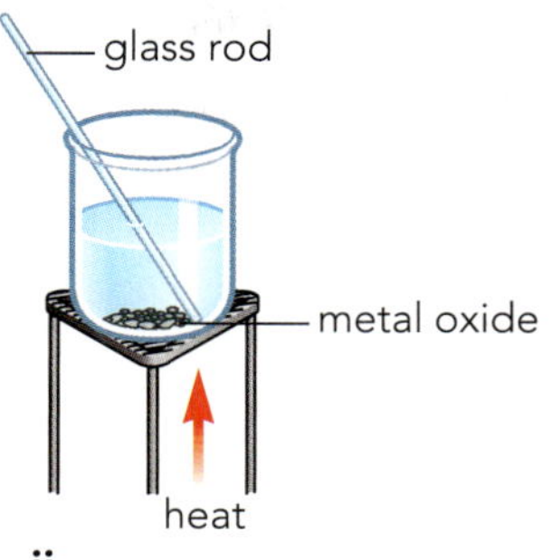

ii
Add an excess of the metal oxide to the acid. Wait until the solution no longer turns blue litmus paper red.

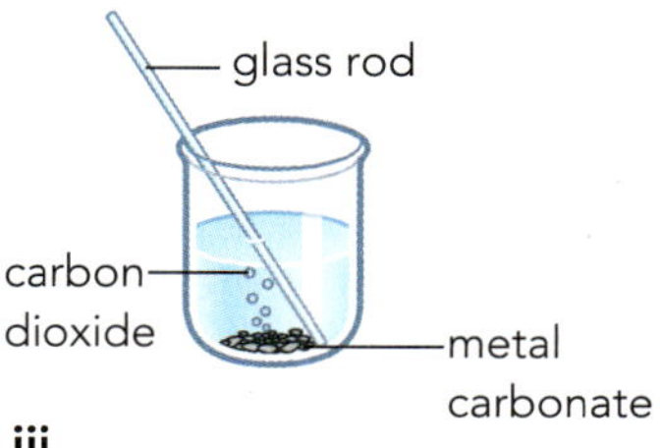

iii
Add an excess of the metal carbonate to the acid. Wait until no more carbon dioxide is given off.

b

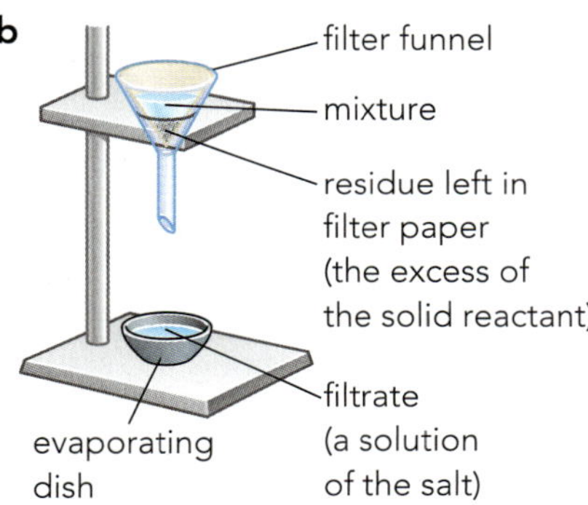

c

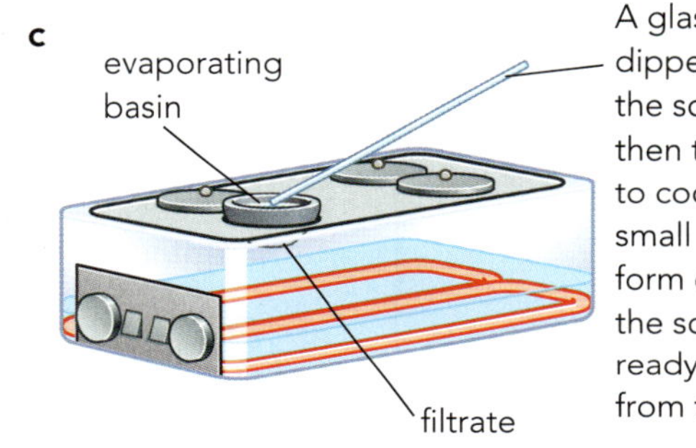

d

Crystals form as solution cools; filter, wash and then dry them.

Figure 12.8: Method A for preparing a soluble salt. **a:** Stage 1: the acid is reacted with either (i) a metal, (ii) a base or (iii) a carbonate. **b:** Stage 2: the excess solid is filtered out. **c:** Stage 3: the solution is carefully evaporated. **d:** Stage 4: the crystals are allowed to form.

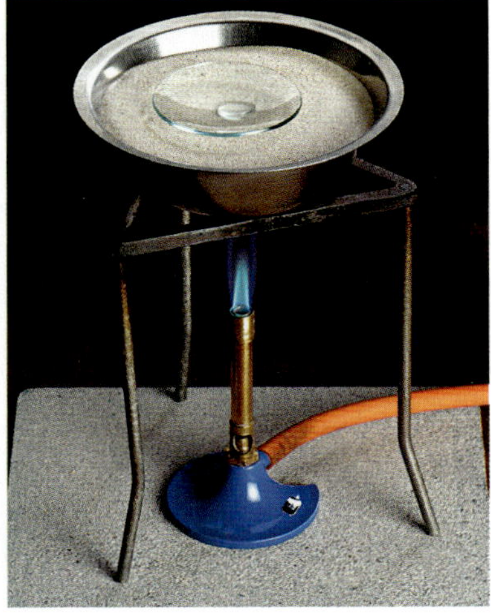

Figure 12.9: Evaporating off the water to obtain salt crystals. Here a sand tray is being used to heat the solution carefully.

Figure 12.10: Crystals of zinc nitrate after drying between filter papers.

EXPERIMENTAL SKILLS 12.1

Quick and easy copper(II) sulfate crystals

This activity is an adaptation of the larger-scale method of preparing a soluble salt (see Figure 12.8). The essential feature of this method is that it is easier to carry out than the method involving multiple stages. The method uses a reaction between a dilute solution of the parent acid (sulfuric acid in this case) and an excess of an insoluble base.

You will need:

- measuring cylinder (25 cm^3)
- boiling tube
- sulfuric acid (2 mol/dm^3)
- beaker (250 cm^3)
- copper(II) oxide
- filter funnel and filter paper
- conical flask (100 cm^3)
- crystallising dish
- Bunsen burner
- tripod and gauze
- heat-resistant mat
- kettle
- balance.

Safety

Wear eye protection throughout. Be careful with chemicals. Never ingest them and always wash your hands after handling them. Note that sulfuric acid is corrosive and an irritant at the concentration used.

Getting started

It is important that you are well organised and have all the apparatus and chemicals that you need before starting the experiment. You should be familiar with the techniques of filtration and crystallisation (Topic 21.2). You may find it helpful to set up the filtration apparatus before starting.

Method

1. Pour 15 cm^3 of 2 mol/dm^3 sulfuric acid into a boiling tube.
2. Place the tube in a beaker half-filled with boiling water from a kettle.
3. Weigh out between 1.8 g and 2.0 g of copper(II) oxide.
4. Add half the copper(II) oxide to the acid in the boiling tube. Agitate the boiling tube and return it to the hot water.
5. When the solid has dissolved, add the remaining portion of copper(II) oxide.
6. Keep the tube in the hot water for five more minutes, taking it out occasionally to agitate the tube.
7. Filter off the unreacted solid, collecting the clear blue solution in a 100 cm^3 conical flask. A fluted filter paper can be used to speed up the filtration.
8. Set up the Bunsen burner and boil the solution for two to three minutes.
9. Pour the hot solution into a clean, dry crystallising dish and watch the crystals grow!

Questions

1. Write word and balanced chemical equations for the reaction taking place.
2. What does the fact that there is some unreacted solid left after the reaction tell you about the proportions of reactants used? Why is it useful that the reaction is carried out with these proportions?

Peer assessment

The practical work to prepare copper(II) sulfate crystal requires careful and coordinated work with your partner. When you have completed the experiment, discuss with your partner how efficiently you worked together. Think about the following questions:

- Were you able to handle the manipulation of the reaction, filtration and crystallisation stages confidently?
- Did you share the work equally between you?
- Did carrying out the experiment help you to understand the importance of the different steps involved in preparing a soluble salt?

REFLECTION

Do you find carrying out practical work a useful way of learning aspects of chemistry?
Is there a way in which you could gain more from performing experiments?

Method B – acid plus alkali by titration

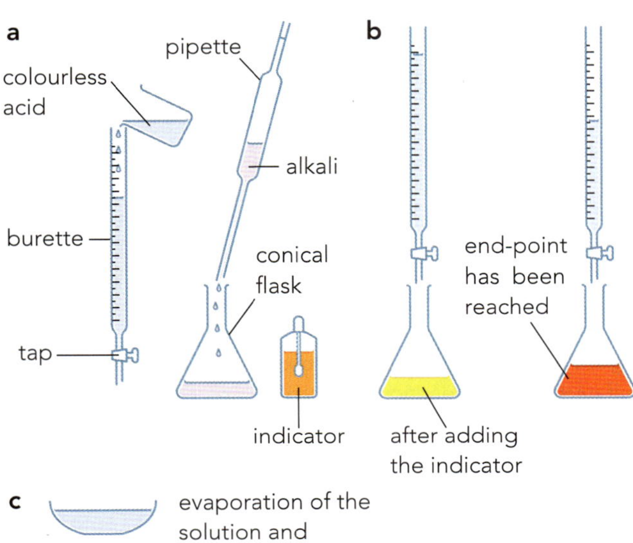

c

evaporation of the solution and crystallisation as in method A

Figure 12.11: Method B (the titration method) for preparing a soluble salt. **a:** Stage 1: the burette is filled with acid and a known volume of alkali is added to the conical flask. **b:** Stage 2: the acid is added to the alkali until the end-point is reached. **c:** Stage 3: the solution is evaporated and crystallised as for method A.

Method B (the **titration** method) involves the neutralisation of an acid with an alkali (e.g. sodium hydroxide) or a soluble carbonate (e.g. sodium carbonate) (Figure 12.11). Since both the reactants and the products are colourless, an indicator is used to find the neutralisation point or end-point.

- Stage 1: The acid solution is poured into a **burette**. The burette is used to accurately measure the volume of solution added. A known volume of alkali solution is placed in a conical flask using a **volumetric pipette**. The pipette delivers a fixed volume accurately. A few drops of an indicator (e.g. thymolphthalein or methyl orange) are added to the flask.

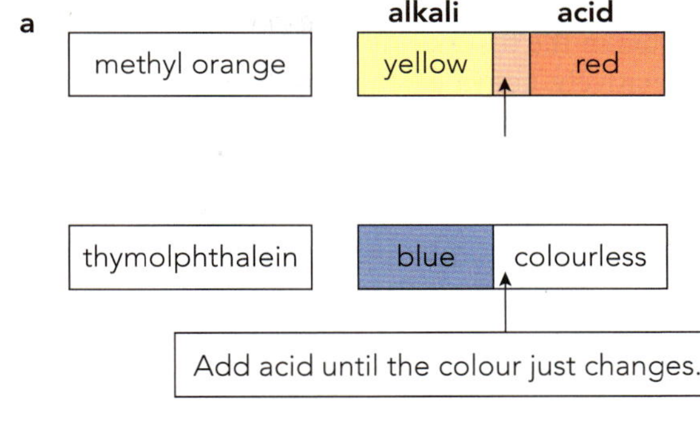

Figure 12.12 a: Colour changes for the indicators methyl orange and thymolphthalein during the titration. **b:** Actual colours of methyl orange in acid and alkali.

KEY WORDS

titration: a method of quantitative analysis using solutions: one solution is slowly added to a known volume of another solution using a burette until an end-point is reached

burette: a piece of glass apparatus used for delivering a variable volume of liquid accurately

volumetric pipette: a pipette used to measure out a volume of solution accurately

end-point: the point in a titration when the indicator just changes colour showing that the reaction is complete

- Stage 2: The acid solution is run into the flask a few drops at a time from the burette until the indicator just changes colour (Figure 12.12). The conical flask must also be swirled after each portion of acid to ensure everything is mixed and the reaction is complete. Having found the **end-point** for the reaction, the volume of acid (titre volume) added is noted. The experiment is then repeated without using the indicator. The same known volume of alkali is used in the flask. The same volume of acid as noted in the first part is then run into the flask. Alternatively, activated charcoal can be added to remove the coloured indicator. The charcoal can then be filtered off.

- Stage 3: The salt solution is evaporated and cooled to form crystals as described in method A.

This titration method is very useful not only for preparing salts but also for finding the concentration of a particular acid or alkali solution (Chapters 5, 11 and 22).

Questions

4 Name the salts formed when:
   a dilute hydrochloric acid reacts with magnesium
   b calcium oxide reacts with dilute nitric acid
   c zinc carbonate reacts with dilute sulfuric acid.

5 a In the methods of preparing a salt using a solid metal, base or carbonate, why is the solid used in excess?
   b In such methods, how is the excess solid removed once the reaction has finished?
   c Name the two important pieces of graduated glassware used in the titration method of preparing a salt.
   d What colour is the indicator methyl orange in alkali?
   e Why should the crystals prepared at the end of experiments to prepare salts not be heated too strongly when drying them?

6 There are two methods of preparing soluble salts depending on the solubility of the reagent reacted with the acid: method **A** (titration using a burette and an indicator) and method **B** (addition of an excess of a base or a metal to a dilute acid and removal of the excess solid). Which method would you use to prepare the soluble salt, zinc sulfate, from the insoluble base, zinc oxide?
   i Write down the method.
   ii Write down the reagent to use.
   iii Write the word equation.

Preparing insoluble salts by precipitation

The reaction between marble chips (calcium carbonate) and sulfuric acid would be expected to produce a strong reaction, with large amounts of carbon dioxide being given off. However, the reaction quickly stops after a very short time. This is caused by the fact that calcium sulfate is insoluble. This insoluble calcium sulfate soon forms a layer on the surface of the marble chips, stopping any further reaction (Figure 12.13).

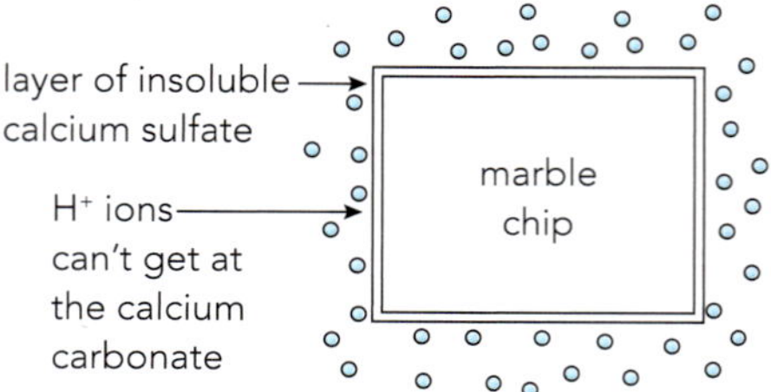

Figure 12.13: Formation of a layer of insoluble calcium sulfate stops the reaction of marble chips with sulfuric acid by forming a protective layer over the surface of the solid.

This reaction emphasises that some salts are insoluble in water, e.g. silver chloride and barium sulfate (Table 12.2). Such salts cannot be made by the crystallisation methods we have described earlier. They are generally made by ionic **precipitation**.

For example, barium sulfate can be made by taking a solution of a soluble sulfate (e.g. sodium sulfate). This is added to a solution of a soluble barium salt (e.g. barium chloride). The insoluble white solid, barium sulfate, is formed immediately. This solid 'falls' to the bottom of the tube or beaker as a precipitate (Figure 12.14). The precipitate can be filtered off. It is then washed with distilled water and dried in a warm oven.

KEY WORD

precipitation: the sudden formation of a solid when either two solutions are mixed or a gas is bubbled into a solution

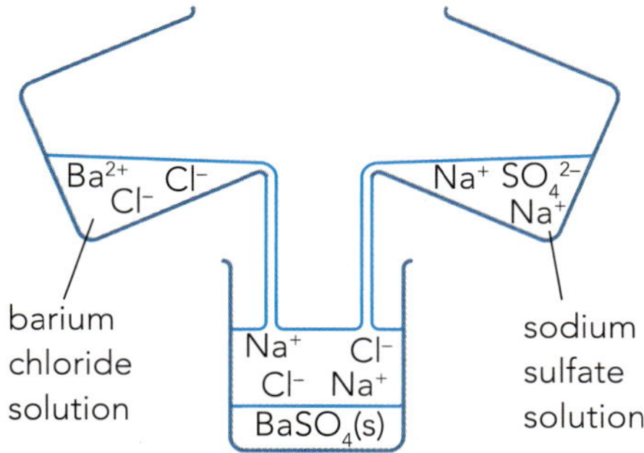

Ba^{2+} and SO_4^{2-} ions combine to form a precipitate of $BaSO_4$; the Na^+ and Cl^- ions stay in solution.

Figure 12.14: Precipitation of barium sulfate. The solid can be collected by filtration or centrifugation.

The equation for this reaction is:

barium chloride	+	sodium sulfate	→	barium sulfate	+	sodium chloride
$BaCl_2(aq)$	+	$Na_2SO_4(aq)$	→	$BaSO_4(s)$	+	$2NaCl(aq)$

This equation shows the importance of state symbols – it is the only way we can tell that this equation shows a precipitation.

The equation can be simplified to show only those ions that take part in the reaction and their products:

$$Ba^{2+}(aq) + SO_4^{2-}(aq) \rightarrow BaSO_4(s)$$

This type of equation is known as an **ionic equation** (Chapter 4). The ions that remain in solution are left out of the equation. The ions that do not take part in the reaction are known as spectator ions.

Figure 12.15 shows the formation of a precipitate of silver chloride when solutions of silver nitrate and sodium chloride are mixed.

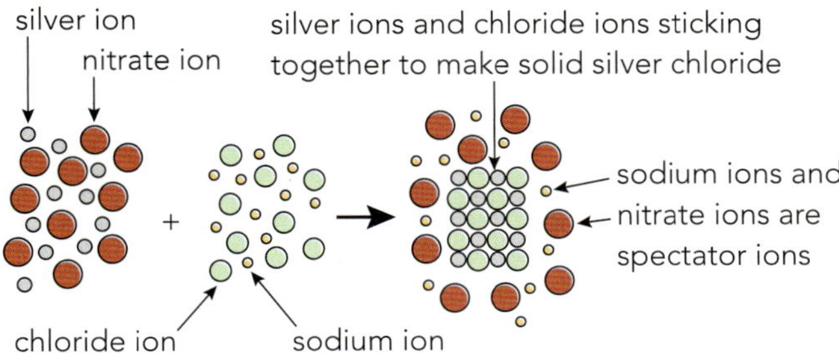

Figure 12.15: Illustration of which ions take part in the precipitation of silver chloride.

The different forms of equations for this precipitation of silver nitrate are:

silver nitrate	+	sodium chloride	→	silver chloride	+	sodium nitrate
$AgNO_3(aq)$	+	$NaCl(aq)$	→	$AgCl(s)$	+	$NaNO_3(aq)$
$Ag^+(aq)$	+	$Cl^-(aq)$	→	$AgCl(s)$		

Precipitation reactions are often used in analysis to identify salts such as chlorides, iodides and sulfates (Chapter 22). Figure 12.16 shows the precipitation of yellow lead iodide by mixing lead nitrate and potassium iodide solutions.

Figure 12.16: Precipitation of lead(II) iodide by mixing two colourless solutions.

The lead(II) nitrate and potassium iodide solutions react to produce solid lead(II) iodide, leaving soluble potassium nitrate in solution. The yellow precipitate of lead(II) iodide (Figure 12.16) can be recovered by filtration. If the precipitate is washed, dissolved in hot water and re-crystallised, some quite spectacular crystals can be obtained.

$$Pb^{2+}(aq) + 2I^-(aq) \rightarrow PbI_2(s)$$

You will note that one easy way to write the ionic equations of these precipitations correctly is to work backwards. First write down the formula for the solid product on the right-hand side. Then, on the left-hand side, write down the ions that combine together to form the precipitate. Finally, write in the state symbols.

KEY WORDS

ionic equation: the simplified equation for a reaction involving ionic substances: only those ions which actually take part in the reaction are shown

Questions

7 What do the following terms mean in connection with chemical reactions to produce salts?

   a precipitation
   b titration
   c an ionic equation

8 Which of these salts can be prepared by precipitation?

   A silver iodide
   B magnesium nitrate
   C lead(II) chloride

9 The insoluble salt, barium sulfate, can be prepared from a solution of barium nitrate by precipitation.

   a Name a soluble salt that could be added to the barium nitrate solution to give a precipitate of barium sulfate.
   b What colour is the precipitate of barium sulfate?
   c Write the ionic equation (with state symbols) for the reaction.

SUMMARY

Salts are an important group of ionic compounds.
Some salts are soluble in water while other salts are insoluble.
There are general solubility rules that apply to the various different types of salt.
A hydrated substance is a substance that is chemically combined with water and an anhydrous substance is a substance containing no water.
The term water of crystallisation indicates the water molecules chemically combined in salt crystals.
Methods are available for the preparation, separation and purification of soluble salts by the reaction of the parent acid with either excess metal, excess insoluble base or excess insoluble carbonate.
A soluble salt can be prepared by titration of an acid with an alkali followed by separation and purification.
An insoluble salt can be prepared by precipitation followed by filtration.

PROJECT

Stop-Start Go – when ions collide!

'Stop-motion' animations are a popular tool for helping to explain concepts that involve movement. The animations are created by linking a series of individually photographed images. In this project, you are going to create a stop-motion animation of a precipitation reaction.

In a precipitation reaction, solutions containing ions are mixed. The ions are moving in the solutions but when certain ions meet, they combine and fall together out of solution (Figures 12.14 and 12.15). The reaction lends itself to animation as we cannot see the individual ions in solution and how they come together.

In a pair or small group, choose a precipitation reaction to focus on. Figure 12.17 shows the precipitation of silver chloride but you can choose one of several different reactions (think about the analysis tests in Chapter 22).

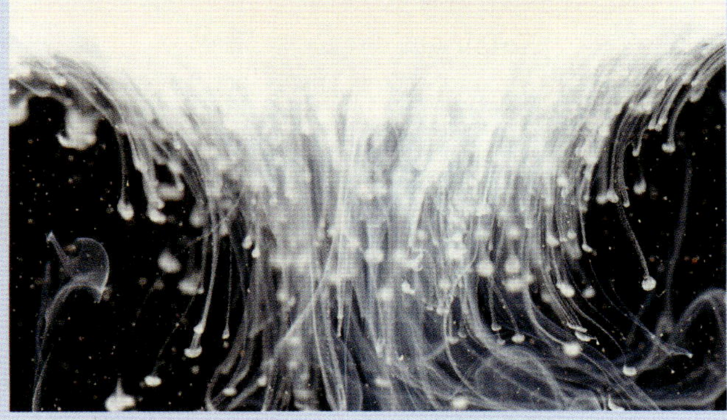

Figure 12.17: Precipitation of silver chloride.

You need to produce a storyboard to help you visualise the final animation. This should explain the movement of the ions during the mixing of the two solutions. Your storyboard must include:

1. an image of the two separate solutions before mixing
2. some images of one solution being poured into the other
3. images of the precipitate settling; possibly of it being stirred up and allowed to settle again.

Having produced your storyboard, you should then produce your animation. This can be done in various ways. For example, you could draw some illustrations similar to Figures 12.15 and 12.16 to show diagrammatically how the precipitate forms. Include some pictures of how an ionic equation for the reaction is constructed. Then take a series of photographs, perhaps adding callouts which you can assemble together as a set of slides. You could use a stop-start animation app that will allow you to put together a series of photographs and add a voiceover to describe the process. In both cases, the more photographs you take and the less the movement between them, the better it will look.

If you do not have access to a camera/laptop then you could produce a flipbook with lots of little images showing a precipitation reaction.

EXAM-STYLE QUESTIONS

1 The method of preparing a soluble salt in the laboratory has certain clear stages. Which three steps shown in the table are needed to make sodium sulfate crystals from sodium hydroxide solution and dilute sulfuric acid?

	First step	Second step	Third step
A	evaporation	crystallisation	neutralisation
B	neutralisation	evaporation	crystallisation
C	neutralisation	crystallisation	evaporation
D	evaporation	neutralisation	crystallisation

[1]

2 A list of salts is shown below:

barium sulfate **copper sulfate** **potassium carbonate**
sodium chloride **zinc nitrate**

Which of these salts:

a cannot safely be made by reacting acid with a metal? **[1]**
b is made by titration? **[1]**
c is insoluble? **[1]**
d reacts with acid to produce a gas? **[1]**

[Total: 4]

3 Excess magnesium carbonate was added to dilute sulfuric acid to make a sample of magnesium sulfate.

a What observation would be made as the reaction took place? **[1]**
b Why was excess magnesium carbonate used? **[1]**
c How would you know when the reaction was complete? **[1]**
d Explain how you would separate a pure sample of magnesium sulfate from the mixture. **[3]**

[Total: 6]

4 Which of the copper salts A–D could be made by the following method?

- mix the solutions of two salts
- filter the mixture
- wash the residue
- dry the residue

A copper carbonate
B copper chloride
C copper nitrate
D copper sulfate **[1]**

COMMAND WORD

explain: set out purposes or reasons/make the relationships between things evident/provide why and/or how and support with relevant evidence

CONTINUED

5 A student wanted to make a sample of the insoluble salt lead(II) chloride.

a Name two substances which could be used to form this salt. [2]

b List the steps the student should use to produce a pure dry sample of lead(II) chloride from these substances. [4]

[Total: 6]

6 A student attempted to make carbon dioxide by reacting lumps of calcium carbonate with dilute sulfuric acid. The reaction produced a small amount of carbon dioxide but then stopped.

a Using your knowledge of the solubility of salts, explain why this reaction did not work. [2]

b What simple change could be made to the method to make carbon dioxide from calcium carbonate? [1]

c Name a carbonate that would react with sulfuric acid to form carbon dioxide successfully. [1]

[Total: 4]

SELF-EVALUATION CHECKLIST

After studying this chapter, think about how confident you are with the different topics. This will help you see any gaps in your knowledge and help you to learn more effectively.

I can	See Topic...	Needs more work	Almost there	Confident to move on
understand that some salts are soluble in water while others are insoluble	12.1			
describe the general rules concerning the solubility of different types of salt in water	12.1			
define a hydrated substance as a substance that is chemically combined with water and an anhydrous substance as a substance containing no water	12.1			
define the term water of crystallisation as the water molecules present in crystals	12.1			
describe the preparation, separation and purification of a soluble salt by reaction of an acid with either excess metal, excess insoluble base or excess insoluble carbonate	12.2			
describe the preparation, separation and purification of a soluble salt by titration of an acid with an alkali	12.2			
describe the preparation and separation of an insoluble salt by precipitation	12.2			

Chapter 13

The Periodic Table

IN THIS CHAPTER YOU WILL:

- learn how the elements are organised into a table of periods and groups based on the order of increasing atomic (proton) number
- describe the relationship between the group number and ionic charge of the elements
- explain the similarities in chemical properties of the elements in a group
- describe the trends in properties of the Group I alkali metals
- describe the trends in properties of the halogens (Group VII)
- describe the noble gases (Group VIII) as unreactive, monatomic gases
- describe the change from metallic to non-metallic character across a period
- describe the key characteristics of the transition elements

- identify trends in groups given information on the elements
- describe the ability of the transition elements to form ions with variable oxidation numbers.

GETTING STARTED

Work in groups of three or four and use a copy of the Periodic Table.

1 Make an estimate of how many more metals there are than non-metals, and where they are placed in the Periodic Table.

2 What are the properties you most obviously associate with a metal? Make a list in your group.

3 Discuss the question 'Are all metals magnetic?'. What do you think?

4 If you have a magnet, try to use it on samples of aluminium, copper and magnesium. Can you name the three most important magnetic, metallic elements?

THE WOMEN BEHIND THE PERIODIC TABLE

The modern Periodic Table has been a major scientific achievement involving both men and women. The breakthrough in the organisation of the elements came in 1869 when the Russian chemist Dmitri Ivanovich Mendeleev put forward his ideas of a Periodic Table. In his first attempt he used 32 of the 61 elements known at that time and stressed the idea there was a periodic pattern to the properties of the elements. Mendeleev drew up his table based on atomic masses, but elements of similar mass and character are difficult to distinguish. Mendeleev was in correspondence with Julia Lermontova, a Russian scientist working in Heidelberg and the first woman to obtain a doctorate in chemistry. Her work on separation methods for the platinum group of elements meant that the elements could be properly placed in the Periodic Table. The success of Mendeleev's Periodic Table was mainly due to him leaving gaps in the table for further possible elements and in predicting the properties of elements that had not yet been discovered.

Many women have been involved in the search for new elements. Marie Curie discovered radium and polonium in the 1890s and was the first woman to be awarded a Nobel Prize. Other notable women involved in the discovery of the elements in the last century include Lise Meitner (protactinium in 1917), Ida Noddack (rhenium in 1925) and Marguerite Perey (francium in 1939). Most of these women worked with male collaborators and their contributions were not always fully acknowledged at the time.

Francium was the last element to be discovered in nature and the search for new elements now requires large teams with particle accelerators and big budgets. The work by American chemists Darleane Hoffman and Dawn Shaughnessy helped to discover the most recent six new 'superheavy' elements (elements 113–118) that were officially recognised in 2015. Clarice Phelps, one of the co-discoverers of tennessine (element 117), is the first African-American woman to discover a new chemical element (Figure 13.1).

Figure 13.1: Clarice Phelps, one of the co-discoverers of tennessine.

The critical role of women in science has been recognised by UNESCO. In 2015 UNESCO established the International Day of Women and Girls in Science, celebrated annually on 11 February, as an opportunity to promote access to and participation in science for women and girls.

Discussion questions

1 Why do you think that much of the early pioneering work by women scientists was not acknowledged at the time?

2 Do you think the situation has changed? Do you think there are obstacles to women and girls participating in science?

13.1 Classifying the elements

Groups and the Periodic Table

All modern versions of the **Periodic Table** are based on the table proposed by Mendeleev in 1869. We have discussed some aspects of the Periodic Table when describing atomic structure in Chapter 2. An example of the Periodic Table is given in Figure 13.2.

In the Periodic Table:

- elements are arranged in order of increasing proton (atomic) number
- vertical columns of elements with similar properties are called **groups**
- horizontal rows are called **periods**.

> **KEY WORDS**
>
> **Periodic Table:** a table of elements arranged in order of increasing proton number (atomic number) to show the similarities of the chemical elements with related electronic configurations
>
> **groups:** vertical columns of the Periodic Table containing elements with similar chemical properties; atoms of elements in the same group have the same number of electrons in their outer energy levels
>
> **period:** a horizontal row of the Periodic Table

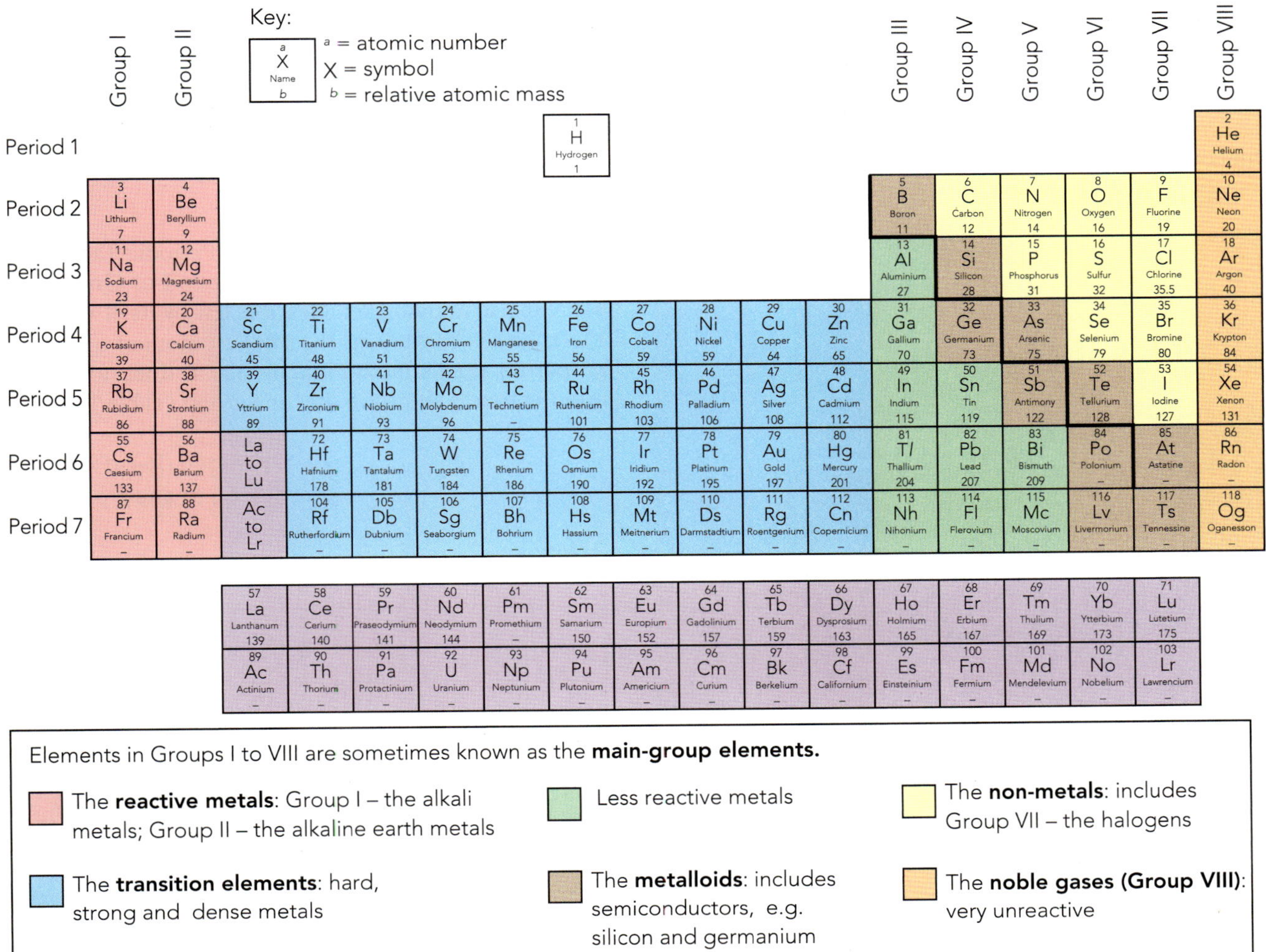

Figure 13.2: Periodic Table showing the major regions.

Metals and non-metals

The main distinction in the Periodic Table is between metals and non-metals. Metals are clearly separated from non-metals. The non-metals are grouped into the top right-hand region of the Periodic Table, above the thick stepped line in Figure 13.2. One of the first uses of the Periodic Table now becomes clear. Although we may never have seen a sample of the element hafnium (Hf), we know from looking at the Periodic Table that it is a metal. We may also be able to predict some of its properties. The change from metallic to non-metallic properties in the elements across a period is not as clear-cut as suggested by drawing the line between the two regions of the Periodic Table. The elements close to the line show properties that lie between these extremes (we will consider the properties of metals and non-metals in more detail in Chapter 14). These elements are now often referred to as **metalloids** (or semi-metals). Such elements have some of the properties of metals and others that are more characteristic of non-metals. There are eight elements that are called metalloids. They often look like metals but are brittle like non-metals. They are neither conductors nor insulators, but make excellent semiconductors. The prime example of this type of element is silicon (Figure 13.3).

Figure 13.3: A sample of the element silicon, the basis of the semiconductor industry.

The Periodic Table allows us to make even more useful subdivisions of elements than simply deciding which are metals and which are non-metals. The elements present in Groups I to VIII of the Periodic Table are sometimes known as the **main-group elements**. These vertical groups show most clearly how elements within the same group have similar chemical and physical properties. Some of these groups have particular names as well as numbers, and we will refer to these names later in the chapter. Between Groups II and III of these main groups of elements is a block of metals known as the **transition elements** (or transition metals). The first row of these elements occurs in Period 4. This row includes such important metals as iron, copper and zinc.

The noble gases, in Group VIII on the right-hand side of the Periodic Table, are the least reactive elements in the Periodic Table. However, the group next to them, Group VII, which are also known as the halogens, and the group on the left-hand side of the Periodic Table, Group I or the **alkali metals**, are the most reactive elements. The more unreactive elements, whether metals or non-metals, are in the centre of the Periodic Table.

Position of hydrogen in the Periodic Table

Hydrogen is difficult to place in the Periodic Table. Different versions place it above Group I or Group VII. More often, in modern tables, it is left by itself (Figure 13.4). This is because, as the smallest atom of all, its properties are distinctive and unique. It does not fit easily into the trends shown in any one group (Table 13.1).

1 H hydrogen — Hydrogen doesn't belong to any group.

Group

I	II	III	IV	V	VI	VII	
							2 He helium
3 Li lithium	4 Be beryllium	5 B boron	6 C carbon	7 N nitrogen	8 O oxygen	9 F fluorine	10 Ne neon
11 Na sodium	12 Mg magnesium	13 Al aluminium	14 Si silicon	15 P phosphorus	16 S sulfur	17 Cl chlorine	18 Ar argon
19 K potassium	20 Ca calcium						

Figure 13.4: Position of hydrogen in the Periodic Table.

Lithium	Hydrogen	Fluorine
solid at room temperature	gas	gas
metal	non-metal; forms diatomic molecules (H_2)	non-metal; forms diatomic molecules (F_2)
atom has one electron in outer shell	atom has one electron in outer shell	atom has seven electrons in outer shell
atom can lose one electron to achieve a noble-gas arrangement (forms a positive ion)	atom can form either a positive or a negative ion; can gain one electron to achieve a noble-gas arrangement, or lose its only electron	atom can gain one electron to achieve a noble-gas arrangement (forms a negative ion)

Table 13.1: A comparison of the properties of hydrogen with lithium (Group I) and fluorine (Group VII).

KEY WORDS

metalloid (semi-metal): element that shows some of the properties of metals and some of non-metals, e.g. boron and silicon

main-group elements: the elements in the outer groups of the Periodic Table, excluding the transition elements (Groups I–VIII)

transition elements (transition metals): elements from the central region of the Periodic Table – they are hard, strong, dense metals, which form compounds that are often coloured

alkali metals: elements in Group I of the Periodic Table; they are the most reactive group of metals

periodic property: a property of the elements that shows a repeating pattern when plotted against proton number (*Z*)

Organisation of the Periodic Table

When the first attempts were made to construct a Periodic Table, the structure of the atom was unknown. The order of the elements was originally provided by their increasing atomic masses. However, it was later found that atomic number provides a better basis for putting the elements in order. Atomic number provides a linear, continuous sequence to listing the elements. But the Periodic Table is a two-dimensional representation obtained by breaking the continuous list at certain points. The resulting fragments of the sequence of elements are then placed underneath each other. We split the sequence of elements according to the structure of the energy levels (shells) of electrons in the atom. A **periodic property** is one that shows a repeating pattern in the Periodic Table.

We can now directly link the properties of an element with its position in the Periodic Table and its electronic configuration (Figure 13.5). The number of outer electrons in the atoms of each element has been found. Elements in the same group of the Periodic Table have the same number of outer electrons. We also know that, as you move across a period in the Periodic Table, a shell of electrons is being filled. Each period in the Periodic Table represents the filling of an electron shell.

There is a clear relationship between electron arrangement and position in the Periodic Table for the main-group elements.

- Elements in the same group have the same number of electrons in their outer shell. For the main-group elements, the number of the group is the number of electrons in the outer shell
- Periods of the Periodic Table also have numbers. This number shows us how many shells of electrons the atom has.

A magnesium atom, for example, has two electrons in its third, outer, shell, and is in Group II and Period 3 (Figure 13.5).

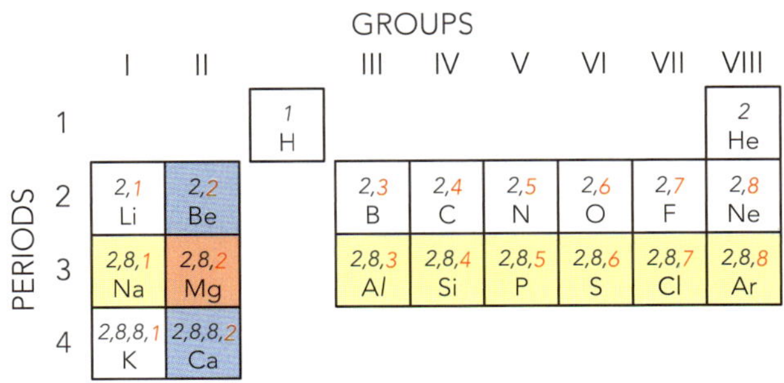

Figure 13.5: Relationship between an element's position in the Periodic Table and the electronic configuration of its atoms.

It is the outer electrons of an atom that are mainly responsible for the chemical properties of any element. Therefore, elements in the same group will have similar properties. Certain electron arrangements are found to be more stable than others. This makes them more difficult to break up. The most stable arrangements are those of the **noble gases**, and this fits in with the fact that they are so unreactive.

There are links between the organisation of particles in the atom and the regular variation in properties of the elements in the Periodic Table. This means that we can see certain broad trends in the Periodic Table (Figure 13.6). These trends become most obvious if we leave aside the noble gases in Group VIII. Individual groups show certain 'group characteristics'. These properties follow a trend in particular groups.

KEY WORDS

noble gases: elements in Group VIII – a group of stable, very unreactive gases

Relationship between group number and ionic charge

The relationship between position in the Periodic Table and electronic configuration means that there is a link between group number and the charge on the ions that atoms form (Figure 13.7).

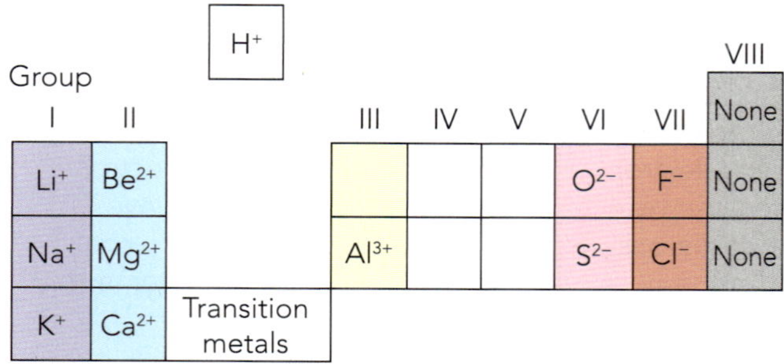

Figure 13.7: Ions formed by the main-group elements of the Periodic Table.

You can see that for the metal ions the positive charge is the same as the group number. These atoms have lost their outer electrons to form the ion. For non-metals the negative charge on the ion is 8 minus the group number. These atoms have gained electrons to form the negative ion. The atoms of elements in the middle of a period usually form covalent rather than ionic compounds. We have discussed this relationship earlier in Chapter 4 when talking about valency and writing formulae.

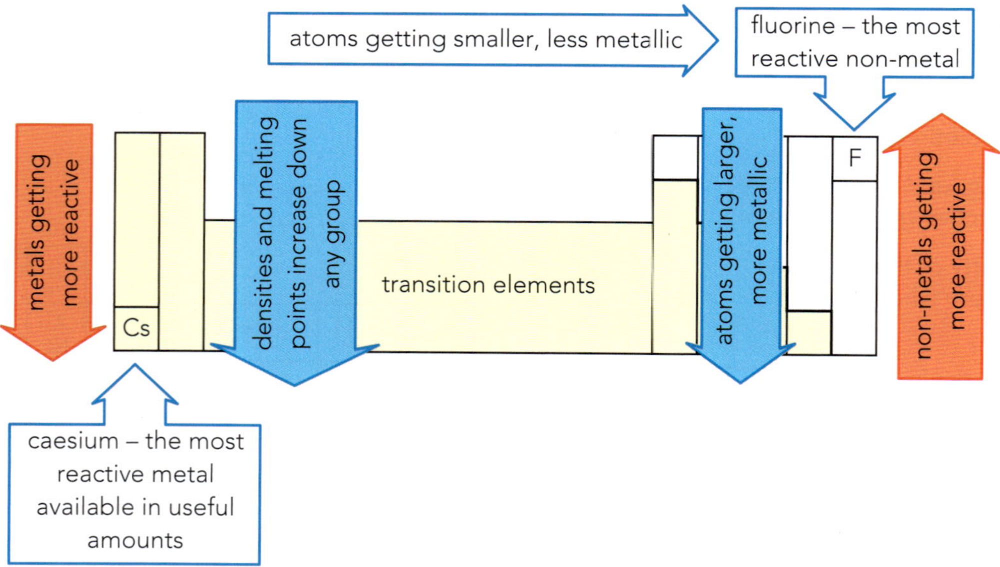

Figure 13.6: General trends in the Periodic Table, leaving asice the noble gases in Group VIII.

ACTIVITY 13.1

Discovering elementary patterns!

In groups, organise yourselves as a team of news reporters working on a breakthrough in organising a table of the elements, or the discovery of new elements. You are putting together a blog post celebrating this scientific progress.

Write a post celebrating the achievement of a scientist of your choice. You could choose a person from the history of chemistry (e.g. Mendeleev, Marie Curie or William Ramsay) or someone involved in the recent discoveries of new elements (e.g. Glenn Seaborg, Dawn Shaughnessy or Clarice Phelps).

Research the scientist of your choice and the context of their work. Remember that this is a news report, so does not need to be too technical.

Once you have written your blog post, get together with another group, swap reports and complete the following peer assessment.

Peer assessment

Having swapped blog reports with another group, discuss the following questions:

Does their blog post include?

1 A catchy/engaging heading.

2 A concise and clear introduction.

3 Specific facts and information about their chosen scientist and their achievement.

4 The group's own opinion on the achievement.

Now, write down the answers to the following questions to give some feedback to the other group:

- What do you think they did well?
- Next time they write a news blog, what could they do to improve?

REFLECTION

Having worked with others on Activity 13.1, consider whether you found group work helped you in your learning. Were you able to contribute confidently to the group discussion and appreciate the input of others to it? What would you do differently to gain more from this type of activity in future?

13.2 Trends in groups

Group I – the alkali metals

The metals in Group I are often called the alkali metals. They are soft solids with relatively low melting points and low densities (Figure 13.8). They are highly reactive and are stored in oil to prevent them reacting with the oxygen and water vapour in the air. When freshly cut with a knife, all these metals have a light-grey, silvery surface, which quickly tarnishes and becomes dull.

Figure 13.8: All alkali metals are soft solids and can be cut with a knife. This is a sample of lithium.

Metal	Electronic configuration	Density / g / cm^3	Melting point / °C	Boiling point / °C	Hardness
lithium	2,1	0.53	181	1342	fairly soft
sodium	2,8,1	0.97	98	883	soft
potassium	2,8,8,1	0.86*	63	760	very soft

** the densities of rubidium and caesium are 1.53 and 1.88, respectively*

Table 13.2: Physical properties of some alkali metals.

Alkali metal	Gas production	Heat produced by reaction	Production of alkaline solution	Production of flame
lithium	fizzes slowly, a few bubbles	moves slowly on the surface, does not melt	metal disappears slowly, solution turns purple	no flame
sodium	fizzes strongly, many bubbles	melts into a ball on surface and moves about	metal disappears quickly, solution turns purple	no flame unless held in place
potassium	fizzes violently, very many bubbles	melts and moves very quickly on surface	disappears very quickly, solution turns purple	hydrogen ignites, a lilac flame produced

Table 13.3: Observations on the reactions of the alkali metals with water (with universal indicator added).

The physical properties of the alkali metals change as we go down the group in Table 13.2. The melting points become lower and the density of the metals increases.

There are several different trends down this group of elements:

- melting points and boiling points decrease down the group
- metals get softer as we go down the group
- densities increase as we go down the group (although this trend is obscured in Table 13.2 because the value for sodium is higher than would be expected; see footnote to table).

Observing the trends in these values for physical properties means we can predict values for other alkali metals. For instance, we would predict the melting point of rubidium to be lower than that of potassium by about 20–30 °C as the gap between melting points is getting smaller at each step. This would suggest a value of between 33–43 °C. The actual value is 39 °C.

Chemical reactivity increases as we go down the group. All Group I metals react with cold water to form hydrogen and an alkaline solution of the metal hydroxide. The production of an alkaline solution can be shown by adding a few drops of universal indicator to the water. The solution will turn the water purple if an alkali is produced. The reactions range from vigorous in the case of lithium to explosive in the case of caesium. The strength of the reaction of each metal can be judged by several observations (Table 13.3). Observations such as the production of a gas, whether the metal melts on contact with the water and whether the gas produced is ignited all give an indication of how vigorous the reaction is.

You might predict that francium, at the bottom of Group I, would be the most reactive of all the metals. However, it is highly radioactive and very rare because it decays with a half-life of five minutes. This means that the amount of francium in any sample is halved in just five minutes. It has been estimated that there are only 30 g of francium in existence on Earth at any one moment in time.

The alkali metals (Group I) are the most reactive metals that occur. They are known as the alkali metals because they react vigorously with water to produce hydrogen and an alkaline solution. The equation for the reaction between sodium and water is:

sodium	+	water	→	sodium hydroxide	+	hydrogen
2Na	+	$2H_2O$	→	2NaOH	+	H_2

We will discuss the reactivity of the alkali metals further in Chapter 14.

Questions

1 What is the name of the alkali formed when potassium reacts with water?

2 Write a word equation for the reaction between lithium and water.

3 Using the information in Tables 13.2 and 13.3:

a suggest a value for the melting point of rubidium

b suggest how the reactivity of rubidium with water compares with potassium.

Group VII – the halogens

The most reactive non-metals are the **halogens** in Group VII of the Periodic Table (Figure 13.9). In contrast with Group I, in Group VII reactivity decreases down the group. For example, fluorine is a dangerously reactive, pale-yellow gas at room temperature. There is a steady increase in melting points and boiling points as we go down the group, and the elements change from gases to solids as the atomic number increases. Interestingly, the lowest element in this group is also a highly radioactive and rare element, astatine. The actual properties of astatine remain a mystery to us, but we could make a good guess at some of them.

> KEY WORDS
>
> **halogens:** elements in Group VII of the Periodic Table – generally the most reactive group of non-metals
>
> **halides:** compounds formed between an element and a halogen, e.g. sodium iodide

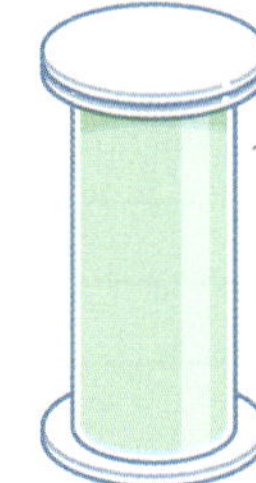

Chlorine (Cl_2)

- dense pale-green gas
- smelly and poisonous
- occurs as chlorides, especially sodium chloride in the sea
- relative atomic mass 35.5

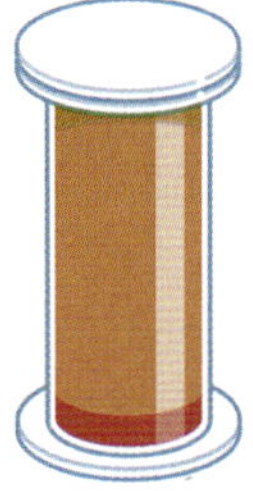

Bromine (Br_2)

- deep-red liquid with red-brown vapour
- smelly and poisonous
- occurs as bromides, especially magnesium bromide in the sea
- relative atomic mass 80

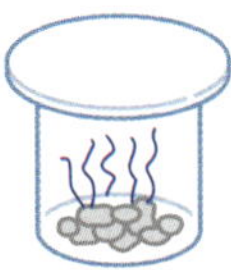

Iodine (I_2)

- grey solid with purple vapour
- smelly and poisonous
- occurs as iodides and iodates in some rocks and in seaweed
- relative atomic mass 127

Figure 13.9: General properties of some of the halogens (Group VII).

The halogen family found in Group VII of the Periodic Table shows clear similarities between the elements of the group. The common properties of the halogens are:

- They are all poisonous and have a similar strong smell.
- They are all non-metals.
- They all form diatomic molecules (e.g. Cl_2, Br_2 and I_2).
- They all have a valency of 1 and form compounds with similar formulae, e.g. hydrogen chloride (HCl), hydrogen bromide (HBr) and hydrogen iodide (HI).
- Their compounds with hydrogen are usually strong acids when dissolved in water, e.g. hydrochloric acid (HCl), hydrobromic acid (HBr) and hydriodic acid (HI).
- They each produce a series of compounds with other elements, e.g. chlorides, bromides and iodides. Together these are known as **halides**.

Halogen	Electronic configuration	Melting point / °C	Boiling point / °C	State at r.t.p	Colour
chlorine	2, 8, 7	−101	−35	gas	pale-green
bromine	2, 8, 18, 7	−7	+59	liquid	reddish–brown
iodine	2, 8, 18, 18, 7	+114	+184	solid	greyish–black

Table 13.4: Physical properties of some halogens.

- The halogens themselves can react directly with metals to form metal halides (or salts).
- They all form negative ions carrying a single charge, e.g. chloride ions (Cl^-), bromide ions (Br^-) and iodide ions (I^-).

There are gradual changes in properties between the halogens (Figure 13.9). As you go down the group, the boiling points increase (Table 13.4). The physical state of the halogens changes from gas to liquid to solid going down the group. The intensity of the colour of the element also increases, from pale to dark. Following these trends, it should not surprise you to know that fluorine is a pale-yellow gas at room temperature.

Chemical reactivity of the halogens

Fluorine and chlorine are very reactive. They combine strongly with both metals and non-metals. A piece of Dutch metal foil (an alloy of copper and zinc) will burst into flames when placed in a gas jar full of chlorine. When chlorine is passed over heated aluminium, the metal glows white and forms aluminium chloride:

$$2Al + 3Cl_2 \xrightarrow{\text{heat}} 2AlCl_3$$

Aluminium also reacts strongly with bromine and iodine. The reaction between a dry mixture of powdered aluminium and iodine can be triggered by adding just a few drops of water. The reaction is highly exothermic and some of the iodine is given off as purple fumes before it has a chance to react.

Hydrogen will burn in chlorine to form hydrogen chloride. Carried out a different way, the reaction can be explosive:

$$H_2 + Cl_2 \rightarrow 2HCl$$

Chlorine dissolves in water to give an acidic solution. This mixture is called chlorine water and contains two acids:

$$\underset{}{Cl_2} + \underset{}{H_2O} \rightarrow \underset{\text{hydrochloric acid}}{HCl} + \underset{\text{hypochlorous acid}}{HClO}$$

Chlorine water acts as an oxidising agent – hypochlorous acid can give up its oxygen to other substances. It also acts as a bleach because some coloured substances lose their colour when they are oxidised. This reaction is used as the chemical test for chlorine gas (Chapter 22). Damp litmus or universal indicator paper is bleached when held in the gas. The halogens become steadily less reactive as you go down the group. Table 13.5 gives some examples of the reactivity of the halogens.

The **halogen displacement reactions** shown in the lower part of Table 13.5 demonstrate the order of reactivity of the three major halogens. For example, if you add chlorine to a solution of potassium bromide, the chlorine displaces bromine (Figure 13.10). Chlorine is more reactive than bromine, so it replaces bromine and potassium chloride is formed. Potassium bromide solution is colourless. It turns orange when chlorine is bubbled through it:

$$Cl_2 + \underset{\text{colourless}}{2KBr} \rightarrow 2KCl + \underset{\text{orange}}{Br_2}$$

Chlorine will also displace iodine from potassium iodide:

$$Cl_2 + \underset{\text{colourless}}{2KI} \rightarrow 2KCl + \underset{\text{yellow–brown}}{I_2}$$

KEY WORDS

halogen displacement reactions: reactions in which a more reactive halogen displaces a less reactive halogen from a solution of its salt

Reaction with	Chlorine	Bromine	Iodine
coloured dyes	bleaches easily	bleaches slowly	bleaches very slowly
iron wool	iron wool reacts strongly to form iron(III) chloride; needs heat to start	iron reacts steadily to form iron(III) bromide; needs continuous heating	iron reacts slowly, even with continuous heating, to form iron(III) iodide
chlorides		no reaction	no reaction
bromides	displaces bromine, e.g. $Cl_2 + 2KBr \rightarrow 2KCl + Br_2$		no reaction
iodides	displaces iodine, e.g. $Cl_2 + 2KI \rightarrow 2KCl + I_2$	displaces iodine, e.g. $Br_2 + 2KI \rightarrow 2KBr + I_2$	

Table 13.5: Some reactions of the halogens.

Figure 13.10: Bromine is displaced by chlorine from a colourless solution of potassium bromide.

Group VIII – the noble gases

When Mendeleev first constructed his table, part of his triumph was to predict the existence and properties of some undiscovered elements. However, there was no indication that a whole group of elements (Group VIII) remained to be discovered! Because of their lack of reactivity, there was no clear sign of their existence. However, analysis of the gases in air (Chapter 17) by John William Strutt (Lord Rayleigh) led to the discovery of argon in 1894. There was no suitable place in the Periodic Table for an individual element with argon's properties. This pointed to the existence of an entirely new group. Helium, which had first been detected by spectroscopy of light from the Sun during an eclipse, and other noble gases in the group (Group VIII) were isolated by Sir William Ramsay in the 1890s. The radioactive gas, radon, was the last to be purified, in 1909.

All of the noble gases are present in the Earth's atmosphere. Together they make up about 1% of the total, though argon is the most common. These gases are particularly unreactive. They were sometimes referred to as the inert gases, meaning they did not react at all. However, since the 1960s, some compounds of xenon and krypton have been made and their name was changed to the noble gases.

The uses of the noble gases depend on their unreactivity. Helium is used in airships and balloons because it is both light and unreactive. Argon is used to fill incandescent light bulbs because it will not react with the filament even at high temperatures. The best-known use of the noble gases is, perhaps, its use in 'neon' lights (Figure 13.11). The brightly coloured advertising lights work when an electric discharge takes place in a tube containing small amount of a noble gas. Different gases give different colours.

Figure 13.11: 'Neon' lights give colour to Shanghai city centre by their use in advertising displays. The different colours are caused by different gases.

The atoms of the noble gases do not combine with each other to form molecules or any other form of structure. Their melting points and boiling points are extremely low (Figure 13.12). Helium has the lowest melting point of any element and cannot be solidified by cooling alone (pressure is also needed).

Figure 13.12: A small piece of rapidly melting 'argon ice'; the melting point is −189 °C.

All these properties point to the atoms of the noble gases being particularly stable:

- the electron electronic configurations of the atoms of the noble gases are energetically very stable
- this means that they do not react readily with other atoms
- in many situations where atoms of other elements bond or react chemically, they are trying to achieve the energetically stable arrangement of electrons found in the noble gases.

The elements of Group VIII lie between the two most reactive groups of elements (Groups I and VII) in the Periodic Table. Indeed, it is their closeness to this group with stable electron arrangements that makes the alkali metals and the halogens so reactive. They can fairly easily achieve a noble-gas electron structure. The Group VII elements gain or share electrons and the Group I elements lose electrons to reach a noble gas electronic configuration (see earlier discussion in Chapter 3).

Questions

4 Which halogen(s) will displace bromine from a solution of potassium bromide?

5 What is the similarity in the electron arrangement of the noble gases?

6 Arsenic is in the same group of the Periodic Table as nitrogen. Arsenic forms a compound with hydrogen just as nitrogen does. A molecule of this compound contains one atom of arsenic and three atoms of hydrogen.

 a What is the formula of this molecule?

  A 3AsH B As_3H

  C AsH_3 D $(AsH)_3$

 b What would you predict to be the formula of phosphine, the compound of phosphorus and hydrogen?

13.3 Trends across a period

Changes across a period

The elements of the vertical groups at either side of the Periodic Table show similar properties within the groups. However, following a period across the Periodic Table highlights the trend from metallic to non-metallic properties within the Periodic Table. This change from metal to non-metal can be explored by looking more closely across each period. The first period of the Periodic Table contains just two elements, hydrogen and helium, both of which are distinctive in different ways. The final period in the Periodic Table is now recently complete with the discovery of elements such as oganesson (Og) (see Chapter 4). Each of the five remaining periods of elements starts with a reactive alkali metal and finishes with an unreactive, non-metallic, noble gas. In Period 3, for example, from sodium to argon, there appears to be a gradual change in physical properties across the period. The change in properties seems to centre around silicon; elements before this behave as metals and those after it as non-metals (Figures 13.13 and 13.14).

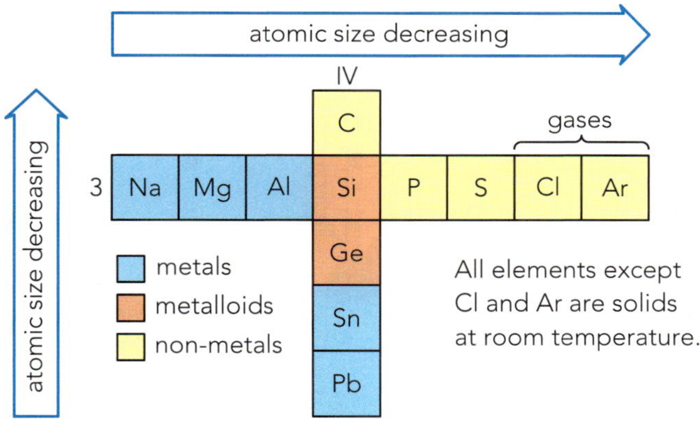

Figure 13.13: Changes in properties of the elements in Period 3 and in Group IV of the Periodic Table.

The changeover in properties is emphasised if we look at Group IV. As we go down this group, the change is from non-metal to metal. The metalloids, silicon and germanium, are in the centre of the group (Figure 13.13).

The different structures of the elements in Periods 2 and 3 are reflected in their boiling points and melting points. Figure 13.14 shows how the values for the melting points and boiling points of the elements of Period 3 increase up to Group IV and then decrease to Group VIII.

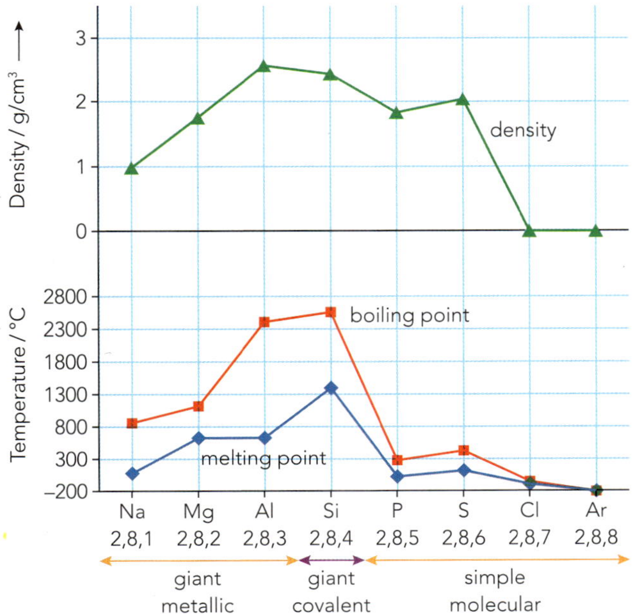

Figure 13.14: Trends in the structure and physical properties of the elements across Period 3.

As you move across a period there are also differences in some chemical properties linked to the metal to non-metal transition. Metal oxides (e.g. magnesium oxide) are usually basic, while those of non-metals (e.g. sulfur dioxide) are acidic. Therefore, there is a change from basic oxides to acidic oxides as we move across a period (see also Chapter 11).

The change from metal to non-metal across a period shows itself in changes in bonding type from metallic bonding to covalent bonding. To be more precise, in Periods 2 and 3 there is a shift from metallic lattice to giant covalent structure and then to simple molecular structure as we move across the Periodic Table. We commented earlier on this when we were discussing the formulae of the elements in Chapter 4.

Transition elements

If we look at Period 4 in the Periodic Table, we see that there is a whole 'block' of elements in the centre of the Periodic Table. This block of elements falls outside the main groups of elements that we have talked about so far. They are best considered not as a vertical group of elements but as a row or block. They are usually referred to as the transition elements (or transition metals). Their properties make them among the most useful metallic elements available to us (Figure 13.15). They are much less reactive than the metals in Groups I and II. Many have excellent corrosion resistance, for example chromium. The very high melting point of tungsten (3410 °C) led to its use in the filaments of incandescent light bulbs.

Many familiar objects are made from transition metals. Figure 13.15 shows a range of these objects, including steel nails, a vanadium spanner, silver cutlery, gold jack plugs, copper pipe joints, an iron magnet, a titanium camera body and chromium-plated balls on the Newton's cradle.

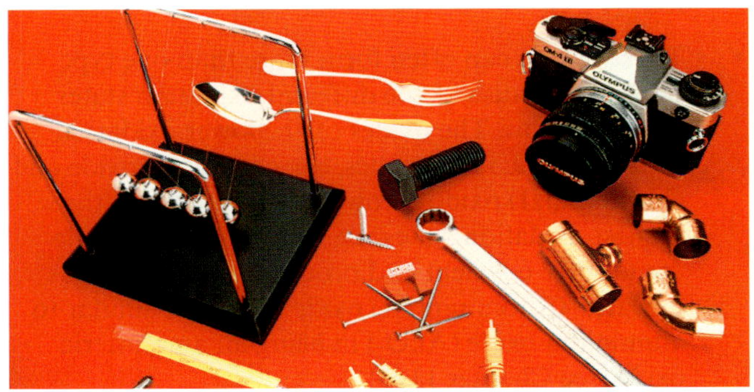

Figure 13.15: Some everyday objects made from transition metals.

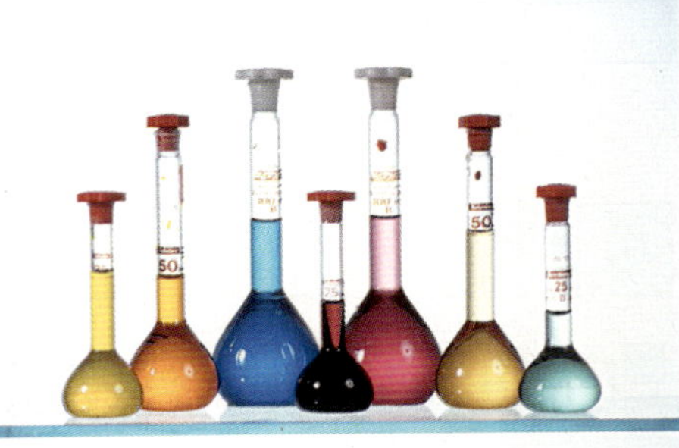

Figure 13.16: Many of the compounds of transition metals are coloured and, when they dissolve, they give coloured solutions.

These general properties mean that the transition metals are useful in a number of different ways. In addition, there are particular properties that make these metals distinctive and useful for more specific purposes. One important feature of transition metals is that their compounds are often coloured (Figure 13.16).

General features of transition metals:

- they are hard and strong
- they have high density
- they have high melting and boiling points.

Distinctive properties of transition metals:

- many of their compounds are coloured
- the transition metals, or their compounds, often act as catalysts. Iron, for example, is used as a catalyst in the Haber process for making ammonia (Chapter 9).
- they often show more than one valency (variable **oxidation number**) – they form more than one type of ion. For example, iron can form compounds containing iron(II) ions (Fe^{2+}) or iron(III) ions (Fe^{3+}).
- These metals can often form more than one type of oxide because of the different oxidation states they can show. The oxides of the lower oxidation state are basic ionic oxides, e.g. copper(II) oxide and chromium(II) oxide . However, the oxides of the highest oxidation states, e.g. chromium(VI) oxide , tend to be covalent and produce acidic solutions in water. Chromium(III) oxide (Cr_2O_3) is similar to aluminium oxide (Al_2O_3) in being an amphoteric oxide (see Chapter 11).

KEY WORDS

oxidation number: a number given to show whether an element has been oxidised or reduced; the oxidation number of a simple ion is simply the charge on the ion

Questions

7 Describe how metallic character changes across Period 2 or 3.

8 a Which metal is the softest and least dense in Period 3?

b Which of the elements in Period 3 has the highest melting point?

9 State four differences between an alkali metal and a transition metal such as nickel.

10 Which phrase (A–D) completes the following statement?

All transition metals in their compounds …

A … are white in colour

B … are magnetic

C … form ions of the type M^{2+} and M^{3+}

D … have variable oxidation states.

SUMMARY

The chemical elements can be organised into the Periodic Table based on increasing atomic (proton) number.
The electronic configuration of an element determines the group in which an element occurs in the Periodic Table and the chemical properties of the element.
The ionic charge of an element relates to the group number of that element.
The chemical and physical properties of the Group I alkali metals (lithium, sodium and potassium) vary as we descend the group and we can predict the properties of other members of the group, given data.
Information on given elements in a group can be used to identify trends within all the elements of that group.
Trends in the data for certain properties of the halogens can be used to predict the properties of other halogens in the group.
The noble gases (Group VIII) are unreactive, monatomic gases because they have eight electrons in their outer shells.
There is a trend from metallic to non-metallic character in the properties of the elements as we move across a period in the Periodic Table.
The transition elements (transition metals) have certain common key characteristics.
Transition metal atoms are able to form ions with variable oxidation numbers.

PROJECT

Elements card game

Symbol: **Li**

Name: **Lithium**

Atomic number: **3**

Mass number: **7**

Group: **I**

Period: **2**

Physical state at r.t.: **Solid**

Figure 13.17: Suggested layout for an elements card – probably not a winning one!

This is a card game in which two people compete with each other on a chosen value in a set of cards featuring a particular collection of objects; in this case the elements. In groups, make a set of cards based on the first 36 elements of the Periodic Table.

Figure 13.17 shows a suggested set of information to enter on the cards, although you could add in further or alternative information (but do make it relevant to your learning).

Having made the cards, deal out a set of cards between a pair of you and challenge each other according to the following criteria:

- for names and symbols – alphabetical precedence wins
- for numbers – the higher number wins
- for physical state – solid beats gas; with the one liquid trumping all.

As you work through the game you can ask each other questions about the cards you are currently challenging with; which element is the more reactive of the two elements, what is the electronic configuration of the atoms and so on.

Use the game to reinforce your familiarity with the elements.

EXAM-STYLE QUESTIONS

1 The figure shows an outline of the Periodic Table with certain elements marked.

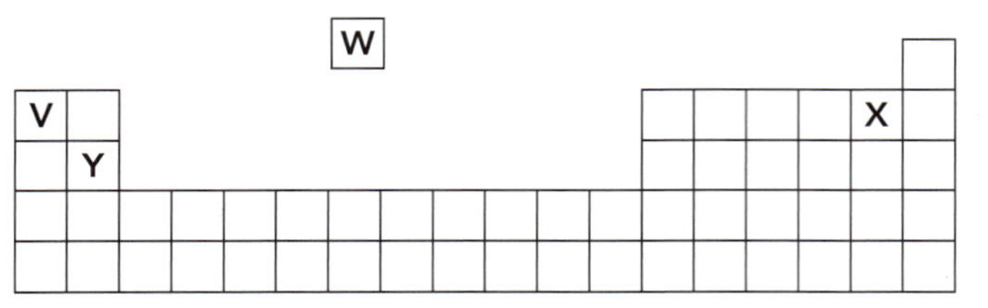

Which combination of the elements **V**, **W**, **X** or **Y** in the table is a metal and a non-metal?

	Non-metal	Metal
A	Y	V
B	Y	X
C	W	X
D	W	V

[1]

2 The reactivity of elements within a group in the Periodic Table changes with their position in the group. Which combination in the table shows the order of increasing reactivity of elements in Group I and Group VII?

	Group I	Group VII
A	Cs → Li	F → I
B	Li → Cs	I → F
C	Li → Cs	F → I
D	Cs → Li	I → F

[1]

3 The table shows part of two groups of the Periodic Table.

Group I	Group VII
lithium, **Li**	
sodium, **Na**	chlorine, **Cl**
potassium, **K**	bromine, **Br**
	iodine, **I**

Choose from the elements given:

a a solid Group VII element [1]

b the Group I element with the lowest density [1]

c an element that is a liquid at room temperature [1]

d the two elements that would react most violently with each other [2]

[Total: 5]

CONTINUED

4 A list of statements about elements of the Periodic Table is given below. **State** whether each of the following statements is true or false.

a Elements are arranged in order of their mass numbers. **[1]**

b All elements in Group VIII, the noble gases, have eight electrons in their outer shell. **[1]**

c Transition elements have high densities. **[1]**

d Transition elements are all metals. **[1]**

e Elements in the same period of the Periodic Table have the same number of electron shells. **[1]**

[Total: 5]

5 a Manganese is an important transition metal. Information about three manganese compounds is given below.

- Manganese(II) sulfate is a pale pink crystalline solid.
- Manganese(IV) oxide is a black solid which speeds up the decomposition of hydrogen peroxide.
- Manganese(VII) oxide is a purple liquid.

Which three characteristic properties of transition metals are shown in this information? **[3]**

b Chromium forms three different metal oxides:

chromium(II) oxide, CrO

chromium(III) oxide, Cr_2O_3

chromium(VI) oxide, CrO_3

CrO is a basic oxide, while CrO_3 dissolves in water to form a strong acid. What does this tell you about the type of bonding present in these two oxides? **[2]**

c Cr_2O_3 has similar bonding to aluminium oxide, Al_2O_3. What type of oxide is it likely to be? **[1]**

[Total: 6]

COMMAND WORD

state: express in clear terms

SELF-EVALUATION CHECKLIST

After studying this chapter, think about how confident you are with the different topics. This will help you see any gaps in your knowledge and help you to learn more effectively.

I can	See Topic...	Needs more work	Almost there	Confident to move on
understand how the elements in the Periodic Table are organised in order of increasing atomic number	13.1			
explain the similarities of elements in the same group in terms of their electronic configuration	13.1			
describe the relationship between group number and ionic charge	13.1			
describe the trend in physical and chemical properties of the alkali metals (Group I)	13.2			
describe the trend in physical and chemical properties of the halogens (Group VII)	13.2			
describe the properties of the noble gases (Group VIII)	13.2			
identify trends in groups given information on the elements	13.2			
describe the change from metallic to non-metallic character of the elements across a period	13.3			
describe the key characteristics of the transition elements (transition metals)	13.3			
describe the ability of transition elements to form ions with variable oxidation numbers	13.3			

Chapter 14

Metallic elements and alloys

IN THIS CHAPTER YOU WILL:

- understand the differences in the physical properties of metals and non-metals
- learn about the reactions of metals with water, acids and oxygen
- explore and appreciate the usefulness of metallic elements
- learn that alloys are mixtures of metals with other elements
- learn how diagrams can be used to represent the physical structure of alloys

- understand the uses of certain alloys in terms of their physical properties
- consider how alloys can be harder or stronger than pure metals.

GETTING STARTED

Metals, and structures made from them, play an important part in our lives.

1 Take a look at the immediate environment around you. What objects do you see that are made using metals?

2 How many different metals can you list that are being used in those objects?

3 Working with a partner, discuss the term 'alloy'.

  a Where have you heard the term before?

  b What do you think an alloy is?

Be ready to share your ideas with others in the class.

SMART ALLOYS REMEMBER THAT SHAPE!

A series of metal alloys with remarkable properties has recently found use in new and exciting applications. These alloys, called shape memory alloys, have great elasticity (they bend without breaking) and a property not usually linked to metals – memory! Ordinary metals do not have any memory of their shape. If you sit on a pair of aluminium spectacles and bend them permanently, it is difficult to get them back to their original shape. Shape memory materials behave differently. They can be bent (deformed) at one temperature but then recover their original, undeformed shape when heated.

The first of these alloys to find widespread practical use was 'nitinol' (a nickel–titanium alloy). There is now a range of these shape memory alloys, with some based on the original nickel–titanium combination mixed with either copper or aluminium. Unlike many traditional metals and alloys, shape memory alloys are both strong and flexible, easy to sterilise, unreactive and biocompatible. Being biocompatible, they are used in medical and health-related equipment, including spectacles frames (Figure 14.1), dental implants and miniaturised stents to hold veins and arteries open and allow blood to flow. They are also lightweight and resistant to corrosion, and have been widely used in engineering applications, such as in robotics and the aerospace industry.

Orbiting satellites use solar panels to generate the power they use to carry out their functions. These solar panels are attached to the satellite by arms made of shape memory metal. The satellites are stored away in the body of the rocket prior to launching. The rocket is launched at a relatively low temperature and the shape memory metal arms are bent inwards with the solar panels safely folded up with the satellite. Once in space, the satellite is released from the rocket. When exposed to the warmth of the Sun's rays, the memory metal arms automatically spring back to their original shape allowing the solar panels to fold out and function.

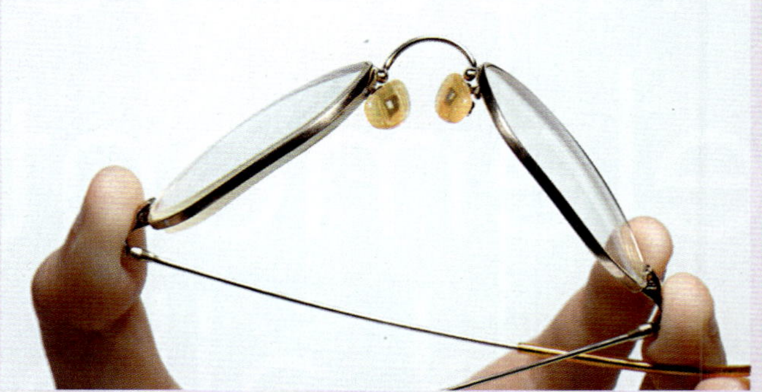

Figure 14.1: A spectacles frame made from nitinol can be flexed, bent or twisted but would return to its original shape afterwards.

Discussion questions

1 Nitinol is the name for the shape memory alloy developed during the 1960s at the Naval Ordnance Laboratory in the USA. Why do you think it was given this name by its discoverer?

2 The study of the properties of material is called material science. Which of the properties of shape memory alloys make them particularly suitable for the different applications?

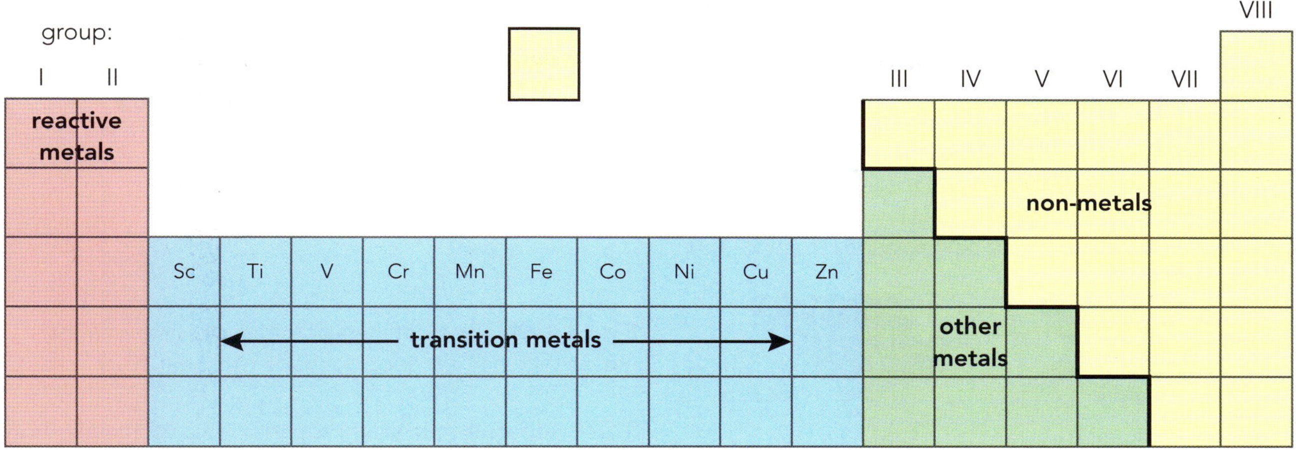

Figure 14.2: Distribution of metals and non-metals in the Periodic Table.

14.1 The properties of metals

Physical properties of metals and non-metals

Most of the elements (70 out of a total of 118 known elements) can be classified as metals. They are positioned to the left in the Periodic Table (Figure 14.2). The metals form a group of elements whose structures are held together by metallic bonding (a form of bonding discussed in detail in Chapter 3). This type of bonding gives rise to many properties that the metals have in common. However, there is wide variation in the level of these properties when we look in detail at different metals. For instance, many metals have a high density (e.g. iron), but the alkali metals (Group I) have low densities (Chapter 13).

Non-metals lie to the right of the table and they tend to have properties that are the opposite of the metals. This transition, from left to right, is one of the major trends present in the structure of the Periodic Table and is discussed in Chapter 13. Non-metals are a less uniform group of elements. Non-metals form either simple molecular structures (e.g. hydrogen, oxygen and sulfur) or giant covalent structure (e.g. carbon). This means that non-metals show a much wider range of properties, reflecting the wider differences in their structure (see Chapter 3 for discussion of the bonding in non-metals).

The best way to identify a metal or non-metal is to check whether it conducts electricity or not. The simple apparatus for doing this was shown as the method for testing the **electrical conductivity** of a solid in Chapter 6. A simple circuit is set up and any current flow is detected by the bulb (or measured by an ammeter).

All metals conduct electricity, even mercury, which is liquid at room temperature. All non-metals, except graphite, do not conduct electricity. Together with electrical conductivity, **malleability**, **ductility** and **thermal conductivity** are the other physical properties that most clearly characterise a metal.

KEY WORDS

electrical conductivity: the ability to conduct electricity

malleability: the ability of a substance to be bent or beaten into shape. Malleable is the word used to describe a substance that can be bent or beaten into shape.

ductility: the ability of a substance to be drawn out into a wire. Ductile is the word used to describe a substance that can be drawn out into a wire.

thermal conductivity: the ability to conduct heat

Metals have a crystalline structure in which the metal atoms are packed in layers that can slide over each other.

This regular lattice structure gives rise to the characteristic properties of metals. Metals are said to be workable (malleable) and are usually easy to beat or bend into shape. This is why metals are so useful to us for making things. They can also be stretched into wires (they are ductile). If a metal is subjected to a small force, it will stretch and then return to its original shape when the force is released. However, if a large force is applied, the layers of atoms slide over each other and stay in their new positions (Figure 14.3).

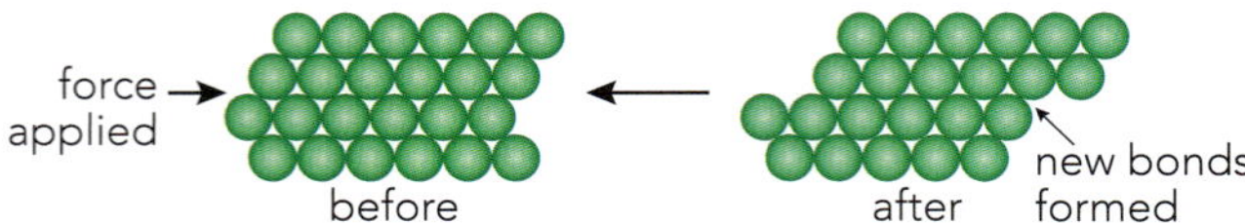

Figure 14.3: The layers of metal atoms in their structure mean that they can slide over each other when a force is applied.

Metals conduct heat well. This good thermal conductivity results from the structure of the metals. When one part of a metal is heated, the metal atoms vibrate more strongly (the vibration of atoms in solids is discussed in Chapter 1). The increased vibration of the metal atoms where the metal is heated is passed throughout the whole metal structure (Figure 14.4).

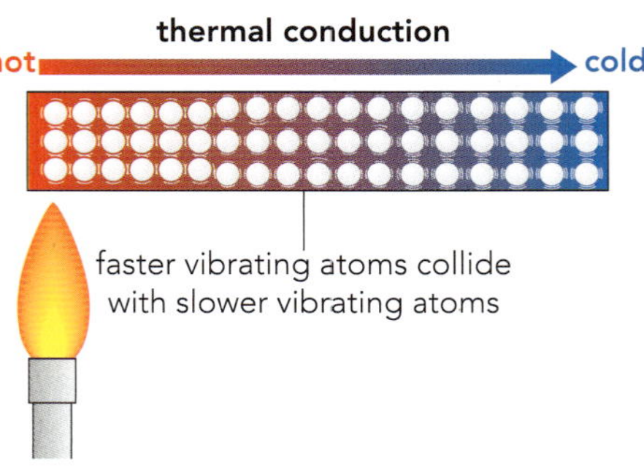

Figure 14.4: The giant metallic lattice structure of metals means that they are good conductors of heat.

These important physical properties that all metals show to a greater or lesser extent are directly related to their structural characteristics. Table 14.1 presents a summary of the differences in the physical properties of metals and non-metals.

KEY WORD

sonorous: a word to describe a metallic substance that rings like a bell when hit with a hammer

Physical property	Metals	Non-metals
melting points and boiling points	Metals are usually solids (except for mercury, which is a liquid) at room temperature. Many metals have high melting and boiling points.	Non-metals are solids or gases at room temperature (except for bromine, which is a liquid). Their melting and boiling points are often low.
electrical conductivity	All metals are good conductors of electricity.	Non-metals are poor conductors of electricity (except for graphite). They tend to be insulators.
thermal conductivity	Metals are good conductors of heat.	Non-metals are generally poor thermal conductors (except diamond, which conducts heat strongly).
malleability and ductility	The shape of a piece of metal can be changed by hammering (they are malleable). They can also be pulled out into wires (they are ductile).	Non-metals are not malleable; they are brittle and break easily when hit.
strength and hardness	Metals are usually strong and dense. When a force is applied, they are hard and do not shatter (they are not brittle).	Most non-metals are softer than metals (but diamond is very hard). Their densities are often low.
ability to produce a sound	Metals usually make a ringing sound when struck (they are **sonorous**).	Non-metals are not sonorous.
colour and appearance	Metals are grey in colour (except gold and copper). They can be polished.	Non-metals vary in colour. They often have a dull surface when solid.

Table 14.1: How the properties of metals and non-metals differ.

Questions

1 Sort the properties in the box into two sets.

   a Those that are characteristic of a metal.

   b Those typical of a non-metal.

is an insulator can be beaten into sheets gives a ringing sound when hit conducts heat well conducts electricity well has a dull surface

2 State the names of the following.

   a A non-metal that conducts electricity.

   b Two metals that are soft.

   c A metal that is liquid at room temperature.

   d A non-metal that conducts heat well.

3 List the three properties that most clearly help you to tell the difference between a metal and a non-metal. Explain how the structure of a metal gives rise to any two of these properties.

REFLECTION

What learning strategies have you used to learn about the link between metallic bonding and metallic properties? Do you find it useful to create summary diagrams or bullet point lists of the key points?

Chemical reactions of metals with air, water and dilute acids

The chemical properties of metals and non-metals are also very different, which makes the distinction between these two types of substances very important. Metals show some characteristic reactions with common reagents such as water, oxygen and acids. However, the strength and rate of these reactions varies greatly depending on the particular metal considered. We have seen in the previous Chapter 13 that there are variations in the reactivity of metals as one goes down a particular group of the Periodic Table. The reactivity of the Group I alkali metals with water increases as we go down the group. Also, we saw that the transition elements (transition metals) are less reactive than the alkali metals.

Sodium reacts violently with both air and water. Iron, on the other hand, reacts much more slowly with air and water, producing rust. Gold remains unchanged in the presence of air and water.

Looking in more detail at these basic reactions reveals a pattern of reactivity among metals. In this chapter and the next, we will add progressively more detail to our understanding of the reactivity of metals.

Table 14.2 summarises the reactions of certain metals with air, water and dilute acid. A gradual decrease in reactivity can be seen moving down the table from the highly reactive (sodium and calcium) to the unreactive (copper, silver and gold). Note that the true reactivity of aluminium is only seen if its protective oxide layer is removed.

Metal	Reaction with ...		
	Air	Water	Dilute HCl
sodium calcium magnesium	burn very strongly in air to form oxide	react with cold water to give hydrogen	react very strongly to give hydrogen
aluminium[(a)] zinc iron	burn less strongly in air to form oxide	react with steam, when heated, to give hydrogen	react less strongly to give hydrogen
lead copper	react slowly to form oxide layer when heated	do not react	do not react
silver gold	do not react	do not react	do not react

[(a)] *These reactions only occur if the protective oxide layer is removed from the aluminium.*

Table 14.2: Reactions of metals with air, water and dilute hydrochloric acid.

It is useful to look in more detail at the reactions of particular metals with air, water and dilute hydrochloric acid.

Reactions of metals with air/oxygen

In discussing the reactions of metals with air, it is important to note that oxygen is by far the most reactive of the gases present. Many metals react directly with oxygen to form oxides. Magnesium, for example, burns brightly in air or oxygen to form a white powder of magnesium oxide.

magnesium	+	oxygen	→	magnesium oxide
$2Mg(s)$	+	$O_2(g)$	→	$2MgO(s)$

Iron only burns in air to give an oxide if it is in powder form (Figure 14.5), or as iron wool.

Figure 14.5: Iron dust ignited in a Bunsen flame.

However, iron exposed to air and water will slowly form an oxide layer (rust) on its surface. This layer of rust (hydrated iron(III) oxide) does not stick to the surface of the metal but flakes off. This rusting process damages the iron object.

Aluminium is normally coated with a thin layer of aluminium oxide (Al_2O_3) by reaction with the oxygen in the air. This oxide layer sticks to the surface of the aluminium and hides the reactivity of the metal.

Reactions of metals with water/steam

Reactive metals such as potassium, sodium and calcium all react with cold water to produce the metal hydroxide and hydrogen gas.

metal + water → metal hydroxide + hydrogen

For example:

sodium	+	water	→	sodium hydroxide	+	hydrogen
$2Na(s)$	+	$2H_2O(l)$	→	$2NaOH(aq)$	+	$H_2(g)$

calcium	+	water	→	calcium hydroxide	+	hydrogen
$Ca(s)$	+	$2H_2O(l)$	→	$Ca(OH)_2(aq)$	+	$H_2(g)$

Magnesium only reacts very slowly in cold water. However, a much more vigorous reaction takes place if steam is passed over heated magnesium. The magnesium glows brightly to form hydrogen and magnesium oxide. The hydrogen given off can be burnt when lit with a splint (Figure 14.6).

magnesium	+	steam	→	magnesium oxide	+	hydrogen
$Mg(s)$	+	$H_2O(g)$	→	$MgO(s)$	+	$H_2(g)$

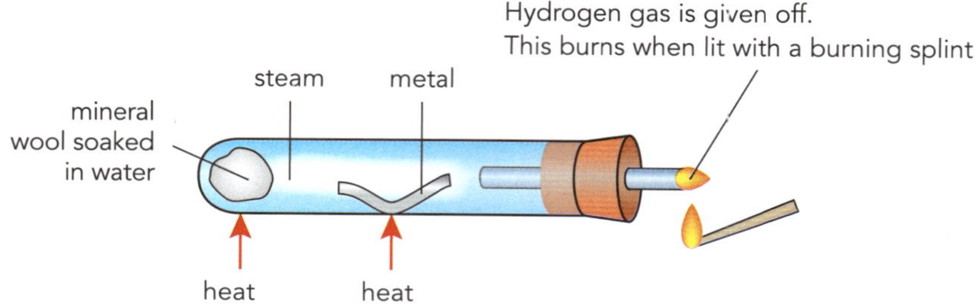

Figure 14.6: Heating magnesium in the presence of steam.

Zinc and iron will react in a similar way if steam is passed over the heated metal.

Iron	+	steam	→	iron oxide	+	hydrogen
$3Fe(s)$	+	$4H_2O(g)$	→	$Fe_3O_4(s)$	+	$4H_2(g)$

Hydrogen gas is produced in all these reactions. You should note that in these cases it is the metal oxide, not the hydroxide, that is formed.

In summary:

- if a metal reacts with cold water, a metal hydroxide and hydrogen are formed
- if a metal reacts only with steam, then a metal oxide is formed.

Reactions of metals with dilute hydrochloric acid

Moderately reactive metals such as magnesium, zinc or iron can be reacted safely with dilute acids to produce hydrogen gas (Figure 14.7)

metal + hydrochloric acid → metal chloride + hydrogen

For example:

iron	+	hydrochloric acid	→	iron(II) chloride	+	hydrogen
$Fe(s)$	+	$2HCl(aq)$	→	$FeCl_2(aq)$	+	$H_2(g)$

zinc	+	hydrochloric acid	→	zinc chloride	+	hydrogen
$Zn(s)$	+	$2HCl(aq)$	→	$ZnCl_2(aq)$	+	$H_2(g)$

Figure 14.7: The reaction between zinc and dilute hydrochloric acid produces hydrogen gas.

The summary of the reactions of metals with these reagents (Table 14.2) allows us to construct an approximate order of reactivity for these metals. We will look in more detail at these and other observations in the next chapter.

Questions

4 A word equation is shown here:

magnesium + hydrochloric acid → magnesium chloride + hydrogen

What is the symbol equation for this reaction?

A $2Mg + HCl \rightarrow MgCl_2 + H_2$

B $Mg + 2HCl \rightarrow MgCl_2 + H_2$

C $Mg + HCl \rightarrow MgCl_2 + H_2$

D $2Mg + 2HCl \rightarrow MgCl_2 + H_2$

5 Select from this list two metals that will not react with hydrochloric acid to produce hydrogen:

zinc, copper, iron, magnesium, silver

6 **a** Some metals react with cold water. Complete the general word equation for this type of reaction.

metal + water → metal ________ + ________

b Other metals do not react with cold water but do react when steam is passed over the heated metal. Complete the general word equation for this type of reaction.

metal + steam → metal ________ + ________

c **i** Write the chemical equation for the reaction when steam is passed over heated magnesium.

ii Balance the following equation for the reaction between iron wire and steam.

$\_\_Fe + \_\_H_2O \rightarrow Fe_3O_4 + \_\_H_2$

14.2 Uses of metals

The production of iron dates back to Middle Eastern cultures in the Middle Bronze Age and then spreads through India and China (Figure 14.8). The skills of the smiths and metalworkers stretch back through history. For example, the 'art' of making the blades of Samurai swords, forged from traditional Japanese steel, involves processes of heat treatment and folding that is truly astounding. Such technology grew out of practical understanding. It gave rise to a whole series of techniques and alloys that produced metals fit for a purpose. This expertise has now been added to as the structure of metals has been increasingly understood. We now have a wide range of steels and other alloys suited to demanding uses.

Figure 14.8: The Iron Pillar of Delhi dates from the beginning of the 5th century and is remarkable for its resistance to rusting.

It is the unreactive, or only moderately reactive, transition metals that have proved most useful to us historically for construction and other purposes. Iron and copper are perhaps the two most important of these transition metals. In addition to the transition metals, aluminium is increasingly used for a variety of purposes.

Uses of iron

The famous bridge at Ironbridge in Shropshire, England (Figure 14.9), marks a historic Industrial Revolution in Europe. Made from cast iron and opened in 1781, it was the first iron bridge in the world.

Figure 14.9: The bridge at Ironbridge was the first ever bridge built of iron and was awarded UNESCO World Heritage status in 1986. It has a span of around 30 m.

We use about nine times more iron than all the other metals put together. Iron is a moderately reactive transition metal. By itself, pure iron is quite soft and weak, and therefore not very useful for construction. Cast iron obtained directly from the blast furnace (Chapter 16) can be used to make very large objects such as the bridge structure at Ironbridge. Cast iron was used in rails, boats, ships, aqueducts and buildings, as well as in iron cylinders in steam engines. Iron production and use was central to the Industrial Revolution using new manufacturing processes in Europe and the United States during the 18th century. Cast iron can even be used in making cooking pots and pans as it conducts heat well. However, the use of iron for smaller or more sophisticated pieces of engineering was restricted as it was too brittle for this widespread use.

The high level of carbon in cast iron (2–4%) disrupts the lattice structure of the iron and makes the metal liable to shatter when too much force is applied (the metal is too brittle). The strength of iron can be improved by carefully controlling the amount of carbon present and by adding other metals to create different forms of steel to match the use intended.

The use of iron was also complicated by the fact that it rusts easily. When exposed to air and water, iron becomes coated with an orange–red powder of hydrated iron(III) oxide (rust), which flakes off and weakens the structure (Chapter 16). This problem has led to various methods to combat rusting and also to the development of stainless steel.

Uses of aluminium

Aluminium was, for a long time, an expensive and little-used metal. In France, around the 1860s, at the Court of Napoleon III (the nephew of Napoleon Bonaparte), honoured guests used cutlery made of aluminium rather than gold. The breakthrough in its usefulness came in 1886 when Charles Hall and Paul Héroult independently found a way to obtain the metal by electrolysis (Chapter 16).

Aluminium is a light, strong metal and has good electrical conductivity. Increasingly, aluminium is used for construction purposes. It is used in the construction of some naval vessels and modern cars. It is commonly used in aeroplanes, where it is usually alloyed with other metals such as copper (Figure 14.10). Its low density and good conductivity have also led to its use in overhead power lines. The aluminium is very resistant to corrosion and its low density prevents sagging of the cables between pylons.

Figure 14.10: The world's first supersonic passenger jet *Concorde* was built out of an alloy containing over 90% aluminium.

Aluminium is particularly useful because it is protected from corrosion by the stable layer of aluminium oxide that forms on its surface. This protective layer stops the aluminium (a reactive metal) from reacting. This makes aluminium foil containers ideal for food packaging because they resist corrosion by natural acids present in the food. Aluminium is also used for external structures such as window frames because the metal resists weathering.

Uses of copper

Copper has a distinctive colour and is one of the least reactive metals in common use. It is used for water pipes and as roofing for buildings. Copper roofs, and indeed statues, become coated with a green layer of basic copper(II) carbonate, known as *verdigris*, when exposed to the atmosphere for a long time.

Copper has a very high electrical conductivity and is very ductile, and so can be easily drawn out as wire. A major use of copper is in electrical cables and domestic wiring (Figure 14.11). It is also used in the circuit boards of smartphones and other electronic equipment. The copper used for wiring and circuit boards must be of high purity (99.99%) or the electrical conductivity will be reduced significantly. This is why copper is refined to very high purity by electrolysis.

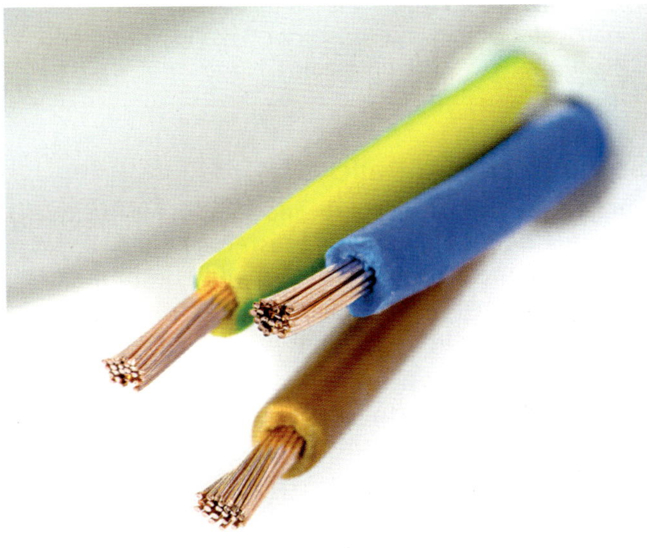

Figure 14.11: A major use for copper is in electrical cabling and domestic wiring.

Metal	Key properties	Uses
iron	hard (but brittle and now replaced by steel) good conductor of heat	construction girders, large castings cooking utensils
copper	strong but easily bent to shape strong with high electrical conductivity	water pipes electrical wiring and cables
zinc	protective coating on iron and steel objects	galvanising iron and steel to prevent rusting
aluminium	low density and strong resistant to corrosion by acids in food	construction, e.g. aircraft and high-voltage power cables food containers

Table 14.3: Properties and uses of some major metals.

Table 14.3 summarises some of the major uses of key metals and the properties that give rise to those uses.

14.3 Alloys

The Periodic Table does not list substances such as steel, bronze and brass, which in everyday use we call metals, and which share the properties listed for metals. They are not elements. They are **alloys**, mixtures of elements (usually metals) designed to have properties that are useful for a particular purpose.

KEY WORD

alloys: mixtures of elements (usually metals) designed to have the properties useful for a particular purpose, e.g. solder (an alloy of tin and lead) has a low melting point

Figure 14.12: Bronze statue of Nelson Mandela (Western Cape, South Africa).

Some important alloys

Pure metals are not always the most useful for practical purposes. Iron is quite soft and rusts easily, and copper is not very strong. However, we can modify the properties of a metal to make it stronger or more resistant to corrosion by making an alloy of it.

The use of alloys such as bronze stretches back into human early history. Modern bronze (Figure 14.12) has a similar composition to the types of bronze developed in Bronze Age cultures (90% copper and 10% tin) about 5000 years ago.

Ironbridge (Figure 14.9) may have seen the construction of the first iron bridge, but modern bridges are now made of steel. Here, iron is alloyed with other transition elements and carbon to make it stronger. The strength of steel means that more visually pleasing and longer designs are structurally possible (Figure 14.13).

Figure 14.13: The Chaotianmen Bridge over the Yangtze River in Chongqing City, China. The bridge is one of the world's longest through-arch steel bridges and has a span of 552 m.

Making an alloy of one metal with another is one of the commonest ways of changing the properties of metals. Alloys are formed by mixing the molten metals together thoroughly and then allowing them to cool and form a solid. Alloying often results in a metal that is stronger, harder and more corrosion resistant than the original individual metals. Other examples of alloys and their properties are given in Table 14.4.

Alloy	Typical composition		Particular properties and uses
brass	**copper**	**70%**	**harder than pure copper; 'gold' coloured / musical instruments, ornaments, electrical connections**
	zinc	**30%**	
bronze	copper	90%	harder than pure copper / statues, bells, machine parts
	tin	10%	
mild steel	iron	99.7%	stronger and harder than pure iron / car bodies
	carbon	0.3%	
stainless steel	**iron**	**74%**	**harder than pure iron; does not rust / cutlery, surgical instruments, reaction vessels in the chemical industry**
	chromium	**18%**	
	nickel	**8%**	
solder	tin	50%	lower melting point than either tin or lead / making electrical connections
	lead	50%	

Table 14.4: Some important alloys.

KEY WORDS

brass: an alloy of copper and zinc; this alloy is hard

stainless steel: an alloy of iron that resists corrosion; this steel contains a significant proportion of chromium which results in the alloy being resistant to rusting

Strength is not the only property to consider when designing an alloy. For example, solder is an alloy of tin and lead. It is useful for making electrical connections because its melting point is lower than that of either of the two separate metals.

Also, steel, which rusts when in contact with oxygen and water, can be prevented from doing so when alloyed with chromium and nickel. The transition metals chromium and nickel are similar to aluminium in that they form a protective oxide layer on their surface and so are resistant to corrosion. The presence of these atoms in the alloy gives the stainless steel formed a resistance to rusting (Table 14.4).

Structure of alloys

As we saw earlier in this chapter (and when discussing metallic bonding in Chapter 3), the atoms in a pure metal are arranged in regular layers. When a force is applied, the layers can slide over each other. When an alloy is made using a mixture of metals, or a metal with a non-metal, the alloy is not simply a mixture of different crystal regions but the atoms of the second element form part of the overall crystal lattice (Figure 14.14).

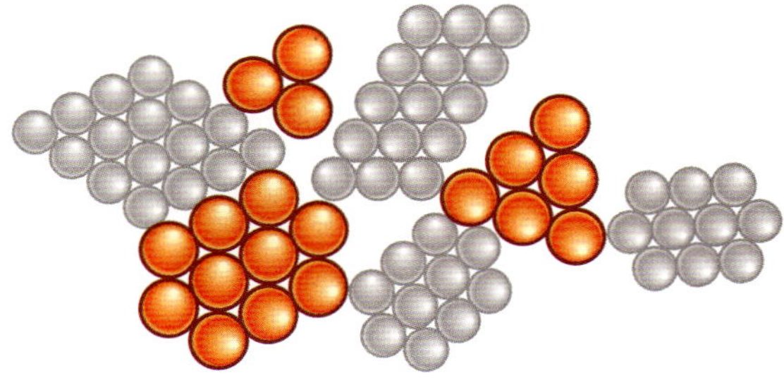

A mixture of metal crystals

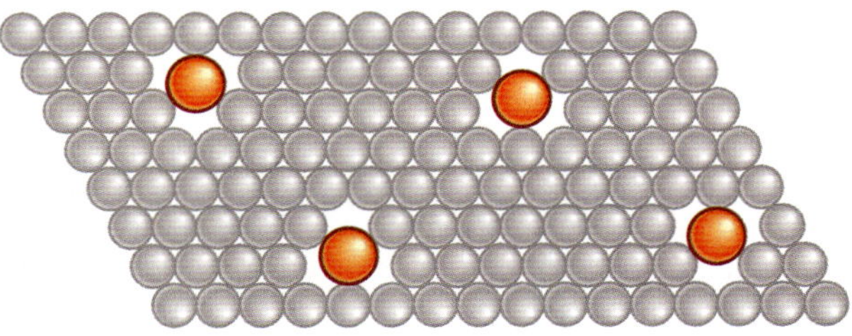

An alloy

Figure 14.14: An alloy is not a mixture of metal crystal regions but a regular lattice of atoms.

When a metal is alloyed with a second metal, the different sized atoms make the lattice structure less regular. The crystal lattice of the main metal present is disrupted. Figure 14.15 shows how the presence of the 'impurity' atoms (the larger atoms in this diagram) makes it more difficult for the metal atoms to slide over each other. This makes the alloy stronger but more brittle than the metals it is made from. The alloy tends to be more brittle than the pure metal as the regularity of the structure has been broken.

The alloy lattice is still held together by metallic bonding with delocalised electrons free to move between the layers of the structure. Alloys, therefore, share the same characteristic physical properties – malleability, ductility, electrical and thermal conductivity – as the metallic elements.

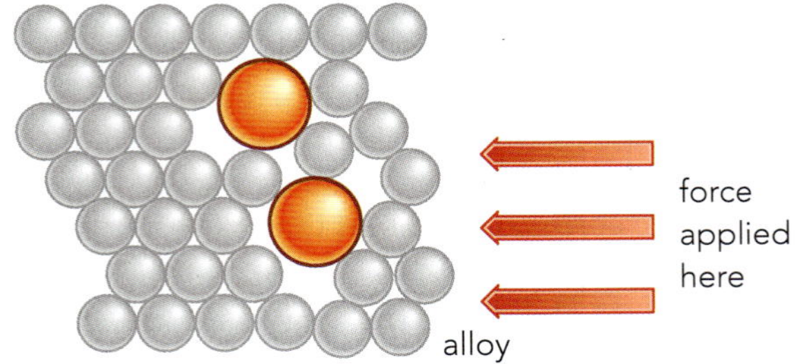

Figure 14.15: Alloys are stronger than pure metals because the layers cannot slide over each other easily as in a pure metal.

Questions

7 a The wiring and connections involved in a domestic electric circuit must be very safe. Figure 14.16 shows the structure of a simple domestic electric plug.

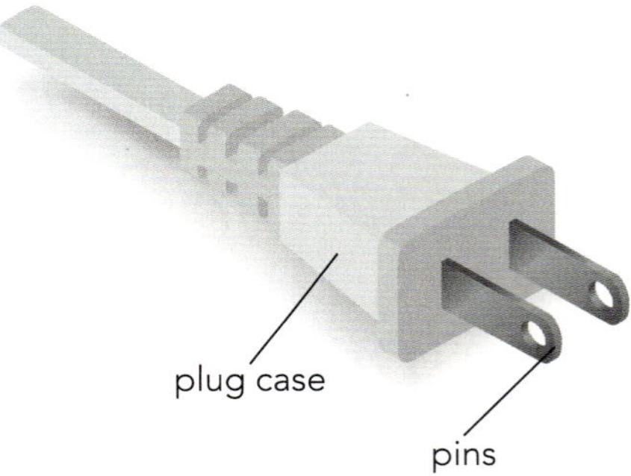

Figure 14.16: An electric plug.

The properties of four materials, some of which could be used in the home, are shown in Table 14.5.

Material	Melting point / °C	Electrical conductivity when solid
W	−39	good
X	−20 to −10	poor
Y	170 to 220	poor
Z	1083	good

Table 14.5: The properties of some materials.

Which of these materials are most suitable to make the case and pins of the electric plug (Figure 14.16)? Choose one of the alternatives A–D:

	Case	Pins
A	X	Y
B	W	X
C	Z	Y
D	Y	Z

b Which alloy is most often used for the pins of such a plug?

8 Electrical wiring requires use of a good conductor and insulation. Figure 14.17 shows the end of some cabling.

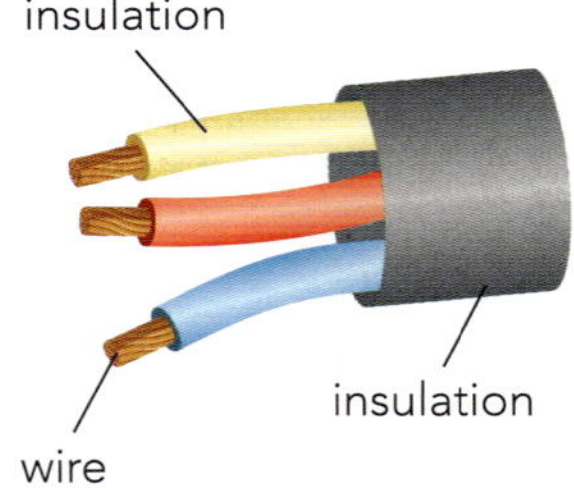

Figure 14.17: Some domestic electric cable.

a Which metal is usually used as the wire in this cable?

b What material is usually used for insulation in the cable?

9 Stainless steel is an alloy often used to make cutlery.

a Name two of the main metal elements in stainless steel.

b What are two of the properties of stainless steel that make it particularly useful in cutlery?

ACTIVITY 14.1

Intriguing alloys!

Alloys are metals designed for a purpose. The properties of the alloy improve on those of the parent metals. By adjusting the composition of the mixture, the properties best suited to a particular purpose are achieved.

If samples of alloys are available in your laboratory, then do examine them. Figure 14.18 shows examples of some of the more important alloys while Figure 14.1 shows a use of a shape memory alloy.

Figure 14.18: Some more important alloys in use (top) pewter, solder; (middle) brass, stainless steel; (bottom) mild steel, copper-nickel alloy.

In groups, discuss the different alloys in the photos and their different uses. If you have samples of the alloys available, pass these around your group as you discuss.

In your discussion, focus on: (You may want to search the internet to help you.)

- the composition of these different alloys
- the property that is being improved, or designed for, in each case
- the uses that these alloys are designed for.

Draw up a table of your findings to share with other groups. Follow the example shown:

Name of alloy	Composition	Properties	What could this be used for?
Brass	Brass is composed of …		

When considering the uses of brass and stainless steel, in particular, think about the following:

- Why is brass not used in electrical wiring?
- Why is stainless steel not used to build very large structures such as bridges?

Peer assessment

Ask your partner the following questions:

- What did I do well during our discussion about the different alloys?
- What is one thing I could improve on?

SUMMARY

Metals and non-metals have different physical properties.
The key physical properties of metals are that they conduct electricity, and are malleable and ductile.
Metals vary considerably in their reactivity with water, oxygen and dilute acids.
Some metals react strongly with cold water to produce alkaline solutions and hydrogen gas.
Other metals react only when heated with steam, while some do not react with water at all.
The most reactive metals can react with oxygen in the air.
A range of metals react with dilute acids to produce hydrogen, but there are some metals so unreactive that they do not react with dilute acids.
Some metals are very useful, with aluminium being particularly useful for its low density and resistance to corrosion.
Alloys are mixtures of metals and other elements, which are often chosen to increase the strength and usefulness of the pure metal.
Stainless steel is a hard alloy created specifically to resist corrosion.
Alloys can be identified in diagrams of metallic structure by noting the different sizes of the atoms in the layered arrangement shown.
Alloys are stronger and harder than pure metals as the layers in the structure can no longer slide over each other.

PROJECT

Making predictions

In this chapter, we have outlined some basic reactions of metals with water and dilute acids. This confirms that some metals are more reactive than others. We can begin to construct an approximate order of reactivity from such results. You can begin to get an idea of such an order using the possible reactions with water and acids.

Figure 14.19: Constructing a flow chart.

Construct a flow chart (Figure 14.19) of the tests you would do to establish how reactive a metal is.

You have access to the following reagents and apparatus to help you:

- dilute hydrochloric acid solution
- cold, distilled water
- dilute sulfuric acid
- a steam generator and apparatus to test for any reaction with steam.

Decide on a series of chemical tests that you would carry out to decide approximately where an unknown metal would be placed in an order of reactivity (or reactivity series). Start with the test that would show whether the metal was very reactive. Then work out a sequence based on the possible results at each stage.

Consider the following questions:

- Are you confident that you have the tests in the best order?
- Would you expect the two acids to give different results?
- Do the tests in your flow chart give you a detailed suggestion of a reactivity series, or is more information required?

Once you've finished, check in groups that you are confident of the products of the different types of reaction you have used and that you can write equations for them. Add these to your flow chart.

EXAM-STYLE QUESTIONS

1 What is produced when sodium metal reacts with water?

A oxygen gas and an acid solution
B hydrogen gas and a neutral solution
C hydrogen gas and an alkaline solution
D oxygen gas and an alkaline solution **[1]**

2 Which of the following statements is true for *all* metals?

A They are grey in colour.
B They are weak and brittle solids.
C They react with water.
D They may be used to make alloys with other elements. **[1]**

3 The table shows some of the properties of metals and non-metals. For each property, write True or False in the space provided to show whether it applies to most metals, most non-metals or to both.

Property	Metals	Non-metals
They conduct electricity		
They react with oxygen in the air		
They are brittle		
They can easily be bent and shaped		
They have high melting points		

[5]

4 Aluminium is used for the following purposes: manufacturing aircraft, overhead electrical power cables and food containers.

a Which property of aluminium is most important for its use in aircraft manufacture? **[1]**
b Which property is most useful for its use in food containers? **[1]**
c Aluminium is a good conductor of electricity, but this is not why it is used in overhead power cables. Why is it used for power cables? **[1]**
d Electrical wiring in the home does not use aluminium. Which metal is used? **Suggest** why this metal is preferred. **[2]**
e Brass is also important in the home for electrical connections. Which of these words correctly describe brass? Choose all correct answers. **[4]**

Alloy **Compound** **Conductor** **Element** **Metal** **Mixture**

f Stainless steel is used to make cutlery. In what ways does stainless steel differ from iron? **[2]**

[Total: 11]

COMMAND WORD

suggest: apply knowledge and understanding to situations where there are a range of valid responses in order to make proposals / put forward considerations

CONTINUED

5 The figure shows an alloy and the metals from which it is made.

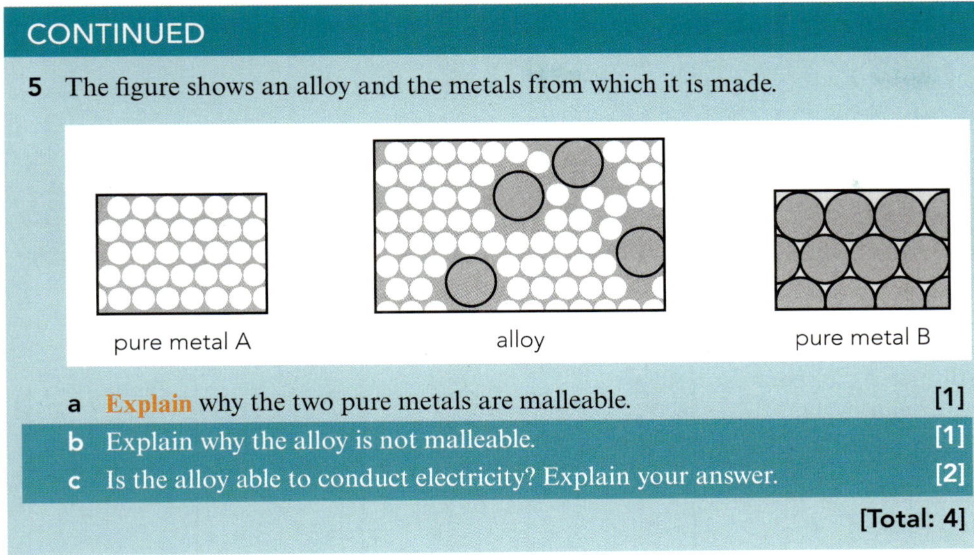

a **Explain** why the two pure metals are malleable. [1]

b Explain why the alloy is not malleable. [1]

c Is the alloy able to conduct electricity? Explain your answer. [2]

[Total: 4]

COMMAND WORD

explain: set out purposes or reasons/make the relationships between things evident/provide why and/or how and support with relevant evidence

SELF-EVALUATION CHECKLIST

After studying this chapter, think about how confident you are with the different topics. This will help you to see any gaps in your knowledge and help you to learn more effectively.

I can	See Topic...	Needs more work	Almost there	Confident to move on
compare the general physical properties of metals and non-metals	14.1			
describe the general chemical properties of metals in relation to their reactions with acids, water and oxygen	14.1			
describe the uses of metals; particularly the uses of aluminium, copper and stainless steel	14.2			
describe an alloy as a mixture of a metal with other elements, referring particularly to brass and stainless steel	14.3			
understand that alloys can be harder or stronger than the pure metals and therefore more useful	14.3			
identify representations of alloys from diagrams of structure	14.3			
describe the use of alloys in terms of their physical properties, including the use of stainless steel	14.3			
explain, in terms of structure, how alloys can be harder or stronger than the pure metals because the layers can no longer slide over each other	14.3			

> Chapter 15

Reactivity of metals

IN THIS CHAPTER YOU WILL:

- establish the concept of a reactivity series
- analyse how different metals show differences in the strength of their reaction with water and dilute acid
- understand that some metals react very strongly with water or acid, while others do not react at all
- learn how this reactivity relates to the reactivity series and predict an order of reactivity given a set of experimental results
- analyse how the different levels of reactivity of metals can be used to organise them into a reactivity series

> understand that the reactivity series relates to the ease with which a metal forms its positive ion by considering the displacement reactions of metals

> understand that these displacement reactions are redox reactions.

GETTING STARTED

You will have seen from the two previous chapters, and possibly also from demonstrations, the reactions of sodium and potassium with water.

- Do all metals react in this way?
- Can you think of some metals that do not react with water?

Discuss in class the different levels of reactivity of metals with water and acid that you have encountered. Can you relate the different levels of reactivity to positions in the Periodic Table and indeed to everyday life?

THE SEARCH FOR LONGER-LASTING BATTERY POWER

The reactivity series is the basis for powering modern technology.

An unusual way to power a simple digital clock is with an electrochemical cell (battery) made using a potato (Figure 15.1). The current is produced by pushing two different metal electrodes (e.g. copper and zinc) into the potato. The potato acts as the electrolyte. Connecting up the electrodes to the clock produces a small current in the circuit, which powers the clock. If you had enough potatoes you could even power your smartphone!

Figure 15.1: A potato-powered digital clock.

This cell works because the two metals have different reactivities. Zinc is more reactive than copper and so forms positive ions more easily. The zinc releases electrons as its atoms become ions, and these electrons give the zinc electrode a negative charge. The electrons move around the circuit to the copper electrode. The electrolyte solution completes the movement of charge in the circuit as the ions present carry the charge between the electrodes. Such a cell converts chemical energy into electrical energy.

A simple electrochemical cell works best if the metals used as the electrodes are far apart in the reactivity series. The further apart the metals are in the reactivity series, the greater the cell voltage. This explains the use of lithium, the very reactive metal at the top of Group I, as one of the electrodes in the modern rechargeable lithium ion batteries found in many portable consumer electronic devices. The aim is to make the difference in reactivity as large as is safely possible. The concern over transporting lithium batteries in the baggage holds of aircraft is precisely because of the reactivity of lithium. There have been many instances of smartphones (and other lithium ion devices) catching fire and exploding!

The importance of portable electric power is increasing as we move away from fossil fuels as our main energy source. Whether it is for powering electric cars or storing solar-generated electrical energy in our homes, there is now a greater need to develop more durable batteries. Research is proceeding strongly with the possibility of solid-state batteries being considered.

Discussion questions

1. What are the environmental concerns over the need to develop more powerful batteries and the fact that many types of battery are considered disposable?
2. What is the one non-metal that is sometimes found in batteries as it conducts an electric current?

15.1 The metal reactivity series

Most of the elements in the Periodic Table are metals. Many of them are useful for a wide variety of purposes; some, such as iron, have an enormous number of uses. The early history of human life is marked by the metals used in making jewellery, ornaments and tools. Early civilisations used metals that could be found 'native' (e.g. gold) for decorative items and then developed alloys such as bronze. Even among the metals that were available, there were obvious differences in resistance to corrosion. The Viking sword in Figure 15.2 emphasises the different reactivities of the gold and silver of the hilt and the iron of the blade. Such observed differences in the reactivity of different metals (Figure 15.3) suggest that there may be an order of reactivity built into the natural world.

Figure 15.2: This Viking sword had a handle made from gold and silver and an iron blade. The blade has corroded badly but the handle is intact.

We have seen in our discussion of Group I of the Periodic Table (Chapter 13) how **reactivity** changes in a particular group. We also referred to the fact that the transition metals are less reactive than the metals in Groups I and II. Many of the important metals that we use in different ways come from more than one group. So, the question arises as to whether there is a broader order of reactivity (a **reactivity series**), which includes all the metals.

At the end of Chapter 14 we suggested that it may be possible to put the metals into an approximate series based on their reactions with oxygen, water or dilute hydrochloric acid. We will now follow that further to establish the reactivity series for metals, looking at the reactions with water and dilute acid. We will then look at metal displacement reactions that allow us to place metals precisely in a series and consider the reasons for the order we observe.

KEY WORDS

reactivity: the ease with which a chemical substance takes part in a chemical reaction

reactivity series (of metals): an order of reactivity, giving the most reactive metal first, based on results from a range of experiments involving metals reacting with oxygen, water, dilute hydrochloric acid and metal salt solutions

Establishing the basis of a reactivity series

In reality a range of reactivity exists for the metals. This range can be observed by simply considering the reaction of metals with air and water. The Group I metals are the most reactive of all the metals and must be kept under oil to avoid reaction with moist air. The trend within any metallic group is for reactivity to increase going down the group.

Reactions with water

All the alkali metals react spontaneously with water to produce hydrogen gas and the metal hydroxide (Table 15.1). The reactions are exothermic. If water is dripped onto a small sample of the metal, the heat produced ignites the hydrogen gas (Figure 15.4). For sodium the flame colour is yellow.

Figure 15.3: The copper in this bronze mask is coated green (verdigris) through reaction with the air.

Figure 15.4: Reaction of sodium with water.

If a piece of sodium (or potassium) is placed on the surface of water in a dish the heat produced is sufficient to melt the sodium (or potassium) as it skips over the surface of the water (Figure 15.5).

The reaction with water is the same in each case and Table 15.1 summarises the order of reactivity for Group I.

Element	Reaction with water	Reaction with air	
lithium	reacts steadily $2Li + 2H_2O \rightarrow 2LiOH + H_2$	tarnishes slowly to give a layer of oxide	increasing reactivity ↓
sodium	reacts strongly $2Na + 2H_2O \rightarrow 2NaOH + H_2$	tarnishes quickly to give a layer of oxide	
potassium	reacts violently $2K + 2H_2O \rightarrow 2KOH + H_2$	tarnishes very quickly to give a layer of oxide	

Table 15.1: Reactions of Group I metals with air and water.

Group II metals shows similar trends in reactivity to Group I. However, they are less reactive than the metals in Group I. Again, reactivity increases going down the group. Magnesium reacts very slowly indeed when placed in cold water. Bubbles of hydrogen form very slowly on the magnesium ribbon. A much stronger reaction is obtained if steam is passed over heated magnesium (Figure 15.6; see also Chapter 14). The magnesium glows brightly to form hydrogen and magnesium oxide:

$$\text{magnesium} + \text{steam} \rightarrow \text{magnesium oxide} + \text{hydrogen}$$

$$Mg(s) + H_2O(g) \rightarrow MgO(s) + H_2(g)$$

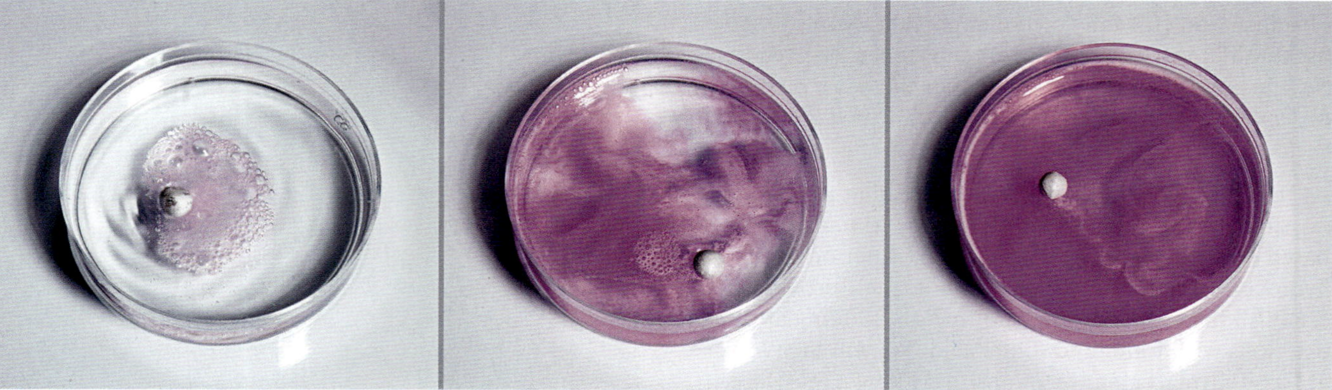

Figure 15.5: Sodium floats in a ball on the surface of water as it reacts. The change in the colour of the indicator solution shows the solution produced is alkaline.

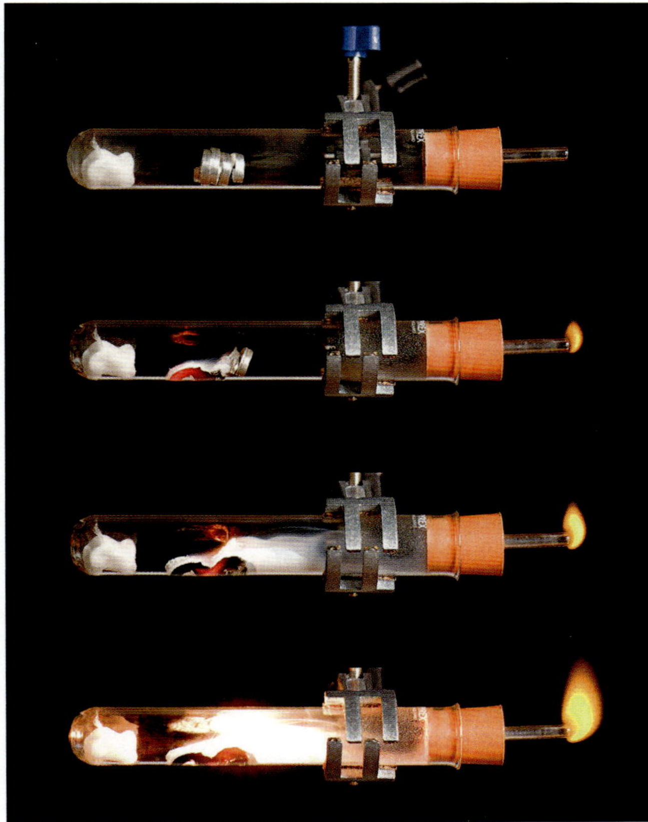

Figure 15.6: Sequence of the reaction between steam and heated magnesium.

Calcium, however, reacts strongly with cold water, giving off hydrogen rapidly:

calcium + water → calcium hydroxide + hydrogen

$$Ca(s) + 2H_2O(l) \rightarrow Ca(OH)_2(aq) + H_2(g)$$

Calcium hydroxide is more soluble than magnesium hydroxide, so an alkaline solution is produced (limewater). As the reaction proceeds, a white suspension is obtained because not all the calcium hydroxide dissolves.

As we have seen in Chapter 14, moderately reactive transition elements such as zinc and iron will also react with steam in a similar way to magnesium. Other transition metals such as copper, silver and gold do not react with either water or steam. Based on the reactions we have seen so far, we can put the metals considered into an order of reactivity:

potassium, sodium, calcium, magnesium, zinc, iron, copper, silver, gold

most reactive → **least reactive**

Reactions with dilute acids

The alkali metals of Group I, and indeed calcium, are too reactive to safely add to even dilute acids. They do produce hydrogen but too vigorously. Other metals show a range of reactivity with dilute acid and this is summarised in Table 15.2.

Metal	Observations with dilute HCl(aq) or dilute $H_2SO_4(aq)$	Equations
magnesium	strong reaction, bubbles of gas, magnesium disappears, colourless solution formed	$Mg(s) + 2HCl(aq) \rightarrow MgCl_2(aq) + H_2(g)$ $Mg(s) + H_2SO_4(aq) \rightarrow MgSO_4(aq) + H_2(g)$
aluminium	slow to react in cold, but bubbles form on heating, aluminium disappears, colourless solution formed	$2Al(s) + 6HCl(aq) \rightarrow 2AlCl_3(aq) + 3H_2(g)$ $2Al(s) + 3H_2SO_4(aq) \rightarrow Al_2(SO_4)_3(aq) + 3H_2(g)$
zinc	bubbles of gas, zinc disappears, colourless solution formed	$Zn(s) + 2HCl(aq) \rightarrow ZnCl_2(aq) + H_2(g)$ $Zn(s) + H_2SO_4(aq) \rightarrow ZnSO_4(aq) + H_2(g)$
iron	bubbles of gas, iron disappears, pale-green solution formed	$Fe(s) + 2HCl(aq) \rightarrow FeCl_2(aq) + H_2(g)$ $Fe(s) + H_2SO_4(aq) \rightarrow FeSO_4(aq) + H_2(g)$
copper	no reaction	
silver	no reaction	
gold	no reaction	

Table 15.2: Reactions of metals with dilute acids.

You will note from Table 15.2 that the reaction of the metal and the acid always produces a salt and hydrogen gas.

metal + acid → a metal salt + hydrogen

The reactions of the metals with dilute acids that can be carried out safely confirm the order of reactivity seen with water/steam.

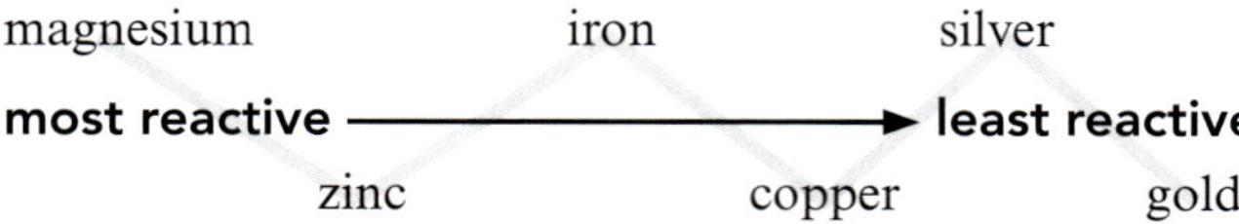

From Table 15.2, you will see that aluminium reacts less strongly than would be expected. This occurs because aluminium reacts easily with oxygen in the air to form a thin layer of aluminium oxide (Al_2O_3) that extends over the whole metal surface.

$$4Al(s) + 3O_2(g) \rightarrow 2Al_2O_3(s)$$

This unreactive oxide layer is very firmly attached to the metal and does not flake off. The layer acts as a protective coating, slowing the reaction. Only when the oxide layer is removed is the true reactivity of aluminium observed. It is possible to remove the protective oxide layer and aluminium then shows a reactivity with acids just below that of magnesium. The reactivity of aluminium (when not covered in a layer of oxide) puts it in a position within the reactivity series just below magnesium, but above zinc (Figure 15.7).

Study of the reactions of metals with water and acids allows us to place hydrogen in the reactivity series (Figure 15.7) even though it is not a metal. This is useful as it demonstrates that metals higher in the series are more reactive than hydrogen. They can displace the gas from water or acids under appropriate conditions.

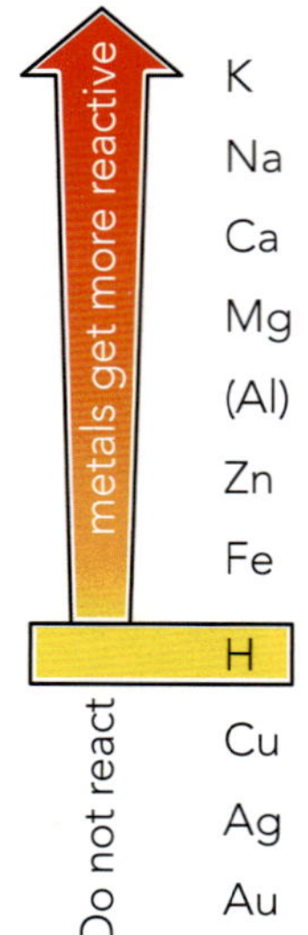

Figure 15.7: A reactivity series of metals with water and dilute acids, showing the position of hydrogen.

Questions

1. Three metals, X, Y and Z, react with dilute hydrochloric acid in different ways.

   Metal X: dissolves slowly with bubbles of gas given off at a slow rate.

   Metal Y: dissolves very quickly and gas is produced very rapidly indeed.

   Metal Z: dissolves rapidly and bubbles of gas are formed rapidly.

   From these reactions, which is the order of increasing reactivity for these three metals?

   A ZXY **B** YXZ
   C XYZ **D** XZY

2. **a** Which gas is given off when the alkali metals are reacted with water?

   b Name the product, other than hydrogen, when potassium is reacted with water.

   c Write a word equation for the reaction of sodium with water.

   d Write a balanced chemical equation for the reaction of potassium with water.

3. **a** Give two characteristic properties of aluminium that make it very useful for construction.

   b Why does aluminium not corrode like some other metals such as iron?

   c Select from this list a metal that will not react with hydrochloric acid to produce hydrogen: magnesium, iron, copper.

ACTIVITY 15.1

Mind-mapping the reactivity series

The development of the metal reactivity series draws information from several different types of experiment. In your groups, you should think of the different reactions that provided evidence for the series.

Construct a mind map to show how the different areas lead from the central idea of reactivity.

Figure 15.8 shows a possible starting structure to the map.

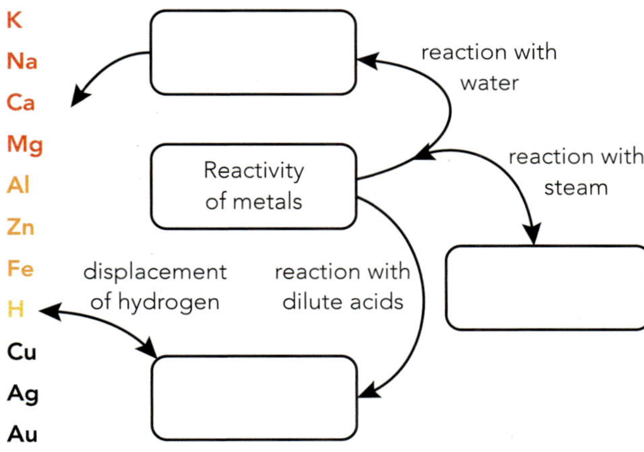

Figure 15.8: A possible framework for a mind map on the reactivity of metals.

Copy the framework shown and use the spaces between the arrows, and alongside them, to fit in information. Use:

- equations
- lists
- comments to increase the usefulness of the information on the map.

You can then discuss how to improve the mind map by devising your own map of the connected ideas. Try to make your maps as interesting as you can in terms of design and present them to the class as part of a discussion of this important concept.

REFLECTION

This chapter discusses how the metal reactivity series is formed by considering the reactivity of metals with air and water. Some of these reactions are not safe to perform in a school laboratory. Do you find images and photographs of experiments useful?

15.2 Metal displacement reactions

A **displacement reaction** can help us to place particular metals more precisely in the reactivity series. We can use this type of reaction to compare directly the reactivity of two metals. In a displacement reaction, a more reactive metal displaces a less reactive metal from solutions of salts of the less reactive metal.

KEY WORDS

displacement reaction: a reaction in which a more reactive element displaces a less reactive element from a solution of its salt

In this type of reaction, the two metals are in direct 'competition'. If a piece of zinc is left to stand in a solution of copper(II) sulfate, a reaction occurs:

zinc	+	copper(II) sulfate	→	zinc sulfate	+	copper
$Zn(s)$	+	$CuSO_4(aq)$	→	$ZnSO_4(aq)$	+	$Cu(s)$
grey		blue		colourless		red–brown

The observed effect of the reaction is that the zinc metal becomes coated with a red–brown layer of copper. The blue colour of the solution fades. The solution will eventually become colourless zinc sulfate (Figure 15.9). Zinc displaces copper from solution, so zinc is more reactive than copper.

Figure 15.9: Zinc is more reactive than copper and displaces copper from copper(II) sulfate solution.

The reverse reaction does not happen. A piece of copper does not react with zinc sulfate solution.

Although copper is an unreactive metal when tested in various ways, there are some metals that are even less reactive. Silver is a metal that is less reactive than copper. If copper metal is placed in a solution of silver nitrate, then a reaction does occur. A deposit of silver grows on the surface of the copper. The solution becomes blue because of the formation of copper nitrate solution:

$$2AgNO_3(aq) + Cu(s) \rightarrow Cu(NO_3)_2(aq) + 2Ag(s)$$

Copper displaces silver from solution, so copper is more reactive than silver. This reaction has been used to produce the 'silver pine tree' by using silver nitrate on a strip of copper folded in the shape of a tree (Figure 15.10).

Figure 15.10: A silver pine tree produced by a displacement reaction.

EXPERIMENTAL SKILLS 15.1

Displacement reactions of metals

In this experiment, you will look at the reactions between metals and solutions of their salts. You will investigate the relationship between the difference in reactivity of the metals involved and how exothermic the reaction is.

You will need:

- three boiling tubes
- measuring cylinder (10 cm^3 or 25 cm^3)
- test-tube rack
- spatula
- zinc sulfate solution (0.5 mol/dm^3)
- copper sulfate solution (0.5 mol/dm^3)
- magnesium powder
- zinc powder
- iron powder
- stopclock
- stirring thermometer (−10 to 100 °C).

Safety

Wear eye protection throughout. Be careful with chemicals. Never ingest them and always wash your hands after handling them.

Getting started

Read through the method for this experiment and create a table in which to record your results. What will you need to record?

What is a line of best fit? Working in pairs, discuss how you would draw a line of best fit.

Method

1. Using a measuring cylinder, pour 10 cm^3 of zinc sulfate solution into a boiling tube.
2. Place the tube in a rack and, using a stirring thermometer, record the temperature of the solution.
3. Add one spatula measure of magnesium powder to the tube, start the stopclock and stir the mixture.
4. Record the temperature every 30 seconds for five minutes, stirring between each reading.
5. Using a fresh tube, repeat the above experiment using copper sulfate solution and zinc powder.
6. Again, record the temperature change over five minutes.
7. Repeat the experiment again, this time using copper sulfate solution and iron powder.
8. Plot three graphs on the same grid showing the temperature change over time for each metal. Draw lines of best fit on each of the three graphs.

Questions

1. What would you expect to happen if the experiment was carried out using iron(II) sulfate solution and zinc powder? Explain your answer.
2. How could you improve the accuracy of your experiment?

Peer assessment

Discuss with your partner how well you worked together during the experiment. What do you think went well? What do you think you could do better if you were asked to repeat the experiment?

Displacement as a redox reaction

Reactive metals are good reducing agents. The most reactive metals are those that can lose electrons most easily. Reactive metals are those most ready to form positive ions. From Chapter 10, you will remember OILRIG:

- oxidation is the loss of electrons
- reduction is the gain of electrons.

The nature of the reaction taking place between zinc and copper sulfate can be explored in more detail by looking at the ionic equation for the reaction:

zinc	+	copper(II) ions	→	zinc ions	+	copper
$Zn(s)$	+	$Cu^{2+}(aq)$	→	$Zn^{2+}(aq)$	+	$Cu(s)$

This shows that the reaction is a redox reaction involving the transfer of two electrons from zinc atoms to copper(II) ions. Zinc atoms are oxidised to zinc ions, while copper(II) ions are reduced (Figure 15.11). In general, the atoms of the more reactive metal lose electrons to become positive ions.

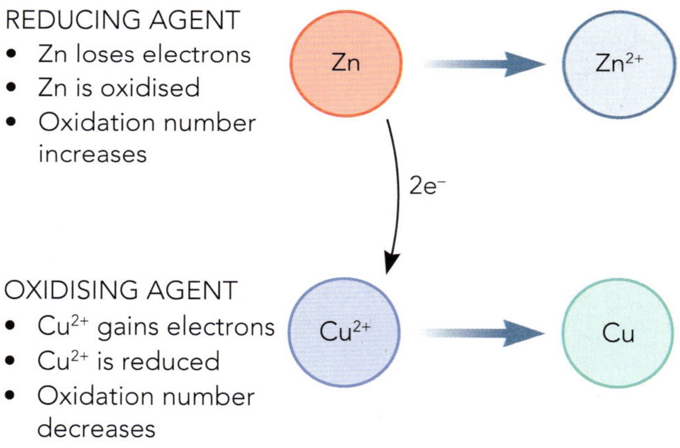

Figure 15.11: The displacement reaction between zinc and copper(II) sulfate is a redox reaction.

Because these are redox reactions, we can write separate half-equations for the two parts of the reaction. The more reactive zinc atoms lose electrons to form ions; they are oxidised.

$Zn(s) \rightarrow Zn^{2+}(aq) + 2e^-$ Oxidation half-equation (loss of electrons)

The ions of the less reactive metal, copper, gain electrons. These copper ions are reduced to metallic copper.

$Cu^{2+}(aq) + 2e^- \rightarrow Cu(s)$ Reduction half-equation (gain of electrons)

In these half-equations, you can see that it is the more reactive metal that loses electrons. The zinc atoms become positive ions. The more reactive a metal is, the more easily it loses its outer shell electrons to form the positive metal ion.

Although we do not often refer to them as such, the reactions in which a metal displaces hydrogen from water or an acid are also redox reactions. The reactive metal forms the positive ion during the reaction (loss of electrons). The hydrogen ions in water or the acid gain electrons and so are reduced. For the reaction between magnesium and an acid, the two half-equations are:

$Mg(s) \rightarrow Mg^{2+}(aq) + 2e^-$ Oxidation half-equation (loss of electrons)

$2H^+(aq) + 2e^- \rightarrow H_2(g)$ Reduction half-equation (gain of electrons)

Consideration of these reactions with water and acids as redox reactions give justification to the inclusion of hydrogen in the reactivity series even though it is not a metal.

Displacement reactions in solution help us to compare directly the reactivity of two metals and to establish the order of the reactivity series.

Questions

4 Write a word equation for the reaction between magnesium and copper(II) sulfate solution.

5 State two observations you would see when a piece of magnesium ribbon is placed in copper(II) sulfate solution.

6 Write a balanced chemical equation and an ionic equation for the reaction between magnesium and copper(II) sulfate solution.

7 We have a list of metals in a particular order in the reactivity series. Can you work out a memory aid (mnemonic) to help remember the order?

An overview of metal reactivity

Through our consideration of trends in the Periodic Table, and the picture of metal reactivity we have developed here, we understand that metals can be placed in a sequence depending on their reactivity. By using specific metal-displacement reactions, we can be clear about the order of metals in the sequence. The reactions of metals with water and acids help us to place hydrogen in the sequence even though it is not a metal (Figure 15.6). Information on the reactivity of metals has been found from various aspects of their chemistry:

- ease of extraction (Chapter 16)
- reactions with air or oxygen
- reactions with water
- reactions with dilute acids
- metal displacement (redox) reactions.

The overall picture that emerges is summarised in Figure 15.12. This is known as the reactivity series of metals.

Evidence for the position of carbon comes from the methods we can use to extract a metal (Chapter 16). Carbon is placed between aluminium and zinc because it can reduce zinc oxide to form zinc but cannot reduce aluminium oxide.

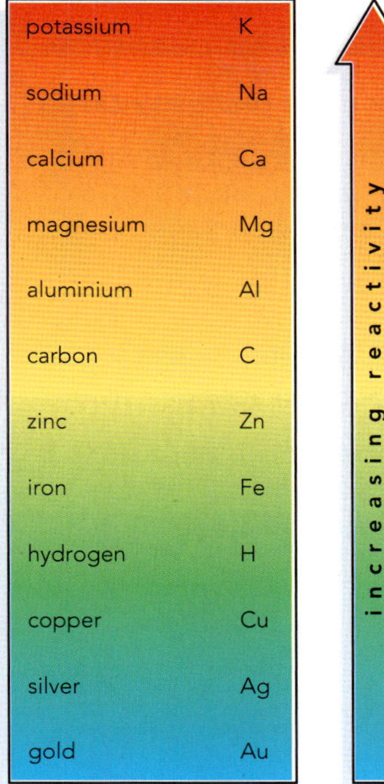

Figure 15.12: Reactivity series of metals.

We can see from these redox reactions that the most reactive metals are those that most readily lose electron(s) to form positive ions. The most reactive metals are the most powerful reducing agents, giving electrons to other ions or compounds.

SUMMARY

Metals show different levels of reactivity and can be arranged in a reactivity series of activity: potassium, sodium, calcium, magnesium, aluminium, carbon, zinc, iron, hydrogen copper, silver, gold.
The different levels of reactivity can be seen in the way the different metals react with such reagents as water and dilute acids.
The elements of Group I (the alkali metals) are the most reactive metals at the top of the series.
Transition metals are more moderately reactive.
Some metals such as copper, silver and gold do not react with water or dilute hydrochloric acid and so are placed at the bottom of the series.
Metal displacement reactions provide evidence that helps to bring detail to the reactivity series.
A more reactive metal will displace a less reactive metal from a solution of its salts.
The position of a metal in the series is related to the tendency of the metal to form its positive ion in reactions.
Metal displacement reactions are redox reactions.

PROJECT

Understanding redox and the reactivity series

Redox reactions are an important type of reaction in Chemistry. You have met ideas on this type of reaction in both this and several earlier chapters (especially Chapter 10).

Together with a partner, write a podcast script (Figure 15.13) that describes the key features of these reactions. In your podcast script:

- Explore the word 'redox' itself. What does it stand for? Redox reactions can be seen as competitions between elements. But competitions for what?
- Discuss the different definitions of the processes involved, including:
  - removal/addition of oxygen
  - gain/loss of electrons.
- How does the idea of redox fit with the reactions used in this chapter to place the metals in a detailed reactivity series? Conclude your presentation by describing a particularly important example of this type of reaction and its significance.

Figure 15.13: Communicating ideas through a podcast.

EXAM-STYLE QUESTIONS

1 Which of these metals will react with cold water to produce hydrogen gas?

A copper
B zinc
C potassium
D aluminium **[1]**

2 A student investigated the reactivity of a number of metals. First, she took a strip of magnesium ribbon and dropped it into a beaker of dilute hydrochloric acid.

a Describe what the student saw as the magnesium ribbon reacted. **[2]**

b She then dropped a different strip of magnesium ribbon into cold water. Describe what she saw. **[1]**

She then tried the same experiments with three other metals: calcium, iron and a third metal that she did not know the name of. The results are shown in the table.

Metal	Reaction with acid	Reaction with water
calcium	reacted violently	reacted rapidly
iron	reacted slowly	no reaction
magnesium	reacted rapidly	no reaction
unknown metal	reacted quite quickly	no reaction

COMMAND WORD

describe: state the points of a topic / give characteristics and main features

CONTINUED

c Put the metals in increasing order of reactivity (least reactive first). [2]

d Suggest the name of the 'unknown metal'. [1]

e Some of these reactions produced a gas. Give a test for this gas.

Test: ____________________

Result: ____________________ [2]

[Total: 8]

3 A student put a strip of a silver/grey metal, A, into a test-tube containing a blue solution. The solution became paler in colour and the strip of metal was covered with a brown coating.

When the same silver metal was added to a solution of zinc sulfate, no reaction took place.

a **Identify** the blue solution. [1]

b Suggest the identity of metal A. [1]

In a separate experiment the student put a strip of copper into a solution of silver nitrate, $AgNO_3(aq)$.

c Write an equation including state symbols for the reaction which took place. [3]

d What observations would the student have made? [2]

e The student repeated this experiment using zinc nitrate solution. **Explain** why no reaction took place. [1]

[Total: 8]

4 Zinc powder reacts with copper(II) oxide when the mixture is heated:

$Zn(s) + CuO(s) \rightarrow ZnO(s) + Cu(s)$

a Which reactant is the reducing agent in this reaction? Explain your answer. [2]

b Describe how electrons are transferred during this reaction. [2]

c Both magnesium and zinc are above copper in the reactivity series. **Examine**, in terms of how readily these metals form ions, why magnesium is more able to remove oxygen from copper(II) oxide than zinc. [2]

[Total: 6]

COMMAND WORDS

suggest: apply knowledge and understanding to situations where there are a range of valid responses in order to make proposals / put forward considerations

give: produce an answer from a given source or recall/memory

identify: name/ select/recognise

explain: set out purposes or reasons/make the relationships between things evident/ provide why and/or how and support with relevant evidence

examine: investigate closely, in detail

SELF-EVALUATION CHECKLIST

After studying this chapter, think about how confident you are with the different topics. This will help you to see any gaps in your knowledge and help you to learn more effectively.

I can	See Topic...	Needs more work	Almost there	Confident to move on
understand the concept of a reactivity series	15.1			
state the order of the reactivity series	15.1			
describe the reactions of potassium, sodium and calcium with cold water	15.1			
describe the reaction of magnesium with steam	15.1			
describe the reactions of metals with dilute hydrochloric acid	15.1			
relate the reactions of metals to their position in the reactivity series	15.1			
deduce an order of reactivity from a set of experimental results	15.1			
explain the apparent unreactivity of aluminium to the presence of a stable oxide layer on the metal	15.1			
relate the reactivity series to the displacement reactions of several metals	15.2			

› Chapter 16

Extraction and corrosion of metals

IN THIS CHAPTER YOU WILL:

- consider the methods of extraction of metals from their ores in the context of the reactivity series of metals
- describe how moderately reactive metals can be extracted by reduction with carbon in a blast furnace, while reactive metals are extracted by electrolysis
- describe how iron is extracted from hematite in a blast furnace
- learn how aluminium is extracted from bauxite by the electrolysis of molten aluminium oxide
- discuss the causes of the rusting and describe some barrier methods of rust prevention

› consider the chemistry of the extraction of iron

› describe the extraction of aluminium from bauxite

› consider galvanisation and sacrificial protection as methods of rust prevention.

GETTING STARTED

In your class groups, discuss which metals are most abundant in the Earth's crust. Extend your ideas to include less abundant metals that are in high demand for applications in modern technology.

What stress do these considerations place on concerns over the environment and the availability of resources?

MINING THE OCEAN FLOOR

Concern over the impact of global warming and climate change has focused considerable attention on the need to reduce our use of fossil fuels. One aspect of this is the potential to use electric battery-powered cars and this has placed an important focus on improved battery technology. Currently, the metal cobalt is an essential ingredient for the rechargeable batteries needed to provide electric cars with the needed power and storage capacity. Demand for cobalt is therefore increasing rapidly and the number of sites where cobalt can be mined worldwide is very limited. Most of the world's cobalt is mined in the Democratic Republic of Congo. Expanding production is not straightforward and this is leading mining companies to consider obtaining cobalt from a very different location – the deep-sea floor!

We are becoming increasingly aware of rocks on the seabed, known as 'polymetallic nodules', that are rich in metals such as manganese, nickel and cobalt (Figure 16.1). The discovery of this potential resource has raised enthusiastic attention over recent years. Different regions of the seabed have been put forward as open to exploitation. Prototype mining machines have been designed and the pressure is growing to develop mining operations given the increasingly pressing need for these rare metals.

But what would be the environmental cost of such deep-sea mining? Gathering these nodules from the seabed would disturb the habitat of the marine life, stirring up clouds of sand and silt. New technology is being designed to reduce the impact of what is effectively open-cast mining, but it surely cannot eliminate all damage. There is significant international interest even though the financial costs are great and there is need for international control of this development.

Figure 16.1: A manganese nodule of the type found on the ocean floor.

Discussion questions

1 What are the environmental advantages and disadvantages of developing seabed mining? Think as broadly as you can about the related issues.

2 What are the problems associated with relying on one major land-based source for a resource that is likely to be in increasing demand?

16.1 Metal extraction and the reactivity series

Some metals exist as pure metals in nature. Others are very difficult to extract from their compounds contained in **ores**. Those metals that do not occur uncombined (native) in the ground can usually be found as the metal oxide, or a compound that can easily be converted to the oxide. To extract the metal from this oxide we need to carry out a reduction (the removal of the oxygen from the metal).

In Chapter 15, we established that there is a reactivity series based on the differences in reactivity of the various metals. The most reactive metals are those that combine most strongly in their compounds or form positive ions most easily. We placed hydrogen in that reactivity series because it is useful to know its reactivity relative to the different metals. We found that hydrogen is less reactive than zinc, but more reactive than copper, for instance. Hydrogen will remove oxygen (reduce) copper oxide to copper metal (Chapter 10); but hydrogen will not reduce zinc oxide to zinc.

At the end of Chapter 15, we commented that it would be useful to place carbon in the reactivity series as it is a strong reducing agent providing a very useful method of extracting metals. Practical experiments have shown that carbon is between aluminium and zinc in the series (Figure 16.2).

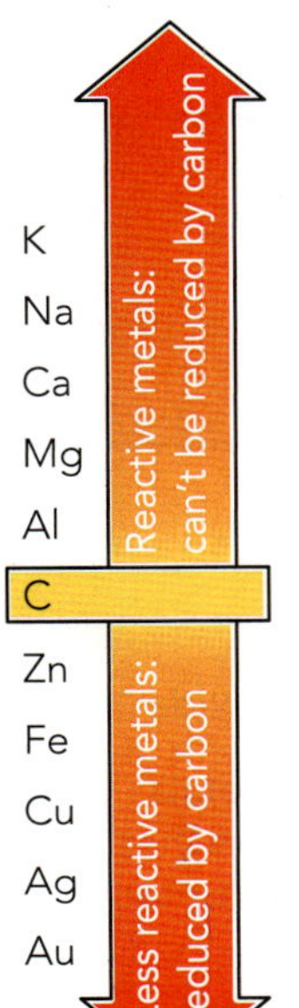

Figure 16.2: Position of carbon in the reactivity series.

Metals below carbon in the series can be extracted by heating their oxides with carbon, but those above carbon must be extracted by **electrolysis**. If a metal is below carbon in the reactivity series the oxygen in its oxide will form covalent bonds more easily with carbon rather than remain as the oxide ion. Carbon is the reducing agent, removing the oxygen from the oxide.

If the metal is more reactive than carbon, then reduction must be achieved by the reaction that takes place at the cathode during electrolysis. The metal ions in the oxide must gain electrons to produce the free metal. This is the type of reaction that takes place in the electrolytic extraction of aluminium.

KEY WORDS

ore: a naturally occurring mineral from which a metal can be extracted

electrolysis: the breakdown of an ionic compound, molten or in aqueous solution, by the use of electricity

Figure 16.3: The blast furnace at a steel-making plant uses carbon to reduce iron ore.

EXPERIMENTAL SKILLS 16.1

Carbon as a reducing agent in metal extraction

In this experiment, you will heat copper(II) oxide with carbon to see whether reduction takes place. If the copper(II) oxide is reduced to copper then it helps place carbon in the reactivity series.

You will need:

- test-tube
- test-tube holder
- test-tube rack
- spatula
- Bunsen burner
- heat-resistant mat
- powdered charcoal
- copper oxide.

Safety

Wear eye protection throughout. Point the test-tube away from you and others while heating it. Be careful with chemicals. Never ingest them and always wash your hands after handling them.

Getting started

You will need the roaring blue flame from the Bunsen burner produced when the air hole is fully open. Practise producing this flame prior to the experiment. When heating the tube, you will need to hold the end of the tube just above the blue cone in the flame.

Method

1 Place one spatula measure of copper oxide into the test-tube.

2 Add one spatula measure of powdered charcoal. DO NOT mix the two.

3 Set up the Bunsen burner on the heat-resistant mat.

4 Place the test-tube in the test-tube holder and heat the two layers strongly for five minutes with the Bunsen burner.

5 Allow the test-tube to cool in the test-tube rack.

6 Look closely at the point where the two powders join.

Questions

1 What do you see at the place where the two powders join?

2 Write an equation for the reaction that is occurring.

Self-assessment

Was your experiment successful? How do you know?

Think about how you did each of the steps in the method. If you were to do this experiment again, is there anything you would do differently?

Extraction of iron from its ore

Iron is the second most common metal in the Earth's crust. The main ore of iron is **hematite** (Fe_2O_3). This ore is rich in iron, containing more than 60% iron.

Production of iron in the blast furnace

The iron is obtained by reduction with carbon in a **blast furnace** (Figures 16.4 and 16.5). The furnace is a steel

KEY WORDS

hematite: the major ore of iron, iron(III) oxide

blast furnace: a furnace for extracting metals (particularly iron) by reduction with carbon that uses hot air blasted in at the base of the furnace to raise the temperature

tower about 30 metres high. It is lined with refractory (heat-resistant) bricks of magnesium oxide that are cooled by water. The furnace is loaded with the 'charge' of raw materials, which consists of iron ore, coke (a form of carbon made from coal) and **limestone** (a **mineral** form of calcium carbonate). The charge is sintered (the ore is powdered and heated with coke and limestone) to make sure the solids mix well, and it is mixed with more coke. Blasts of hot air are sent in to the furnace through holes near the bottom of the furnace. A series of chemical reactions takes place within the furnace to produce molten iron.

Figure 16.4: A worker in protective clothing takes a sample from a blast furnace in a steel works.

KEY WORDS

limestone: a form of calcium carbonate ($CaCO_3$)

mineral: a naturally occurring rock containing a particular compound

The carbon (coke) burns in the air blast and the furnace gets very hot from the heat from this exothermic reaction.

carbon + oxygen → carbon dioxide

As carbon dioxide rises through the furnace it reacts with more carbon and is reduced to carbon monoxide.

carbon dioxide + carbon → carbon monoxide

The most important reaction that then occurs is the reduction of the ore by carbon monoxide:

iron(III) oxide + carbon monoxide → iron + carbon dioxide

The iron produced flows to the bottom of the furnace where it can be 'tapped off' as a liquid because the temperature at the bottom of the furnace is higher than the melting point of iron (Figure 16.4).

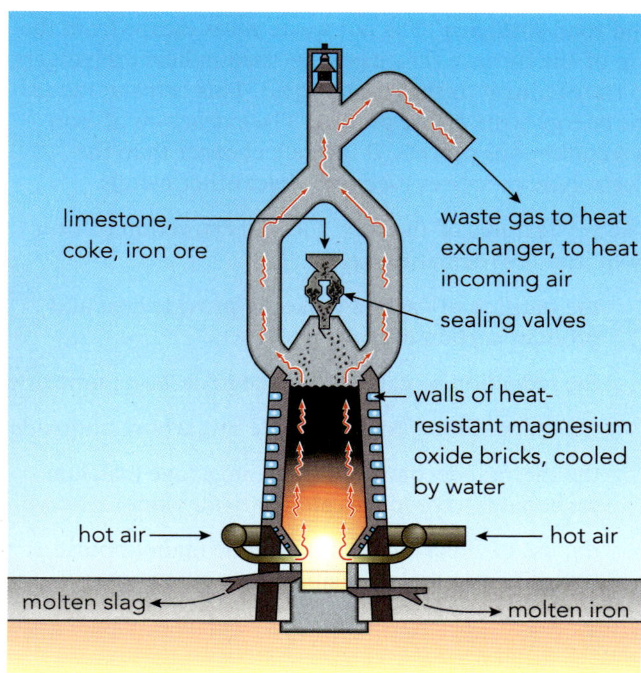

Figure 16.5: Reduction of iron ore to iron in the blast furnace.

One of the major impurities in iron ore is sand (silicon(IV) oxide, SiO_2 – also called silica). The limestone added to the furnace helps to remove this impurity. The limestone (calcium carbonate) undergoes a thermal decomposition reaction in the furnace produce calcium oxide and carbon dioxide.

calcium carbonate → calcium oxide + carbon dioxide

The lime produced then reacts with the silica to form calcium silicate **slag**.

calcium oxide + silicon(IV) oxide → calcium silicate

KEY WORD

slag: a molten mixture of impurities, mainly calcium silicate, formed in the blast furnace

The calcium silicate formed is also molten. It flows down the furnace and forms a molten layer of slag on top of the iron (Figure 16.5). It does not mix with the iron, and it is less dense. The molten slag is 'tapped off' separately. When solidified, the slag is used in concrete for buildings and road surfacing. The hot waste gases escape from the top of the furnace. The gases are used in heat exchangers to transfer heat to the incoming air. This helps to reduce the energy costs of the process. The extraction of iron is a continuous process. It is much cheaper than the electrolytic processes used to extract other metals.

The key features of the blast furnace extraction of iron from iron ore (hematite) are:

- the burning of carbon (coke) to provide heat and produce carbon dioxide
- the reduction of carbon dioxide to carbon monoxide
- the reduction of iron(III) oxide by carbon monoxide
- the thermal decomposition of limestone (calcium carbonate) to produce calcium oxide (lime)
- the use of limestone to remove the main impurity (sand) as slag (calcium silicate).

Chemistry of the blast furnace

A number of different reactions take place in the blast furnace. The reaction that takes place depends on the position in the furnace and the temperature. Figure 16.6 shows the regions of the furnace where the key reactions take place and the chemical equations for those reactions.

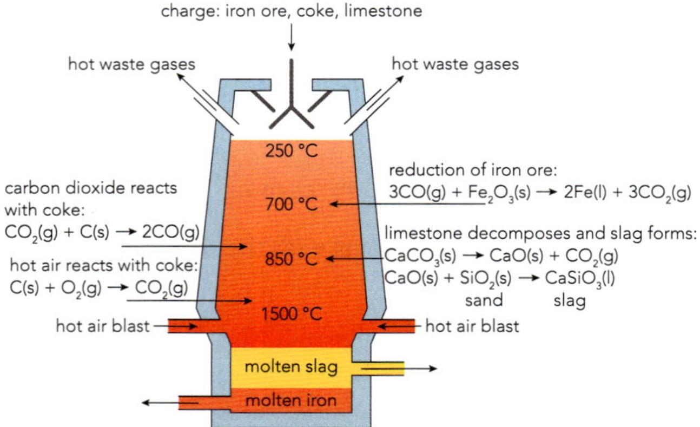

Figure 16.6: Key reactions of the extraction of iron and the regions of the blast furnace where they take place.

The iron produced in the blast furnace is only about 95% pure. The impurities are mainly carbon, but they also include sulfur, silicon and phosphorus. Most of the iron produced in the furnace is taken and purified further and then turned into various forms of steel. Steels are alloys of iron and a defined amount of carbon, often with other transition metals added depending on the type of steel required. The addition of chromium and nickel, for example, make the steel hard and resistant to **corrosion** (rusting). This alloy is known as stainless steel (Chapter 14).

KEY WORD

corrosion: the process that takes place when metals and alloys are chemically attacked by oxygen, water or any other substances found in their immediate environment

Questions

1. Why is limestone added to the blast furnace? Write the balanced symbol equation for the thermal decomposition of the limestone to lime.
2. What is the reducing agent in the extraction of iron from iron(III) oxide? Give the word equation and the balanced symbol equation of the reduction reaction.
3. Write the word and symbol equation for the production of the calcium silicate slag.

Extraction of aluminium from its ore

We have seen that reduction with carbon does not work for more reactive metals. The metals are held in their compounds (oxides or chlorides) by stronger bonds that need a lot of energy to break them. This energy is best supplied by electricity. Extracting metals in this way is a three-stage process:

- mining the ore
- purification of the ore
- electrolysis of the molten ore.

The extraction of a metal by electrolysis is expensive. Energy costs to keep the ore molten and to separate the ions can be very high. Because of this, many of these metals are extracted in regions where hydroelectric power from dams is available. Aluminium plants (*smelters*) are the most important examples. They produce sufficient aluminium to make it the second most widely used metal after iron. Although aluminium is the most common metal found in the Earth's crust, it is too reactive with other elements to occur naturally as the free metal.

Figure 16.7: The major ore of aluminium is bauxite. Bauxite usually contains some iron(III) oxide, which gives the ore its brown colour.

Bauxite (Figure 16.7) is the major form of aluminium oxide. Up to 25% of bauxite consists of impurities such as iron(III) oxide and sand. The iron(III) oxide gives it a red–brown colour.

Aluminium is extracted from bauxite by electrolysis. The bauxite is converted to alumina (aluminium oxide) using sodium hydroxide. Aluminium oxide is insoluble in water and has a very high melting point (2030 °C) making it difficult to change it into a suitable **electrolyte**. However, in 1886, the Hall–Héroult electrolytic method for extracting aluminium was invented independently by the American chemist Charles Martin Hall and the French scientist Paul Héroult.

KEY WORDS

bauxite: the major ore of aluminium; a form of aluminium oxide, Al_2O_3

electrolyte: an ionic compound that will conduct electricity when it is molten or dissolved in water; electrolytes will not conduct electricity when solid

Hall–Héroult process

The Hall–Héroult electrolytic process involves the following stages.

1 The bauxite is treated with sodium hydroxide in a refinery to obtain pure aluminium oxide (alumina). The alumina produced is shipped to the electrolysis plant.

2 The purified aluminium oxide (Al_2O_3) is dissolved in molten **cryolite** (another ore of aluminium). Dissolving the aluminium oxide in cryolite lowers the melting point of the oxide and therefore the working temperature of the cell. The cryolite thus provides a considerable saving in energy costs.

3 The molten mixture of aluminium oxide and cryolite is electrolysed in a cell fitted with graphite electrodes (Figure 16.8).

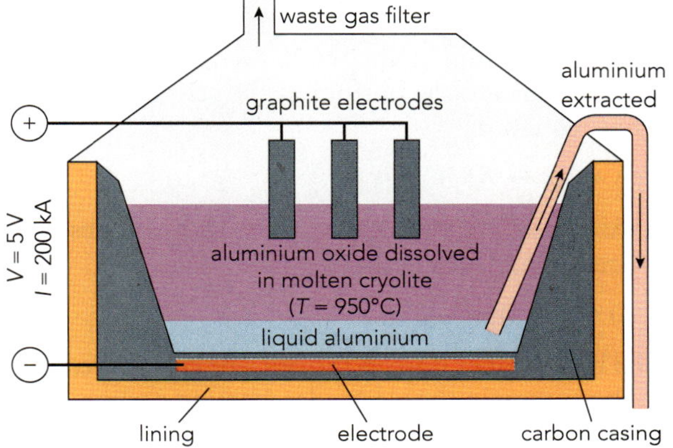

Figure 16.8: Electrolytic cell for extracting aluminium.

KEY WORD

cryolite: sodium aluminium fluoride (Na_3AlF_6), an ore of aluminium used in the extraction of aluminium to lower the operating temperature of the electrolytic cell; now replaced by synthetic sodium aluminium fluoride produced from the common mineral fluorite

Chemistry of the Hall–Héroult process

The electrolyte in the **electrolytic cell** is a solution of alumina in cryolite as shown in Figure 16.8. Cryolite (sodium aluminium fluoride, Na_3AlF_6) is a mineral found naturally in Greenland. It is no longer mined commercially there, and all the cryolite now used is made synthetically. Cryolite is used to lower the working temperature of the electrolytic cell. The melting point of aluminium oxide is 2030 °C. This is reduced to 900–1000 °C by dissolving the aluminium oxide in cryolite.

The process is run continuously. Liquid aluminium is siphoned out and more alumina is added as necessary. The voltage used is quite low (5 volts) but the current flowing is very large (200 kiloamps). The large current flowing through the electrolyte produces a lot of heat, which keeps the mixture molten.

Aluminium ions are attracted to the cathode. The cathode is effectively the carbon lining of the cell. Aluminium (melting point 660 °C) is discharged as a liquid and sinks to the bottom of the cell, as shown in this half equation.

$$Al^{3+}(l) + 3e^- \rightarrow Al(l)$$

The oxide ions are attracted to the carbon anodes where they are discharged as oxygen gas.

$$2O^{2-}(l) \rightarrow O_2(g) + 4e^-$$

At the high temperature in the cell, the oxygen reacts with the carbon in the electrodes forming carbon dioxide. The carbon anodes slowly burn away and have to be replaced frequently.

$$\text{carbon} + \text{oxygen} \xrightarrow{\text{heat}} \text{carbon dioxide}$$

$$C(s) + O_2(g) \xrightarrow{\text{heat}} CO_2(g)$$

This means that the anodes need to be replaced regularly. The cryolite is not used up by the electrolysis and so only alumina needs to be added to keep the process running.

The Hall–Héroult process uses a great deal of energy and it is also costly to replace the anodes. These costs emphasise why the recycling of aluminium is important as it is much cheaper to recycle the metal than to manufacture it. The energy requirement for recycling is about 5% of that needed to manufacture the same amount of 'new' aluminium metal.

KEY WORDS

electrolytic cell: a cell consisting of an electrolyte and two electrodes (anode and cathode) connected to an external DC power source where positive and negative ions in the electrolyte are separated and discharged

REFLECTION

What features helped you understand why different methods of extraction are used for different metals? Do you think you learnt more from the written text or from the figures? Would a video or a visit in person help you to understand this topic further?

Questions

4 Why is aluminium expensive to extract?

5 Why is cryolite added to the electrolytic cell as well as alumina?

6 Why does the anode in the electrolytic cell need replacing regularly?

16.2 Corrosion of metals

The problem of corrosion

When a metal is attacked by air, water or other surrounding substances, it is said to corrode. Most metals corrode. They undergo a chemical reaction with oxygen and other gases in the air to form compounds that collect on the surface of the metal. Some very reactive metals such as sodium and potassium need to be stored under oil to keep them away from air and water. Others such as calcium will slowly corrode away to powder over time.

In the case of iron and steel, the corrosion process is also known as **rusting**. Rusting is a serious economic problem. Large sums of money are spent each year replacing damaged iron and steel structures or protecting structures from such damage. Rust is a red–brown powder consisting mainly of hydrated iron(III) oxide ($Fe_2O_3{\cdot}xH_2O$). Water and oxygen are essential for iron to rust (Figure 16.10). The problem is made worse by the presence of salt; seawater increases the rate of corrosion as can be seen on countless shipwrecks around the world (Figure 16.9). Acid rain also increases the rate at which iron objects rust.

Figure 16.9: Rusted shipwreck on rocky beach.

> **KEY WORD**
>
> **rusting:** the corrosion of iron and steel to form rust (hydrated iron(III) oxide)

Aluminium is more reactive than iron, but it does not corrode in the damaging way that iron does. Both metals react with air. In the case of aluminium, a very thin single layer of aluminium oxide forms, which sticks strongly to the surface of the metal. This micro-layer seals the metal surface and protects it from further attack. Aluminium is a useful construction material because it is protected by this layer. The protective layer of aluminium oxide can be made thicker by electrolysis (anodised aluminium).

In contrast, when iron corrodes, the rust forms in flakes. It does not form a single layer. The attack on the metal can continue over time as the rust flakes come off and holes can be formed in iron sheets by the rusting process. Chromium is another metal, similar to aluminium, that is protected by an oxide layer. If chromium is alloyed with iron, a 'stainless' steel is produced. However, it would be too expensive to use stainless steel for all the objects built out of iron. Electroplating a layer of chromium on steel is used to protect some objects from rusting (Chapter 6).

Rust prevention

Many methods have been devised to protect iron and steel from rusting.

Barrier methods

Rust can be prevented by a coating of material that prevents the iron or steel from coming into contact with water and oxygen. These methods are known as barrier methods.

Painting

Painting is a very common method of protection, and is used for objects ranging in size from ships and bridges to garden gates. Some paints react with the metal to form a stronger protective layer. However, generally, painting only protects the metal for as long as the paint layer is unscratched. Regular repainting is often necessary to keep this protection intact.

Oiling and greasing

The oiling and/or greasing of the moving parts of machinery forms a protective film, preventing rusting. Again, the treatment must be repeated to continue the protection.

Plastic coatings

Plastic coatings are used to form a protective layer on items such as refrigerators and garden chairs. The plastic poly(vinyl chloride), PVC, is often used for this purpose.

ACTIVITY 16.1

Causes of rusting

The results of an experiment to investigate rusting are shown in Figure 16.10. The following tasks are based on the experiment. Discuss each of them with a partner before you attempt to write anything down.

- What three things do the results of this experiment tell you about rusting?
- Make a list of the apparatus you would need to carry out this experiment.
- Write an instruction sheet explaining, to someone else, how they should carry out the experiment.
- Think of a possible fifth tube you could set up to show a way of preventing rusting.

Discuss your answers with a partner and comment on how each of you completed the task. Were you able to evaluate the different causes clearly?

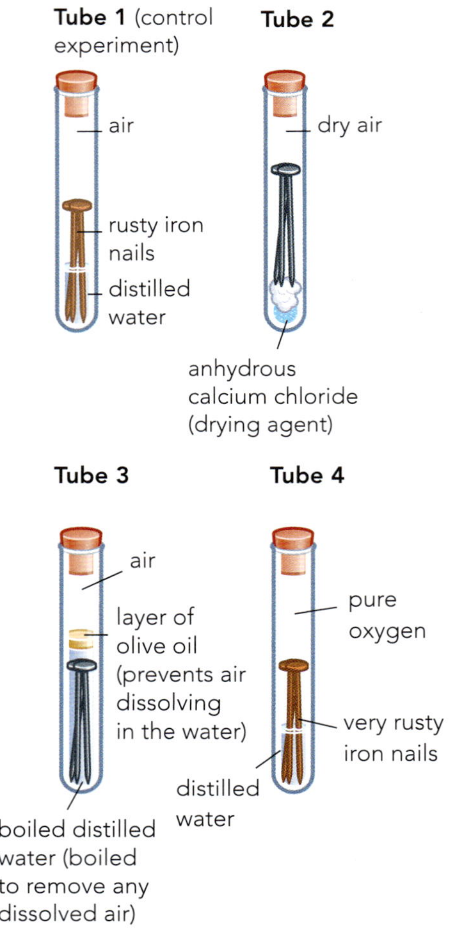

Figure 16.10: The results of an experiment to investigate the factors that are involved in rusting. In tube 2, the air is dry, so the nails do not rust. In tube 3, there is no oxygen in the water, so the nails do not rust. In tube 4, pure oxygen and water are present, so the nails are very rusty.

Electroplating

An iron or steel object can be electroplated with a layer of chromium or tin to protect against rusting (Chapter 6). A 'tin can' is made of steel coated on both sides with a fine layer of tin. Tin is used because it is unreactive and non-toxic. However, this does raise a problem. With both these metals, if the protective layer is broken, then the steel beneath will begin to rust.

Galvanising and sacrificial protection

Galvanising

An object may be coated with a layer of the more reactive metal, zinc. This is called **galvanising**. It has the advantage over other barrier methods in that the protection still works even if the zinc layer is badly scratched. Galvanising is in fact both a barrier method of protection, while the zinc layer is unbroken, and a form of sacrificial protection, if the zinc layer is scratched or broken. If the zinc layer is broken it is still corroded away in preference to the iron as zinc is a more reactive metal than iron.

The zinc layer (Figure 16.11) can be applied by several different methods. These include electroplating or dipping the object into molten zinc. The bodies of cars are dipped into a bath of molten zinc to form a protective layer.

Figure 16.11: A photograph of zinc grains on a galvanised fence post.

Sacrificial protection

Sacrificial protection is a method of rust prevention in which blocks of a reactive metal are attached to the iron surface. Zinc or magnesium blocks are attached to oil rigs and to the hulls of ships (Figure 16.12). These metals are more reactive than iron and will be corroded in preference to it: they 'sacrifice' themselves so the iron does not rust. Underground gas and water pipes are connected by wire to blocks of magnesium to obtain the same protection. In all cases, an electrochemical cell is set up. The metal blocks lose electrons in preference to the iron and so prevent the iron forming iron(III) oxide. The more reactive metal oxidises more readily than iron, so it 'sacrifices' itself while the iron does not rust.

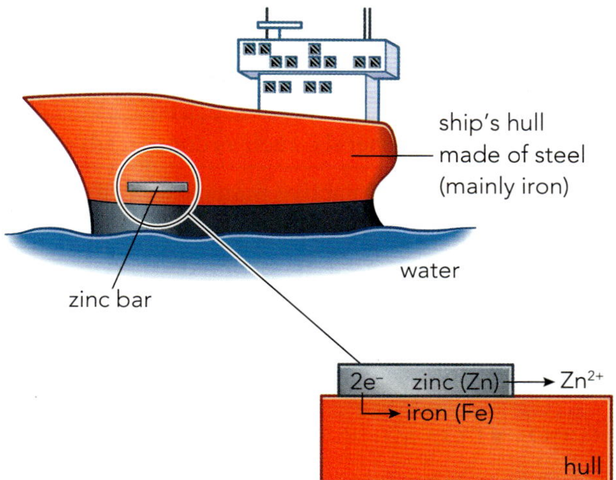

Figure 16.12: Blocks of zinc (or magnesium) are used for the sacrificial protection of the hulls of ships.

KEY WORDS

galvanising: the protection of iron and steel objects by coating with a layer of zinc

sacrificial protection: a method of rust protection involving the attachment of blocks of a metal more reactive than iron to a structure; this metal is corroded rather than the iron or steel structure

Questions

7 What two substances must be present before iron can rust? What other factors can increase the rate of rusting?

8 What simple barrier methods can be used to prevent rusting?

9 Steel is often plated with another metal. Chromium and zinc are both used. Suggest why zinc is more effective at stopping rusting.

SUMMARY

The method of extracting a metal from its ore depends on the reactivity of the metal.
The least reactive metals, such as copper, silver and gold, can be found uncombined in the Earth's crust.
Metals that are less reactive than carbon can be extracted by heating with carbon in a blast furnace.
The main ore of iron is hematite.
Iron is extracted from hematite by reduction in a blast furnace.
The extraction of iron from hematite in a blast furnace can be described in a series of different stages.
The balanced symbol equations for the extraction of iron from hematite in a blast furnace are: $C + O_2 \rightarrow CO_2$ $CO_2 + C \rightarrow 2CO$ $Fe_2O_3 + 3CO \rightarrow 2Fe + 3CO_2$
The main ore of aluminium is bauxite.
Aluminium is more reactive than carbon and must be extracted by electrolysis.
Aluminium oxide is prepared from bauxite and is the molten electrolyte used to produce aluminium.
Aluminium oxide is dissolved in molten cryolite to lower its melting temperature and aluminium ions are discharged at the cathode during the electrolysis to produce molten aluminium.
The carbon anodes need to be regularly replaced during the process because they slowly burn away during electrolysis.
The reactions at the electrodes in the electrolytic cell can be represented by half-equations involving the loss or gain of electrons.
Rusting of iron and steel requires the presence of oxygen and water, and is increased by the presence of salts or acid rain.
Corrosion of iron and steel is a major problem and various methods have been devised for rust protection.
Barrier methods of rust prevention include painting, the use of oil or grease and coating the object in plastic.
Galvanising an object by coating with zinc is both a barrier method of rust prevention and offers protection even if the zinc layer is broken.
Sacrificial protection is a form of rust protection using blocks of a more reactive metal attached to an iron or steel structure; electrons are lost from the more reactive metal to the iron or steel to prevent the formation of iron(III) oxide.

PROJECT

Less than super superyachts: changing the design

The owners of superyachts have reported that the hulls of their expensive vessels are rusting. The rusting occurs as any exposed iron can react with water and oxygen to from hydrated iron(III) oxide, more commonly known as rust. It is known that salt in the sea water accelerates the rusting process because salt water is a better electrolyte than pure water. With the yachts selling for very high prices, the company making them needs to:

- suggest solutions to existing owners that will prevent further rusting
- work with its current team of designers to ensure the next generation of yachts do not experience the same problems.

You have been asked to research as many solutions as possible and give the strengths and weaknesses of each.

1 Copy and complete the tables shown. Include additional rows as necessary.

2 Use your tables to prepare a presentation that will be given to the company's board of directors. The presentation should explain why the problem has arisen, give the options available and then justify which proposal you would recommend using to help fix the current problem and which proposal you would recommend to ensure such issues do not arise in the future.

Figure 16.13: A luxury superyacht.

To aid your work, you might want to research metals or alloys that do not corrode, but you should also consider costs and any weight penalties due to increased density. Your research could include novel metal alloys or even non-metallic materials.

Options for reducing rusting from the hulls of current superyachts

Suggested fix	Scientific reason why this approach would work	The benefits of this approach	Possible risks

Options for preventing rusting from the hulls of future (next-generation) superyachts

Suggested fix	Scientific reason why this approach would work	The benefits of this approach	Possible risks

EXAM-STYLE QUESTIONS

1 The figure shows a blast furnace for extracting iron from hematite.

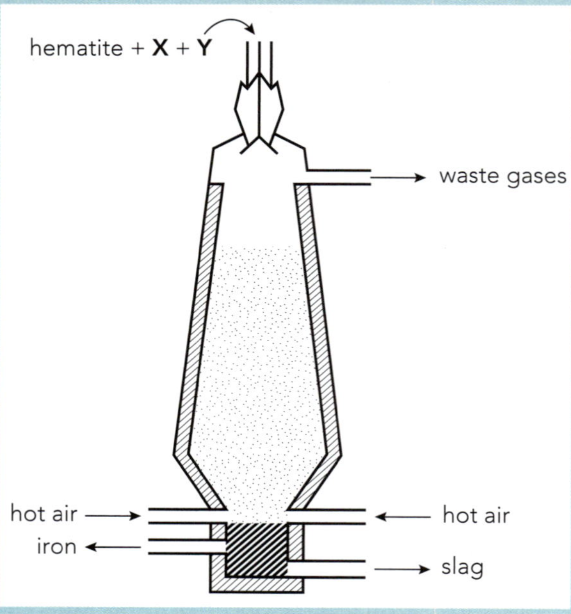

What are **X** and **Y**?

	X	Y
A	lime	coal
B	limestone	coal
C	lime	coke
D	limestone	coke

[1]

2 Which treatment is used to prevent the rusting of an iron girder of a bridge?

A electroplating
B painting
C galvanising
D coating with grease **[1]**

3 The equation shows how iron(III) oxide is reduced to form iron in the blast furnace.

iron(III) oxide(s) + carbon monoxide(g) → iron(l) + carbon dioxide(g)

a What is the main ore of iron containing iron(III) oxide? **[1]**
b Which substance in the equation is the reducing agent in the reaction? **[1]**
c Why is the iron produced as a liquid? **[1]**
d Slag is also formed in the blast furnace. How is slag produced? **[2]**

[Total: 5]

CONTINUED

4 Iron and aluminium are two of the most important metals for industry. They are both extracted from their oxides.

This reaction of carbon occurs during both extraction processes.

$C(s) + O_2(g) \rightarrow CO_2(g)$

a Where does the oxygen come from when this reaction occurs in the extraction of iron? [1]

b **Explain** why this reaction is important in the extraction of iron. [2]

c Where does the oxygen come from during the extraction of aluminium? [2]

d Explain why this reaction is a problem during the extraction of aluminium. [2]

Carbon dioxide is also produced from the thermal decomposition of limestone, $CaCO_3$.

$CaCO_3(s) \rightarrow CaO(g) + CO_2(g)$

e Explain how the calcium oxide produced is important in the extraction of iron. [2]

[Total: 9]

5 When iron is left in the open air it rusts. Iron can be protected from rusting by painting or coating with oil. This stops air and water reaching the surface of the iron.

Coating the iron with zinc works more effectively than these methods.

a What name is given to the process of coating iron with zinc? [1]

b **Suggest** why this method is more effective than coating with paint. [1]

c If an iron object is connected to a block of a metal such as zinc or magnesium, it doesn't rust, even though it is exposed to air and water. Explain how this protection works. [3]

d Explain why aluminium, a more reactive metal than iron, doesn't corrode. [2]

[Total: 7]

COMMAND WORDS

explain: set out purposes or reasons/make the relationships between things evident/provide why and/or how and support with relevant evidence

suggest: apply knowledge and understanding to situations where there are a range of valid responses in order to make proposals / put forward considerations

SELF-EVALUATION CHECKLIST

After studying this chapter, think about how confident you are with the different topics. This will help you see any gaps in your knowledge and help you to learn more effectively.

I can	See Topic...	Needs more work	Almost there	Confident to move on
understand how the ease of extraction of a metal from its ore is related to its position in the reactivity series of metals	16.1			
describe the extraction of iron from hematite in the blast furnace by heating with carbon	16.1			
describe the chemical changes taking place in the blast furnace extraction of iron, including the reduction of iron(III) oxide by carbon monoxide	16.1			
describe the process of removing impurities from the furnace as a molten slag	16.1			
state the balanced symbol equations for the reactions taking place in the blast furnace extraction of iron	16.1			
state that the main ore of aluminium is bauxite and that aluminium is extracted from it by electrolysis	16.1			
describe the extraction of aluminium by electrolysis, including the role of cryolite and the electrode reactions	16.1			
state the conditions required for the rusting of iron and steel	16.2			
describe barrier methods for rust prevention, including painting, greasing and coating with plastic	16.2			
describe the use of galvanising with zinc as both a barrier and sacrificial protection method of rust prevention	16.2			
explain sacrificial protection in terms of the reactivity of the metals used and the loss of electrons	16.2			

Chapter 17
Chemistry of our environment

IN THIS CHAPTER YOU WILL:

- learn how to describe the composition of clean dry air
- investigate common air pollutants and their adverse effects
- describe how carbon dioxide and methane are greenhouse gases linked to global warming and climate change
- describe and state the photosynthesis reaction
- describe the tests for the presence and purity of water, and why distilled water is used in experiments
- discuss the sources of substances in water from natural sources and outline their impacts on human health and the environment
- describe the main steps needed to purify the domestic water supply

- explain how oxides of nitrogen form in car engines and are removed by catalytic converters
- understand how greenhouse gases cause global warming
- state the symbol equation for photosynthesis.

GETTING STARTED

Air and water are being contaminated with substances produced because of human activities.

In a small group, discuss the importance of clean air and water to human life. Can you think of different types of air and water pollution (contamination with other substances) in and around the area where you live or go to school?

There is a wide range of pollutants generated by manufacturing industries, power plants, farms and transportation. Make a list of some of the pollutants you already know of and try to give their sources. Having done this, discuss the impact these pollutants can have on people's daily lives and whether your lifestyle choices contribute to local air and water pollution.

THE GROWING PROBLEM OF AIR POLLUTION

The air we breathe is a mixture of different gases. There are the non-toxic gases such as nitrogen and argon, which we simply breathe in and out, there is the oxygen we require for respiration, but there are also some gases that have been linked to health problems, quality of life and environmental issues (Figure 17.1). These pollutant gases are generally the result of human activity, with most being produced either by burning fossil fuels or because of modern intensive farming methods. These pollutants include oxides of nitrogen, sulfur dioxide, carbon monoxide, methane, particulates and carbon dioxide. Governments around the world are aware of the seriousness of these pollutants and are gradually working together to set strict limits on how much of the pollutants can be released into the environment.

Figure 17.1: High levels of industrial atmospheric pollution are causing environmental problems.

To meet these targets and improve air quality, governments need to change in part how their citizens lead their daily lives. Steps to do this are already being made with cities banning high-polluting vehicles or setting high tariffs (congestion charges) to drive into city centres. Some governments are increasing taxes on companies or those technologies that use fossil fuels. Governments can also choose to promote more sustainable methods of farming and encourage people to eat less meat in their diets.

It is important to consider whether decisions about environmental effects should be left to governments, companies or individuals. These are difficult decisions and many different factors have to be considered. Individuals may feel powerless to act in isolation or believe that changes in lifestyle do not have a significant impact. Companies may resist changes because of concerns over increased prices. Governments face problems as they are linked by large global trade agreements and increasing tariffs to meet a greener future in one country may simply result in increased imports from places that have less-strict environmental standards.

Discussion questions

1 Think about how the air quality where you live compares with other parts of the world. What might be the reasons for the differences in air quality?

2 What factors might encourage or prevent an individual choosing more environmentally friendly options in their day-to-day lifestyle?

17.1 Air quality

Composition of air

The **atmosphere**, a layer approximately 480 km thick that surrounds the Earth, is made up of a mixture of air and water vapour. Levels of water vapour are highly variable and can range from 0.2% in high mountain regions to 4.0% in tropical rainforests. To compare samples of air, the water vapour is removed to produce **clean dry air**.

Clean dry air consists mainly of 78% nitrogen (N_2) and 21% oxygen (O_2) (Figure 17.2). This leaves only 1% for the other gases. Argon makes up 0.9% so the other gases are only present in very small quantities. The remainder includes carbon dioxide (CO_2), other noble gases (neon, krypton and helium), methane, oxides of nitrogen and sulfur dioxide.

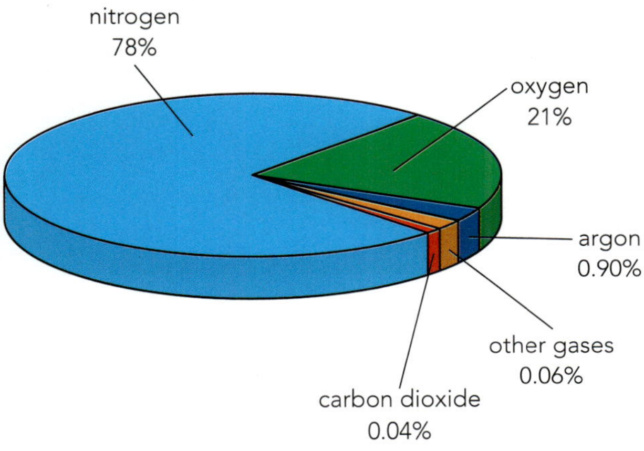

Figure 17.2: Composition of clean dry air.

Oxygen is essential for **respiration** and so life on Earth is dependent on it. We can extract oxygen from the air to treat people with respiratory problems in hospital and for industrial welding.

Nitrogen, as found in the atmosphere, is unreactive. Some species of bacteria can use ('fix') nitrogen directly from the air to produce amino acids, but most living organisms are unable to convert gaseous nitrogen into useful products. Nitrogen can also be extracted from the air and is used as an **inert** atmosphere for food packaging or combined with hydrogen to produce ammonia. Ammonia is the starting material for most fertilisers (Chapter 9).

Carbon dioxide, although used in **photosynthesis** and produced by respiration, makes up just 0.04% of the Earth's atmosphere.

Air pollution

Human activity has contributed significantly to atmospheric pollution and increased levels of oxides of nitrogen, sulfur dioxide, methane, carbon monoxide and carbon dioxide. Carbon dioxide occurs naturally in the air but, when present in larger quantities, it is classed as a **pollutant** because it is partly responsible for **global warming** and the associated **climate change**.

KEY WORDS

atmosphere: the layer of air and water vapour surrounding the Earth

clean dry air: containing no water vapour and only the gases that are always present in the air

respiration: the chemical reaction (a combustion reaction) by which biological cells release the energy stored in glucose for use by the cell or the body; the reaction is exothermic and produces carbon dioxide and water as the chemical by-products

inert: a term that describes substances that do not produce a chemical reaction when another substance is added

photosynthesis: the chemical process by which plants synthesise glucose from atmospheric carbon dioxide and water giving off oxygen as a by-product: the energy required for the process is captured from sunlight by chlorophyll molecules in the green leaves of the plants

pollutants: substances, often harmful, which are added to another substance

global warming: a long-term increase in the average temperature of the Earth's surface, which may be caused in part by human activities

climate change: changes in weather patterns brought about by global warming

The following section will look at sulfur dioxide, oxides of nitrogen (nitrogen oxides, NO_x), carbon monoxide and **particulates**. The section will focus on the sources of these pollutants and the problems they create, and will highlight possible ways of minimising their production. Given the importance of global warming and climate change, the sources and impacts of carbon dioxide and methane will be addressed separately in Topic 17.2.

> **KEY WORDS**
>
> **particulates:** very tiny solid particles produced during the combustion of fuels
>
> **acid rain:** rain that has been made more acidic than normal by the presence of dissolved pollutants such as sulfur dioxide (SO_2) and oxides of nitrogen (nitrogen oxides, NO_x)

Sulfur dioxide

Fossil fuels such as coal and crude oil often contain sulfur as an impurity. When the fuel is burnt, the sulfur combines with oxygen from the air to release sulfur dioxide (SO_2).

$$S(s) + O_2(g) \rightarrow SO_2(g)$$

Sulfur dioxide has been linked to breathing difficulties because it irritates the lining of the respiratory tract and so is especially dangerous for people with asthma. Sulfur dioxide, along with nitrogen oxide (see next), is also known to react with oxygen and water vapour in the air to form **acid rain**. Acid rain is harmful to life both on land and in the water. Increased acidity levels in lakes can kill fish and other aquatic life, while many plants are extremely sensitive to pH levels. Some building materials react with acid rain over time leading to increased rates of damage and corrosion (Figure 17.3).

Figure 17.3: Medieval sculpture damaged by acid rain.

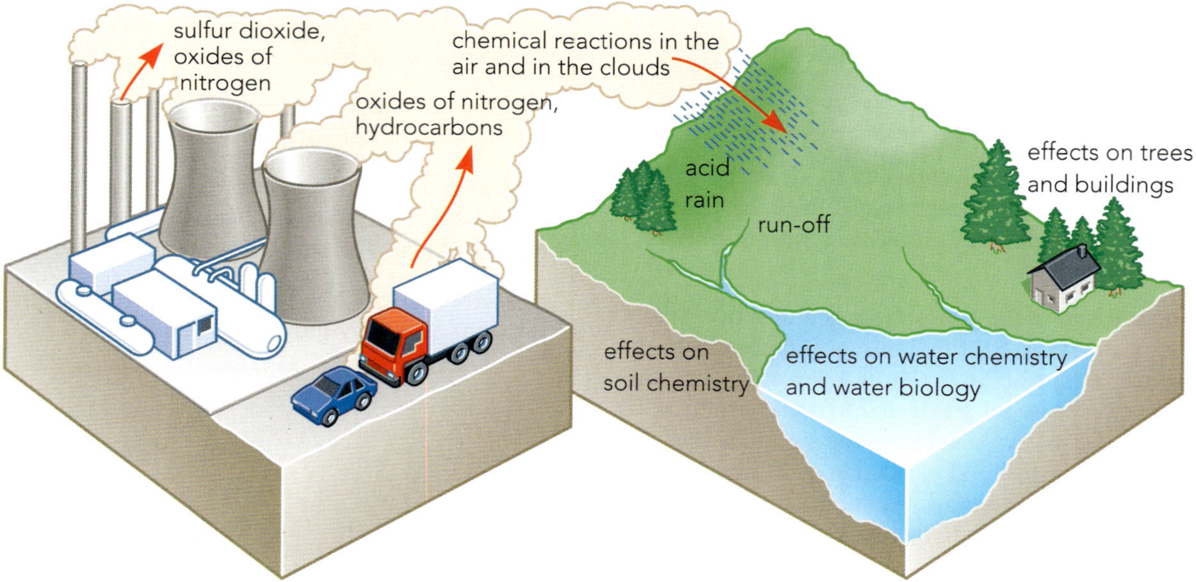

Figure 17.4: Formation of acid rain.

The main sources of sulfur dioxide are electricity generation using either coal or oil and burning fossil fuels such as petrol (gasoline) and diesel in vehicle engines (Figure 17.4).

To prevent the harmful effects of sulfur dioxide, scientists have researched ways to reduce the amounts being released into the air. Removal of SO_2 from combustion gases is done in a chimney ('flue'). The flue gases react with water to form an acidic solution that can then be neutralised by the reaction with calcium oxide in a process known as **desulfurisation** or 'scrubbing' (this is an **acid–base reaction**). To prevent the formation of SO_2, it is possible to remove most of the sulfur containing compounds before combustion. This gives the ultra low sulfur petrol and diesel which are required in most countries across the world.

Oxides of nitrogen

Oxides of nitrogen, also referred to as nitrogen oxides (NO_x), is the general name given to represent several different oxides including nitric oxide (NO) and nitrogen dioxide (NO_2). Oxides of nitrogen form when nitrogen and oxygen from the air react at high temperatures. An example of this is:

$$N_2(g) + O_2(g) \rightarrow 2NO(g)$$

Like sulfur dioxide, oxides of nitrogen have been linked to the formation of acid rain (Figure 17.4). An example of a reaction showing how nitrogen dioxide reacts with water to form nitric acid (HNO_3) is:

$$3NO_2(g) + H_2O(l) \rightarrow 2HNO_3(aq) + NO(g)$$

A second problem linked to increased levels of oxides of nitrogen is the formation of **photochemical smog**. Smog results when several air pollutants react with sunlight to form the characteristic brown haze seen over many large cities.

KEY WORDS

desulfurisation: the removal of sulfur dioxide from the fumes of power stations

acid–base reaction: (see **neutralisation**)

photochemical smog: a form of local atmospheric pollution found in large cities in which several gases react with each other to produce harmful products

Photochemical smog is harmful to human health, and has been linked to respiratory disease and increased numbers of asthma attacks.

Although oxides of nitrogen can form naturally, most come from human activity. These include vehicle emissions, power production and other industrial processes. The extent to which human activity produces NO_x was shown by the dramatic reduction in airborne nitrogen oxide pollution during the lockdowns enforced as part of the response to the 2020 worldwide coronavirus pandemic.

One way to reduce levels of NO_x emitted by vehicles is to use a **catalytic converter** connected as part of the exhaust system (Figure 17.5). The toxic gases produced by the engine are converted into less harmful gases before they are emitted into the atmosphere.

A catalytic converter uses a rare transition metal catalyst (e.g. platinum, palladium or rhodium), which is coated as a thin layer onto a honeycomb support. Within a catalytic converter, carbon monoxide will react with nitrogen oxide to release carbon dioxide and nitrogen:

carbon monoxide + nitrogen oxide → carbon dioxide + nitrogen

$$2CO(g) + 2NO(g) \rightarrow 2CO_2(g) + N_2(g)$$

Catalytic converters can also catalyse other reactions to further reduce the emissions of pollutant gases.

Carbon monoxide

Incomplete combustion occurs when a hydrocarbon fuel is burned in a limited supply of oxygen. When this happens, one possible product is the pollutant carbon monoxide (CO). The incomplete combustion of octane (an important constituent of petrol) is shown by:

octane + oxygen → carbon monoxide + water

$$2C_8H_{18} + 17O_2 \rightarrow 16CO + 18H_2O$$

This should be compared to **complete combustion** of hydrocarbon fuels that occurs in a plentiful supply of oxygen and releases only carbon dioxide and water.

Carbon monoxide is toxic to humans as it binds irreversibly with haemoglobin in red blood cells preventing them from carrying oxygen around the body.

Carbon monoxide can be produced when there is a lack of oxygen getting into an engine. It can also be made when the air inlets get blocked in a gas central heating system as this prevents oxygen from entering the system.

To reduce emissions of carbon monoxide, heating systems should be regularly checked, and any inlets cleaned. In vehicles, the catalytic converter also prevents harmful emissions of carbon monoxide from entering the atmosphere (Figure 17.5).

KEY WORDS

catalytic converter: a device for converting polluting exhaust gases from cars into less dangerous emissions

incomplete combustion: a type of combustion reaction in which a fuel is burnt in a limited supply of oxygen; the incomplete combustion of hydrocarbon fuels produces carbon, carbon monoxide and water (see also **complete combustion**)

complete combustion: (see also **incomplete combustion**) a type of combustion reaction in which a fuel is burnt in a plentiful supply of oxygen; the complete combustion of hydrocarbon fuels produces only carbon dioxide and water

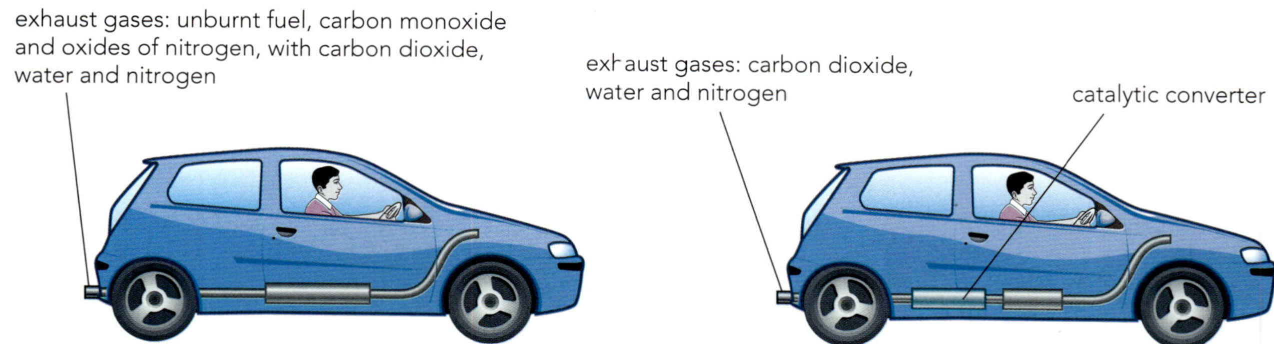

Figure 17.5: A catalytic converter converts harmful exhaust gases into safer gases.

Particulates

Like carbon monoxide, carbon particulates ('soot' particles) are formed as a result of incomplete combustion of fuel. The incomplete combustion of octane to form particulates is shown by:

octane	+	oxygen	→	particulate (soot)	+	water
$2C_8H_{18}$	+	$9O_2$	→	$16C$	+	$18H_2O$

Particulates are linked to increased respiratory disease and there is also evidence that they can cause cancer.

An important source of particulates is from diesel vehicles where they are produced because of a lack of oxygen getting into an engine. To reduce particulate emissions, diesel vehicles are fitted with particulate traps (fine mesh filters) that remove the particles from the exhaust gas.

Questions

1 What are the percentages of nitrogen, argon and oxygen in clean dry air?

2 Describe some of the problems caused by acid rain.

3 Fuels undergo either complete or incomplete combustion in oxygen.

  a State the word equation for the incomplete combustion of methane to produce carbon monoxide and water.

  b Balance this symbol equation:
  $\_\_C_6H_{14} + \_\_O_2 \rightarrow \_\_CO + \_\_H_2O$

4 Explain how oxides of nitrogen are formed and give some of the problems they are linked to.

5 Give a balanced symbol equation to show how a catalytic converter can remove carbon monoxide (CO) and nitrogen oxide (NO) from the exhaust gases to produce only nitrogen and carbon dioxide.

17.2 Carbon dioxide, methane and climate change

Greenhouse gases

Carbon dioxide, although produced naturally by respiration, is considered a particularly important pollutant. It is one of the **greenhouse gases** and increased amounts in the atmosphere have resulted in global warming. Global warming causes an increase in average temperatures and, because of this, is leading to climate change. There are several greenhouse gases including water vapour, carbon dioxide, methane, nitrous oxide, sulfur hexafluoride and chlorofluorocarbons (CFCs). In this section we will only look at two main greenhouse gases: carbon dioxide and methane.

KEY WORDS

greenhouse gas: a gas that absorbs thermal energy reflected from the surface of the Earth, stopping it escaping the atmosphere

Release of carbon dioxide into the atmosphere

Carbon dioxide (CO_2) is produced during the complete combustion of fossil fuels. Fossil fuels are used to produce electricity and are the basis of many forms of transport. For example, methane (natural gas) produces carbon dioxide when burnt in a plentiful supply of oxygen:

methane	+	oxygen	→	carbon dioxide	+	water
CH_4	+	$2O_2$	→	CO_2	+	$2H_2O$

Release of methane into the atmosphere

Like carbon dioxide, methane (CH_4) is a greenhouse gas with levels in the atmosphere increasing over recent years. Some of the reasons for rising levels of methane in the atmosphere are linked to increased cattle farming and more waste being generated by larger populations. Cattle emit large amounts of methane as part of their digestive system. A single cow can produce over 1000 times the mass of methane compared with a human. The decomposition of food waste under anaerobic conditions by bacteria at landfill sites also releases large amounts of methane into the atmosphere.

Global warming

Carbon dioxide and methane occur naturally in the atmosphere and play an important role in maintaining a constant temperature on Earth. This relatively constant temperature is due to a natural phenomenon that scientists call the **greenhouse effect**. By trapping thermal energy reflected from the Earth's surface, the greenhouse gases maintain an average surface temperature of around 15 °C. Without the greenhouse effect the average temperature would be much lower, possibly only −18 °C. At this temperature, life of the type we know would not exist. Over the last 200 million years, levels of greenhouse gases have remained relatively constant but because of human activity this is now starting to change. Increased use of fossil fuels and changes in farming have caused levels of carbon dioxide and methane to rise, and this in turn has resulted in an increase in the average surface temperature of the Earth. The increase is known as global warming and it has resulted in changing weather patterns (climate change). Scientists have linked global warming to increased levels of greenhouse gases because of the strong correlations between the concentration of the gases in the atmosphere and the average recorded temperatures.

KEY WORDS

greenhouse effect: the natural phenomenon in which thermal energy (heat) from the Sun is 'trapped' at the Earth's surface by certain gases in the atmosphere (greenhouse gases)

Greenhouse effect

The greenhouse gases allow high-energy, short wavelength radiation from the Sun to pass through the atmosphere and reach the Earth's surface. Some of this thermal energy is absorbed and heats the oceans and land, and some is radiated (reflected) back into the atmosphere. The thermal energy radiated by the Earth has a lower energy and a longer wavelength. The actual wavelength of this reflected radiation falls within the infrared part of the electromagnetic spectrum. The greenhouse gases such as carbon dioxide and methane can absorb this infrared radiation and then reradiate (re-emit) it in all directions. As it is re-emitted in all directions, some comes back towards the Earth's surface. This reduces the thermal energy loss to space and increases the temperature of the lower atmosphere. This phenomenon is called the greenhouse effect because the absorption and reflection of thermal energy that warms the atmosphere works in a similar way to a greenhouse (Figure 17.6).

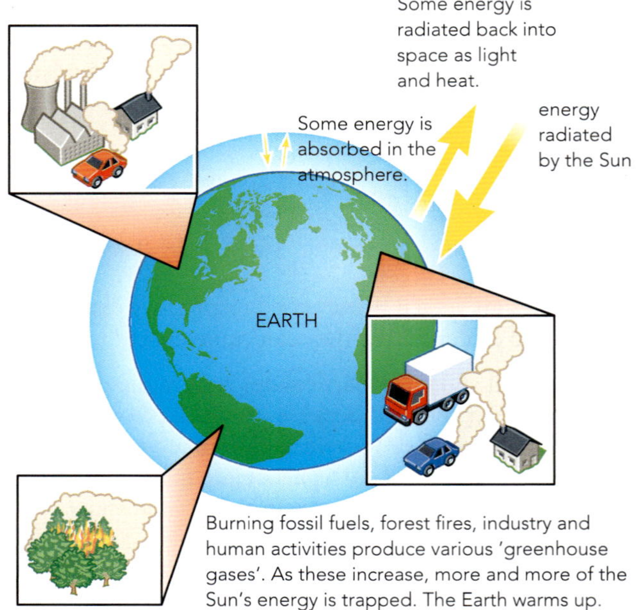

Figure 17.6: The greenhouse effect.

Climate change

Global warming has brought about a range of consequences for the Earth's climate. The impacts of global warming differ from country to country and general changes in weather patterns are known collectively as climate change.

An important impact of climate change is that the increase in average temperature has caused quicker rates of melting of the Earth's polar ice caps and glaciers. This has led to rising sea levels and so there has been increased flooding in some low-lying countries and faster rates of coastal erosion. The melting of ice caps is also causing changes to the life cycles and migratory patterns of animals and birds. Polar bears who rely on sea ice as they travel between different hunting grounds have been particularly badly affected.

An increase in temperature can also lead to more severe droughts. With higher temperatures, the soil dries out more quickly, and this is then compounded by changes in rainfall. Very dry soil and low rainfall increase the chances of crop failure and, in the longer term, arable land can become so arid it turns to desert. In regions such as California and parts of Australia, drying out of grasslands as a result of climate change has increased the frequency and severity of wildfires.

The problems linked to climate change are not limited to increased temperatures. For many countries, climate change means more frequent, more extreme weather patterns. For some this has included more severe storms and associated flooding and landslides.

While the effects of climate change can be disruptive to human ways of living, they can be catastrophic for animals and plants. Many living organisms are extremely sensitive to even slight changes in average temperature. For example, changes in temperature can lead to differences in the seasons when plants bud and produce fruit, which then has consequences for the wildlife whose life cycles depend on these plants. Changes in sea temperature can lead to bleaching of coral reefs and the associated loss of marine life.

Reducing the amount of CO_2 released into the atmosphere

There are many steps that can be taken to reduce the release of greenhouse gases at an individual, national and global level. In recent years, governments from around the world have tried to take steps together to reduce their greenhouse gas emissions. There have been important climate change conferences where agreements have been made, including the Kyoto Protocol of 2005 and the Paris Agreement of 2016. Increased public awareness of the impact of climate change and a stronger presence of environmental groups has placed pressure on governments to react.

An important step towards decreased greenhouse gas emissions of carbon dioxide is to reduce our reliance on fossil fuels for transportation and electricity generation. This can be done by turning to **renewable** sources of energy such as wind and solar. Many countries have started to remove coal power stations and replace them with renewable energy sources.

Methods of transport are starting to change, and several countries have committed themselves to the phasing out of diesel engines. There is significant interest in moving away from petrol cars to electric cars, although there are still issues about how the electricity used to power these vehicles is created. Some manufacturers are also looking at developing fuel cell vehicles. Hydrogen fuel cells are of particular interest as the only chemical product is water (Chapter 6).

As well as reducing the CO_2 being produced, steps are also being taken to remove CO_2 from the atmosphere and so reverse the impact of global warming. At electricity generating plants, it may be possible to capture the CO_2 before it is released into the atmosphere and then store it underground. An alternative method being used by many countries is to plant additional trees (afforestation) which will capture the CO_2 through photosynthesis.

Photosynthesis is the process of taking in carbon dioxide and water, and then using energy from sunlight in the presence of **chlorophyll** to produce glucose and oxygen. Glucose is necessary as the starting material for other carbohydrates including starch and cellulose. The word equation for photosynthesis is:

carbon dioxide + water → glucose + oxygen

The balanced symbol equation for photosynthesis is:

$$6CO_2 + 6H_2O \rightarrow C_6H_{12}O_6 + 6O_2$$

KEY WORDS

renewable (resources): sources of energy and other resources that cannot run out provided they are managed sustainably, or that can be made at a rate faster than our current rate of use

chlorophyll: a green pigment in plants which traps energy from the Sun in photosynthesis

Reducing the amount of methane released into the atmosphere

Methane is released by both rotting vegetation and livestock. Innovative approaches have been considered to reduce methane emissions from livestock, including changes in their diet and even trying to capture the gases they produce.

In the short term, a more realistic option for reducing methane emissions from cattle is to better educate people about the harmful effects of an excessively meat-rich diet. There is evidence that encouraging greater dependence on plant-based food is being successful, with increasing numbers of vegetarians and vegans around the world.

Methane is also released by decomposing waste in landfill. To reduce the production of methane, better separation of household waste might lower the amount of food waste going to landfill and so lower the amount of methane being produced by bacteria. Landfill methane can be trapped and burnt as a clean energy source for generating electricity or heat.

Questions

6 Name two greenhouse gases and give a source for each.

7 Why are levels of carbon dioxide in the atmosphere increasing?

8 What are some of the problems linked to climate change and the melting of the ice caps? List some of the strategies that could be used to reduce these environmental issues.

9 The greenhouse gases allow short length, high-energy radiation from the Sun to pass through the atmosphere. What do they do to the longer wavelength radiation reflected into space from the Earth's surface?

REFLECTION

How easy do you find it to apply your scientific understanding to issues such as global warming and climate change? Could you explain these environmental issues to someone else? Are there any other factors you would need to consider?

How can the scientific knowledge you have learnt benefit people and the environment?

17.3 Water

Water is vital to life and plays an important role in many industrial processes. In this section, we will consider how to test for the presence of water, look at the differences between distilled water and drinking water ('potable' water), and consider some of the different substances found in water from natural sources.

Tests for the presence and purity of water

There are many colourless liquids, including water, ethanol and cyclohexane, so how do we know whether water is present? Either of two simple chemical tests can be used to show the presence of water (Chapter 9). Blue cobalt chloride paper turns pink in the presence of water (Figure 17.7). Anhydrous cobalt(II) chloride is converted to the hydrated salt.

$$\underset{\text{blue}}{CoCl_2} + 6H_2O \rightarrow \underset{\text{pink}}{CoCl_2{\cdot}6H_2O}$$

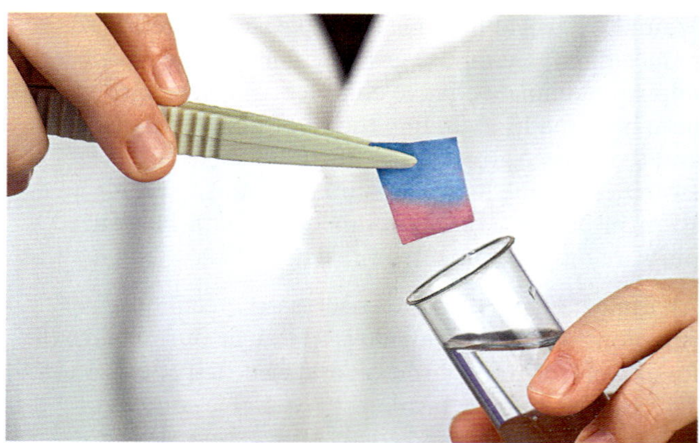

Figure 17.7: Test for the presence of water: blue cobalt chloride paper turns pink. Note the use of tweezers when handling the test paper.

Alternatively, solid white anhydrous copper(II) sulfate forms blue hydrated copper(II) sulfate ($CuSO_4{\cdot}5H_2O$) if water is added to it (see Figure 9.3 in Chapter 9).

$$\underset{\text{white}}{CuSO_4} + 5H_2O \rightarrow \underset{\text{blue}}{CuSO_4{\cdot}5H_2O}$$

The purity of any substance, including water, can be tested by recording its melting and boiling points (Chapters 1 and 21). This is because pure water has a fixed melting point of 0 °C and boiling point of 100 °C.

Distilled water

Tap water is a mixture of natural minerals (dissolved salts) and substances added due to human activity. Experiments that require water as a solvent should use distilled water rather than tap water. This is because substances dissolved in the tap water may interfere with experiments giving unwanted side reactions. They would also interfere with the results of standard analysis, e.g. for metal ions. Distillation (Chapter 21) removes the impurities and uses the key idea that different substances have different boiling points.

Substances in natural water

Water from natural sources contains many dissolved substances. Some of these substances are beneficial and some are potentially harmful.

Substances that are beneficial

Dissolved oxygen

One of the most important substances to be found dissolved in water is oxygen. Oxygen is needed to support animal and plant life. The levels of oxygen present will determine the number and variety of living organisms that can be supported.

Oxygen enters the water either as a result of photosynthesis by aquatic plants or through diffusion of oxygen from the air (Figure 17.8). Oxygen is removed from the water by respiration in plants and animals. The amount of oxygen dissolved in water is dependent on temperature and whether it is saltwater or freshwater.

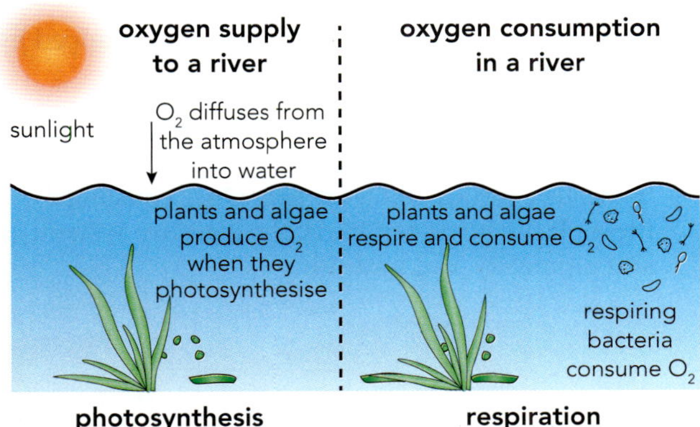

Figure 17.8: Oxygen in river water.

Metallic compounds

As water passes over and through different types of rock it can dissolve metallic compounds called minerals, e.g. calcium and magnesium salts. There are a large range of metals that are needed in trace amounts to support good health. These include Group I metal ions (sodium and potassium), Group II metals (calcium and magnesium) and a range of transition metal ions (iron, cobalt, nickel, copper, zinc and chromium). Calcium, for example, supports the health of teeth and bones while iron is needed in the production of haemoglobin (for use in red blood cells).

Substances that are harmful

Some metallic compounds

Alongside the useful metallic compounds, water can be contaminated with a range of potentially harmful metals. Heavy metals such as lead and mercury can enter water systems from a variety of sources including mining, metal smelting, waste disposal, corrosion and metal processing plants. Lead can cause liver and kidney damage, and mercury has been linked to damage of the nervous system.

Sewage

Wastewater (sewage) produced by humans includes a range of contaminants. Sewage is usually carried by underground pipes (sewers) and taken to wastewater treatment plants to remove the harmful materials. Any solids can be filtered or digested and the treated liquid (effluent) returned into rivers or the sea. Leaks of sewage into drinking water can happen during natural disasters such as extreme weather events or earthquakes. When this happens, harmful microbes enter the drinking water spreading diseases such as diarrhoea, cholera, dysentery, typhoid and polio.

Nitrates and phosphates

NPK fertilisers are used to increase crop yields by adding three essential plant nutrients nitrogen (N), phosphorous (P) and potassium (K) to the soil (Chapter 9). They are made from water-soluble compounds (salts) that are easily absorbed through the roots of plants. These water-soluble compounds create problems if there is heavy rain after the fertiliser has been spread onto the crops. Under these conditions, instead of being taken in by the plants the fertiliser will be washed over the surface of the soil and into waterways. This process is called **run-off**.

> **KEY WORD**
>
> **run-off:** water that travels over the surface of the land before entering waterways such as rivers and lakes; run-off from farmland may contain dissolved substances such as fertilisers

Figure 17.9: Rapid algae growth in a waterway possibly contaminated with phosphates and nitrates.

When fertilisers enter waterways, such as streams and rivers, they cause the rapid growth of algae. These algae form huge blooms that cover the surface of the water (Figure 17.9) and block out sunlight. Without sunlight, aquatic plants that live below the surface are unable to photosynthesise and die. This leads to a drop in oxygen levels in the water and so causes the death of many aquatic animals.

The most common forms of pollutant that enter water because of run-off are nitrates and phosphates. Phosphates can also enter water because of washing; for example, pentasodium triphosphate ($Na_5P_3O_{10}$) is widely used as an ingredient in biological washing powders.

Plastics

Plastics are polymers (Chapter 20) and they are used throughout our daily life. Unlike the other forms of pollutant, plastics are insoluble in water and so can be easily removed. The issues being caused by plastics are due to the volume of waste being released.

Poor disposal of these materials combined with a lack of biodegradability has resulted in polymers being released into waterways. As they do not breakdown (biodegrade), they can rapidly accumulate. The impacts of accumulated plastics in our oceans and waterways on wildlife can be catastrophic.

There are several aspects to the problem of plastic pollution of the oceans and waterways:

- Larger sea creatures and sea birds can be trapped by discarded fishing nets.
- Large scale debris such as intact plastic bags can be confused as prey such as jelly fish and so consumed by whales, turtles and large fish. Once consumed these plastic items can block the digestive system and ultimately lead to death.

- Microplastic debris can accumulate in the surface layer of the ocean. These small pieces of plastic can easily be consumed by fish and damage the digestive systems of the animals.

The nature of plastics and the problems caused by their pollution of the oceans are described in further detail in Chapter 20.

The growing awareness of plastic pollution has resulted in several initiatives to minimise the problem. Methods of removing plastics from water are being developed and chemists are also designing new polymers that are **biodegradable**.

KEY WORD

biodegradable: a substance that can be broken down, or decomposed, by microorganisms

Purification of domestic water

Tap water, the domestic water supplied to our houses, undergoes several purification steps from the point it is collected to the point it is delivered. In many countries, domestic water is taken from lakes and reservoirs.

The first step in purification is to remove large insoluble objects such as rocks, plastic bags and branches, in a process known as screening. The water is then taken to a sedimentation tank. In the sedimentation tank the soil and sand will drop to the bottom of the tank (as sediment).

The next step in the process is to filter the water to remove smaller insoluble particles. The water is often passed through a very fine sand to filter out these particles.

Water may contain dissolved organic compounds that can cause the water to have an unwanted odour or taste. The organic compounds, often present in exceptionally low amounts, can be removed by using an activated carbon filter.

Before water is distributed to homes, the final step is disinfection. Disinfection is needed to kill harmful waterborne microbes such as bacteria that can cause disease. Different countries use different methods to disinfect water but one of the most common and effective methods is to add small amounts of chlorine to the water. Typically, chlorine is added at a concentration of 2–3 mg/dm^3.

A summary of the main stages involved in water treatment is shown in Figure 17.10.

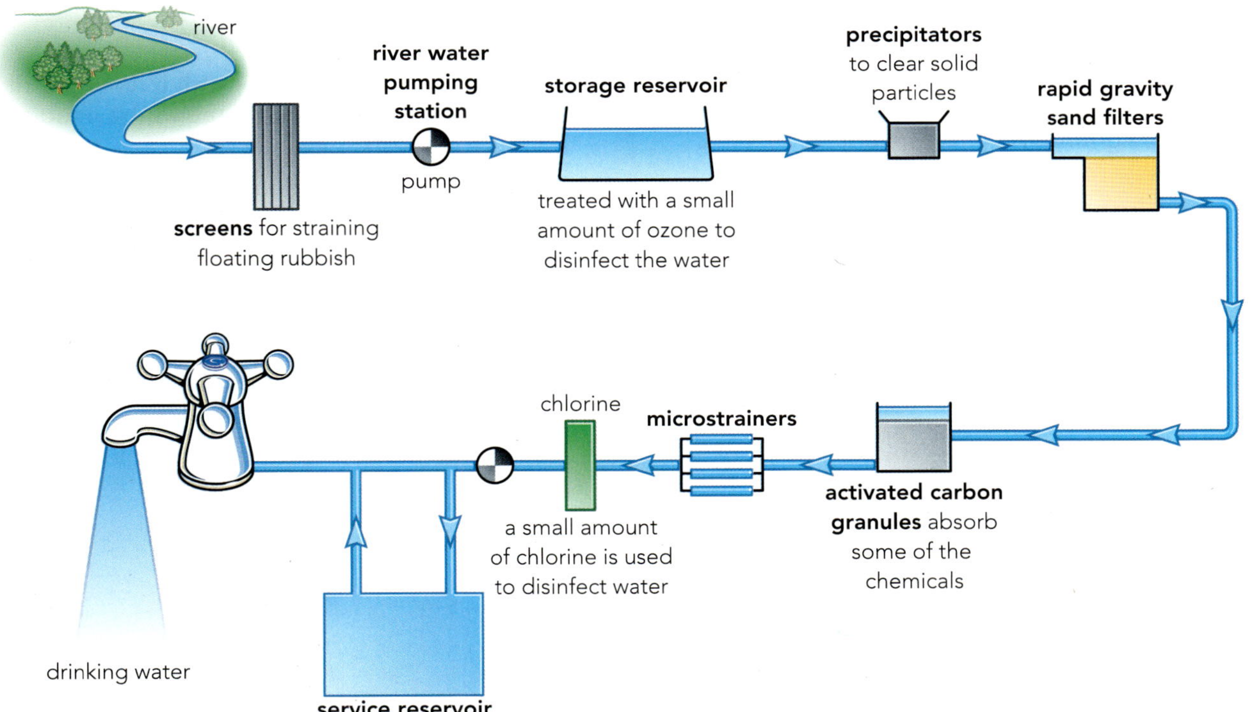

Figure 17.10: Purifying water for the domestic water supply.

Questions

10 State a simple test, giving its result, for detecting the presence of water.

11 How could you prove that a sample of water was pure?

12 Nitrates and phosphates can enter waterways as a result of the run-off of fertiliser from fields. Describe the problems that can be caused when this happens.

13 Name two useful metals found in water samples and state the reasons why they are useful. Name two metals found in water that are not useful and state the problems they can create.

14 Why is it important that scientists develop new biodegradable plastics rather than continuing to use the current types of plastic?

15 Water from a lake contained microplastics, traces of dissolved organic compounds and microbes linked to diarrhoea. How would these three contaminants be removed to enable the water to be sold as domestic water?

ACTIVITY 17.1

A letter of concern!

Evidence from a recent survey suggests one country has given permits to mining companies who have allowed dangerous metal compounds to contaminate local drinking water. The same country has opened themselves up to offer plastic disposal for other countries and the evidence indicates that significant amounts of this plastic waste are not being recycled but are instead ending up in waterways and being transported out to sea.

As an expert in water pollution, you have been asked to write a letter to the president of this country highlighting the environmental concerns. The letter should have the following structure.

- In the first paragraph, you should state the importance of ensuring water does not become contaminated by unwanted metal compounds. You need to give examples of the problems that metal compounds cause and then suggest steps to ensure that they do not enter the water.
- In the second paragraph, you need to explain some of the problems associated with plastic waste and microplastics.
- You should finish your letter with a reminder of the importance of clean water and the need for countries to work together.

Self-assessment

List three criteria for a good explanation of an environmental concern. How well do you think your explanations about the effects of metal and plastic pollutants have met these criteria? Can you rewrite one of your explanations adding further improvements?

SUMMARY

The atmosphere is made up of a mixture of gases and includes a variable amount of water vapour.
Clean dry air (air without water vapour) is a mixture of nitrogen (78%), oxygen (21%), noble gases and carbon dioxide.
The main air pollutants include carbon dioxide (CO_2), methane (CH_4), oxides of nitrogen (NO_x), sulfur dioxide (SO_2), carbon monoxide (CO) and carbon particulates.
Oxides of nitrogen, NO_x, are formed when nitrogen and oxygen from the air react at high temperature. They can cause respiratory problems, acid rain and photochemical smog. NO_x can be removed by the use of a catalytic converter.
In a catalytic converter, CO and NO_x react to produce nitrogen and carbon dioxide.
SO_2 is formed when sulfur in fossil fuels reacts with oxygen during combustion. SO_2 reacts with water vapour to form acid rain. Levels of SO_2 can be reduced by using ultra-low sulfur fuels (desulfurisation of fossil fuels) or by neutralisation of flue gases.
CO and carbon particulates are products of incomplete combustion of fossil fuels and can cause health problems. CO can be removed from exhaust gases using a catalytic converter and carbon particulates are removed by using a particulate trap.
CO_2 is a greenhouse gas produced by the combustion of fossil fuels; CH_4 is a greenhouse gas released by cattle and from rotting vegetation at landfill sites.
Greenhouse gases allow short wavelength energy from the Sun to reach the Earth's surface but trap and re-emit the longer wavelength radiation reflected from the Earth.
Plants remove CO_2 from the atmosphere by photosynthesis.
The symbol equation for photosynthesis is $6CO_2 + 6H_2O \rightarrow C_6H_{12}O_6 + 6O_2$.
Increased levels of greenhouse gases in the atmosphere have caused an increase in average temperatures (global warming), which has a range of important environmental effects.
Strategies to reduce the effects of these environmental issues include using renewable sources of energy, decreasing the use of fossil fuels, planting more trees and a reduction in livestock farming.
The presence of water can be determined by using anhydrous copper(II) sulfate, which turns from white to blue with water, or anhydrous cobalt(II) chloride, which turns from blue to pink with water; the purity of water can be assessed by its melting point (0 °C) and boiling point (100 °C).
Distilled water is used in practical experiments because it contains fewer impurities than tap water.
Water from natural sources contains a variety of substances; some of these substances are beneficial (oxygen for aquatic life and some metal compounds for essential nutrients); others are harmful (some metal compounds are toxic, sewage contains harmful microbes and plastic are hazardous to aquatic life).
Domestic water supplies go through several purification steps, including screening, sedimentation and filtration (to remove solids), carbon filtration (to remove odours and taste) and chlorination (to kill harmful microbes).

PROJECT

The great energy debate

Your government needs to plan its energy resources for future use (Figure 17.11). In a group of three, organise a debate, putting forward different ideas for the plan.

- Person 1. You work for the oil industry. In your presentation you must give reasons why it is necessary to continue burning fossil fuels. You should think about the ways in which fossil fuels are used, e.g. transport and heating and the problems in swapping to alternatives. You might also want to state ways in which you have made improvements in how fossil fuels are being used.
- Person 2. You represent suppliers of renewable energy. In your presentation you need to explain the benefits of moving to renewable fuels. Think about the emissions benefits but also any logistical concerns.
- Person 3 (the chair or judge). You will need to listen to the presentations from each person and then ask them additional questions.

At the end of the debate, the group should produce a brief summary of the arguments and reach a consensus in order to produce an outline of a recommended plan for your government to use.

Figure 17.11: Green energy versus polluting energy.

EXAM-STYLE QUESTIONS

1 The domestic water supply is treated in various ways to make it safe for use. Which of the following treatments is designed to kill any bacteria present in the water?

A sedimentation
B filtration through gravel beds
C treatment with chlorine
D passage through a carbon filter **[1]**

2 The pie chart shows the composition of clean dry air.

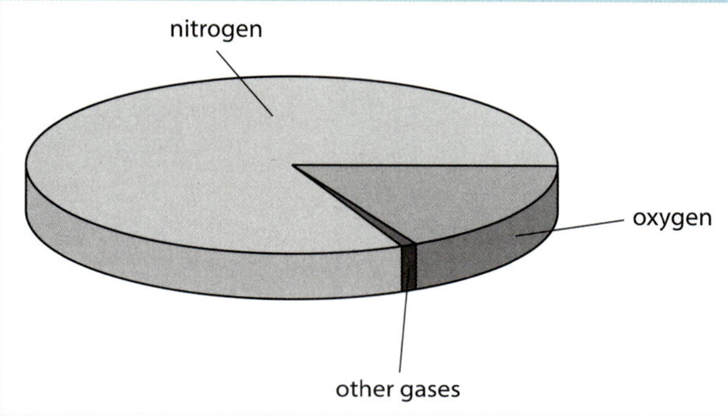

a What is the percentage of nitrogen in clean dry air? **[1]**

b Apart from nitrogen and oxygen, **state** the names of two other gases present in clean dry air. **[2]**

c The following questions relate to these four gases, which are all found as air pollutants.

A carbon monoxide
B methane
C oxides of nitrogen
D sulfur dioxide

i Name the gas given off from decomposition in landfill waste disposal sites. **[1]**

ii Name the gas produced when fuels burn incompletely. **[1]**

iii State which gases cause acid rain. **[2]**

iv State which gases are a harmful component of photochemical smog. **[1]**

[Total: 8]

3 The burning of fossil fuels is the main source of air pollution. Fuels are burnt in vehicles and in power stations to make energy.

a When coal is burnt in power stations, sulfur dioxide is one of the pollutants formed. How can power stations prevent sulfur dioxide from entering the atmosphere? **[2]**

b Oxides of nitrogen are produced in motor car engines. Where do the oxygen and nitrogen come from? **[1]**

COMMAND WORD

state: express in clear terms

CONTINUED

c How can the burning of fossil fuels cause pollution of water in rivers and lakes? [2]

d Carbon monoxide is produced in motor car engines.

i Why is carbon monoxide produced instead of carbon dioxide? [1]

ii Why is carbon monoxide harmful? [1]

e Carbon monoxide and oxides of nitrogen can be removed from exhaust gases by using a catalytic converter. **Explain** how, using a balanced symbol equation in your answer. [2]

[Total: 9]

4 The reaction shown occurs in catalytic converters.

$2NO_2(g) + 4CO(g) \rightarrow 4CO_2(g) + N_2(g)$

Which statement about this reaction is true?

A Carbon monoxide is an oxidising agent and is oxidised to carbon dioxide.

B Carbon monoxide is a reducing agent and is oxidised.

C Nitrogen dioxide is an oxidising agent and oxidises nitrogen.

D Nitrogen dioxide is a reducing agent and reduces carbon monoxide. [1]

5 a Carbon dioxide in the atmosphere can cause global warming. Name a natural process that removes carbon dioxide from the atmosphere and give the balanced symbol equation for the reaction involved in this process. [3]

b Name another different gas that also causes global warming and name one natural source of this gas. [2]

c Explain how the gases you have named caused global warming. [3]

[Total: 8]

COMMAND WORD

explain: set out purposes or reasons/make the relationships between things evident/provide why and/or how and support with relevant evidence

SELF-EVALUATION CHECKLIST

After studying this chapter, think about how confident you are with the different topics. This will help you to see any gaps in your knowledge and help you to learn more effectively.

I can	See Topic...	Needs more work	Almost there	Confident to move on
state the composition of clean dry air	17.1			
give the sources of the main pollutants: NO_x, sulfur dioxide, carbon monoxide, particulates, carbon dioxide and methane	17.1/17.2			
state the problems caused by NO_x, sulfur dioxide, carbon monoxide and particulates	17.1			
describe ways of reducing sulfur dioxide and NO_x	17.1			
state that carbon dioxide and methane are greenhouse gases, which at higher concentrations lead to global warming and the associated climate change	17.2			
describe how the greenhouse gases cause global warming	17.2			
give ways in which levels of greenhouse gas emissions can be reduced	17.2			
describe photosynthesis and give the word equation linked to the process	17.2			
state the symbol equation for photosynthesis	17.2			
describe tests for the presence and purity of water	17.3			
explain why distilled water is used in practical chemistry	17.3			
explain the importance of dissolved oxygen in water	17.3			
state that water can contain dissolved metal compounds, some of which are beneficial and others that are harmful to human health	17.3			
explain that sewage can pollute water leading to possible disease	17.3			
give the problems caused when nitrates and phosphates enter water by run-off	17.3			
describe the problems caused by plastic contamination in water	17.3			
describe the main stages in the purification of the domestic water supply	17.3			

Chapter 18

Introduction to organic chemistry

IN THIS CHAPTER YOU WILL:

- learn that a homologous series of organic compounds has the same general formula and similar chemical properties
- state that alkanes are saturated compounds as all the carbon–carbon bonds are single covalent bonds
- understand that compounds containing at least one carbon–carbon double or triple bond are unsaturated molecules
- distinguish between saturated and unsaturated molecules by testing
- learn that alcohols and carboxylic acids are further homologous series and that the type of organic compound present is given a systematic name
- distinguish different organic compounds by their molecular and displayed formulae

- understand how to write the structural formulae of molecules from various homologous series
- define structural isomers
- describe the general characteristics of a homologous series
- understand how to draw different formulae of unbranched members of different homologous series.

GETTING STARTED

What do you understand by the word 'organic'? The word has been linked to ideas related to the natural world. Organic molecules were once believed to have distinctive chemical properties, making them unique to living organisms. It was believed that there was a separate chemistry that took place in living things.

Discuss your understanding of the word in groups and make a list of any substances that you would think of as being organic. What element do they all contain? What molecules do we search for on other planets and in the dust of comets that would be evidence for life other than on Earth? Then come together as a class to share your ideas.

BUILDING A SPACE ELEVATOR

Carbon is a non-metal in Group IV of the Periodic Table. The uniqueness of carbon lies in the different ways in which it can form bonds. This shows even in the element itself. Carbon exists in several different forms (Chapter 3). The fullerenes, carbon nanotubes and graphene, have been discovered relatively recently and the exploitation of these types of carbon is one of the major features of the exciting new area of nanotechnology.

One important property of carbon nanotubes and graphene is their amazing tensile strength. These structures are far stronger than steel or titanium. Carbon nanotubes and graphene also show very high electrical conductivity. The prospect of using these new materials for the elevator cable has given new belief in the idea of constructing a space elevator (Figure 18.1).

Figure 18.1: Computer artwork of a future space elevator viewed looking down the length of the elevator cable towards the Earth.

The idea of a space elevator to carry humans into orbit is not new – the Russian scientist Konstantin Tsiolkovsky envisaged something like it at the end of the 19th century. However, today technology is being developed to move the idea towards reality. The elevator would comprise a huge (50 km) base tower on Earth, with a cable connected to it that stretches into space to a point beyond geostationary orbit, at an altitude of more than 35 000 km. Tracks and platforms along the tower and cable would allow vehicles to move to different levels, including into orbit without the need to use large rockets.

Preliminary experiments on aspects of the design of a space elevator have been conducted from the International Space Station. Companies in Japan and China have expressed an interest in developing such an elevator and an International Space Elevator Consortium has been set up. One development that must be explored is that of generating nanotube or graphene structures for the cable that are much longer than those currently available. However, the possibilities created by these new materials are quite staggering.

Discussion questions

1 What do you understand of the purposes and advantages of developing a space elevator?

2 Discuss the structural similarities between graphite, graphene and the fullerenes to explain why they are able to conduct electricity even though they are non-metallic.

18.1 Names and formulae of organic compounds

Carbon's unique properties

The remarkable versatility and complexity of the structures that carbon atoms can form is the very basis of the different forms of life here on Earth. Amino acids, simple sugar molecules and even fats may be relatively simple molecules, but the construction of complex molecules, such as long-chain carbohydrates and proteins, shows the versatility of carbon-containing compounds. DNA (deoxyribonucleic acid), the molecule that makes life possible, is a complex organic molecule (Figure 18.2).

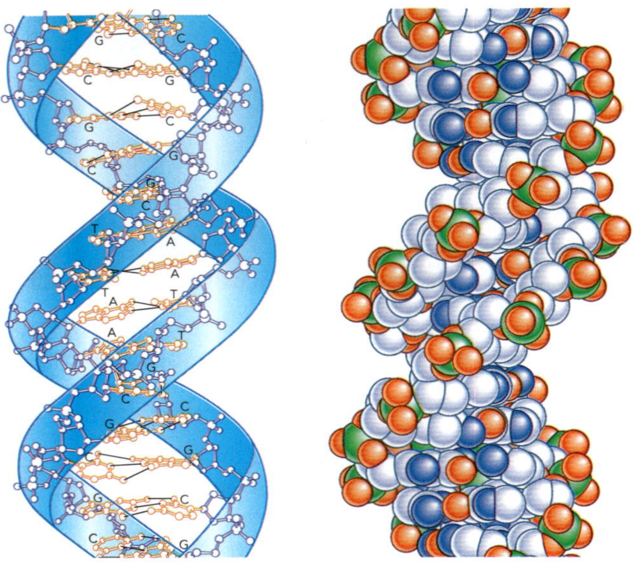

Figure 18.2: Two ways of showing a section of the complex molecule DNA.

Carbon is unique in the variety of molecules it can form. The chemistry of carbon-containing compounds is known as **organic chemistry**. There are three special features of covalent bonding involving carbon (Figure 18.3):

- carbon atoms can join to each other to form long chains; atoms of other elements can then attach to the chain
- the carbon atoms in a chain can be linked by single, double or triple covalent bonds
- carbon atoms can also arrange themselves in rings.

a Carbon can form four bonds, and carbon atoms can join to one another to form long chains.

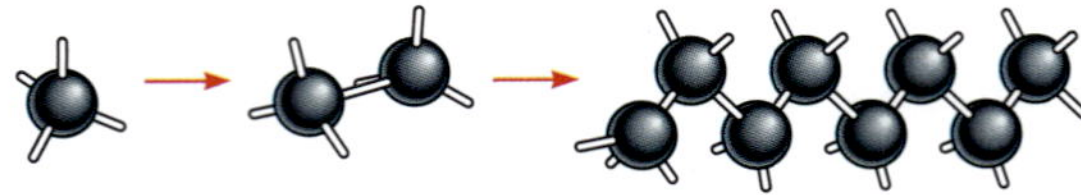

b In alkanes, only hydrogen atoms are joined to the side positions on the chains. Other atoms can be attached instead, forming other families of organic compounds.

c Double bonds can occur in simple molecules and in the long chains.

d Carbon atoms can also join to form ring molecules, for example glucose, as shown here.

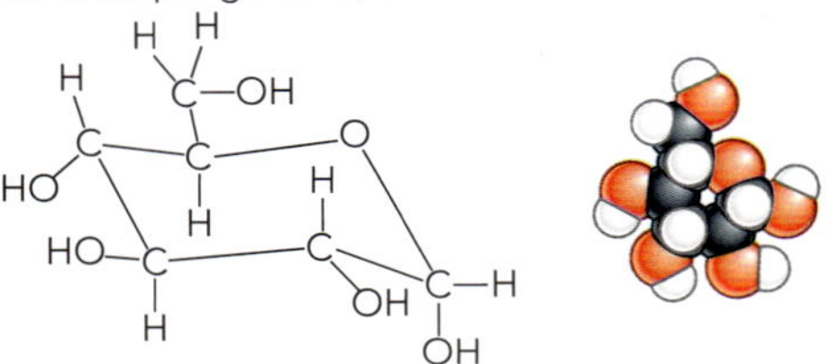

e Long-chain fat molecules can be formed, as well as numerous other molecules.

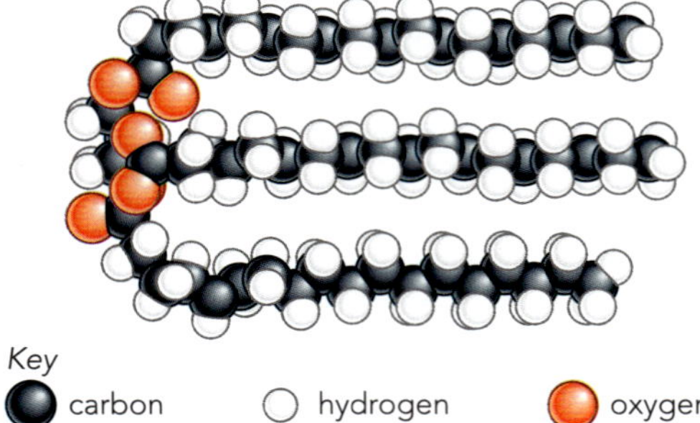

Key

carbon hydrogen oxygen

Figure 18.3: Carbon is very versatile and able to form chain and ring structures.

KEY WORDS

organic chemistry: studies on the structure, properties and reactions of organic compounds that contain carbon in covalent bonding

Only carbon can achieve all these different bonding arrangements to the extent that we see. Indeed, there are more compounds of carbon than of all the other elements put together.

Figure 18.3 gives some idea of how these bonding arrangements can produce different types of molecule.

Hydrocarbons

One of the simplest types of organic compound is the **hydrocarbons**. Hydrocarbons are compounds that contain carbon and hydrogen *only*.

Alkanes

The hydrocarbons studied at this level can be subdivided into two 'families' of molecules. Some hydrocarbons are saturated. These molecules contain only single covalent bonds between carbon atoms. Since carbon has a valency of 4, the bonds not used in making the chain are linked to hydrogen atoms (Figure 18.4). No further atoms can be added to molecules of these compounds, which is why we say they are saturated. This family of **saturated hydrocarbons** is known as the **alkanes**.

> **KEY WORDS**
>
> **hydrocarbons:** organic compounds that contain carbon and hydrogen only; the **alkanes** and **alkenes** are two series of hydrocarbons
>
> **saturated hydrocarbons:** hydrocarbons molecules in which all the carbon–carbon bonds are single covalent bonds
>
> **alkanes:** a series of hydrocarbons with the general formula C_nH_{2n+2}; they are saturated compounds as they have only single bonds between carbon atoms in their structure

The simplest alkane contains one carbon atom and is called methane. We discussed the covalent bonding in methane in Chapter 3. The series of alkanes compounds is built up by simply adding an extra carbon atom to the chain (Figure 18.4 and Table 18.1). Note that the names of this series of hydrocarbons all end in *-ane*. The first part of the name (the prefix) tells you the number of carbon atoms in the chain. We will see later that these prefixes are used consistently in naming organic compounds.

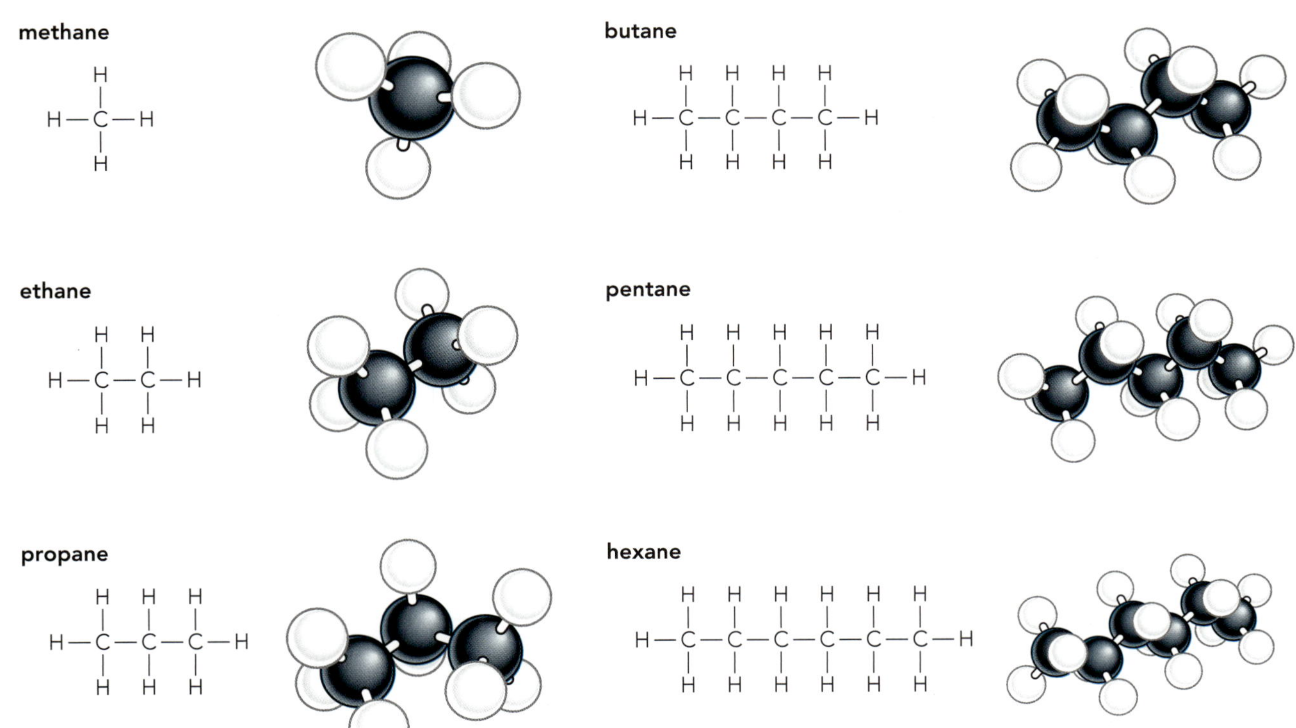

Figure 18.4: Displayed formulae and structures of the first six alkanes.

Alkane	Number of carbon atoms	Prefix to name	Molecular formula (C_nH_{2n+2})
methane	1	meth-	CH_4
ethane	2	eth-	C_2H_6
propane	3	prop-	C_3H_8
butane	4	but-	C_4H_{10}
pentane	5	pent-	C_5H_{12}
hexane	6	hex-	C_6H_{14}

Table 18.1: Names and molecular formulae of the alkanes.

The formulae given in Table 18.1 are the **molecular formulae** of the compounds (Chapter 4). These formulae show the total numbers of each type of atom in a molecule of the compound. Each molecule increases by a $-CH_2-$ group as the chain gets longer (Figure 18.4). The formulae of these molecules all fit the general formula C_nH_{2n+2} (where n is the number of carbon atoms present). Therefore, if you know the number of carbon atoms in an alkane molecule, then you can always work out the molecular formula.

In organic chemistry, the structure of a molecule is also very important. Figure 18.4 shows the displayed formulae of the first six alkanes in the series. A **displayed formula** shows the bonds between all the atoms in a molecule. The displayed formulae of the alkanes emphasise the fact that these molecules are saturated; all the bonds between carbon atoms are single bonds.

KEY WORDS

molecular formula: a formula that shows the actual number of atoms of each element present in a molecule of the compound

displayed formula: a representation of the structure of a compound that shows all the atoms and bonds in the molecule

alkenes: a series of hydrocarbons with the general formula C_nH_{2n}; they are unsaturated molecules as they have a C=C double bond somewhere in the chain

Alkenes

The ability of carbon atoms to form double bonds with each other gives rise to the **alkenes**. The alkenes are another family of hydrocarbons. Alkenes have the general formula C_nH_{2n}. Note that there are two less hydrogen atoms compared with a similar alkane because of the presence of the carbon–carbon double bond. The simplest alkene *must* contain two carbon atoms (which are needed for one carbon–carbon double bond to be formed) and is called ethene. Figure 18.5 shows a dot-and-cross diagram of the bonding in ethene and also the structures of the first three alkenes. You will note that the names of all the alkenes end with *-ene*.

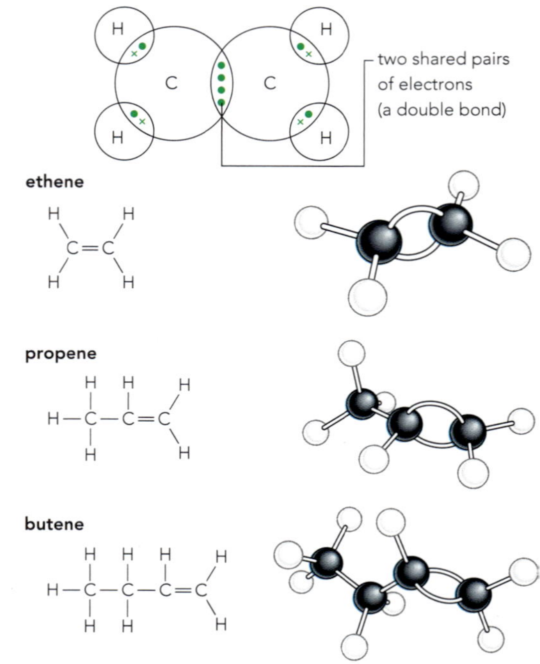

Figure 18.5: Bonding in ethene and structures of the first three alkenes.

Table 18.2 shows the names and molecular formulae of the first five alkenes. Note that the same prefixes used for the alkanes are also used to indicate the number of carbon atoms in the molecules of the alkenes.

Alkene	Number of carbon atoms	Prefix to name	Molecular formula (C_nH_{2n})
ethene	2	eth-	C_2H_4
propene	3	prop-	C_3H_6
butene	4	but-	C_4H_8
pentene	5	pent-	C_5H_{10}
hexene	6	hex-	C_6H_{12}

Table 18.2: Names and molecular formulae of the first five alkenes.

The presence of the carbon–carbon double bond in an alkene molecule makes alkenes much more reactive than alkanes (alkanes contain only carbon–carbon single bonds). Other atoms can add on to alkene molecules when the double bond breaks open. The alkenes are **unsaturated hydrocarbons**. This difference is the basis for a simple test for unsaturation in a hydrocarbon molecule.

Alkynes are a third family of hydrocarbons. The alkynes are also unsaturated hydrocarbons as the molecules contain a carbon–carbon triple bond. The simplest member is ethyne (C_2H_2). We do not study the alkynes any further at this level.

Chemical test for unsaturation

If an alkene, such as ethene, is shaken with a solution of bromine in water, the bromine loses its colour. Bromine has reacted with ethene in an **addition reaction**, producing a colourless compound (Figure 18.6b).

> **KEY WORDS**
>
> **unsaturated hydrocarbons:** hydrocarbons whose molecules contain at least one carbon–carbon double or triple bond
>
> **addition reaction:** a reaction in which a simple molecule adds across the carbon–carbon double bond of an alkene

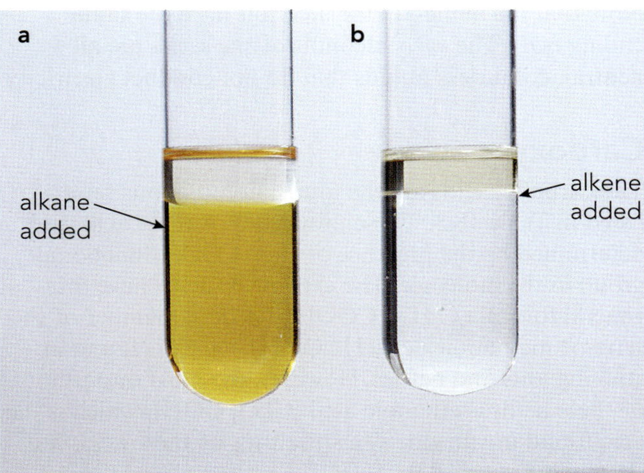

Figure 18.6: Test for unsaturation. Aqueous bromine with **a:** an alkane and **b:** an alkene.

Alkanes are already fully saturated and cannot take part in addition reactions. An alkane would give no reaction with aqueous bromine; the solution would stay orange–brown (Figure 18.6a). The only way to introduce a new atom into an alkane is by substitution (Chapter 19).

Questions

1 Write down the names and molecular formulae of the first four alkanes.

2 Draw the displayed formulae of methane and butane.

3 Draw a diagram showing the arrangement of electrons in the bonding of ethene. Only show the outer (valency) electrons.

4 What do you observe if ethene is bubbled through aqueous bromine? Explain the difference between this observation and what would occur if ethane was bubbled through aqueous bromine.

Homologous series of organic compounds

The presence of the carbon–carbon double bond in an alkene molecule gives this family of compounds their characteristic properties. The alkenes have the properties of unsaturated molecules, as distinct from the alkanes that only have carbon–carbon single bonds. A family of organic compounds with similar chemical properties are called a **homologous series**. The atom, or group of atoms, that gives the series its particular characteristic properties is called the **functional group**. For this reason, the carbon–carbon double bond is known as the functional group of the alkenes. Two other homologous series are the alcohols and the carboxylic acids. The functional group for the alcohols is the hydroxyl group (–OH), and that of the carboxylic acids is the carboxyl group (–COOH). All the members of a particular homologous series contain the same functional group (Table 18.3).

> **KEY WORDS**
>
> **homologous series:** a family of similar compounds with similar chemical properties due to the presence of the same functional group
>
> **functional group:** the atom or group of atoms responsible for the characteristic reactions of a compound

Homologous series	Name ending	Functional group	General formula	First member of series
alkanes	-ane	$-\overset{\vert}{\underset{\vert}{C}}-\overset{\vert}{\underset{\vert}{C}}-$	C_nH_{2n+2}	methane (n = 1)
alkenes	-ene	>C=C<	C_nH_{2n}	ethene (n = 2)
alcohols	-ol	–O–H	$C_nH_{2n+1}OH$	methanol (n = 1)
carboxylic acids	-oic acid	–C(=O)–O–H	$C_nH_{2n+1}COOH$	methanoic acid (n = 0)

Table 18.3: Different homologous series.

Alcohols

The alcohols are a homologous series of compounds with the –OH functional group and the general formula $C_nH_{2n+1}OH$. The simplest member of the series is methanol (CH_3OH). Methanol has one carbon atom per molecule and its covalent bonding was discussed in Chapter 3 (Figure 18.7).

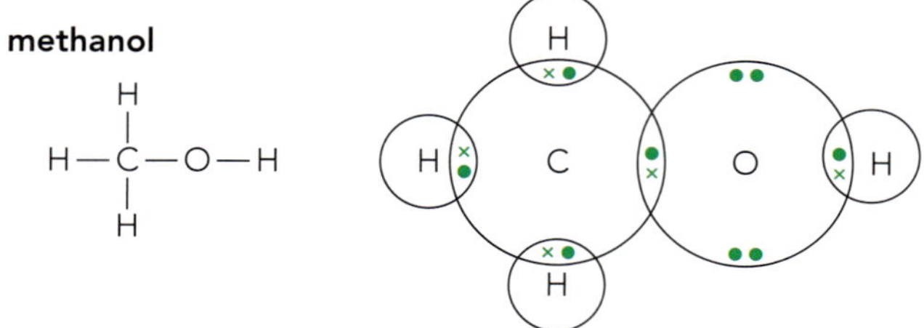

Figure 18.7: The displayed formula and bonding of methanol (CH_3OH).

The most important member of the alcohol series, however, is ethanol (C_2H_5OH). Ethanol is the second member of the series with two carbon atoms in each molecule (Figure 18.8).

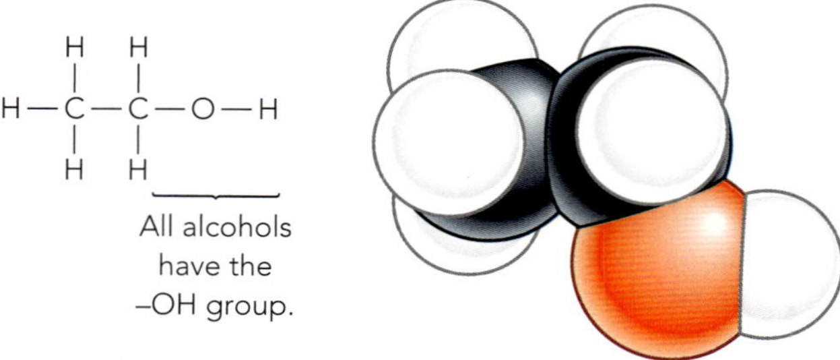

Figure 18.8: Displayed formula and model of ethanol (C_2H_5OH).

Note that the names all the alcohols have the same ending (*-ol*). The early alcohols of the series are all neutral, colourless liquids that do not conduct electricity.

Carboxylic acids

The carboxyl acids are the fourth homologous series listed in Table 18.3. These properties of these acids are determined by the presence of the –COOH functional group in the molecule. The carboxylic acids have the general formula $C_nH_{2n+1}COOH$. The first member of the series is methanoic acid (HCOOH), which is found in stinging ants and nettles. However, the most important of these acids is ethanoic acid (CH_3COOH), which is the acid found in vinegar. The structures of these acids are shown in Figure 18.9.

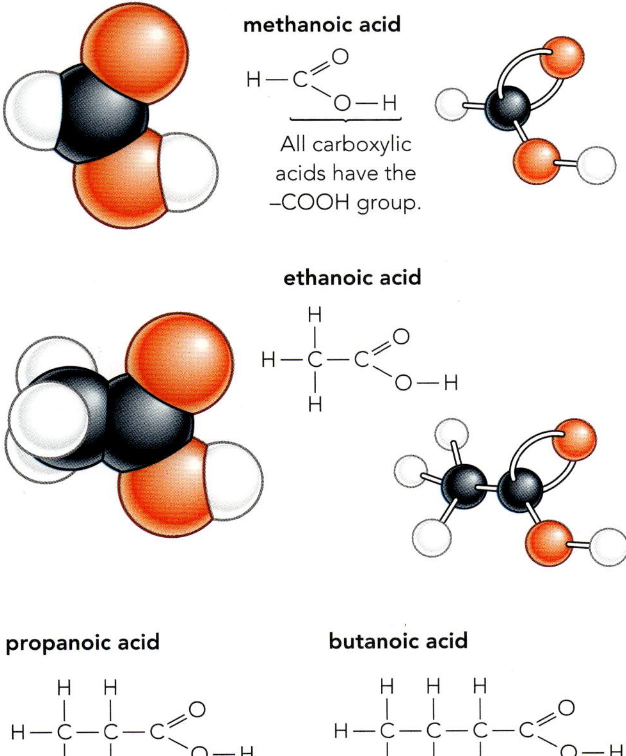

Figure 18.9: The structures of methanoic acid, ethanoic acid, propanoic acid and butanoic acid; the first four carboxylic acids.

Naming organic compounds

The alkanes are a homologous series of saturated hydrocarbons. Their names all end in *-ane*, no matter how long the chain. Figure 18.10 shows a model of tetradecane ($C_{14}H_{30}$). The names of the first six alkanes were given earlier in Table 18.1.

Figure 18.10: Model of a straight-chain alkane ($C_{14}H_{30}$).

The prefixes to the names of the alkanes are standard and indicate the number of carbon atoms in the chain (see Table 18.1). Therefore, a compound in any homologous series with just one carbon atom will always have a name beginning with *meth-*, one with two carbon atoms *eth-* and so on. Hence the names of the early members of the alcohol and carboxylic acid series are as shown in Figures 18.8 and 18.9. When a halogen atom is introduced into a chain, the name of the compound contains a prefix indicating which halogen is present.

The different homologous series all have particular endings to their names (Table 18.3).

Many different organic compounds are formed when a hydrogen in the original alkane 'backbone' is replaced by another group. The product formed when ethene reacts with bromine in solution in hexane (Figure 18.11) illustrates the system of naming organic compounds.

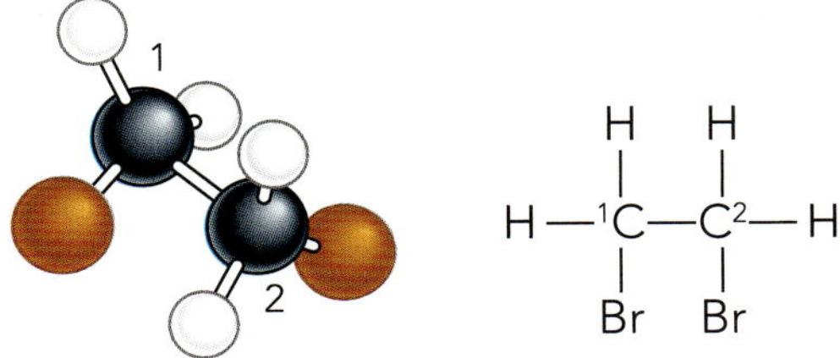

Figure 18.11: Naming the product of adding bromine to ethene (1,2-dibromoethane).

The process of naming the addition product of ethene and bromine is as follows:

- The product has two carbon atoms joined by a single bond. So, it is named after ethane.
- The molecule contains two bromine atoms. It is called dibromoethane.
- The bromine atoms are not both attached to the same carbon atom. One bromine atom is bonded to each carbon atom. The carbon atoms are numbered 1 and 2. The full name of the compound is 1,2-dibromoethane.

The naming of the early members of the alkenes, alcohols and carboxylic acids is systematic and uses the prefixes and name endings we have met earlier (Table 18.4). You will note that in naming ethanoic acid, that the carbon atom of the –COOH functional group is counted as part of the carbon chain. This is true when naming any member of the carboxylic acid series.

Compound (where *n* = 2)	Prefix	Name ending	Displayed formula
ethene	eth-	-ene	$H_2\overset{2}{C}=\overset{1}{C}H_2$
ethanol	eth-	-ol	H–$\overset{2}{C}$H₂–$\overset{1}{C}$H₂–O–H
ethanoic acid	eth-	-oic acid	H–$\overset{2}{C}$H₂–$\overset{1}{C}$(=O)–O–H

Table 18.4: Names and displayed formulae for members of different series with two carbon atoms in the chain.

Questions

5 What are the functional groups of the following homologous series?

- **a** alkenes
- **b** alcohols
- **c** carboxylic acids.

6 What are the molecular and displayed formulae of propanoic acid?

7 What are the correct names for the following straight-chain organic compounds?

- **a** C_4H_{10}
- **b** C_3H_6
- **c** C_3H_7OH
- **d** C_3H_7COOH

ACTIVITY 18.1

Organic flash cards

Organic chemistry involves several very specific words and definitions. In groups, get together to devise a set of flash cards to aid your understanding and help with revision.

The flash cards should have a question on the front of the card and the corresponding answer on the back of the card.

Questions covered could range from the following suggestions:

- What is a hydrocarbon?
- What is a homologous series?
- What are the formulae and structures of the different functional groups?

Through to questions on specific structures:

- What is the displayed formula of propene? Use different cards to put the questions from both directions (name → structure / structure → name).

Different groups could make cards covering different types of molecule and structure. Construct a set of cards and use them to quiz each other in the group. Then compare notes with the rest of the class to refine the content covered.

You could also investigate the wide range of online software available for creating and using virtual flash cards.

REFLECTION

The naming and organisation of organic compounds and their molecules is in many ways highly systematic. How do you visualise the different types of compound and remember their differences? Could you develop a written approach to help you?

18.2 Structural formulae, homologous series and isomerism

Structural formulae and homologous series

In organic chemistry, the structure of a molecule is very important. Figure 18.5 shows the displayed formulae of the first three alkenes in that series. So far, we have concentrated on showing the molecular formulae and the displayed formulae for the members of the different homologous series. A displayed formula is important because it shows clearly all the bonds present in a molecule.

However, there is a further way of representing the formula of an organic compound and this is the **structural formula** of the molecule. This type of formula is not as complex as drawing all the bonds of a structure but is an unambiguous way of representing the way the atoms of a molecule are arranged. Ethane has the molecular formula C_2H_6 and the molecule can be clearly represented by its structural formula CH_3CH_3. Table 18.5 shows the structural formulae of some early members of different homologous series.

Homologous series	Compound	Molecular formula	Structural formula
alkanes	ethane	C_2H_6	CH_3CH_3
	propane	C_3H_8	$CH_3CH_2CH_3$
alkenes	ethene	C_2H_4	$CH_2{=}CH_2$
	propene	C_3H_6	$CH_3CH{=}CH_2$
alcohols	ethanol	C_2H_5OH	CH_3CH_2OH
	propanol	C_3H_7OH	$CH_3CH_2CH_2OH$
carboxylic acids	ethanoic acid	CH_3COOH	CH_3COOH
	propanoic acid	C_2H_5COOH	CH_3CH_2COOH

Table 18.5: Structural formulae of representative molecules from different homologous series.

KEY WORDS

structural formula: the structural formula of an organic molecule shows how all the groups of atoms are arranged in the structure; ethanol, CH_3CH_2OH, for example

So far, we have only considered molecules made up of an unbranched, straight hydrocarbon chain. However, it is possible to have molecules where the chain is branched. In butane, all four carbon atoms are arranged in one 'straight' main chain. However, the atoms do not have to be arranged in this way. The fourth carbon atom can go off from the main chain to give the branched structure of 2-methylpropane (Table 18.6).

	butane	2-methylpropane
Molecular formula	C_4H_{10}	C_4H_{10}
Displayed formula	H H H H H–C–C–C–C–H H H H H	H H–C–H] methyl group This carbon atom is the second in the chain. H H H–¹C–²C–³C–H H H H
Structural formula	$CH_3CH_2CH_2CH_3$	$CH_3CH(CH_3)CH_3$
Properties	alkane, burns to give CO_2 and H_2O, boiling point 0 °C	alkane, burns to give CO_2 and H_2O, boiling point −12 °C

Table 18.6: Two different alkanes with the same molecular formula.

Two factors are important when naming the molecule of 2-methylpropane. The first is the name of the side-chain, in this case a CH_3– group. When we remove a hydrogen from an alkane chain, we are left with a group known as an alkyl group. Alkyl groups are named after the hydrocarbon by changing the *-ane* ending to *-yl*. So we call CH_3– a methyl group, C_2H_5– an ethyl group and so on.

The second consideration is numbering the carbon atom to which the side-chain is attached. In this case, it is simple as the methyl group is attached to the middle carbon in a chain of three carbon atoms; the second carbon, numbered 2. In general, where a hydrocarbon side-chain has replaced a hydrogen to produce a more complex molecule we number the carbon atoms in the chain. The numbering always starts at one end of the chain. The counting starts at the end which keeps the number of the side-chain position as low as possible, and we can then indicate where the side-chain is in the name.

Butane and 2-methylpropane are both alkanes and they have very similar chemical properties as they have the same functional group. The difference between them shows itself mainly in their melting points and boiling points. Hydrocarbons containing branched chains have lower melting points and lower boiling points than straight-chain compounds with the same number of carbon atoms.

Figure 18.12 shows a graph of the increasing boiling points of the straight-chain alkanes with the length of the hydrocarbon chain. As the length of the hydrocarbon chain increases, the strength of the weak forces between the molecules (**intermolecular forces**) is increased. This shows itself in the increasing boiling points of the

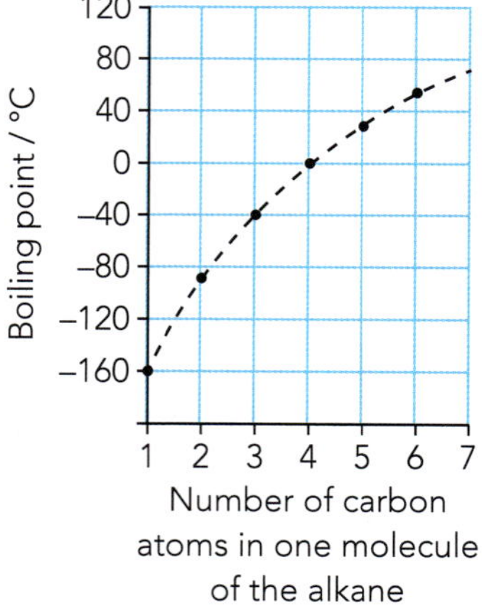

Figure 18.12: Plot of the boiling points of the first six straight-chain alkanes (methane to hexane).

members of the series. Under normal conditions, the first four members of the straight-chain alkanes are gases, and those between C_5H_{12} and $C_{16}H_{34}$ (which are called C_5 to C_{16} alkanes) are liquids. The compounds in the alkane family with 17 or more carbon atoms are waxy solids.

Similar graphs can be plotted for the melting points and densities of the alkanes. The values of these physical properties show a progressive increase in value with increasing chain length. The same trends in physical properties are observed in other homologous series such as the alkenes or alcohols. This adds a further general feature to the characteristics of a homologous series. A homologous series is therefore a series of organic molecules having:

- the same functional group
- the same general formula
- members that differ from one to the next by a $-CH_2-$ unit
- a consistent trend in physical properties with increasing molecular size
- similar chemical properties.

Structural isomerism of hydrocarbons and alcohols

The system of naming compounds emphasises the importance of molecular structure. Molecules with the same molecular formula can have different structures. We have seen in Table 18.6 that two members of the same homologous series can have the same number of atoms in their molecules, but these atoms can be connected together in different ways. Butane and 2-methylpropane have the same molecular formula (C_4H_{10}) but different structural formulae ($CH_3CH_2CH_2CH_3$ and $CH_3CH(CH_3)CH_3$, respectively). This is known as **structural isomerism**, and butane and 2-methylpropane are **structural isomers** of each other.

KEY WORDS

intermolecular forces: the weak attractive forces that act between molecules

structural isomerism: a property of compounds that have the same molecular formula but different structural formulae; the individual compounds are known as **structural isomers**

Structural isomerism can be shown by members of any homologous series and can arise from changing the position of the functional group in a molecule. In the alkenes, the two forms of butene (C_4H_8) are the first examples of this. The two isomers are but-1-ene and but-2-ene (Figure 18.13).

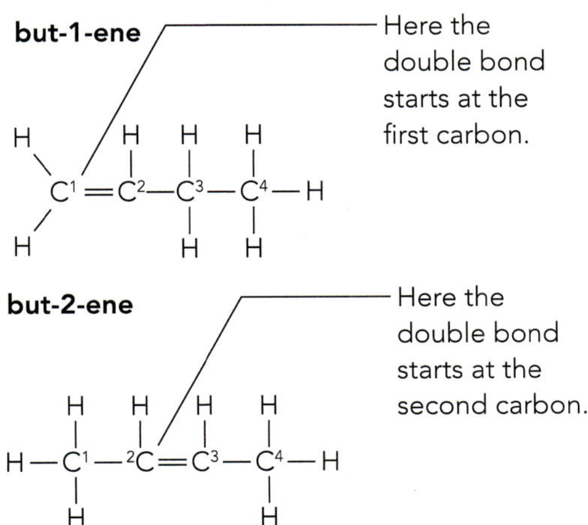

Figure 18.13: But-1-ene and but-2-ene are structural isomers of each other.

The structures in Figure 18.13 are different. Again, the carbon atoms are numbered. The number added to the formula indicates the position of the double bond. In but-1-ene the double bond is between carbon atoms 1 and 2, whereas in but-2-ene it is between carbon atoms 2 and 3. This type of structural isomerism is also known as position isomerism.

This type of structural isomerism where the position of the functional group is moved within the molecule also takes place in the alcohols. Here the carbon atom to which the –OH group attaches changes. The alcohol concerned must have three or more carbon atoms in the chain for isomerism to take place. There are two structural isomers of propanol:

- propan-1-ol, $CH_3CH_2CH_2OH$
- propan-2-ol, $CH_3CH(OH)CH_3$

Here the –OH group is attached to the terminal carbon atom in propan-1-ol, but to the central carbon atom in propan-2-ol. A similar situation can arise with butanol. Figure 18.14 shows the two straight-chain position isomers of butanol (C_4H_9OH).

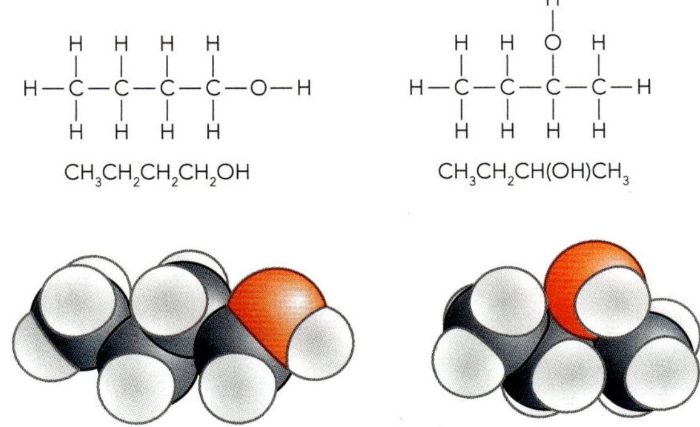

Figure 18.14: Structural isomers of butanol (C_4H_9OH).

In this case, we must be careful as it would be easy to think that there should be a third isomer with the alcohol group attached to the third carbon atom. This is not the case as that structure is identical to butan-2-ol turned over on itself. There are just the two straight-chain structural isomers of butanol. We will discuss these isomers of alcohols further in Chapter 19.

Esters of carboxylic acids and isomerism

There is a further homologous series of organic compounds that we have not discussed yet. This series of molecules are the **esters** formed when an alcohol reacts with a carboxylic acid. For example, ethyl ethanoate is formed by the reaction between ethanol and ethanoic acid:

ethanoic acid + ethanol → ethyl ethanoate + water

$$CH_3COOH(l) + C_2H_5OH(l) \rightarrow CH_3COOC_2H_5(l) + H_2O(l)$$

This type of reaction is known as esterification. Figure 18.15 shows the displayed formula of ethyl ethanoate and indicates how the ethyl group from ethanol has attached to the acid grouping.

KEY WORD

esters: a family of organic compounds formed by esterification, characterised by strong and pleasant tastes and smells

$CH_3COOCH_2CH_3$

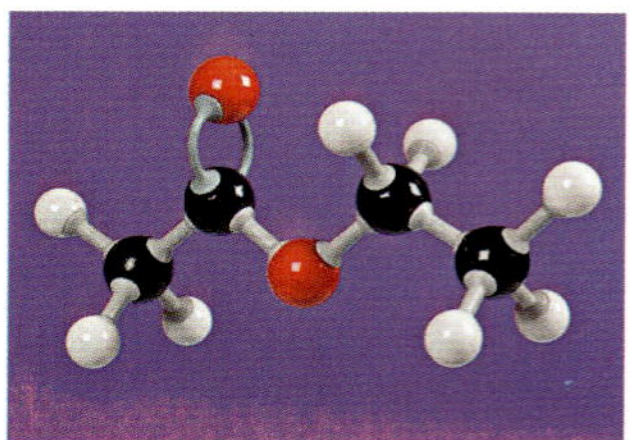

Figure 18.15: Displayed formula and molecular model of the ester, ethyl ethanoate ($CH_3COOC_2H_5$).

Esters are named after the acid from which they are formed; the acid is indicated by the second, most important, part of the name. This means that *ethanoate* is the second part of this name. The alkyl group from the alcohol, *ethyl* in this case, is the first part of this name. Ethyl ethanoate is just one example of a series of esters. Different combinations of alcohol and carboxylic acid can produce different esters that are structural isomers of each other. Table 18.7 shows an example of such isomerism formed from different unbranched alcohols and carboxylic acids. Propyl butanoate and butyl propanoate are structural isomers of each other as they have the same molecular formula.

Questions

8 Give the structural formulae of the following compounds:

- **a** propan-2-ol
- **b** but-2-ene
- **c** propyl ethanoate.

9 Define the term structural isomerism.

10 Plot a graph of the boiling points of the first six alcohols (Table 18.8) against the number of carbon atoms in the molecule. Comment on the shape of the graph.

Name of alcohol	Boiling point / °C
methanol	65
ethanol	78
propan-1-ol	97
butan-1-ol	117
pentan-1-ol	137
hexan-1-ol	158

Table 18.8: Boiling points of the early members of the homologous series of alcohols.

Ester	Molecular formula	Alcohol	Carboxylic acid	Displayed formula of ester
propyl butanoate	$C_7H_{14}O_2$	propan-1-ol	butanoic acid	$(C_3H_7COOC_3H_7)$
butyl propanoate	$C_7H_{14}O_2$	butan-1-ol	propanoic acid	$(C_2H_5COOC_4H_9)$

Table 18.7: Propyl butanoate and butyl propanoate are structural isomers of each other.

Formulae and information

Every organic compound can be represented by several different types of formulae. These formulae are designed to give us information about the composition and nature of the compound. The first and most crucial is the *molecular formula*. This represents the actual number of atoms present in the molecule; thus, for methane it is CH_4, for ethane it is C_2H_6, and so on.

As we have seen in this chapter, the molecular formula does not indicate the structure of the molecule. It is the displayed formula for the compound that gives us that information. A displayed formula shows the structure and all the bonding in the molecule. Examples of displayed formula can be seen in Figure 18.4 and Table 18.7.

Drawing the displayed formula each time we refer to a compound can be unnecessary. A structural formula is designed to show information on structure without showing all the bonds. The two straight-chain structural isomers of butene can be written as $CH_3CH_2CH{=}CH_2$ (but-1-ene) and $CH_3CH{=}CHCH_3$ (but-2-ene), for instance.

The final type of formula relevant to all compounds including organic substances is the empirical formula (Chapter 4). This formula is the simplest possible whole-number ratio of the atoms in a compound; thus, for methane it is CH_4, but for ethane it is CH_3.

SUMMARY

Homologous series of organic compounds (e.g. alkanes or alkenes) have the same general formula and similar chemical properties because they contain the same functional group.
Alkanes are a homologous series of saturated hydrocarbons as all the carbon–carbon bonds are single covalent bonds.
Compounds containing at least one carbon–carbon double or triple bond are said to be unsaturated (e.g. alkenes are unsaturated as they contain a carbon–carbon double bond).
Saturated and unsaturated compounds can be distinguished by testing with aqueous bromine; unsaturated compounds will decolourise aqueous bromine.
Alcohols and carboxylic acids are further examples of homologous series and that organic compounds are given a systematic name depending on which functional group is present and the number of carbon atoms present in their structure.
Different organic compounds can be distinguished from each other by their molecular and displayed formulae.
The molecules from various homologous series, including the esters of carboxylic acids, can be represented and distinguished by their structural formulae.
Structural isomers are compounds with the same molecular formula but different structural formulae.
The general characteristics of a homologous series include increasing chain length, similar chemical properties and related trends in physical properties.
Structural and displayed formulae of unbranched members of the different homologous series containing up to four carbon atoms, including the unbranched esters formed from such alcohols and carboxylic acids, can be drawn to represent the bonding in the different molecules.

PROJECT

Modelling different homologous series

Molecular modelling is an important aid to understanding the series of molecules studied in this chapter and the nature of a homologous series.

Work in groups and use modelling kits (or polystyrene spheres and cocktail sticks) to create models of:

- the structures of members of different series having the same number of carbon atoms (Figure 18.16 shows hydrocarbon molecules from different series containing two carbon atoms)
- the early members of a particular homologous series (e.g. methanol, ethanol, propanol)
- the possible isomers with a certain molecular formula (propan-1-ol and propan-2-ol).

Take photos of your models (or draw sketches) as you produce them. Working in your groups, use these photos or sketches to help you to create a classroom display that explains the homologous series. Before you start, write a list of success criteria for a good chemistry display.

Figure 18.16: Molecular models showing the carbon–carbon single, double and triple bonds of ethane, ethene and ethyne.

Peer assessment

Look at another group's classroom display. Use the success criteria you created to write down two things that the group have done well, and one thing they could improve next time.

EXAM-STYLE QUESTIONS

1 Organic compounds have names that correspond to the homologous series they belong to. Four types of organic compound have names ending in *-ane*, *-ene*, *-ol* or *-oic acid*. How many of these types of compound contain oxygen?

A three
B two
C one
D four **[1]**

2 The displayed formula of five organic compounds **A–E** are shown in the figure.

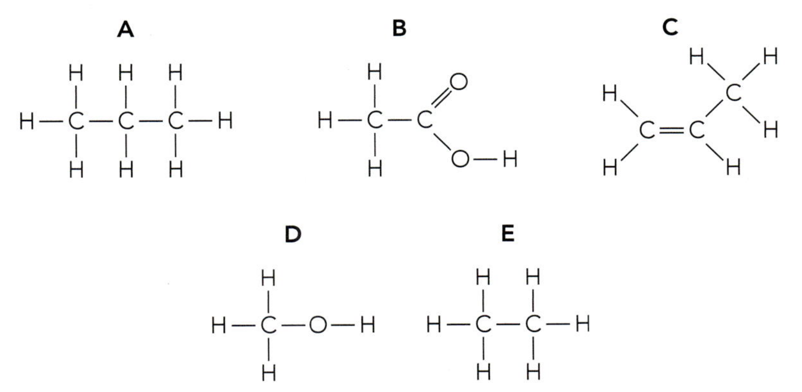

Use these compounds to answer the questions which follow. Each compound **A–E** can be used once, more than once or not at all.

a State which compound is an alkene. **[1]**
b State which two compounds belong to the same homologous series. **[2]**
c State the functional group in the alkene homologous series. **[1]**
d Alkenes are referred to as unsaturated compounds. Describe what is meant by the term unsaturated. **[1]**
e State the molecular formula and the empirical formula of compound **C**. **[2]**

[Total: 7]

COMMAND WORDS

state: express in clear terms

describe: state the points of a topic / give characteristics and main features

CONTINUED

3 One feature of a homologous series is that there is a regular change in the values of certain physical properties. The graph shows the way in which the boiling points of some alkanes depend on the number of carbon atoms in their molecules.

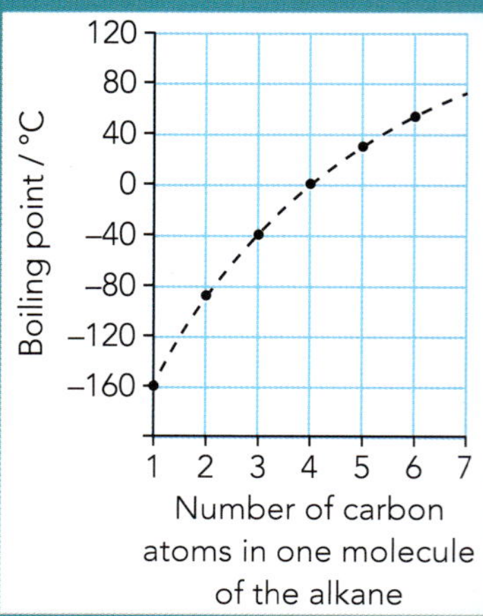

From the graph, **deduce** the boiling point of propane.

A −80 °C

B −40 °C

C 0 °C

D +40 °C **[1]**

COMMAND WORD

deduce: conclude from available information

4 **a** One possible displayed formula for the C_4H_8 molecule is shown in the figure. This compound is called cyclobutane.

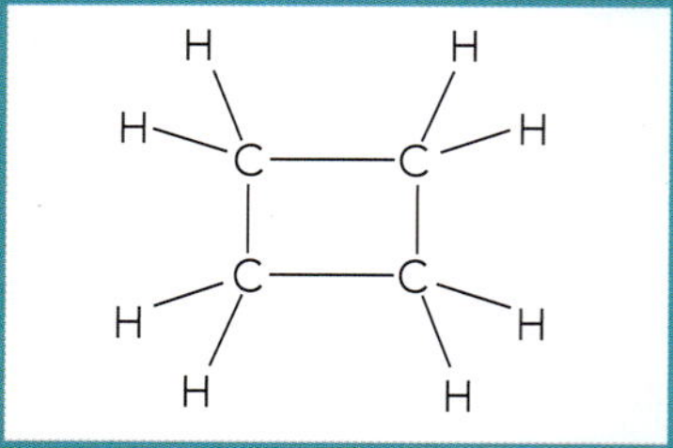

i Draw the displayed formulae of two other isomers of this molecular formula which are both alkenes and state their names. **[4]**

ii Why is cyclobutane not an alkene? **[1]**

b Butane, C_4H_{10}, has two structural isomers. Draw the displayed formula and structural formula of each. **[4]**

c Alkanes are a homologous series of compounds; give the general formula for an alkane. **[1]**

d **Give** three characteristics of a homologous series. **[3]**

[Total: 13]

COMMAND WORD

give: produce an answer from a given source or recall/memory

SELF-EVALUATION CHECKLIST

After studying this chapter, think about how confident you are with the different topics. This will help you see any gaps in your knowledge and help you to learn more effectively.

I can	See Topic...	Needs more work	Almost there	Confident to move on
understand that a homologous series is a family of compounds with the same general formula and similar chemical properties because they contain the same functional group	18.1			
state that the alkanes are a series of saturated hydrocarbons as all the carbon–carbon bonds are single covalent bonds	18.1			
understand that compounds containing at least one carbon–carbon double or triple bond are unsaturated molecules	18.1			
describe the test to distinguish unsaturated and saturated molecules using aqueous bromine	18.1			
understand that alcohols and carboxylic acids are further homologous series and that organic compounds are given a systematic name depending on their functional group and the number of carbon atoms they contain	18.1			
distinguish the different organic compounds by their molecular and displayed formulae	18.1			
understand how to write the structural formulae of molecules from the different homologous series, including the esters of carboxylic acids	18.2			
define structural isomers as compounds with the same molecular formula but different structural formulae	18.2			
describe the general characteristics of a homologous series, including increasing chain length, similar chemical properties and related trends in physical properties	18.2			
understand how to draw the structural and displayed formulae of unbranched members of different homologous series containing up to four carbon atoms, including the unbranched esters formed from such alcohols and carboxylic acids	18.2			

Chapter 19

Reactions of organic compounds

IN THIS CHAPTER YOU WILL:

- describe the alkanes as a series of generally unreactive compounds that burn readily and undergo substitution reactions with chlorine
- understand that alkenes can be obtained by catalytic cracking
- understand that ethanol is manufactured either by fermentation or by the catalytic hydration of ethene, and that it can be used as a solvent and as a fuel
- describe the reactions of ethanoic acid with metals, bases and metal carbonates

- understand that the substitution reactions of alkanes with chlorine are photochemical reactions
- describe how alkenes take part in addition reactions
- understand the advantages and disadvantages of the two methods of manufacturing ethanol
- understand how ethanoic acid can be formed by the oxidation of ethanol

CONTINUED

- describe the reaction of a carboxylic acid and an alcohol to form an ester.

GETTING STARTED

The combustion of organic compounds is an important feature of this chapter. In groups, discuss your understanding of the meaning of the terms combustion, burning and fuels. Which fuels are you familiar with? What are the dangers of their use? Are you familiar with the ideas of complete and incomplete combustion of organic fuels?

MOLECULAR ENVELOPES

Nanotechnology has recently brought a new dimension to the idea of test-tube chemistry. The world's smallest test-tube has been made from a carbon nanotube built out of sheets of the latest revolutionary form of carbon known as graphene (Chapter 3).

The 'nano test-tube' was used to polymerise carbon-60 epoxide ($C_{60}O$) molecules (Figure 19.1a). The spherical molecules of $C_{60}O$ join together in a straight line linked through the oxygen atoms. Without the restriction of the nano test-tube, random branching and tangling of the polymer chains occurs. However, with the monomers lined up in the test-tube, the polymer is unbranched and linear; the tube controls the direction of polymerisation.

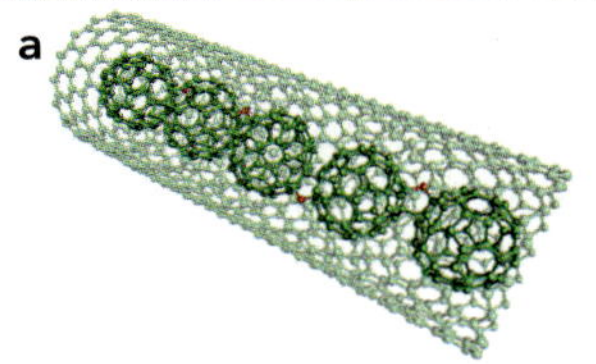

Figure 19.1 a: The smallest possible test-tube – an experiment carried out in a nanotube. **b:** Professor Ijeoma Uchegbu, an expert on pharmaceutical drug delivery using nanotubes.

The use of a nanotube as a test-tube is just one example of what can be done with this technology. Professor Ijeoma Uchegbu is directly involved in developing medical uses for nanotubes (Figure 19.1b). One key development has been the design of polymers that can form themselves into nanoparticles able to carry substances to targeted locations in the body, such as transporting specific drugs to the site of a tumour. Many of the most useful medicinal drugs do not like a water environment (they are hydrophobic = water-hating). This makes it difficult to deliver them across cell membranes to the place in the body where they are needed. The delocalised electrons present in the structure of the graphene sheets means that water-liking molecules can be attached to the surface of the sheets. The designed nanoparticles have a water-liking outer surface and the drug is enclosed in these capsules. The medical compounds are then delivered to the place where they are needed.

Professor Uchegbu is not only developing ways of using nanoparticles in the absorption of hydrophobic drugs, she is also researching potential uses of nanoparticles in the treatment of brain tumours, including how they could first identify, then target and destroy diseased brain cells. Her university and pharmaceutical industry research places her at the very forefront of the development of revolutionary medical work.

Discussion questions

1. Each layer of graphite, and of graphene, is made up of carbon atoms covalently bonded together. It is considered to be a giant covalent structure. How would the approximately spherical molecules of the fullerenes (C_{60} and C_{70}) be classified?
2. What properties of the nanotubes make them useful for a drug delivery device? Does the fact that they have delocalised electrons present help in this use?

19.1 Characteristic reactions of different homologous series

Chemistry of the alkanes

The alkanes are quite unreactive compounds. Alkanes are saturated molecules, so they cannot take part in addition reactions. They are unaffected by acids or alkalis. However, they can take part in substitution reactions, particularly with chlorine.

Alkanes as fuels

The simplest alkane is methane (CH_4), which is the major component of natural gas. Other alkanes are obtained from the fractional distillation of petroleum. One chemical property that all these alkanes have in common is that they burn very exothermically. Natural gas and various alkanes have therefore proved very useful fuels.

Compressed natural gas (CNG) has been used as a cleaner alternative fuel to petrol (gasoline) for vehicles ranging from long-distance motor coaches (Figure 19.2a) to urban autorickshaws (tuk-tuks) (Figure 19.2b). Propane and butane burn with very hot flames and are sold as liquefied petroleum gas (LPG). LPG can also be used as a fuel in autorickshaws and minibuses.

Figure 19.2 a: India's first long-range CNG buses, which can travel up to 1000 kilometres in a single fill, in New Delhi. **b:** A TVS autorickshaw in Chennai, India.

LPG has various compositions of propane and butane depending on its use. The gases are kept as liquids under pressure, but they vaporise easily when the pressure is released. In areas where there is no mains supply of natural gas, propane tanks supply the fuel for heating systems. Cylinders of butane gas are used in portable gas fires in the home. Butane is also used in portable camping stoves, blowtorches and gas lighters (Figure 19.3).

Figure 19.3: A butane portable camping stove.

Combustion of the alkanes

Alkanes burn readily (Figure 19.4). When they burn in a good supply of air, the products are carbon dioxide and water vapour:

methane + oxygen → carbon dioxide + water

$$CH_4(g) + 2O_2(g) \rightarrow CO_2(g) + 2H_2O(g)$$

ethane + oxygen → carbon dioxide + water

$$2C_2H_6(g) + 7O_2(g) \rightarrow 4CO_2(g) + 6H_2O(g)$$

butane + oxygen → carbon dioxide + water

$$2C_4H_{10}(g) + 13O_2(g) \rightarrow 8CO_2(g) + 10H_2O(g)$$

The same products are obtained whichever alkane is burnt, so long as there is a sufficient oxygen supply. Note that the key to balancing the symbol equations for these reactions is to balance the oxygen correctly.

Figure 19.4: Excess natural gas (mainly methane) is sometimes flared off at oilfields.

However, if the air supply is limited, then the poisonous gas carbon monoxide can also be formed. Carbon monoxide is the product of incomplete combustion of a hydrocarbon. For example:

methane + oxygen → carbon monoxide + water

$2CH_4(g) + 3O_2(g) \rightarrow 2CO(g) + 4H_2O(g)$

Carbon monoxide (CO) is toxic because it interferes with the transport of oxygen around our bodies by our red blood cells.

Incomplete combustion can also produce fine particles of carbon (particulates or 'soot'). These particles of carbon have not even reacted to produce carbon monoxide. It is these fine carbon particles that can glow yellow in the heat of a flame. They give a candle flame or the 'safety' flame of a Bunsen burner its characteristic yellow colour (Figure 19.5).

The incomplete combustion of fossil fuels is a major cause of air pollution (Chapter 17).

Figure 19.5: 'Safety' flame of the Bunsen burner. The air supply to the flame is restricted.

Substitution reactions with chlorine

One reaction that alkanes will take part in is a **substitution reaction** with chlorine. In these reactions a hydrogen atom of the alkane is replaced by a chorine atom.

methane + chlorine → chloromethane + hydrogen chloride

ethane + chlorine → chloroethane + hydrogen chloride

A substituted chloroalkane is produced together with hydrogen chloride gas. The hydrogen chloride gas can be detected using moist blue litmus paper, which turns red.

KEY WORDS

substitution reaction: a reaction in which an atom (or atoms) of a molecule is (are) replaced by different atom(s), without changing the molecule's general structure

The substitution reaction of an alkane with chlorine is interesting because it is a **photochemical reaction**:

methane + chlorine $\xrightarrow{\text{sunlight}}$ chloromethane + hydrogen chloride

$CH_4(g) + Cl_2(g) \longrightarrow CH_3Cl(g) + HCl(g)$

KEY WORDS

photochemical reaction: a chemical reaction where the activation energy required to start the reaction is provided by light, usually of particular wavelength, falling on the reactants

Methane and chlorine react in the presence of sunlight or a source of ultraviolet light. There is no reaction between an alkane and chlorine in the dark. The ultraviolet light provides the activation energy (E_a) for the reaction and splits chlorine molecules into separate energised atoms. These atoms then react with methane. So, the overall result is that a chlorine atom replaces (substitutes for) a hydrogen atom in a methane molecule to give chloromethane (CH_3Cl) (Figure 19.6).

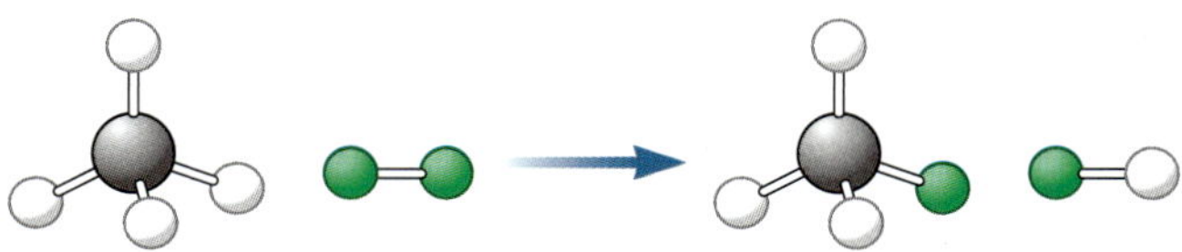

Figure 19.6: Modelling the reaction between methane and chlorine.

The reaction can continue further as more hydrogen atoms are substituted. Compounds such as dichloromethane (CH_2Cl_2), trichloromethane ($CHCl_3$) and tetrachloromethane (CCl_4) are formed in this way. Substituted alkanes are also good organic solvents. 1,1,1-trichloroethane is used frequently in dry-cleaning, for example.

Other alkanes will react in a similar way with chlorine:

ethane	+ chlorine	→	chloroethane	+	hydrogen chloride
$C_2H_6(g)$	+ $Cl_2(g)$	→	$C_2H_5Cl(g)$	+	$HCl(g)$

butane	+ chlorine	→	chlorobutane	+	hydrogen chloride
$C_4H_{10}(g)$	+ $Cl_2(g)$	→	$C_4H_9Cl(l)$	+	$HCl(g)$

Ethane forms just one monosubstituted product CH_3CH_2Cl (chloroethane). Longer chain alkanes will produce a mixture of monosubstituted products as hydrogen atoms attached to different carbons can be replaced. Propane will react to produce a mixture of the 1-chloropropane ($CH_3CH_2CH_2Cl$) and 2-chloropropane ($CH_3CHClCH_3$) isomers of C_3H_7Cl. Butane will similarly produce a mixture of two isomeric products. Figure 19.7 shows the displayed and structural formulae, together with molecular models, of the two **isomers** formed. 1-Chlorobutane and 2-chlorobutane are the two isomers. Remember that the counting starts at the end of the molecule that keeps the number of the substituent position as low as possible.

KEY WORD

isomers: compounds that have the same molecular formula but different structural arrangements of the atoms – they have different structural formulae

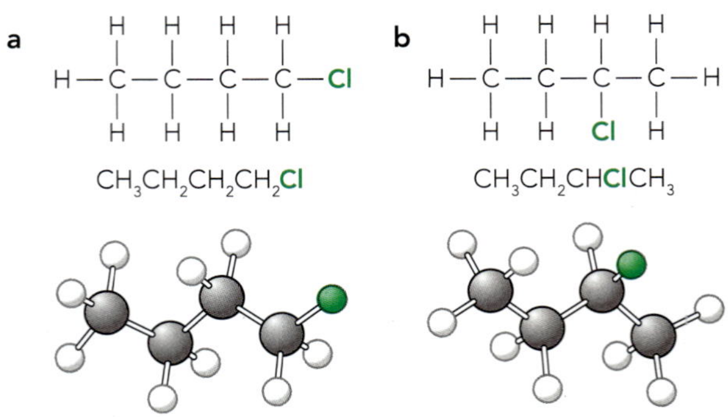

Figure 19.7: Structures of **a:** 1-chlorobutane and **b:** 2-chlorobutane: the isomers of C_4H_9Cl.

Questions

1. The hydrocarbon propane is an important constituent of liquified petroleum gas (LPG). For the burning of propane in an excess of air, give:
   - **a** a word equation
   - **b** a balanced symbol equation.
2. - **a** Write a word equation for the incomplete combustion of methane.
   - **b** What is the formula of carbon monoxide?
   - **c** Why is carbon monoxide toxic?
3. Bromine reacts with alkanes in a similar way to chlorine. Hydrogen bromide is made in the substitution reaction between propane and bromine:

   propane + bromine → bromopropane + hydrogen bromide

   - **a** Draw the structure of propane.
   - **b** Draw the structure of a form of bromopropane.
   - **c** The reaction between propane and bromine is a photochemical reaction. Suggest what is meant by photochemical.

Chemistry of alkenes

Alkenes are unsaturated hydrocarbons and are much more reactive than alkanes. Under suitable conditions, molecules such as bromine, hydrogen and water (steam) will add across the C=C double bond. The reactivity of the alkenes makes them important for the synthesis of other organic molecules. The most important source of alkenes is from the **catalytic cracking** of long-chain alkanes from petroleum.

Catalytic cracking as a source of alkenes

The major source of hydrocarbons industrially is from refining oil (Chapter 20). However, the demand for the various fractions does not necessarily match their supply from the **fractional distillation** process (Figure 19.8). For lighter fractions such as petrol, the demand is greater than the supply. The opposite is true for heavier fractions such as kerosene (paraffin) and diesel oil (gas oil). Larger molecules from these heavier fractions can be broken into smaller, more valuable, molecules. This process is called catalytic cracking ('cat cracking').

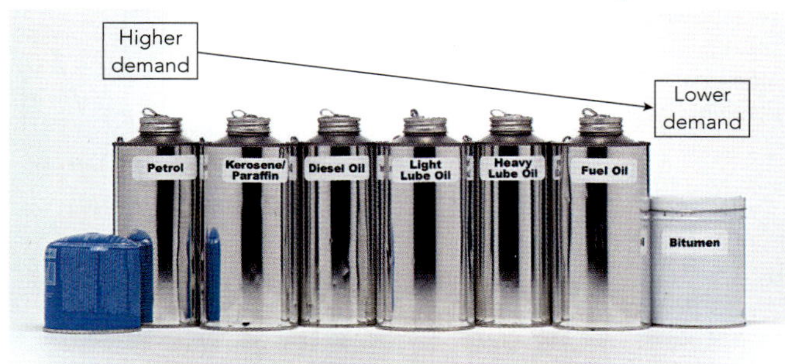

Figure 19.8: There is not the same economic demand for all the fractions from petroleum.

KEY WORDS

catalytic cracking: the decomposition of long-chain alkanes into alkenes and alkanes of lower relative molecular mass; involves passing the larger alkane molecules over a catalyst heated to 500 °C

fractional distillation: a method of distillation using a fractionating column, used to separate liquids with different boiling points

Cracking takes place in a huge reactor (Figure 19.9). In this reactor, particles of catalyst (made of powdered minerals such as silica, alumina and zeolites) are mixed with the hydrocarbon fraction at a high temperature around 500 °C. The cracked vapours containing smaller molecules are separated by distillation.

Figure 19.9: A cracking plant in an oil refinery.

The shortened hydrocarbon molecules are produced by the following type of reaction:

$$\text{decane} \xrightarrow{\text{heat}} \text{octane} + \text{ethene}$$

$$C_{10}H_{22} \xrightarrow{\text{heat}} C_8H_{18} + C_2H_4$$

Figure 19.10 uses the displayed formulae to show the nature of the cracking reaction.

```
    H   H   H   H   H   H   H   H   H   H
    |   |   |   |   |   |   |   |   |   |
H — C — C — C — C — C — C — C — C — C — C — H
    |   |   |   |   |   |   |   |   |   |
    H   H   H   H   H   H   H   H   H   H

                        ↓

    H   H   H   H   H   H   H   H            H   H
    |   |   |   |   |   |   |   |            |   |
H — C — C — C — C — C — C — C — C — H   +    C = C
    |   |   |   |   |   |   |   |            |   |
    H   H   H   H   H   H   H   H            H   H
```

Figure 19.10: Cracking decane to produce octane and ethene.

This is just one of the possible reactions when decane is cracked. The molecules may not all break in the same place. The alkene fragment is not always ethene: propene and but-1-ene may also be produced. Some hydrogen is also broken off from the long-chain alkanes being cracked.

All cracking reactions give:

- an alkane with a shorter chain than the original and a short-chain alkene
- or two or more alkenes and hydrogen.

All the products of cracking are useful. The shortened alkanes can be blended with the gasoline fraction to enrich the petrol. The alkenes are useful as raw materials for making several important products. Figure 19.11 shows the various important uses for the ethene produced, while other alkenes are also useful:

- propene can be polymerised to polypropene (also referred to as polypropylene); while butene polymerises to produce synthetic rubber
- the hydrogen produced can be used for the synthesis of ammonia in the Haber process (Chapter 9) or as a fuel.

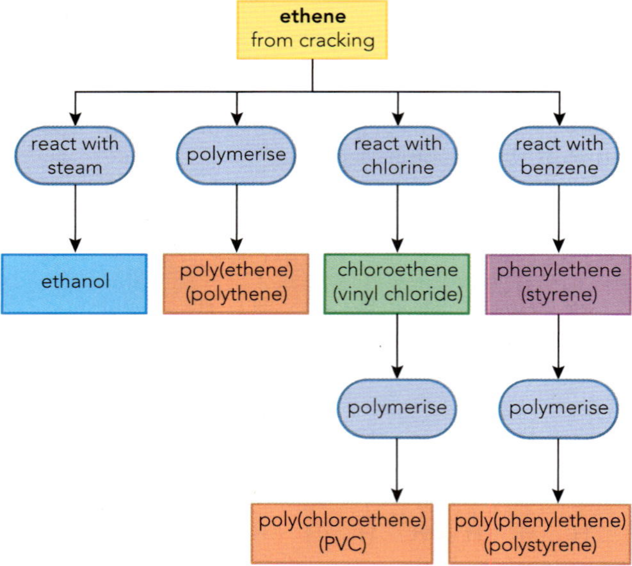

Figure 19.11: Important products can be made from ethene produced by catalytic cracking.

The cracking reaction can be carried out in the laboratory using medicinal paraffin oil (Figure 19.12). The catalyst is heated strongly first and then the paraffin. The paraffin vapour passes over the catalyst that is kept hot. The gases produced in the cracking reaction can be collected in test-tubes. These gases produced can then be tested to see if they are alkenes using the aqueous bromine test (Chapter 18). For safety, the delivery tube can be fitted with a safety valve (Bunsen valve) to prevent cold water being drawn back into the hot apparatus.

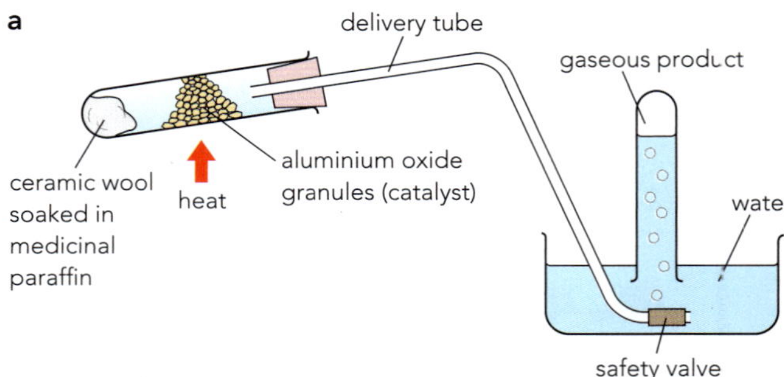

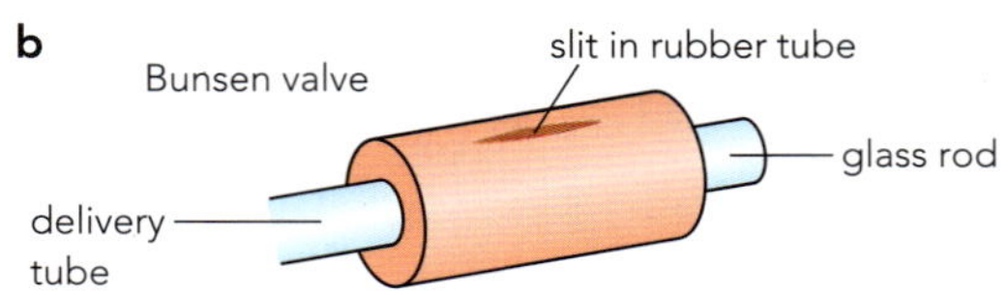

Figure 19.12 a: Cracking of a long-chain alkane in the laboratory. **b:** Use of a Bunsen safety valve to prevent 'suck back'.

Combustion of alkenes

Alkenes are similar to other hydrocarbons when burnt. They give carbon dioxide and water vapour as long as the air supply is sufficient:

$$\text{ethene} + \text{oxygen} \rightarrow \text{carbon dioxide} + \text{water}$$

$$C_2H_4(g) + 3O_2(g) \rightarrow 2CO_2(g) + 2H_2O(g)$$

The presence of the carbon–carbon double bond in an alkene molecule makes these molecules much more reactive than alkanes (alkanes contain only carbon–carbon single bonds). Alkenes take part in addition reactions in which two substances react to form just one single product.

Bromination

Alkenes can take part in addition reactions. Other atoms can add on to alkene molecules when the double bond breaks open. This difference in reactivity between unsaturated alkenes and saturated alkanes produces a simple test for unsaturation using aqueous bromine

(Chapter 18). Aqueous bromine is decolorised when shaken with an alkene. The addition of bromine to an alkene can be carried out using a solution of the halogen in an organic solvent such as ethanol.

The addition of bromine to an alkene can be carried out using a solution of the halogen in an organic solvent such as ethanol. In this case, the bromine atoms add across the double bond in the alkene. The double bond in ethene breaks open and forms new bonds to the bromine atoms (Figure 19.13). The red–brown colour of the bromine is lost during the reaction as the product in each case is colourless. This type of reaction, where a double bond breaks and adds two new atoms, is an addition reaction.

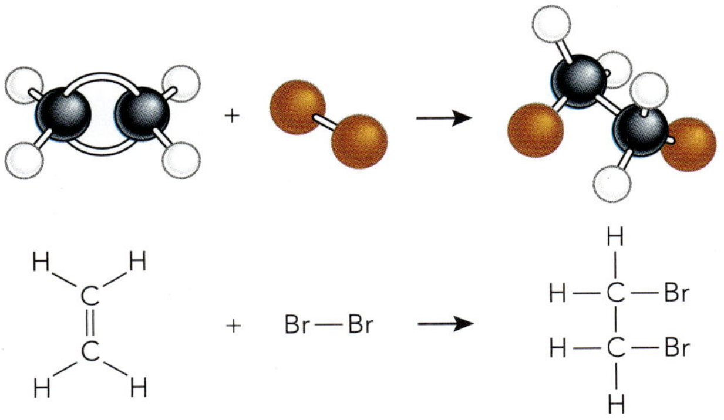

ethene + bromine → 1,2-dibromoethane

propene + bromine → 1,2-dibromopropane

$CH_3CH—CH_2 + Br_2 \rightarrow CH_3CHBrCH_2Br$

Figure 19.13: Bromination of ethene and propene showing displayed and structural formulae.

The product formed if the reaction is carried out with aqueous bromine is more complex. The bromine is still decolourised, but the colourless product includes an added –OH group from the water (Figure 19.14).

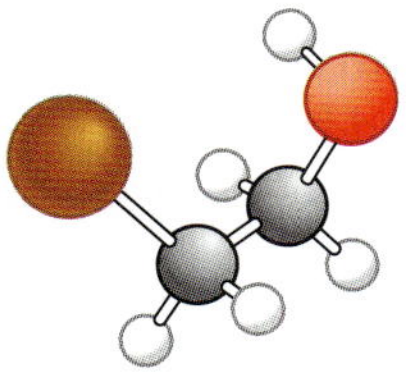

$C_2H_4 + Br_2 + H_2O \rightarrow CH_2BrCH_2OH + HBr$

bromoethanol

Figure 19.14: Addition product (2-bromoethanol) of the reaction between ethene and aqueous bromine.

Hydrogenation

Hydrogen reacts with alkenes to form alkanes. The unsaturation of the alkene is removed to produce the corresponding alkane. The addition of hydrogen across a carbon–carbon double bond is known as **hydrogenation**. Ethene reacts with hydrogen if the heated gases are passed together over a nickel catalyst. The unsaturated ethane is the product (Figure 19.15):

ethene + hydrogen $\xrightarrow[\text{nickel}]{150\text{–}300\,°C}$ ethane

$CH_2{=}CH_2 + H_2 \longrightarrow CH_3CH_3$

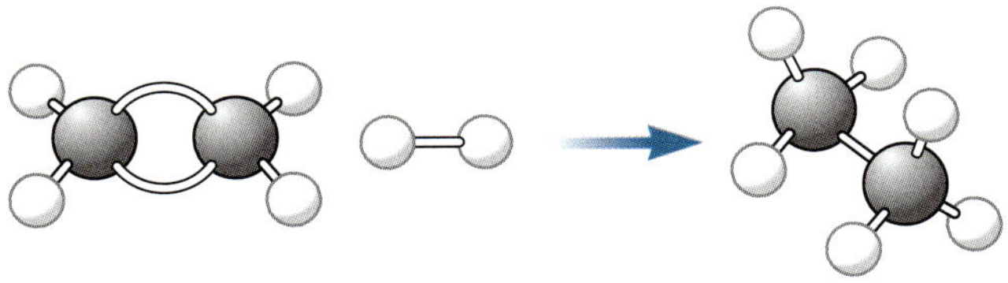

Figure 19.15: Hydrogenation of ethene.

Other alkenes also react in a similar way to produce the saturated alkene. Note that the isomers but-1-ene and but-2-ene both produce the same product on the addition of hydrogen.

$CH_3CH_2CH{=}CH_2 + H_2 \rightarrow CH_3CH_2CH_2CH_3 \leftarrow H_2 + CH_3CH{=}CHCH_3$

but-1-ene butane but-2-ene

Hydrogenation reactions similar to the reaction with ethene are used in the manufacture of margarine from vegetable oils, such as corn oil and sunflower oil. These are edible oils and contain long-chain carboxylic acids. The hydrocarbon chains of these acids contain one or more carbon–carbon double bonds; they are unsaturated molecules (Figure 19.16).

KEY WORD

hydrogenation: an addition reaction in which hydrogen is added across the double bond in an alkene

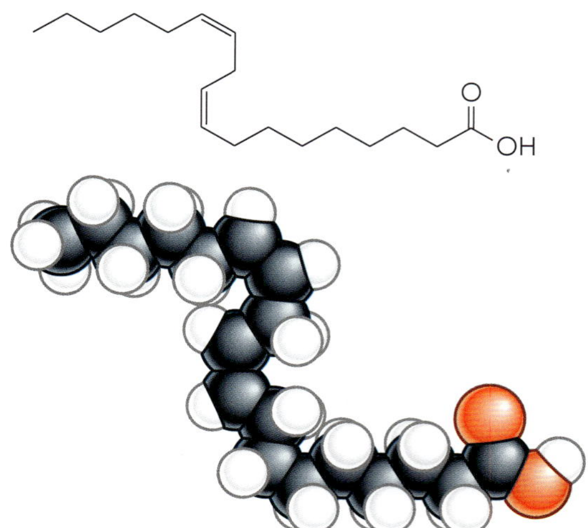

Figure 19.16: Sunflower oil and its products are rich in fats containing unsaturated molecules (note the carbon–carbon double bonds in the chain).

Hydrogen reacts with the vegetable oil when it passes over a nickel catalyst at about 60 °C. By hydrogenating some, but not all, of the carbon–carbon double bonds, the liquid vegetable oil can be made into a solid but spreadable fat (margarine).

Catalytic addition of steam

Another important addition reaction is the reaction used in the manufacture of ethanol. Ethanol is an important industrial chemical and solvent. It is formed when a mixture of steam and ethene is passed over a catalyst of immobilised phosphoric(V) acid (the acid is adsorbed on silica pellets) at a temperature of 300 °C and a pressure of 6000 kPa:

$$\text{ethene} + \text{steam} \xrightarrow[\text{phosphoric acid}]{300\,°\text{C},\ 6000\,\text{kPa}} \text{ethanol}$$

$$C_2H_4(g) + H_2O(g) \rightarrow C_2H_5OH(g)$$

This reaction produces the ethanol of high purity needed in industrial organic chemistry.

Other alkenes can be hydrated by similar reactions. Care is needed in identifying the product when the carbon–carbon double bond is located at the end of the alkene molecule. Propene ($CH_3CH{=}CH_2$) and but-1-ene ($CH_3CH_2CH{=}CH_2$) give a mixture of products as the water molecule can add across the double bond in two different ways. Propene produces a mixture of propan-1-ol ($CH_3CH_2CH_2OH$) and propan-2-ol ($CH_3CH(OH)CH_3$), with propan-2-ol the major product. Butan-1-ol also produces a mixture of two isomers as the product (Figure 19.17) with butan-2-ol the major isomer formed:

```
    butan-1-ol                      butan-2-ol
   H   H   H   H                 H   H   H   H
   |   |   |   |                 |   |   |   |
H—C—C—C—C—O—H             H—C—C—C—C—H
   |   |   |   |                 |   |   |   |
   H   H   H   H                 H   H   OH  H
```

Figure 19.17: Products of hydration of but-1-ene, with butan-2-ol the major product.

The **hydration** of but-2-ene produces just one product, butan-2-ol.

> **KEY WORD**
>
> **hydration:** the addition of the elements of water across a carbon–carbon double bond; H– adds to one carbon, and –OH to the other

Questions

4 What are the molecular and displayed formulae of 1,2-dibromoethane?

5 One source of hydrogen for the Haber process to make ammonia is the catalytic cracking of ethane. Write the word and symbol equations for this reaction.

6 Unsaturated hydrocarbons take part in addition reactions.

   a Write a word equation for the reaction between propene and hydrogen.

   b Write a symbol equation for the reaction between butene and steam.

19.2 Chemistry of ethanol

Manufacture and uses of ethanol

Ethanol is one of the best-known organic compounds. It is just one of a whole family of compounds called **alcohols**. The alcohols are a homologous series of compounds that contain –OH as the functional group (Figure 19.18).

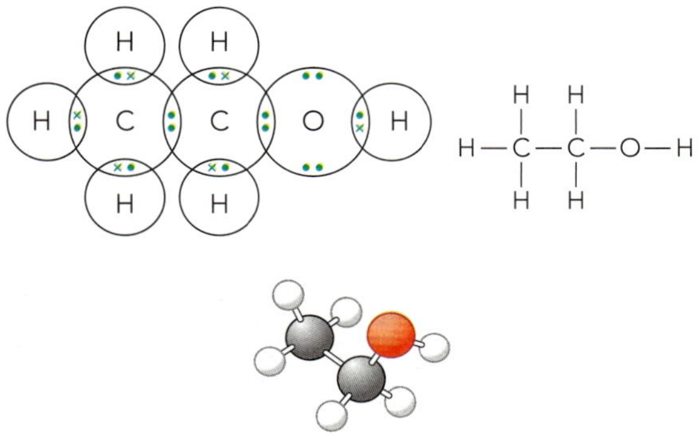

Figure 19.18: Bonding and structure of ethanol (CH_3CH_2OH).

Fermentation

Ethanol can be produced industrially by **fermentation**. Ethanol and carbon dioxide are the natural waste products of yeasts when they ferment sugar. Sugar is present in all fruit and grains, and in the sap and nectar of all plants. Yeasts are single-cell, living fungi found in many places including the surfaces of leaves and in the soil and air. Yeasts contain enzymes that catalyse the fermentation of sugar by anaerobic respiration to gain energy; there is no oxygen present during the process. When we produce ethanol by fermentation, we use specific forms of yeast to avoid unwanted reactions taking place. As ethanol is toxic to yeast, fermentation is self-limiting. Once the ethanol concentration has reached about 14%, or the sugar runs out, the multiplying yeast die and fermentation ends. The best temperature for carrying out the process is between 25 and 35 °C. The reaction is catalysed by enzymes in the yeast:

$$\text{glucose} \xrightarrow{\text{yeast}} \text{ethanol} + \text{carbon dioxide}$$

$$C_6H_{12}O_6(aq) \xrightarrow{\text{enzymes}} 2C_2H_5OH(aq) + 2CO_2(g)$$

Fermentation is an important chemical reaction in many different industries, for example in the production of biofuels and the baking of bread.

Fermentation can be carried out in the laboratory using the apparatus in Figure 19.19. The airlock allows gas to escape from the vessel but prevents airborne bacteria entering. Fermentation is an anaerobic process. It takes place under conditions where there is no air or oxygen available.

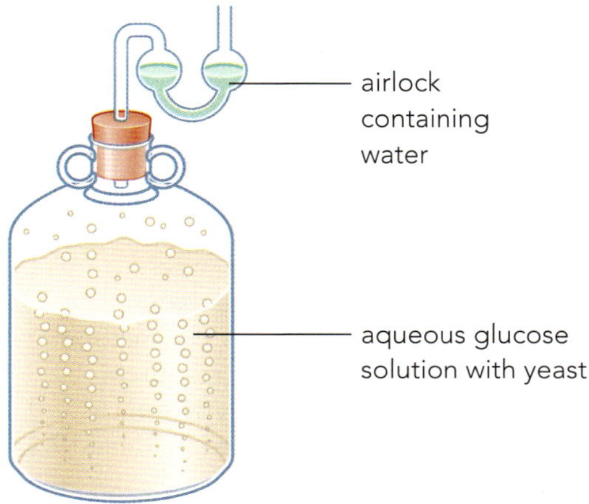

Figure 19.19: A laboratory fermentation vessel.

KEY WORDS

alcohols: a series of organic compounds containing the functional group –OH and with the general formula $C_nH_{2n+1}OH$

fermentation: a reaction carried out using a living organism, usually a yeast or bacteria, to produce a useful chemical compound; most usually refers to the production of ethanol

Hydration of ethene

The alternative industrial method of making ethanol involves the catalytic addition of steam to ethene that we saw earlier. In this method, ethene and steam are compressed to 6000 kPa (60 atmospheres) and passed over a catalyst of immobilised phosphoric(V) acid at 300 °C, and high purity ethanol is produced.

Ethanol by the catalytic hydration of ethene	Ethanol by fermentation
originates from a non-renewable resource (petroleum)	made from readily renewable resources
small-scale equipment capable of withstanding pressure	relatively simple, large vessels
a continuous process	a batch process – need to start process again each time
a fast reaction rate	a relatively slow process
yields highly pure ethanol	ethanol must be purified by subsequent distillation – though fermented product can be used as it is for some purposes
a sophisticated, complex method	a simple, straightforward method

Table 19.1: Comparison of the industrial methods of ethanol production.

Comparing the methods of ethanol production

The two different methods of producing ethanol have their own advantages and disadvantages. The method chosen will depend on the availability of resources and the main purpose for producing the ethanol. A comparison of the methods is summarised in Table 19.1.

The ethanol produced by fermentation comes from a renewable resource. When used as a fuel, the ethanol produced in this way is potentially 'carbon-neutral'. The carbon dioxide released during fermentation and by burning the fuel is balanced by the carbon dioxide absorbed from the atmosphere by the crop, usually sugar cane, as it grows.

Uses of ethanol

Ethanol is an important solvent and a raw material for making other organic chemicals. Many everyday items use ethanol as a solvent. These include paints, glues, perfumes, aftershave and printing inks.

Ethanol burns with a clear flame, giving out a lot of heat:

$$\text{ethanol} + \text{oxygen} \rightarrow \text{carbon dioxide} + \text{water}$$

$$C_2H_5OH(l) + 3O_2(g) \rightarrow 2CO_2(g) + 3H_2O(g)$$

On a small scale, ethanol can be used as methylated spirit (ethanol mixed with methanol or other compounds) in spirit lamps and stoves. However, ethanol is such a useful fuel that some countries have developed it as a fuel for cars, usually when blended with petrol.

Brazil was one of the first countries to produce ethanol fuel, in 1973. With a climate ideally suited for growing sugar cane, the country remains a world leader in ethanol fuel production (Figure 19.20). Ethanol produced by fermentation of sugar from sugar cane has been used as an alternative fuel to petrol or mixed with petrol (gasoline) to produce 'gasohol'.

Figure 19.20: An ethanol and petrol station in Sao Paulo, Brazil.

Ethanol is a renewable resource and has the potential to reduce petroleum imports. Countries in Asia, Europe and the Americas have developed various blends of 'gasohol'; the most common being an E10 blend of 10% ethanol and 90% petrol. 'Gasohol' and other oxygenated fuels have the advantage of reducing the emissions of carbon monoxide from cars. With responsible development and control, biofuels have the potential to reduce future environmental pollution. However, issues of sustainability must be considered too. Land is needed to grow biofuel crops, so overuse of biofuels could encourage unsustainable levels of deforestation (Chapter 17).

ACTIVITY 19.1

The profile of a molecule

Ethanol (Figure 19.21) is a highly important organic compound with important industrial and social implications.

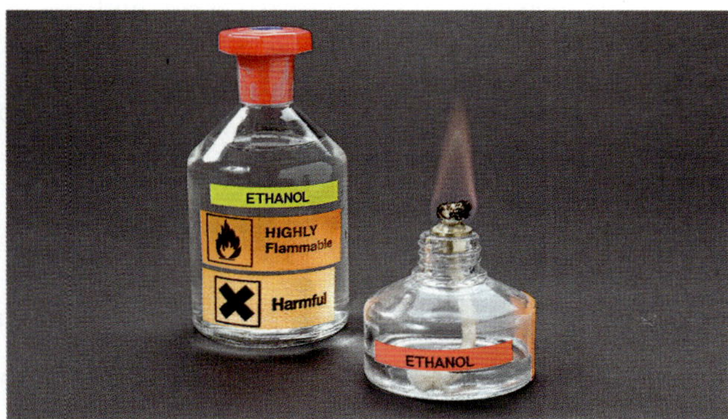

Figure 19.21: Ethanol: an important laboratory chemical and fuel.

In a group, produce an information leaflet that highlights and summarises these different aspects of ethanol. Include:

- a depiction of its molecular structure
- the methods of industrial production
- its usefulness as a fuel and solvent.

Illustrate your leaflet with colourful images from magazines or downloaded from the internet.

When complete, come together as a class to compare leaflets and the different ways of illustrating the impact of this compound.

Peer assessment

Use the time you are in a group producing the poster on ethanol to question and challenge each other on the details of the structure, reactions and uses of ethanol. Include questions on how the compound fits into the general series of compounds known as the alcohols.

19.3 Carboxylic acids and esters

The **carboxylic acids** are another homologous series of organic compounds. All these acids have the functional group –COOH attached to a hydrocarbon chain. Table 19.2 shows the molecular formulae and models of the first two members of the series.

The compounds have the general formula $C_nH_{2n+1}COOH$ (or $C_nH_{2n+1}CO_2H$). When naming a carboxylic acid, remember that the carbon atom of the acid group is counted as the first carbon in the chain. That is why CH_3COOH is the formula of ethanoic acid: there are two carbon atoms in the molecule.

Carboxylic acid	Molecular formula $C_nH_{2n+1}COOH$	Ball-and-stick models
methanoic acid	HCOOH	
ethanoic acid	H–C(H)(H)–C(=O)–O–H CH_3COOH	

Table 19.2: The first two members of the carboxylic acid homologous series.

KEY WORDS

carboxylic acids: a homologous series of organic compounds containing the functional group –COOH (–CO_2H), with the general formula $C_nH_{2n+1}COOH$

The first two acids in the series are liquids at room temperature, although ethanoic acid will solidify if the temperature falls only slightly. The acids dissolve in water to produce solutions that are weakly acidic. Methanoic acid is present in nettle stings and ant stings, while ethanoic acid (once called acetic acid) is well known as the acid in vinegar.

Formation of ethanoic acid by oxidation of ethanol

Vinegar is a weak solution of ethanoic acid (previously called acetic acid). It is produced commercially from wine by biochemical oxidation using bacteria (*Acetobacter*). The production of wine vinegar is an example of traditional biotechnology. The bacteria used are naturally present in the air and wine can simply become 'vinegary' if it is left open to the air.

The same oxidation can be achieved quickly by powerful oxidising agents such as warm acidified potassium manganate(VII):

ethanol + oxygen from oxidising agent → ethanoic acid + water

$$C_2H_5OH + 2[O] \rightarrow CH_3COOH + H_2O$$

Figure 19.22 shows the apparatus used for this oxidation. The condenser is arranged vertically to prevent the volatile alcohol escaping; this would be dangerous as ethanol is highly flammable. The use of apparatus set up in this way is known as **refluxing**. The colour of the potassium manganate(VII) solution turns from purple to colourless.

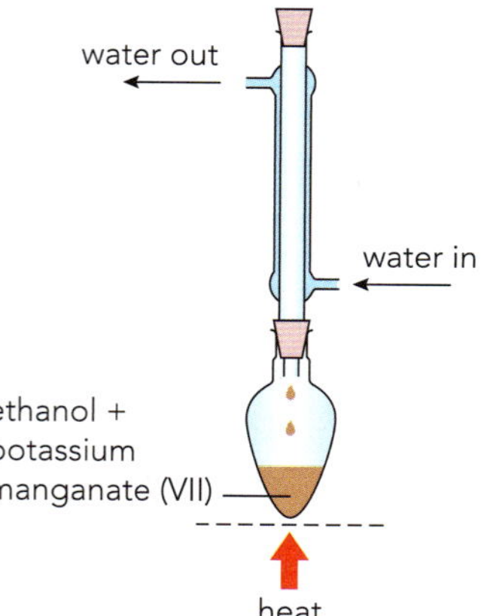

Figure 19.22: Oxidation of ethanol with acidified potassium manganate(VII) under reflux.

Reactions of ethanoic acid as an acid

A solution of ethanoic acid will show the characteristic reactions of an acid.

- Ethanoic acid solution will react with reactive metals such as magnesium to give a salt and hydrogen gas:

ethanoic acid + magnesium → magnesium ethanoate + hydrogen

$$2CH_3COOH + Mg \rightarrow (CH_3COO)_2Mg + H_2$$

- It will react with bases to form salts:

ethanoic acid + sodium hydroxide → sodium ethanoate + water

$$CH_3COOH + NaOH \rightarrow CH_3COONa + H_2O$$

- Vinegar can be used as a 'descaler' in hard water areas. The ethanoic acid in vinegar reacts with limescale (calcium carbonate), producing carbon dioxide and dissolving the scale:

calcium carbonate + ethanoic acid → calcium ethanoate + water + carbon dioxide

$$CaCO_3 + 2CH_3COOH \rightarrow (CH_3COO)_2Ca + H_2O + CO_2$$

Commercial descalers are often based on weak acids (e.g. methanoic acid, HCOOH) or moderately strong acids (e.g. sulfamic acid, H_3NSO_3).

Ethanoic acid is a weak acid (Chapter 11). Whereas a strong acid such as hydrochloric acid is completely split into ions, ethanoic acid only partially dissociates into ions in water. A dynamic equilibrium is set up in the solution. The solution does contain an excess of hydrogen ions (H^+) over hydroxide ions (OH^-), so the solution is weakly acidic:

ethanoic acid ⇌ ethanoate ions + hydrogen ions

$$CH_3COOH(aq) \rightleftharpoons CH_3COO^-(aq) + H^+(aq)$$

KEY WORD

refluxing: a practical technique using a (reflux) condenser fitted vertically to condense vapours from an experiment back into a flask

Esterification

Alcohols react with organic acids to form sweet-smelling oily liquids known as *esters*. Ethanoic acid will react with ethanol, in the presence of a few drops of concentrated sulfuric acid, to produce ethyl ethanoate. This type of reaction is known as **esterification**. Esterification is a reversible reaction and concentrated sulfuric acid is a catalyst for the forward reaction:

ethanoic acid + ethanol $\xrightarrow{\text{conc. } H_2SO_4}$

$CH_3COOH(l) + C_2H_5OH(l) \rightleftharpoons$

ethyl ethanoate + water

$CH_3COOC_2H_5(l) + H_2O(l)$

Figure 19.23 shows the displayed formulae for the organic reactants and product. The water given off comes in part from the acid and partly from the alcohol.

Ethyl ethanoate is just one example of an ester. The naming and structures of the esters formed by the straight-chain alcohols and acids of up to four carbon atoms are discussed in Chapter 18. The ester family of compounds have strong and pleasant smells. Many of these compounds occur naturally. They are responsible for the flavours in fruits and for the scents of flowers; butyl butanoate, for instance, gives the flavour to pineapples. We use them as food flavourings and in perfumes. The ester group or linkage is also found in complex molecules such as natural fats and oils, and in man-made polyester fibres.

KEY WORD

esterification: the chemical reaction between an alcohol and a carboxylic acid that produces an ester; the other product is water

H—C(H)(H)—C(=O)—O—H + H—O—C(H)(H)—C(H)(H)—H ⇌ H—C(H)(H)—C(=O)—O—C(H)(H)—C(H)(H)—H + H_2O

$CH_3COOC_2H_5$
ethyl ethanoate

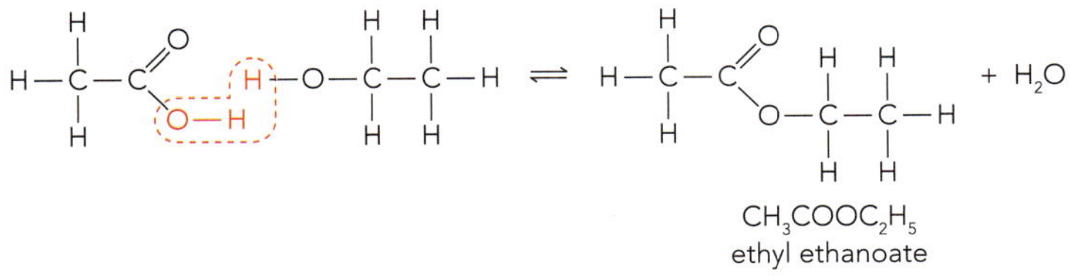

Figure 19.23: The water released during esterification is partly from the acid and partly from the alcohol.

Questions

7 Give two advantages of using ethanol as a fuel.

8 a What are the essentials needed for the production of ethanol by fermentation?

b Name the gas produced during fermentation.

c The molecular formula of ethanol is C_2H_5OH. Draw the displayed formula for ethanol.

d When ethanol is heated with an excess of acidified potassium manganate(VII), it is converted to ethanoic acid:

ethanol (C_2H_5OH) → ethanoic acid (CH_3COOH)

What type of chemical reaction is this?

9 The flavour and smell of foods are partly due to esters. An ester can be made from ethanol and ethanoic acid.

a Name this ester.

b Write a word equation for the reaction between ethanol and ethanoic acid.

SUMMARY

The alkanes are a generally unreactive series of hydrocarbons that burn readily and are used as fuels; they also undergo substitution reactions with chlorine.

The substitution reactions of alkanes with chlorine are photochemical reactions in which ultraviolet light provides the activation energy for the reaction.

Alkenes can be obtained from the catalytic cracking of long-chain alkane molecules present in the fractions obtained from the fractional distillation of petroleum.

Alkenes are more reactive than alkanes and take part in addition reactions with bromine, hydrogen and steam.

Ethanol can be manufactured by fermentation or the catalytic hydration of ethene and that it can be used as a solvent and a fuel.

The two methods of manufacturing ethanol have their own distinctive advantages and disadvantages.

Ethanoic acid reacts as a characteristic acid with metals, bases and metallic carbonates.

Ethanoic acid is formed by the bacterial or chemical oxidation of ethanol.

The reaction of a carboxylic acid with an alcohol produces an ester and water.

PROJECT

Charting linked reactions

When studying the different types of reaction that a compound, or group of compounds, can take part in, it is often useful to draw a reaction summary. Figure 19.24 shows such a chart for the characteristic reactions of acids.

Working in a group, use a large sheet of paper to create a similar chart for an alkene of your choice.

In your class, make sure each group is creating a chart for a different alkene.

Include as many reactions as you can.

Once all groups have finished, explain your chart to the rest of the class. Make notes while other groups are explaining their alkene reaction chart.

Working together as a class, can you link all the charts together to make one global chart for all the reactions covered in this chapter?

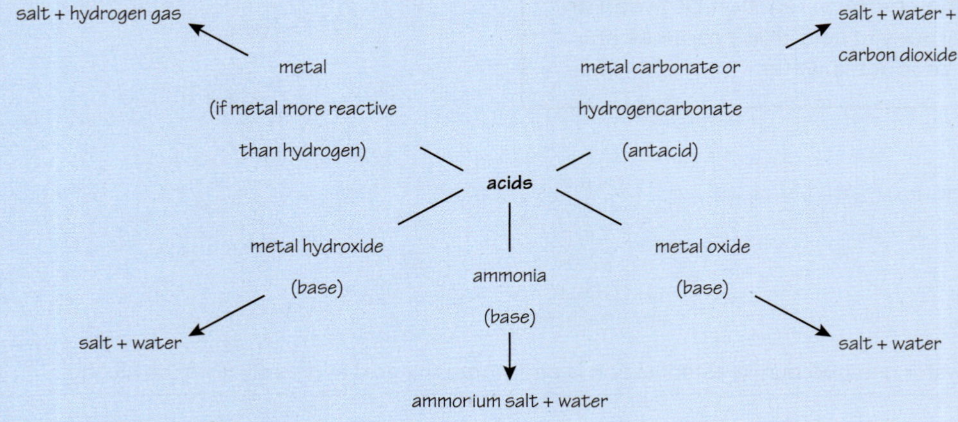

Figure 19.24: Chart of the reactions of acids.

REFLECTION

Devising a chart of reactions such as that for ethanoic acid or the alkenes can be a helpful tool to organise your overall understanding of a topic. Did drawing up the chart help you understand the relationships of ideas within the topic?

EXAM-STYLE QUESTIONS

1 The two compounds whose structures are shown in the figure have similar chemical properties.

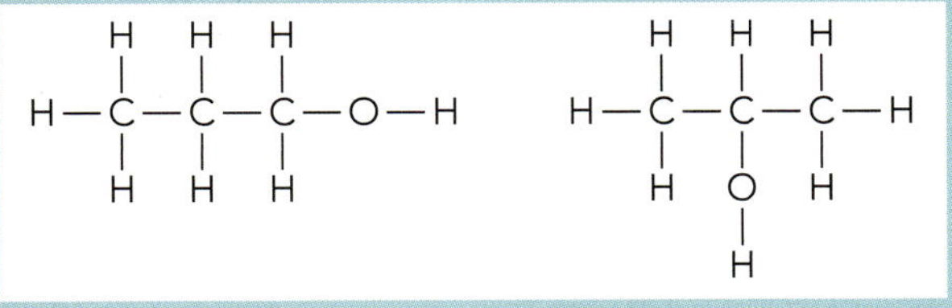

```
    H   H   H                  H   H   H
    |   |   |                  |   |   |
H — C — C — C — O — H      H — C — C — C — H
    |   |   |                  |   |   |
    H   H   H                  H   O   H
                                   |
                                   H
```

What is the reason for this?

Their molecules have the same:

A relative molecular mass

B functional group

C number of oxygen atoms

D number of carbon atoms **[1]**

2 The names of some organic compounds are listed below.

A ethane

B ethanoic acid

C ethanol

D methanoic acid

E methanol

Use the organic compounds A–E to answer the questions that follow. Each compound can be used once, more than once or not at all.

a Which compound is found in vinegar? **[1]**

b Which two compounds react together to form the ester ethyl methanoate? **[2]**

c Which compound is a gas at room temperature and pressure? **[1]**

[Total: 4]

3 Ethane and ethene are both hydrocarbon gases.

a What do you understand by the term 'hydrocarbon'? **[1]**

b Aqueous bromine can be used to distinguish between the two gases. What result would be seen if ethane was bubbled through aqueous bromine? **[1]**

c Ethene reacts with steam to produce ethanol. What conditions are needed to make this reaction happen? **[2]**

CONTINUED

d Give two uses of ethanol. **[2]**

e Ethene is used as a monomer in the formation of poly(ethene). What is the name of the process by which poly(ethene) is made? **[1]**

[Total: 7]

4 A chemical reaction was carried out using the apparatus shown in the figure (ethanoic acid and ethanol).

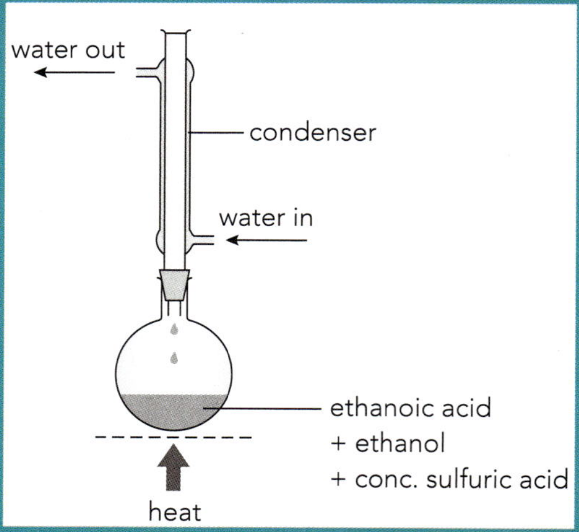

What substance would be produced by this reaction?

A alkene

B condensation polymer

C ester

D salt **[1]**

5 Ethane and ethene are both hydrocarbon gases. They react with halogens in different ways.

a What type of reaction occurs when ethane reacts with chlorine? **[1]**

b What product is formed when ethene reacts with bromine? **[1]**

Ethanol can be formed from ethene by reaction with steam and also by fermentation.

c Give an advantage and a disadvantage of using the fermentation method. **[2]**

d Ethanol can be oxidised to form ethanoic acid. Give the oxidising agent and the conditions for this reaction. **[2]**

e Ethanoic acid and ethanol react together in the presence of a catalyst to form an ester. What is the name and displayed formula of the ester formed? **[3]**

[Total: 9]

COMMAND WORD

give: produce an answer from a given source or recall/memory

SELF-EVALUATION CHECKLIST

After studying this chapter, think about how confident you are with the different topics. This will help you to see any gaps in your knowledge and help you to learn more effectively.

I can	See Topic...	Needs more work	Almost there	Confident to move on
describe alkanes as a series of generally unreactive hydrocarbons that are used as fuels and undergo substitution reactions with chlorine	19.1			
understand that the substitution reactions of the alkanes are photochemical reactions using ultraviolet light to provide activation energy	19.1			
describe how alkenes are obtained from long-chain alkanes present in fractions obtained from the distillation of petroleum	19.1			
describe how alkenes take part in addition reactions with bromine, hydrogen and steam	19.1			
describe how ethanol can be manufactured by fermentation or the catalytic hydration of ethene and can be used as a solvent or a fuel	19.2			
understand the advantages and disadvantages of the two methods of producing ethanol	19.2			
understand how ethanoic acid can be formed by the bacterial or chemical oxidation of ethanol	19.2			
describe the reactions of ethanoic acid with metals, bases and metal carbonates	19.2			
describe the reaction of a carboxylic acid and an alcohol to form an ester	19.3			

Chapter 20

Petrochemicals and polymers

IN THIS CHAPTER YOU WILL:

- learn that the major fossil fuels are coal, natural gas and petroleum
- describe how fractional distillation can be used to separate petroleum into a range of fractions
- define polymers as long-chain molecules built up from smaller molecules (monomers)
- describe how plastics are made from synthetic polymers
- describe the environmental challenges posed by plastics

- identify the structure of an addition polymer or its repeat unit
- understand that polymers can also be formed by condensation reactions
- identify the structure of a condensation polymer or its repeat unit from given monomers
- describe the differences between addition and condensation polymerisation
- describe proteins as natural condensation polyamides, and describe the general structure of amino acids and proteins.

GETTING STARTED

You will be aware of the importance of fossil fuels to the way we currently live. Discuss in groups how the use of these natural resources has affected us economically and practically. What are the three major fossil fuels? How has our use of these resources benefitted our lives? Environmental concerns mean that we need to explore alternatives to some uses of these resources. What changes do you feel are the most crucial?

ENZYMES THAT RECYCLE PLASTICS

In 2016, something remarkable was discovered in a Japanese recycling plant. Among the discarded bottles, bacteria were found that could break down the plastic, PET. Researchers found that bacteria called *Ideonella sakaiensis* can use PET, one of the most common forms of plastic used in making drinks bottles and food containers, as its main source of energy and carbon. The bacteria were isolated from a sample of PET-contaminated sediment near the plastic bottle recycling facility. The researchers found the bacteria had evolved an enzyme (a biological catalyst) capable of turning PET back into its original components. These bacteria, and the enzyme they contain, could revolutionise plastic recycling.

The PET-digesting enzyme, given the name PETase, has been isolated and its protein structure determined (Figure 20.1). The computer model shows how complex the protein is. The protein chain folds in a specific way so that it can catalyse a particular reaction. Remarkably, by manipulating the structure of the protein, researchers improved its ability to digest PET by 20%. Further studies have improved the efficiency of the enzyme even more and made it stable at 72 °C, close to the perfect temperature for fast degradation. This modified enzyme was able to break down the vast majority of one tonne of waste PET bottles in only ten hours.

Enzymes are non-toxic, biodegradable and can be produced in large quantities by microorganisms. More development is needed but it appears enzyme technology has the potential to help society deal with plastic waste.

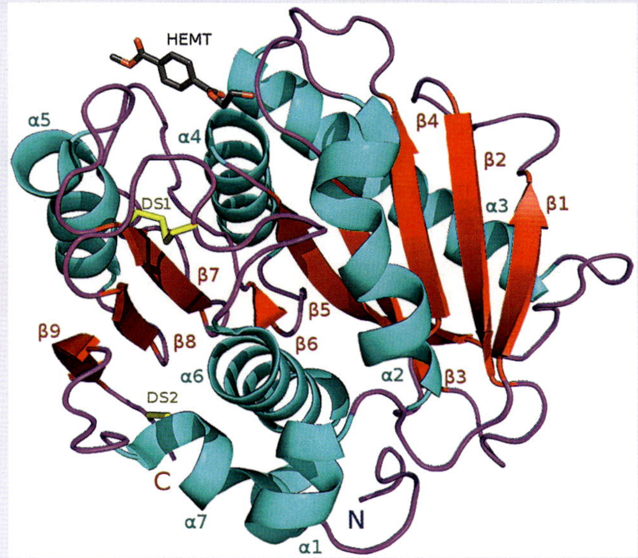

Figure 20.1: Computer model of the structure of the PETase enzyme.

Discussion questions

1 What are the issues that make the recycling of plastics so important to the management of the environment?

2 What are the key features of catalysts, and enzymes in particular, that make them particularly useful in plastic recycling?

20.1 Petroleum and its products

Fossil fuels

Fossil fuels were formed in the Earth's crust from material that was once living. There are three major fossil fuels:

- **coal**
- **natural gas**
- **petroleum**.

These fossil fuels all contain hydrocarbons. Coal comes from fossil plant material. Petroleum (sometimes called crude oil) and natural gas are formed from the bodies of marine microorganisms. Methane is the main component of natural gas but petroleum is a complex mixture of many different hydrocarbons. The formation of these fuels took place over many millions of years. These fuels are therefore a **non-renewable** and **finite resource**.

KEY WORDS

fossil fuels: fuels, such as coal, oil and natural gas, formed underground over geological periods of time from the remains of plants and animals

coal: a black, solid fossil fuel formed underground over geological periods of time by conditions of high pressure and temperature acting on decayed vegetation

natural gas: a fossil fuel formed underground over geological periods of time by conditions of high pressure and temperature acting on the remains of dead sea creatures; natural gas is more than 90% methane

petroleum (or **crude oil**)**:** a fossil fuel formed underground over many millions of years by conditions of high pressure and temperature acting on the remains of dead sea creatures

non-renewable (finite) resources: sources of energy, such as fossil fuels, and other resources formed in the Earth over many millions of years, which we are now using up at a rapid rate and cannot replace

chemical feedstock: a chemical element or compound which can be used as a raw material for an industrial process making useful chemical products

fractional distillation: a method of distillation using a fractionating column, used to separate liquids with different boiling points

Formation of petroleum

Petroleum is one of the Earth's major natural resources, the result of a process that began up to 400 million years ago. When prehistoric marine creatures died, they sank to the seabed and were covered by mud. The change into petroleum and natural gas was brought about by high pressure, high temperature and bacteria acting over millions of years. The original organic material was broken down into hydrocarbons.

Geological movements and pressure created reservoirs of oil and gas. These reservoirs are *not* lakes of oil or pockets of gas. Instead, the oil or gas is spread throughout the pores in coarse rocks such as sandstone or limestone, in much the same way as water is held in a sponge. Oilfields and gas fields are detected by a series of geological and seismic surveys. Once a field is established, production oil rigs can be set up, on land or at sea (Figure 20.2).

Figure 20.2: An oil rig in the Caspian Sea.

Fractional distillation of petroleum

Petroleum is a mixture of many different hydrocarbon molecules. Most of the petroleum that is extracted from the ground is used to make fuel, but around 10% is used as a **chemical feedstock**, or raw material, in the chemical industry. Before the petroleum can be used, the various hydrocarbon molecules are separated by refining. This is done by **fractional distillation** at an oil refinery.

Fraction	Approximate number of carbon atoms in hydrocarbons		Approximate boiling range / °C	
refinery gas	1–4	C_1–C_4	below 25	chain length, b.p. and viscosity decrease from the bottom to the top of the fractionating column; volatility increases
petrol (gasoline)	4–12	C_4–C_{12}	40–100	
naphtha	7–14	C_7–C_{14}	90–150	
kerosene (paraffin)	12–16	C_9–C_{16}	150–240	
diesel oil (gas oil)	14–18	C_{14}–C_{18}	220–300	
fuel oil	19–25	C_{19}–C_{25}	250–320	
lubricating oil fraction	20–40	C_{20}–C_{40}	300–350	
bitumen	over 70	$>C_{70}$	above 350	

Table 20.1: Various petroleum fractions (note the different terms used in different parts of the world for the same fraction).

At a refinery, petroleum is separated into different groups of hydrocarbons that have different boiling points. These groups are known as **fractions**. These different boiling points are roughly related to the number of carbon atoms (chain length) in the hydrocarbons (Table 20.1).

Separation of the hydrocarbons takes place by fractional distillation using a **fractionating column** (or tower). At the start of the refining process, petroleum is preheated to a temperature of 350–400 °C and pumped in at the base of the tower. As it boils, the vapour passes up the tower. It passes through a series of bubble caps, and cools as it rises further up the column. The different fractions cool and condense at different temperatures, and therefore at different heights in the column. The fractions condensing at the different levels are collected on trays. Fractions from the top of the tower are called 'light' and those from the bottom are called 'heavy'. Each fraction contains a number of different hydrocarbons. The individual single hydrocarbons can then be obtained by further distillation. Figure 20.3 shows the separation into different fractions and some of their uses.

KEY WORDS

fractions (from distillation): the different mixtures that distil over at different temperatures during fractional distillation

fractionating column: the vertical column that is used to bring about the separation of liquids in fractional distillation

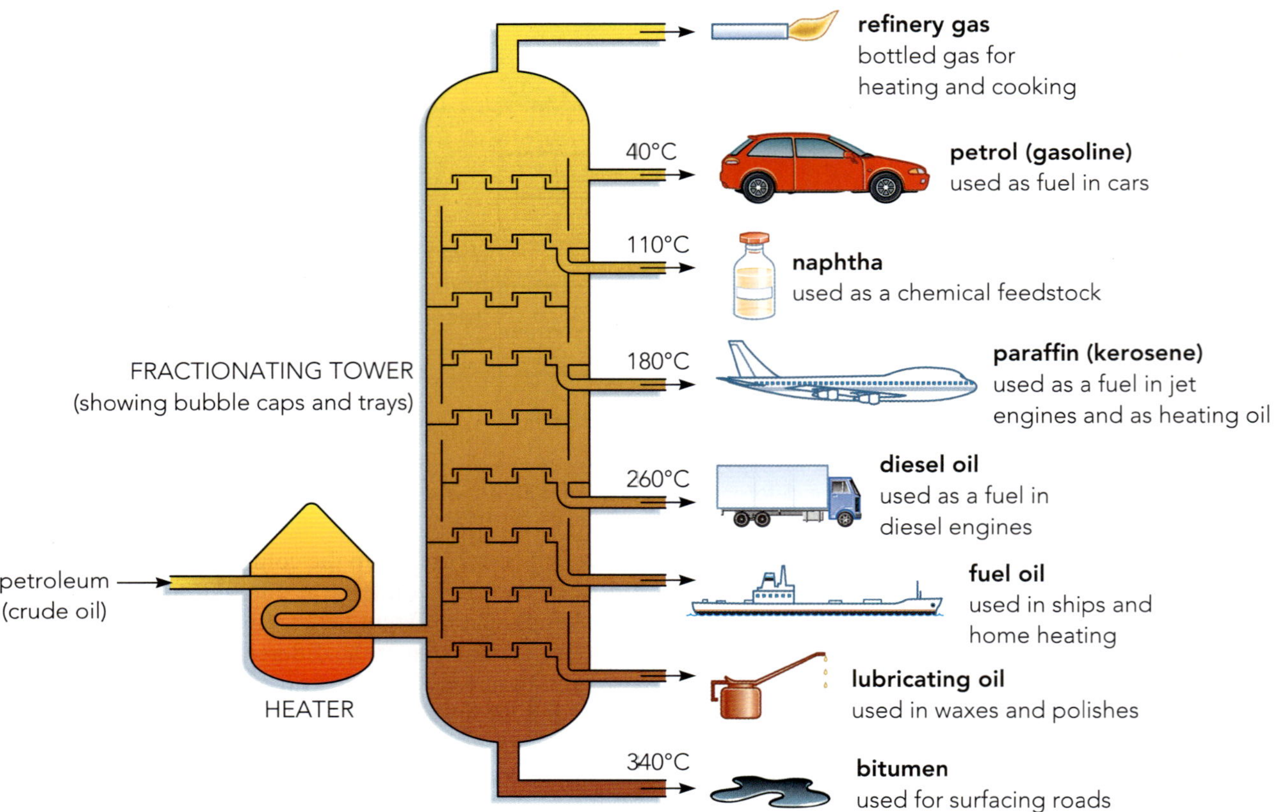

Figure 20.3: Fractional distillation of petroleum in a refinery.

Catalytic cracking

All of the fractions obtained from the distillation of petroleum are useful (see Figure 20.3). However, some are more useful than others, and there is a greater demand for those fractions.

The demand for the various fractions from the refinery does not necessarily match with their supply from the oil (Figure 20.4). The petroleum from different oilfields varies in composition so the values shown here are variable, but the general trend is true. There is a greater demand for the lighter fractions such as petrol that have shorter chain length. The opposite is true for heavier fractions such as paraffin and diesel.

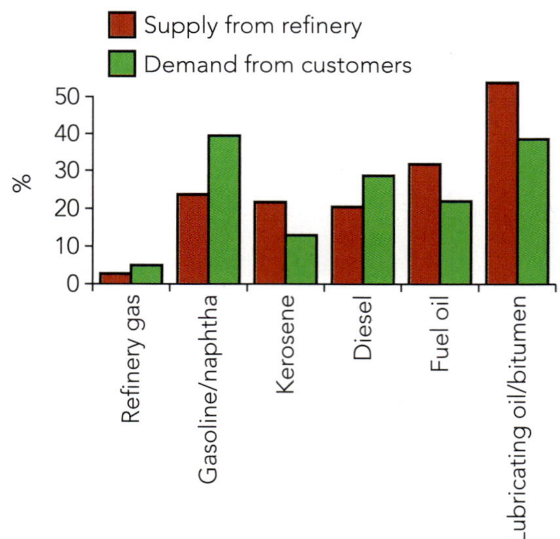

Figure 20.4: Differences in supply and demand for the refinery fractions.

The longer molecules from these heavier fractions can be broken into shorter, more valuable, alkane molecules. Short alkenes and hydrogen are also produced. We introduced this process, known as catalytic cracking ('cat cracking'), in Chapter 19 where we discussed saturated and unsaturated hydrocarbons. All cracking reactions give:

- an alkane with a shorter chain than the original, and a short-chain alkene
- or two or more alkenes and hydrogen.

For example:

decane → pentane + propene + ethene

$C_{10}H_{22} \rightarrow C_5H_{12} + C_3H_6 + C_2H_4$

butane → butene + hydrogen

$C_4H_{10} \rightarrow C_4H_8 + H_2$

Several different reactions can take place during cracking and the mixture of products must be separated by fractional distillation. All the products are useful. The shortened alkanes can be blended with the petrol fraction to enrich the petrol. Hydrogen can be used in the Haber process for making ammonia (Chapter 9) and potentially as a fuel for vehicles. The alkenes are useful as raw materials (a chemical feedstock) for making several important products including addition polymers such as poly(ethene) and poly(propene).

Questions

1. Put the following fractions in order of increasing boiling point: paraffin, diesel, petrol, refinery gas, bitumen, naphtha.
2. State a use for the following fractions from the distillation of petroleum (crude oil): bitumen, fuel oil, diesel, kerosene.
3. Explain what is meant by 'catalytic cracking', and write word and balanced symbol equations to show how decane can be cracked to give octane and ethene.

ACTIVITY 20.1

Refined sudoku

Work in pairs. The aim is to devise and use this logic puzzle to help you learn, and test each other on, the different fractions produced from the distillation of petroleum. Many of you will have solved these puzzles with the numbers 1 to 9. Here we are creating our puzzle using the names of the fractions from the distillation of petroleum and, because we need nine items, another related fossil fuel, natural gas (Table 20.2; Figure 20.5). Alternatively, you can set up a puzzle using the uses of the fractions (the right-hand column).

Figure 20.5: Technicians inspect pipelines at natural gas storage plant.

Natural gas	Methane gas
Refinery gas	Heating / cooking
Gasoline	Fuel – cars
Naphtha	Chemical feedstock
Kerosene	Jet fuel
Diesel oil	Diesel engines
Fuel oil	Fuel – ships
Lubricating oil	Lubricants
Bitumen	Roads

Table 20.2: Fossil fuel products and their uses.

The puzzle is based on a grid and you need to work out the contents of the blank squares (Table 20.3). The rules for filling in the grid are:

- no entry into the horizontal squares can be repeated
- no entry in the vertical squares can be repeated
- there can be no repeat entries in each of the 3 × 3 boxes in the grid.

CONTINUED

Table 20.3 is an example set up for you to try. Complete the puzzle using the names of the fractions. This puzzle is a relatively easy one.

Bitumen	Natural gas	Refinery gas	Lubricating oil	Gasoline	Kerosene			
		Gasoline	Bitumen			Lubricating oil		Kerosene
Lubricating oil								Bitumen
		Natural gas	Fuel oil	Refinery gas				
Gasoline								Naphtha
				Lubricating oil	Naphtha	Fuel oil		
Diesel oil								Natural gas
Refinery gas		Lubricating oil			Natural gas	Kerosene		
			Diesel oil	Fuel oil	Refinery gas	Naphtha	Bitumen	Lubricating oil

Table 20.3: A sudoku puzzle grid using the petroleum refinery fractions.

Having completed the first puzzle, each use a blank grid to set up a puzzle for your partner to try. You can use the fraction names again, or the uses. You can change the difficulty by altering the number of squares you fill in at the start (Table 20.3 has 32 spaces filled; a harder puzzle could have as few as 28 filled spaces).

While working on the puzzle, discuss with your partner the basis of the separation that gives these different fractions:

- Which fractions have the highest boiling points?
- How do properties such as the chain length of the molecules and the viscosity of the fractions vary between the fractions?

A third column of data, such as the length of the carbon chains in the fractions, could be included in the puzzle. Would it be useful for you to set up a puzzle based on that data?

Peer assessment

Once you've completed the puzzle and finished your discussion, use the following checklist to peer assess:

My partner ...

- discussed which fractions had the highest boiling point
- could explain how the chain length of molecules vary
- could explain how the viscosity of the fractions vary.

Is there anything your partner could do to improve their knowledge on this topic?

20.2 Polymers

All living things contain **polymers**. **Proteins**, carbohydrates, wood and natural rubber are all polymers. Synthetic polymers, often called plastics, are to be found everywhere in modern technological societies, ranging from car and aircraft components to packaging and clothing.

Polymers are large organic molecules and are made up of many small repeating units known as **monomers** (Figure 20.6) joined together by **polymerisation**. These units are repeated any number of times from about a hundred to more than a million.

Figure 20.6: Making a chain of beads is similar to joining monomers together to make a polymer.

Some polymers are homopolymers, containing just one monomer. Poly(ethene), poly(propene) and poly(chloroethene) are three examples of homopolymers. Other polymers are copolymers, made of two or more different types of monomer. For example, nylon is made from two monomers and biological proteins are made from 20 different monomers (**amino acids**).

KEY WORDS

polymer: a substance consisting of very large molecules made by polymerising a large number of repeating units, or **monomers**

proteins: polymers of amino acids formed by a condensation reaction; they have a wide variety of biological functions

monomer: a small molecule, such as ethene, which can be polymerised to make a **polymer**

polymerisation: the chemical reaction in which molecules (monomers) join together to form a long-chain polymer

amino acids: naturally occurring organic compounds that possess both an amino ($-NH_2$) group and an acid ($-COOH$) group in the molecule; there are 20 naturally occurring amino acids and they are polymerised in cells to make proteins

Addition polymerisation

The alkene fragments from the catalytic cracking of petroleum fractions produced the starting monomers for the first plastics. Alkenes such as ethene contain a C=C double bond. These molecules can take part in addition reactions (Chapter 19) where the double bond is broken and other atoms attach to the carbons. The double bond in ethene enables many molecules of ethene to join to each other to form a large molecule called poly(ethene) (Figure 20.7), where *n* is a very large number. Note that when you draw out a section of the polymer chain, you must show the open continuation bonds at the end of the section (Figure 20.7).

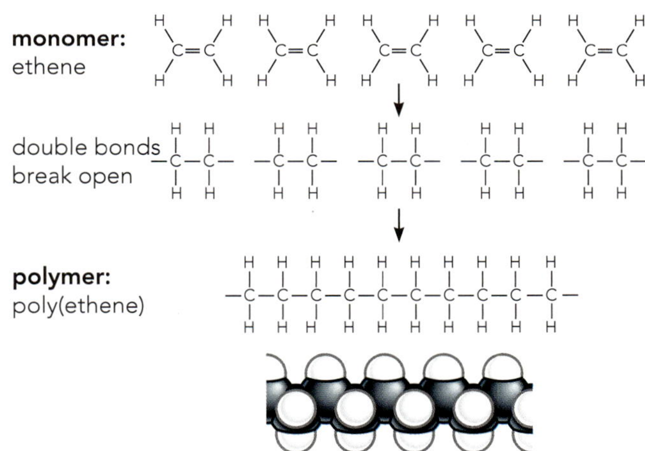

Figure 20.7: The polymerisation of ethene produces poly(ethene).

Various conditions can be used to produce different types of poly(ethene). Generally, a high pressure, a temperature at or above room temperature and a catalyst are needed. The reaction can be summarised by the equation shown in Figure 20.8.

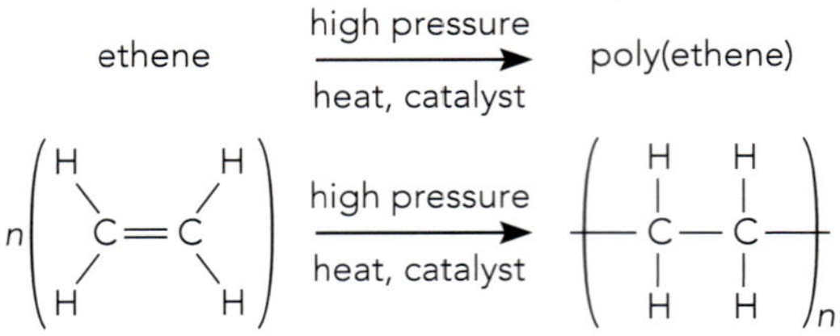

ethene $\xrightarrow[\text{heat, catalyst}]{\text{high pressure}}$ poly(ethene)

$$n\,(\mathrm{CH_2{=}CH_2}) \xrightarrow[\text{heat, catalyst}]{\text{high pressure}} \mathrm{{-}\!\left(CH_2{-}CH_2\right)\!_n{-}}$$

Figure 20.8: Polymerisation of ethene; a very large number of monomers (n) are joined together.

Poly(ethene) was found to be a chemically resistant material that was very tough and durable, and a very good electrical insulator.

Other alkene molecules can also produce addition polymers. Propene will polymerise to produce poly(propene) (Figure 20.9):

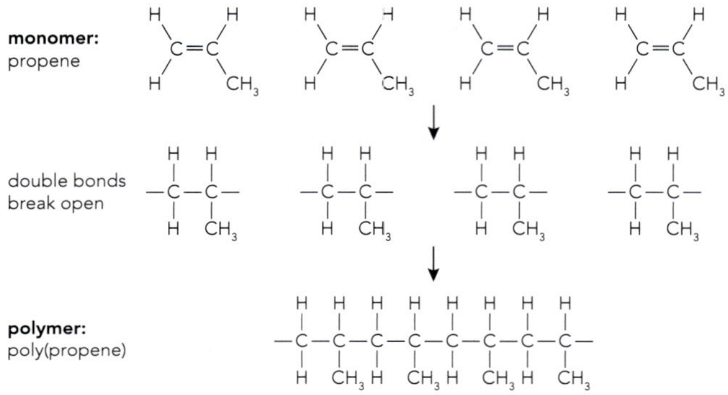

monomer: propene — four molecules of $\mathrm{CH_2{=}CH(CH_3)}$

↓

double bonds break open — four units of $\mathrm{{-}CH_2{-}CH(CH_3){-}}$

↓

polymer: poly(propene) — $\mathrm{{-}CH_2{-}CH(CH_3){-}CH_2{-}CH(CH_3){-}CH_2{-}CH(CH_3){-}CH_2{-}CH(CH_3){-}}$

Figure 20.9: The polymerisation of propene showing four repeating units joining together.

As we saw earlier for the polymerisation of ethene, the reaction to form poly(propene) can be summarised by the equation in Figure 20.10.

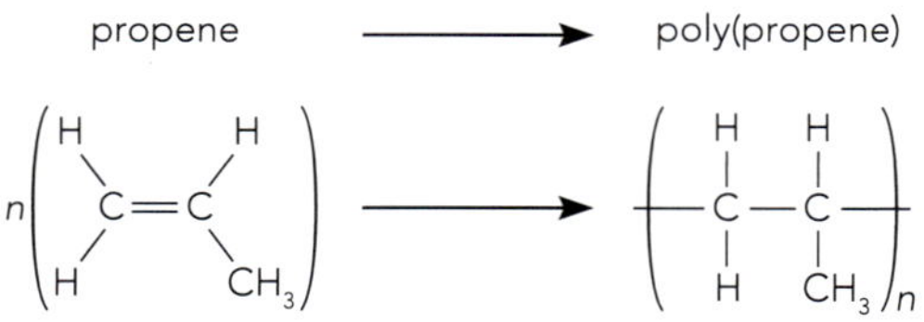

propene ⟶ poly(propene)

$$n\,(\mathrm{CH_2{=}CH(CH_3)}) \longrightarrow \mathrm{{-}\!\left(CH_2{-}CH(CH_3)\right)\!_n{-}}$$

Figure 20.10: The polymerisation of propene monomers to form poly(propene).

The long-chain molecule of poly(propene) is similar in structure to poly(ethene) but with a methyl ($-CH_3$) group attached to every other carbon atom in the chain (Figures 20.9 and 20.10). Be careful, the diagram of the structure of poly(propene) is quite easy to get wrong. It is important to realise that the $-CH_3$ group is a side-group here; the methyl group does not become part of the chain. The chain is formed by the carbon atoms that are joined by the C=C bond in the monomer. Poly(propene) is commonly referred to as polypropylene. Figure 20.11 shows models of poly(propene), poly(chloroethene) and poly(tetrafluoroethene).

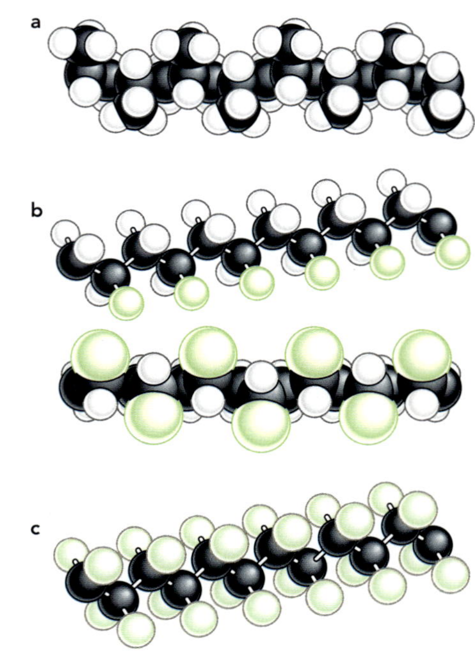

Figure 20.11 a: Poly(propene), **b:** poly(chloroethene) and **c:** poly(tetrafluoroethene).

The equations that summarise the formation of poly(chloroethene) and poly(tetrafluoroethene) are shown in Figure 20.12.

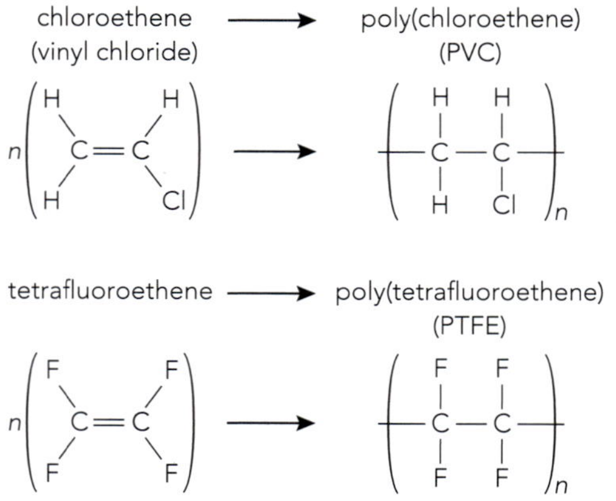

chloroethene (vinyl chloride) ⟶ poly(chloroethene) (PVC)

$$n\,(\mathrm{CH_2{=}CHCl}) \longrightarrow \mathrm{{-}\!\left(CH_2{-}CHCl\right)\!_n{-}}$$

tetrafluoroethene ⟶ poly(tetrafluoroethene) (PTFE)

$$n\,(\mathrm{CF_2{=}CF_2}) \longrightarrow \mathrm{{-}\!\left(CF_2{-}CF_2\right)\!_n{-}}$$

Figure 20.12: The equations for the formation of PVC and PTFE.

You can see from this discussion that there are different ways to represent the formation and structure of an addition polymer. Table 20.4 illustrates how to represent the structure of the polymer formed from a given monomer, and also the reverse process of deducing the monomer used from the structure of the polymer.

Poly(chloroethene) (PVC) is stronger and harder than poly(ethene) and therefore good for making pipes for plumbing. PTFE has some unusual properties: it is very stable at high temperatures and forms a very slippery surface. The properties of some addition polymers are given in Table 20.5.

From monomer to polymer	From polymer to monomer
• draw the monomers with their double bonds next to each other • remove the double bonds and draw single bonds in the space between the monomers	• identify the repeating unit – it will have two carbon atoms with a single bond between them; put a bracket round it • draw that unit as a separate molecule with a double bond in it
monomer: $CH_3(H)C{=}CH_2$ + $CH_3(H)C{=}CH_2$ → polymer: $-CH(CH_3)-CH_2-CH(CH_3)-CH_2-$	polymer: $-CH(CH_3)-CH_2-CH(CH_3)-CH_2-$ → monomer: $CH_3(H)C{=}CH_2$

Table 20.4: How to deduce the relationship between monomer and addition polymer.

Polymer (and trade name(s))	Monomer	Properties	Examples of use
poly(ethene) (polyethylene, polythene, PE)	ethene $CH_2{=}CH_2$	tough, durable	plastic bags, bowls, bottles, packaging
poly(propene) (polypropylene, PP)	propene $CH_3CH{=}CH_2$	tough, durable	crates and boxes, plastic rope
poly(chloroethene) (polyvinyl chloride, PVC)	chloroethene $CH_2{=}CHCl$	strong, hard (not as flexible as polythene)	insulation, pipes and guttering
poly(tetrafluoroethene) (polytetrafluoroethylene, Teflon, PTFE)	tetrafluoroethene $CF_2{=}CF_2$	non-stick surface, withstands high temperatures	non-stick frying pans, non-stick taps and joints
poly(phenylethene) (polystyrene, PS)	phenylethene (styrene) $C_6H_5CH{=}CH_2$	light, poor conductor of heat	insulation, packaging (foam)

Table 20.5: Examples of some widely used addition polymers.

Such synthetic polymers are very versatile. Many, for example poly(propene), are easy to shape by melting and moulding. Poly(propene) is therefore used to make sturdy plastic objects such as crates (Figure 20.13a). However, it can also be drawn out into long fibres for making ropes (Figure 20.13b). Poly(propene) is particularly suited to rope manufacture as it is elastic, stronger than alternative natural materials and sufficiently lightweight that it will float.

Figure 20.13 a: Plastic crates made of poly(propene) on a conveyer belt at a distribution centre. **b:** Different forms of poly(propene) fibres.

The main features of **addition polymers** can be summarised as follows:

- all polymers are long-chain molecules made by joining together a large number of monomer molecules
- addition polymerisation involves monomer molecules that contain a C=C double bond
- addition polymers are homopolymers, made from a single monomer
- during addition, the double bonds open up and the monomer molecules join to themselves to make a molecule with a very long chain (remember to put in the n when representing the structure).

KEY WORDS

addition polymer: a polymer formed by an addition reaction – the monomer molecules contain a C=C double bond

Questions

4 State what is meant by addition polymerisation and give an equation for the formation of poly(ethene) from ethene.

5 Draw the structure of the repeating unit in the following polymers:
   a poly(propene)
   b poly(chloroethene) (PVC)

6 a What common bonding feature must all monomers, whether substituted or not, contain?
   b Draw the structure of the addition polymer made from styrene monomers (show at least three repeating units in your structure).

Condensation polymerisation

A condensation polymer is one that is formed when a condensation reaction takes place between the monomers to join them together in the polymer chain. A **condensation reaction** is one in which two molecules react together to form a new, larger molecule with the elimination of a small molecule, usually water. We will now consider two types of condensation polymer: **polyamides** and **polyesters**.

KEY WORDS

condensation reaction: a reaction where two or more substances combine together to make a larger compound, and a small molecule is eliminated (given off)

polyamide: a polymer where the monomer units are joined together by amide (peptide) links, e.g. nylon and proteins

polyester: a polymer where the monomer units are joined together by ester links, e.g. PET

The reaction between a carboxylic acid and a molecule containing an amine group (–NH_2) produces a compound known as an amide. The link between the two parts of the newly formed molecule is an amide link (Figure 20.14a).

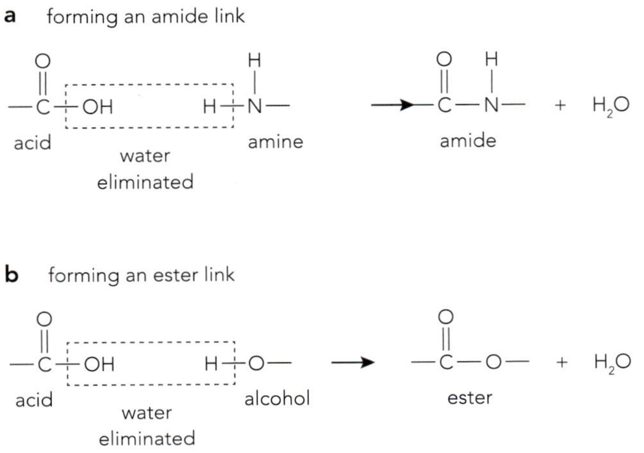

a forming an amide link

acid + amine → amide + H_2O (water eliminated)

b forming an ester link

acid + alcohol → ester + H_2O (water eliminated)

Figure 20.14: Condensation reactions involved in forming polyamides and polyesters. **a:** Forming an amide link. **b:** Forming an ester link.

The reaction between an alcohol and a carboxylic acid (esterification) forms an ester. The link between the two parts of an ester molecule is an ester link (Figure 20.14b).

The polymerisation of a large number of appropriate monomers using one or other of these condensation reactions produces a polyamide or a polyester, respectively.

Nylon (a polyamide)

In the early 1930s, the American company DuPont was conducting research into artificial fibres. Knowledge of silk and wool gave clues as to how protein molecules are built. Wallace Carothers, the leader of organic chemistry at DuPont, imitated the linkage in proteins and produced the first synthetic fibre, 'nylon'. Nylon is a solid when first formed, but it can then be melted and forced through small holes. The long filaments cool, and the fibres produced are stretched to align the polymer molecules and then dried. The fibres can be woven into fabric to make shirts, ties, sheets and so on, or turned into ropes, nets (Figure 20.15) or racket strings. However, nylon is not just made into fibres. It has proved to be a very versatile material and can be moulded into strong plastic items such as gearwheels.

Figure 20.15: Nylon fishing nets.

Nylon is a copolymer of two different monomers, a diamine and a dicarboxylic acid (Figure 20.16). Each monomer consists of a chain of carbon atoms (which are shown here simplified as blocks). At both ends of the monomers are functional groups. An amine group (–NH_2) on the first monomer reacts with a carboxylic acid group (–COOH) on the second monomer to make a link between the two molecules. Each time a link is made, a water molecule is lost.

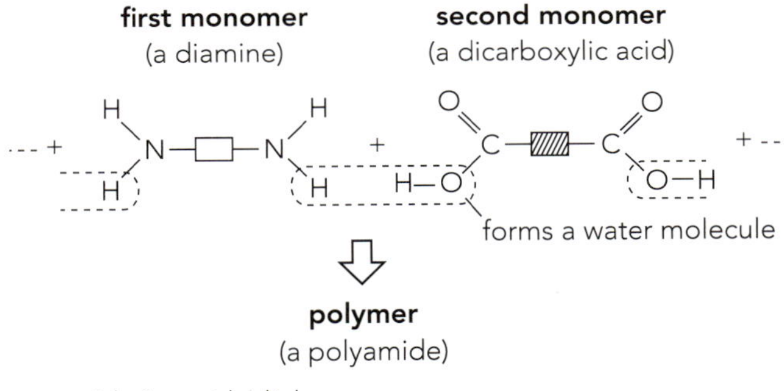

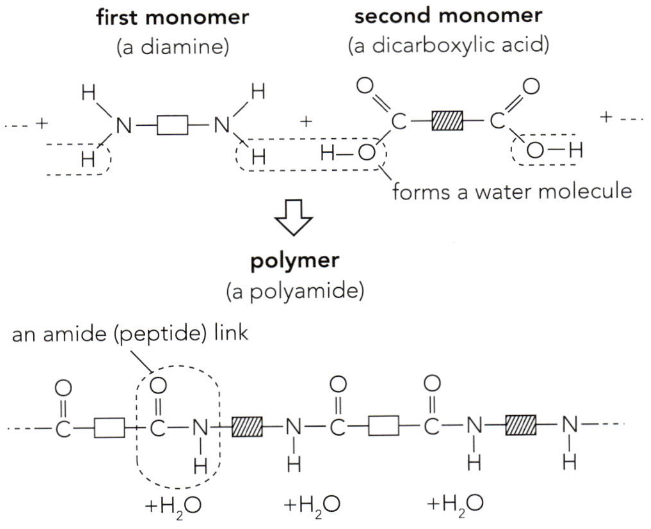

Figure 20.16: The formation of a polyamide (nylon) by condensation polymerisation.

As a result, this type of polymer is known as a **condensation polymer**. Because an **amide link** (or peptide link) is formed during condensation polymerisation, nylon is known as a polyamide.

Take careful note of the structure of the two monomers used to make the synthetic polyamide, nylon (Figure 20.15). The fact that each has the same functional group at either end of the molecule results in an important feature of the polymer produced. This feature is that the direction of each amide link is the opposite of the two link regions immediately on either side of it. It is important to remember this when drawing the structure of a section of the nylon chain (see Figure 20.16).

A version of nylon polymerisation can be carried out in the laboratory (Figure 20.17).

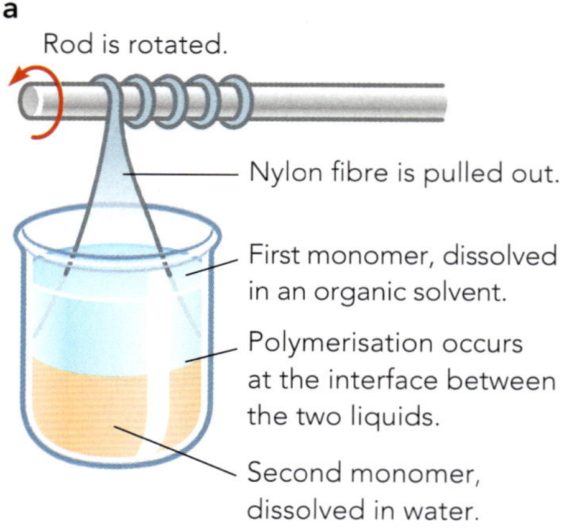

Figure 20.17: Nylon is a polyamide and can be made in the laboratory.

This demonstration shows the production of nylon at the interface between two reactant layers. A solution of the first monomer (the dicarboxylic acid in cyclohexane) is carefully floated on an aqueous solution of the second monomer (the diamine). Nylon forms at the interface and can be pulled out as fast as it is produced, forming a long thread – the 'nylon rope' (Figure 20.17). Remember that you do not need to know the names of the monomers, just the type of functional groups they contain.

PET (a polyester)

Condensation polymerisation can also be used to make other polymers with properties different from those of nylon. *Polyesters* are condensation copolymers made from two monomers. One monomer has an alcohol group (–OH) at each end; this monomer is a diol. The other monomer has a carboxylic acid group (–COOH) at each end; it is a dicarboxylic acid. When the monomers react in condensation polymerisation, an **ester link** is formed, with water being lost each time (Figure 20.18). Note that the directions of the ester links in the synthetic polyester chain are alternating. This feature of synthetic polymers made from bifunctional monomers that each have the same functional group at either end is true for both polyamides and polyesters (see Figures 20.16 and 20.18).

KEY WORDS

condensation polymer: a polymer formed by a condensation reaction, e.g. nylon is produced by the condensation reaction between 1,6-diaminohexane and hexanedioic acid; this is the type of polymerisation used in biological systems to produce proteins, nucleic acids and polysaccharides

amide link (or peptide link): the link between monomers in a protein or nylon, formed by a condensation reaction between a carboxylic acid group on one monomer and an amine group on the next monomer

ester link: the link produced when an ester is formed from a carboxylic acid and an alcohol; also found in polyesters and in the esters present in fats and vegetable oils

first monomer has alcohol functional groups

second monomer has carboxylic acid functional groups

forms a water molecule

polymer (polyester)

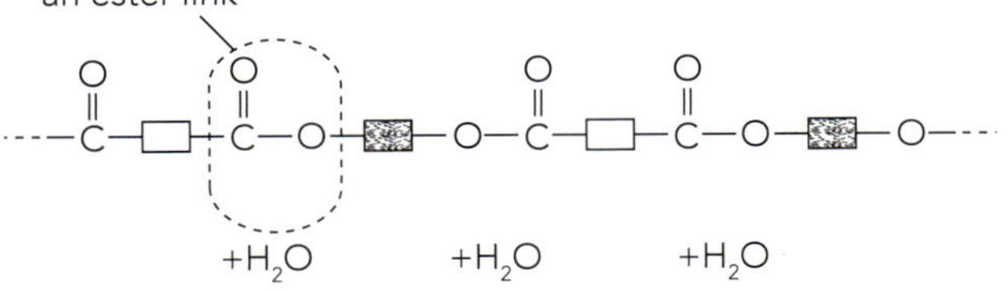

Figure 20.18: The formation of a polyester (PET) by condensation.

PET is the most common polyester polymer used in products all over the world. It can be found in a wide variety of packaging for both food and drink, as well as in clothing fibres (Figure 20.19). Most PET produced globally is for use in synthetic fibres (in excess of 60%) with bottle production accounting for about 30% of global demand.

Figure 20.19: A polyester shirt made from recycled PET bottles.

Comparing addition and condensation polymerisation

Some very useful synthetic polymers have been made by the two types of polymerisation. Both methods take small molecules and make long repeating chains from them. However, there are differences between the two methods. These are summarised in Table 20.6.

	Addition polymerisation	Condensation polymerisation
monomers used	usually many molecules of a single monomer	molecules of two monomers usually used
	monomer is unsaturated, usually contains a C=C double bond	monomers contain reactive functional groups at ends of each molecule
reaction taking place	an addition reaction – monomers join together by opening the C=C double bond	condensation reaction with loss of a small molecule (usually water) each time a monomer joins the chain
nature of product	only a single product – the polymer	two products – the polymer plus water (or some other small molecule)
	non-biodegradable	can be biodegradable (nylon takes 40–50 years)
	resistant to acids	PET can be hydrolysed back to monomers by acids or alkalis

Table 20.6: A comparison of the processes of making synthetic polymers.

Natural polymers (proteins)

All living organisms rely on polymers for their existence. These polymers range from the very complex DNA that makes life itself possible to the more straightforward proteins and carbohydrates that keep living things functioning. The tissues and organs of our bodies are made up of protein. In addition, enzymes, which are responsible for controlling the body's chemical reactions, are proteins. Proteins are built from amino acid monomers. There are 20 different amino acids used, and they each contain two functional groups: an amino group ($-NH_2$) and a carboxylic acid group (–COOH) (Figure 20.20).

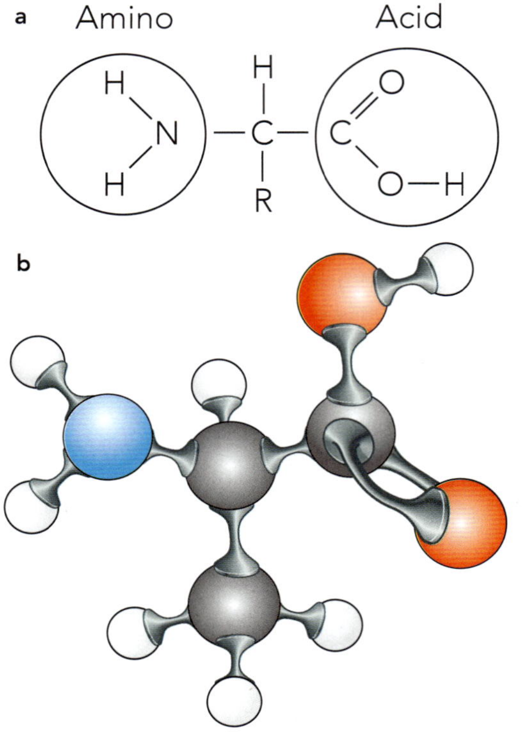

Figure 20.20 a: The structure of an amino acid showing the two functional groups; R represents the different types of side-chain. **b:** The structure of alanine, where R = $-CH_3$ (a methyl group).

The difference between the 20 amino acids used to make proteins lies in the different side-chains (R–). Alanine, shown in Figure 20.20b has a methyl group as the side-chain (R = $-CH_3$). The simplest amino acid is glycine, where R = –H.

Glycine and alanine are the two simplest amino acids. When they react together, an amide linkage (or peptide linkage) is formed to produce a dipeptide (two amino acids joined together) (Figure 20.21). This reaction is a condensation reaction and a water molecule is eliminated in forming the amide link.

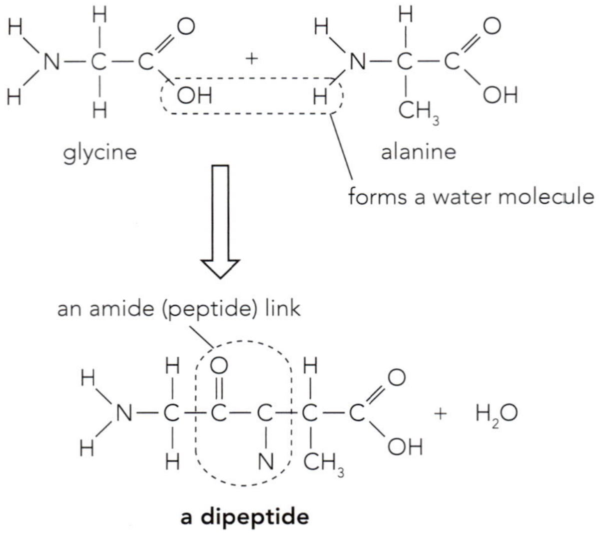

glycine

alanine

forms a water molecule

an amide (peptide) link

a dipeptide

Figure 20.21: Condensation reaction between glycine and alanine.

When this is repeated many times using the different amino acids, a polymer is formed. Short polymers (up to 15 amino acids) are known as peptides. Longer chains are called polypeptides or proteins. The structure of a protein molecule can be drawn where the different amino acids are represented by differently shaded boxes (Figure 20.22).

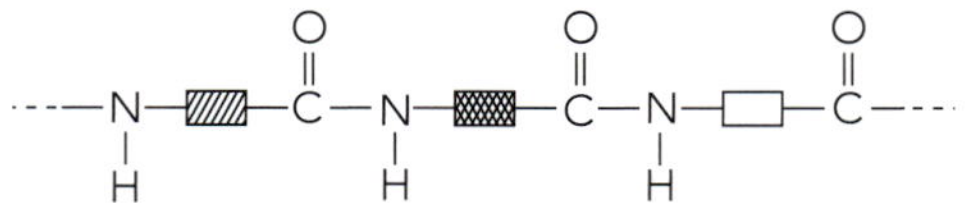

Figure 20.22: Diagram representing the structure of a protein chain, showing the links between three different amino acids.

Questions

7 **a** What are the essential features of condensation polymerisation?

b Name two artificial condensation polymers, and specify the type of linkage present in each.

8 Draw a schematic diagram representing the formation of the polyester, PET. Show the linkage between the monomers (the structure of the monomer is not required and can be represented as a block).

9 Nylon is a synthetic polymer that is held together by the same linkage as protein molecules.

   a What is the name of this linkage?

   b Draw a diagram of the structure of nylon (again, the structure of the monomer is not required and can be represented as a block).

   c Give a major difference between the structure of nylon and of a protein.

20.3 Plastics

Plastics are a group of polymeric materials characterised by their plasticity – their ability to be moulded or shaped under heat and pressure. This property makes plastics incredibly versatile. Many plastics have other advantageous properties, too. They can be extremely strong but also low-density, so are useful for products that need to be tough but lightweight such as soft-drinks bottles. Low conductivity means that plastics are ideal for heat-retaining products, including food containers and insulating foam. Some plastics are also very flexible, which makes them useful for items ranging from plastic bags to garden hoses. Plastics are also increasingly being used in 3D printing, which has applications in many important areas, including the production of respirator masks and the creation of prosthetic limbs (Figure 20.23).

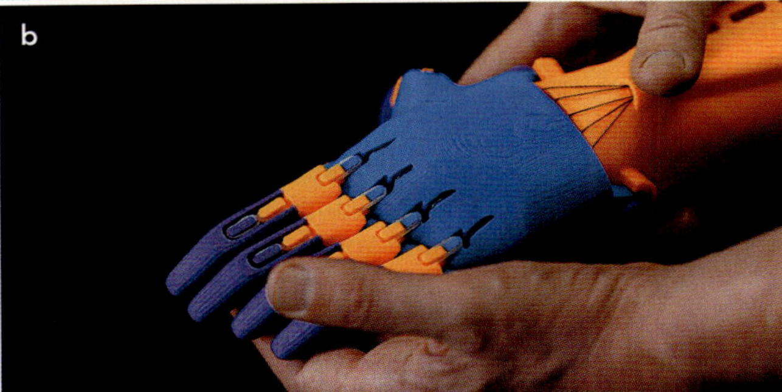

Figure 20.23 a: Respirator masks and **b:** a prosthetic hand made from recycled plastics by 3D printing.

KEY WORD

plastics: polymers that can be moulded or shaped by the action of heat and pressure

The reuse, recycling and disposal of plastic waste

Plastic waste is a common sight all over the world today. Plastic products are light, cheap and resistant to corrosion – characteristics that have resulted in a significant rise in the use of plastics over the past 30 years, as they have been introduced to replace traditional materials in a wide variety of applications. Increased use means increased waste. The problem with plastic waste arises because most plastics are not biodegradable, although research is taking place to find microorganisms that can break them down.

Not all modern plastics are single-use and go straight to waste, however. For example, PET can be used to make items such as soft-drinks bottles, and these can be re-used or recycled. PET also creates bottles that are lightweight, strong and mouldable, which are all benefits for manufacturers. To encourage reuse and recycling of PET drinks bottles, many countries operate deposit schemes, where consumers can return used bottles to 'reverse vending' machines or collection points (Figure 20.24).

Figure 20.24: Commuters in Rome, Italy, trade used plastic bottles for transit credit at a reverse vending machine.

Unfortunately, reuse practices like these are not available or suitable for most plastics. So what should we do with our waste plastic? Recycling is clearly preferable to disposal, both economically and environmentally, as simply throwing plastics away means they go into landfill. However, recycling is not always easy. Most plastic products are made of several different types of plastic, which may need to be recycled in different ways – or which may not be recyclable at all. Most products now have identification numbers and/or symbols on them that indicate their type and whether or not they can be recycled. Optical scanners are also being introduced at recycling plants to help sort the waste, alongside manual sorting methods. Once separated into its different types, the plastic waste can be recycled using the most suitable method. Granulated plastic produced from PET plastic bottles can be supplied to manufacturers for melting and re-moulding into useful products (Figure 20.25).

Figure 20.25: Plastic pellets (PET) produced at a recycling centre; these pellets (nurdles) will be sent for processing and reuse.

As mentioned earlier (see Table 20.6) the ester linkage that joins the monomer units in the polyester (PET) can be broken down by acid or alkaline **hydrolysis**. Hydrolysis is the reverse of the condensation reaction that produced the polymer. The monomers produced by this hydrolysis could be re-polymerised and the new polymer moulded into new objects.

KEY WORD

hydrolysis: a chemical reaction between a covalent compound and water; covalent bonds are broken during the reaction and the elements of water are added to the fragments; can be carried out with acids or alkalis, or by using enzymes

Environmental challenges

The use of plastics and, at times, our casual disposal of them poses a number of environmental problems. In recent years, there has been increasing awareness of the need to adapt our use of these advantageous materials in the following ways:

- to reduce the level of plastic packaging wherever possible
- to avoid the use of 'single-use plastic' (and the subsequent littering with it)
- to reuse and recycle wherever possible.

There are alternatives to recycling for the disposal of plastic waste but all have their detrimental effects.

- Incineration can be used to burn plastic waste, although care must be taken not to release toxic fumes into the air. Incineration of PVC, for instance, can release acidic fumes of hydrogen chloride. Open-air burning of plastic occurs at lower temperatures, and normally releases toxic fumes containing dioxins and furans. Controlled high-temperature incineration, above 850 °C, must be carried out to break down these toxins. Municipal solid waste incinerators also normally include flue gas treatments to reduce pollutants further.
- Disposal in landfill sites suffers from the problem that most plastics are not biodegradable and therefore the plastic waste increasingly fills the space available. This imposes expansion problems on the site and uses up natural resources. Research is being carried out to produce plastics that are biodegradable or photodegradable (broken down by sunlight). However, in all these cases there is the problem of degradation products leaching into the groundwater of a locality used for a landfill site.
- Accumulation in oceans. Plastic pollution of the oceans is a major problem that has been highlighted in recent years (and is discussed in the context of other types of pollution in Chapter 17). Increased awareness of this problem has been part of the background to the drive to reduce the level of 'single-use plastics' in our shopping and packaging. Images of ocean wildlife and sea birds harmed by interaction with waste plastic debris in the oceans have highlighted the problem. The importance of the problem means that projects to collect plastic waste from the oceans have been started in various regions (Figure 20.26).

Figure 20.26: Debris removal from the ocean around the Hawaiian Islands (NOAA, National Ocean and Atmospheric Administration, USA).

The debris can be most evident when it is washed up on beaches after a storm. However, concern has been developed further with the realisation of the presence of 'ocean garbage patches' in various areas of the major oceans of the world. The major currents that occur in our oceans cause circulation effects or 'gyres' in various regions. There are five major ocean-wide gyres – the North Atlantic, South Atlantic, North Pacific, South Pacific and Indian Ocean gyres.

One unwanted effect of these major current rotations is that a great amount of the plastic waste in the oceans is collected into what have been called ocean 'garbage patches (or trash vortices)'. Figure 20.27 shows the location of the two Great Pacific garbage patches, the Western Garbage Patch (left), and the Eastern Garbage Patch (right), in the North Pacific Ocean. These patches contain exceptionally high concentrations of plastics, chemical sludge, wood pulp and other debris that have become trapped by the flows of the North Pacific Current. As the wind and waves stir up this huge mix of debris, it spreads widely – not only for thousands of kilometres across the surface of the ocean, but polluting a large region immediately below the surface, too. Since the discovery of the Eastern Garbage Patch in 1987 similar regions have been found in the South Pacific, North Atlantic and Indian Ocean gyres.

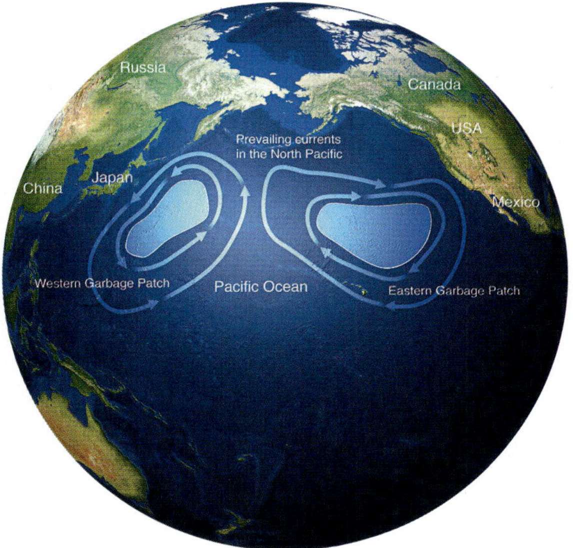

Figure 20.27: The location of the Great Pacific garbage patches.

While high concentrations of items of litter – bottles and plastic bags – can be found in the ocean, much of the debris is actually small pieces of floating plastic not immediately visible to the naked eye. This type of pollution arises in the form of **microplastics**. Some of the small pieces of plastic in the ocean arise from the break-up of larger objects by the oceans themselves but much of them originate from processes in the plastics industry. These microplastics take the form of **nurdles** and **microbeads** (Figure 20.28a and b).

KEY WORD

microplastics: small pieces of plastic less than 5 mm in length that enter natural ecosystems from a variety of sources including cosmetics, clothing and industrial processes: **nurdles** and **microbeads** are different types of microplastic

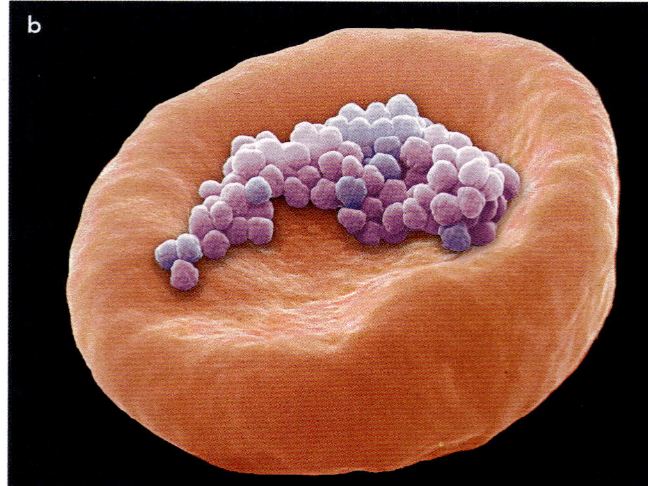

Figure 20.28 a: Nurdles (plastic beads) found on a beach. **b:** A scanning electron microscope photograph of plastic microbeads from cosmetics on a single red blood cell to show the scale.

Nurdles are pre-production plastic pellets and resin materials (usually under 5 mm in diameter). Nurdles serve as raw material in the production of plastic products and these small plastic beads find their way to coastal waterways and oceans.

Microbeads are the much smaller plastic spheres (less than 1 mm in diameter) found in cosmetic facial scrubs, shower gels and toothpastes. As with nurdles, they can accumulate in the world's oceans, lakes and estuaries, harming marine life and entering the food chain. They are an avoidable form of plastic pollution and in recent years many countries have introduced a ban on the use of microbeads in cosmetics.

REFLECTION

The use and misuse of plastics is a major environmental issue. Do you feel confident that you can develop strategies that will enable you to form reasoned judgements and take actions where necessary?

Questions

10 What are the dangers in simply burning plastics to dispose of them?

11 Explain the difference between the reuse and the recycling of plastics.

12 What are two problems that need to be overcome in the effective recycling of plastics?

SUMMARY

The major fossil fuels are coal, natural gas and petroleum, with methane being the main constituent of natural gas and petroleum being a mixture of hydrocarbons.
Fractional distillation can be used to separate petroleum into a range of useful fractions, with these fractions progressively varying in chain length, volatility, viscosity and boiling point.
The fractions from the distillation of petroleum each have distinctive uses.
The longer chain alkanes of the heavier fractions can be shortened by catalytic cracking, and this process also yields hydrogen and alkenes for use in other industrial processes.
Polymers are long-chain molecules built up from a large number of monomer molecules, with poly(ethene) as an example of the addition polymerisation of ethene monomers.
The structure of an addition polymer or its repeat unit can be identified from the formula of a given alkene.
Polymers can also be formed by condensation reactions, with polyamides (nylon) and polyesters (PET) being two major forms of condensation polymer.
The structure of a condensation polymer or its repeat unit can be identified from given monomers.
Addition and condensation polymers depend on different types of monomers, linkage regions and means of formation.
PET can be converted back to its monomers and re-polymerised.
Proteins are natural condensation polyamides formed from amino acid monomers.
Plastics are made from synthetic polymers and the properties of plastics raise problems for their disposal.
There are environmental challenges posed by the disposal of plastics in landfill sites, toxic fumes from incineration and the accumulation of plastic debris in the oceans.

PROJECT

Making sense of plastics

Figure 20.29: A technician using a 3D printer.

Figure 20.29 shows a 3D printer in use; a positive use of plastics to produce an object for a specific use.

Work in pairs to design a campaign brochure that outlines the advantages and disadvantages of using plastics. You can search the internet to help you.

Try to distinguish the advantages/disadvantages that are:

- part of the nature of the plastics themselves
- the result of our misuse of the materials.

When everyone has finished, come together as a class to discuss the content of the brochures you have created. Consider the following question:

To what extent do you agree that there should be a ban on single-use plastic items?

EXAM-STYLE QUESTIONS

1 Butane, ethanol and hydrogen are all used as fuels or being developed as such. Which of these substances produce both carbon dioxide and water when used as fuels (see the table)?

	butane	ethanol	hydrogen
A	yes	yes	yes
B	yes	no	yes
C	no	yes	no
D	yes	yes	no

[1]

2 Some of the fractions separated from petroleum are:

A bitumen
B kerosene
C petrol (gasoline)
D refinery gas

Use the fractions A–D to answer the following questions.

Each fraction may be used once, more than once or not at all.

a Which fraction is not used as a fuel? [1]

b Which fraction has the highest boiling point? [1]

c Which fraction is used to power motor vehicles? [1]

[Total: 3]

3 a Petroleum is a mixture of compounds from the same homologous series. What is the name of this series? [1]

b Petroleum can be separated into fractions by fractional distillation. Which physical properties of the compounds in each fraction make this separation possible? [1]

c What is the main use of the kerosene fraction? [1]

d The heavier fractions are sometimes cracked to form alkenes.

i What conditions are needed for this process? [2]

ii What product is formed during cracking which is not a hydrocarbon? [1]

e A long-chain molecule from petroleum was cracked to form two different alkenes.

Complete the equation.

$C_{10}H_{22} \rightarrow C_4H_8 + 2\_\_\_\_\_ + \_\_\_\_$ [2]

[Total: 8]

4 Plastics are used in every aspect of life. Many thousands of tonnes of plastics are used every day.

Plastics cause a pollution problem both on land and in water unless they are biodegradable.

a What is the meaning of the term 'biodegradable'? [2]

b What happens to most plastics that are discarded in rubbish? [1]

CONTINUED

c Why can plastics cause problems if they get into the oceans? [2]

d Some plastics are burnt to produce energy. Why does this cause pollution problems? [1]

[Total: 6]

5 Nylon is a polymer. Which pair of words in the table correctly describes the type of polymer nylon is?

	Type of polymer	Type of polymerisation
A	polyamide	addition
B	polyester	condensation
C	polyamide	condensation
D	polyester	addition

[1]

6 a The two compounds shown in the figure can be reacted together to form a polyester.

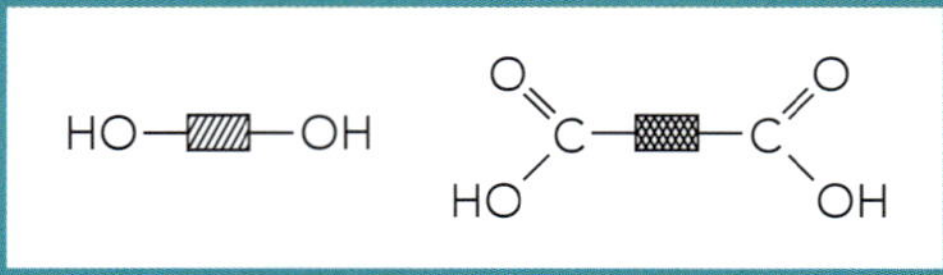

HO—▨—OH   HOOC—▩—COOH

i What other compound is formed in addition to the polyester? [1]

ii What type of polymerisation is this? [1]

b The compound shown is an amino acid.

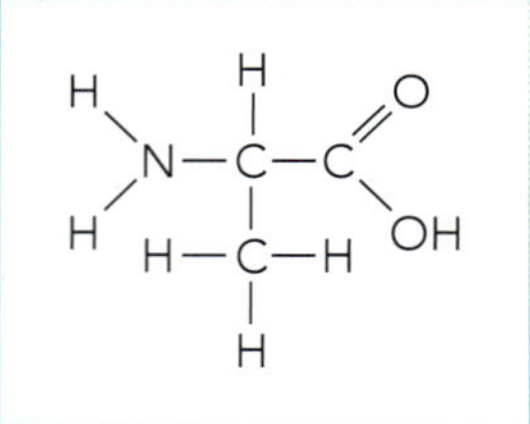

H_2N—CH(CH_3)—COOH

i What is the name of polymers formed from amino acids? [1]

ii Why are they known as natural polymers? [1]

[Total: 4]

SELF-EVALUATION CHECKLIST

After studying this chapter, think about how confident you are with the different topics. This will help you see any gaps in your knowledge and help you to learn more effectively.

I can	See Topic...	Needs more work	Almost there	Confident to move on
name the major fossil fuels as coal, natural gas and petroleum, with methane the main constituent of natural gas and petroleum being a mixture of hydrocarbons	20.1			
describe how petroleum can be separated into useful fractions by fractional distillation, with these fractions progressively varying in chain length, volatility, viscosity and boiling point	20.1			
describe how each different fraction from the distillation has its distinctive uses	20.1			
understand that alkenes and hydrogen can be obtained from catalytic cracking of the longer alkane molecules in certain fractions, with the shortened alkanes being used to enrich the economically more important fractions	20.1			
define polymers as long-chain molecules built up from large numbers of monomer molecules, with poly(ethene) as an example of addition polymerisation of ethene monomers	20.2			
identify the structure of an addition polymer or its repeating unit from the formula of a given alkene	20.2			
understand that polymers can also be formed by condensation reactions, with polyamides (nylon) and polyesters (PET) being two major forms of condensation polymer	20.2			
identify the structure of a condensation polymer or its repeat unit from given monomers and draw diagrams of the structures of nylon and PET	20.2			
describe the differences between addition and condensation polymerisation	20.2			
describe proteins as natural condensation polyamides formed from amino acid monomers, and describe the general structure of amino acids and proteins	20.2			
describe how plastics are made from synthetic polymers and how their properties raise problems for their disposal	20.3			
identify the environmental challenges posed by the disposal of plastics in landfill sites or by incineration, and the accumulation of plastic debris in the oceans	20.3			

› Chapter 21

Experimental design and separation techniques

IN THIS CHAPTER YOU WILL:

- develop an understanding of experimental design
- learn how to select the most appropriate methods and apparatus to use in an experiment together with their possible advantages and disadvantages
- name appropriate apparatus for measuring different variables
- explore and identify techniques to separate and purify different substances
- discover how melting and boiling points can be used to identify and assess the purity of a substance
- describe how chromatography can be used to separate mixtures of soluble coloured substances

CONTINUED

state and use the equation to determine the R_f value.

describe how paper chromatography can be used to separate mixtures of soluble colourless substances using a locating agent.

GETTING STARTED

Chemists need to be able to design experiments carefully and safely. Work in groups of three of four to consider the scenario below and discuss the questions that follow.

You have been asked to plan an experiment to analyse how temperature affects the rate at which a new drug can be produced. The drug is produced as a white precipitate (insoluble salt) when solutions of two chemicals A and B are mixed together.

1 What should you include as you plan your experiment?

2 Can you think of any general hazards that would need to be taken into consideration?

3 How would you reduce any risk?

4 What measurements would you need to make?

SAFE DRINKING WATER FOR ALL

Safe drinking water is considered a basic human right by the United Nations. Many people take access to clean drinking water for granted. However, it is estimated that around 800 million people lack even a basic drinking-water service. The problems are being compounded with growing populations and the pressures caused by climate change increasing the demand for clean water. Traditional supplies such as reservoirs and wells cannot meet the demand and so research has focused on alternative methods to provide clean water. Many of these approaches are based on simple separation techniques.

For small-scale production it is possible to distil water. This approach involves boiling and then condensing the water. It is therefore energy intensive, and if fossil fuel is used it is also unsustainable. This problem can be overcome by using small solar-powered distillation units. The stills are simple black-bottomed containers filled with water and covered with clear glass or plastic. Sunlight is absorbed by the black material and increases the rate of evaporation. The evaporated water is trapped by the clear cover and funnelled away. Most pollutants do not evaporate and remain in the container. The stills are usually used in remote areas with limited freshwater, but only produce small volumes of water and require access to sunlight.

An alternative, larger-scale approach is desalination, which removes salt and other minerals from water using distillation or membrane filtration (reverse osmosis). Desalination is energy intensive and has an environmental impact as the toxic brine produce can damage coastal and marine ecosystems. The process works well for countries with limited freshwater but with access to seawater. The Middle East has most of the world's desalination capacity, with Oman producing over nearly 90% of its drinking water from desalination (Figure 21.1).

Figure 21.1: Chemists working in a laboratory at a desalination plant in Oman.

A novel approach to supplying clean water is to use mini personal water filtration systems ('filter straws') to drink directly from natural water sources. The filter straws contain many hollow fibre micro-tubes that trap contaminants while allowing clean water to pass through. The straws are highly effective at removing not only small particles of mud and sand but also filter out harmful microbes.

CONTINUED

Discussion questions

1 What are the strengths and weaknesses of using portable distillation units and desalination to produce drinking water?

2 What does the filter straw need to be able to remove from the water? Would you drink untreated river water using one of these straws?

21.1 Experimental design

Planning an experimental investigation

There are several steps to consider when planning an experimental investigation (Figure 21.2).

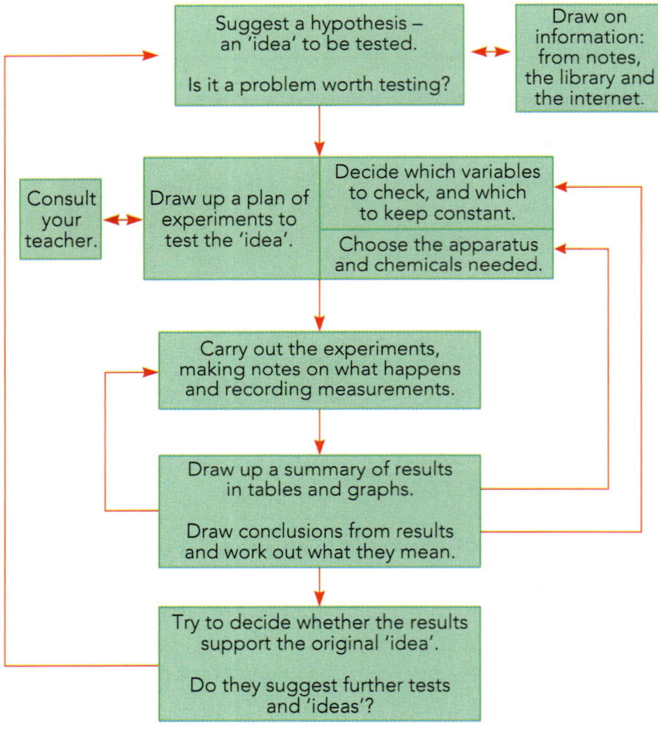

Figure 21.2: Steps involved in planning an experimental investigation.

The choice of apparatus requires careful thought. It is also necessary to consider the level of accuracy, how to control variables, sources of error and any steps to minimise safety concerns.

At the design stage you should plan which equipment to use and write a method detailing how to use it. You need to evaluate alternative approaches, taking into account any possible advantages and disadvantages of each approach, that may help you to improve the method.

Selecting the correct apparatus

The type of apparatus used in an experiment will depend on the scale and the **accuracy** of the results required. The scale of an experiment is a measure of the amount of product needed. A small-scale experiment as carried out in a laboratory may produce milligram or gram amounts of a product, whereas large-scale experiments of the type used in industry will produce many tonnes of material. The accuracy is a measure of how close a result is to its true value. The apparatus chosen for an experiment should give **precise** results (results that are close to each other).

You need to be able to identify common pieces of laboratory equipment (Figure 21.3), and should be aware of the purpose and accuracy of the apparatus.

KEY WORDS

accuracy: how close a value is to the true value

precision: the degree to which repeat measurements are consistent (close to each other)

Equipment for measuring time, mass, temperature and pH

Time is measured using a stopclock or stopwatch. Measurement of time is particularly useful when determining the rate of a chemical reaction (Chapter 8). Digital stopwatches are capable of reading to two decimal places, i.e. to within 1/100 second (0.01 seconds).

Mass can be measured approximately in spatula amounts, e.g. add one spatula of a compound to $2\,cm^3$ of acid, or more accurately by using a top-pan balance.

Figure 21.3: Common laboratory equipment.

A simple balance may have a low **resolution**, e.g. 0.1 g, whereas more complex balances may have a higher resolution of 0.001 g or greater. The resolution is the smallest change in a measurement that can be detected.

Temperatures in practical work are measured in degrees Celsius (°C) using a thermometer. Often these are simple liquid thermometers (spirit or mercury), which can be read to the nearest degree. It is also possible to use a digital temperature probe, which can provide a higher level of resolution, e.g. to within 0.1 °C.

pH can be measured using either universal indicator paper or more accurately with a digital probe (Figure 21.4). A digital probe may read to 0.1 or even 0.01 on the pH scale, giving significantly higher resolution. However, **calibration** may be required before using a digital meter; this involves using known standards to check values.

Figure 21.4: Digital pH meter being used by a research scientist to test water quality.

KEY WORDS

resolution: the smallest division on the instrument, e.g. this could be 1 mm on a ruler or 0.1 g on a balance

calibration: the process of checking that the device gives accurate values by using it to read samples with known values

Equipment for measuring volume

Volume can be measured using a beaker, a measuring cylinder, a syringe, a volumetric pipette or a burette. The equipment used depends on the accuracy needed. For example, a beaker could be used to prepare a water-bath containing 100 cm^3 of water, but it could not be used to

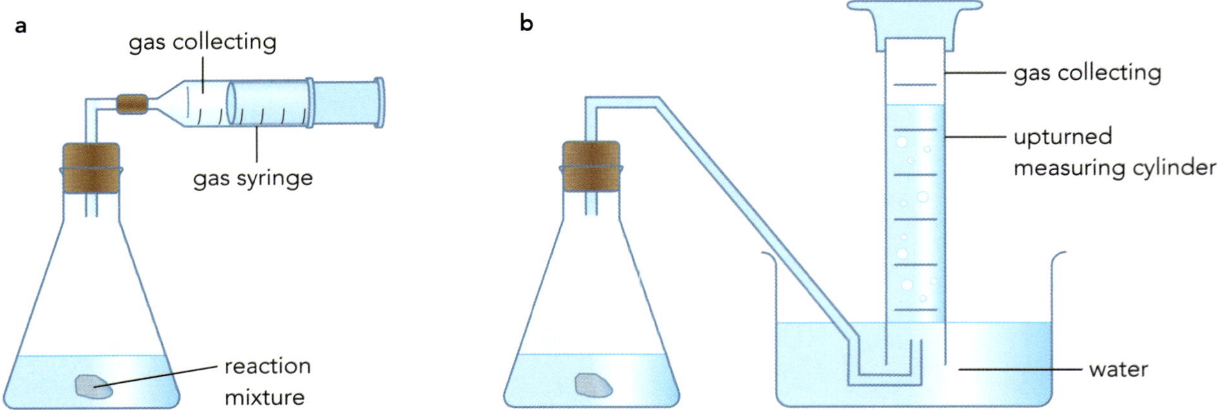

Figure 21.5: Collecting and measuring the volume of a gas: **a:** in a gas syringe and **b:** over water in an inverted measuring cylinder.

accurately measure 25.0 cm^3 of alkali for use in an acid–base titration (Chapter 22). For general use, a measuring cylinder will often provide a sufficient level of accuracy. A volumetric pipette is the most accurate piece of apparatus for measuring a fixed volume of liquid (typically volumetric pipettes are either 10.0 cm^3 or 25.0 cm^3). A burette is the most accurate piece of apparatus for measuring a variable volume (often between 0 and 50.0 cm^3).

Some experiments produce a gas. The volume of gas released during a reaction can give useful information for quantitative purposes as well as when investigating rates of reactions (Chapter 8). Most experiments use one of two standard methods for capturing a gas: direct capture into a gas syringe or by displacement of water in a measuring cylinder (Figure 21.5). Care must be taken if the gas is bubbled through water as some gases (e.g. CO_2) are sparingly soluble in water.

Accuracy and precision of the data recorded

The data obtained from an experiment should be accurate; this means that the value obtained is close to the true value. For example, in an acid–base titration, the true volume of acid needed to neutralise 25.0 cm^3 of alkali is 14.10 cm^3. When carried out, one student recorded values of 12.45 cm^3, 12.55 cm^3 and 12.5 cm^3. This data is precise as the repeated values show only a small difference but is not accurate. Another student recorded values of 14.30 cm^3, 14.00 cm^3 and 14.15 cm^3. This data is not precise as the repeated values show larger differences, but it is more accurate as the mean is closer to the true value. In the diagrams in Figure 21.6, the true value is represented as the centre spot of the circle.

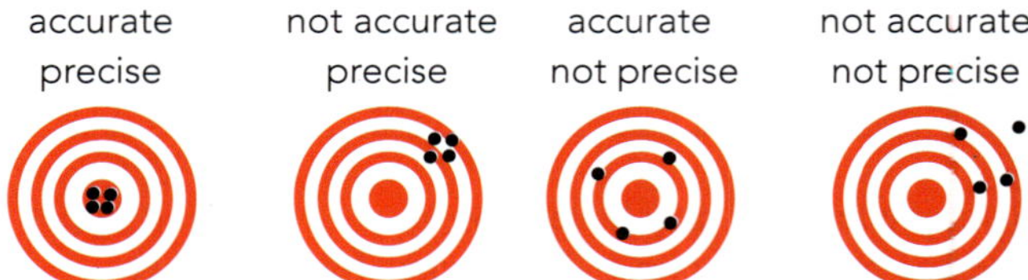

Figure 21.6: Accuracy versus precision.

Repeatable and reproducibility

When chemists carry out reactions, they should consider the different factors that influence their results. Data needs to be both **repeatable** (reliable) and **reproducible**. Data is repeatable when the same person has conducted an experiment several times and obtained similar results. Data can only be reproducible when several different people have performed an experiment with similar variables and obtained similar results.

KEY WORD

repeatability: where an experiment (or series of experiments) can be repeated using the same method and obtain **reproducible** (similar) results

Variables

To ensure experimental work is both repeatable and reproducible, the variables need to be considered when planning an experiment. There are three main types of variable:

- controlled variable
- independent variable
- dependent variable

KEY WORDS

control variable: a variable that is kept the same during an investigation

independent variable: the variable that is altered during a scientific investigation

dependent variable: the variable that is measured during a scientific investigation

A controlled variable is a variable that is kept constant throughout an experiment. As an example, consider a student who wanted to investigate how the type of metal changed the rate of reaction when added to sulfuric acid. To give repeatable results, as the rate of a reaction is dependent on both concentration and temperature, they would have to use the same starting temperature and concentration of the acid in each experiment. Temperature and concentration are therefore controlled variables.

The independent variable is the variable that is changed during an experiment. In the experiment investigating how the type of metal changed the rate of reaction, the type of metal is the independent variable. The independent variable forms the *x*-axis (horizontal axis) of a graph. The dependent variable is the variable that is measured during an experiment, e.g. the rate of the reaction, and this will form the *y*-axis (vertical axis) of a graph.

Sources of error and display of observations

As part of the experimental design, it is useful to consider the sources of error. Almost every measurement has some degree of error or uncertainty in it. Some pieces of apparatus are more accurate than others.

There are two main types of sources of error: random errors and systematic errors. The impact of random errors can be reduced by repeating an experiment several times, removing any anomalous data and taking an average. Systematic errors mean the data is wrong by the same amount each time. An example of this is a zero error, i.e. the apparatus is producing a value when it should not. For example, if there is no mass on a top-ban balance the reading should be 0.00 g. Systematic errors can be reduced by 'taring' the apparatus or subtracting the zero error from all results.

KEY WORDS

random errors: these are unpredictable variations in results caused by factors such as human errors

systematic errors: these are consistent errors that may arise because of a problem with the experimental design or in a piece of equipment being used

anomalous: something that is unusual or unexpected and deviates from the normal; one of a series of repeated experimental results that is much larger or smaller than the others is an anomalous result

zero error: a type of systematic error in a measuring instrument, e.g. the reading on a balance may not reset to zero when there is nothing on the balance

An awareness of accuracy and sources of error is important in evaluating the results of an experiment. Tables and graphs of results should be checked for results that do not fit the pattern. A typical graph is shown in Figure 21.7.

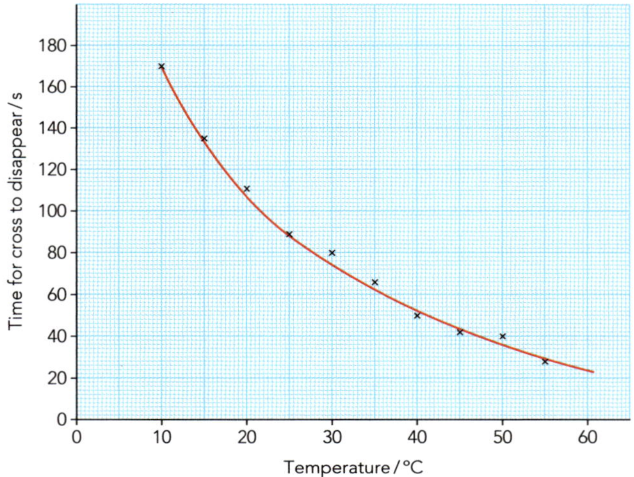

Figure 21.7: Plotting a graph is important to get the most from experimental data. This sample graph is from an experiment such as the one in Chapter 8 (Figure 8.12).

When plotting graphs, the line through the points should be a 'best-fit' line. Do not try to include points that are obviously out of place. The line you draw, after carefully plotting the points, should show the general trend or pattern of the results. Very often this will be a straight line or a gentle curve. Try to draw the line so that the points are evenly scattered on either side of the line. If a curve seems to be the best fit, then make the line as smooth as possible.

The key points when drawing graphs are:

- plot the independent variable ('Temperature' in Figure 21.7) on the x-axis (horizontal axis)
- plot the dependent variable ('Time' in Figure 21.7) on the y-axis (vertical axis)
- make the scales as large as possible: the scales do not have to start at zero
- label each axis with the name of the variable and its units
- give the graph a title
- plot the points with a cross (or a dot in a small circle) using a sharp pencil
- draw the best-fit line, which does not have to pass through all the points and which may be a straight line or a curve.

A point that does not fit the pattern is probably due to a random error in a particular reading. A line that should pass through the origin but does not do so could be due to a systematic error.

Safety

Safety is of great importance in experiments. You should be aware of the possible dangers associated with the apparatus and with those chemicals that can pose a risk.

When you plan your investigation, it is important to assess the risks associated with each part of the experiment and to take the relevant safety precautions.

Questions

1 A student wanted to accurately measure the temperature change for the reaction between a solution of sodium carbonate and hydrochloric acid. What equipment would you advise they select to measure the temperature and volumes?

2 The true time for a reaction is 1.06 min. A student measured the time for the experiment and recorded results of 1.23 min, 1.22 min and 1.23 min. The teacher said that this data was accurate but not precise. Was the teacher correct? Explain your answer.

3 A student used a balance with a resolution of 0.1 g to measure 1.3 g of magnesium ribbon. They added this to 10.0 cm^3 of 1.0 mol/dm^3 hydrochloric acid using a 20 cm^3 measuring cylinder with a resolution of 0.5 cm^3. Which of these readings would have given the largest source of error? How could any random errors associated with this experiment be reduced?

21.2 Separation and purification

Standard separation techniques

A mixture can be separated using physical processes. Different types of mixture require different standard techniques, including:

- using a suitable solvent
- **filtration**
- **crystallisation**
- **simple distillation**
- **fractional distillation**
- **chromatography**.

KEY WORDS

filtration: the separation of a solid from a liquid, using a fine filter paper which does not allow the solid to pass through

crystallisation: the process of forming crystals from a saturated solution

simple distillation: a distillation method for separating the liquid solvent from a solution containing dissolved solids

fractional distillation: a method of distillation using a fractionating column, used to separate liquids with different boiling points

chromatography: a technique employed for the separation of mixtures of dissolved substances, which was originally used to separate coloured dyes

This section focuses on techniques using a suitable solvent, filtration, crystallisation and distillation. Chromatography will form the basis of Topic 21.3.

The method of separation selected depends on the states of the chemicals involved. To separate a mixture containing a solid and a liquid, e.g. fine mud particles from water, requires filtration. To separate a mixture of two or more liquids with different boiling points requires distillation. To remove water from a solution to give a salt (ionic compound) requires crystallisation. It may be possible to separate two solids using a suitable solvent that will dissolve one solid and not the other. An example of this is the separation of salt and sand. The salt would dissolve into the water giving a **solution**, whereas the sand is insoluble and could then be removed by filtration.

Separation using different solvents

Differences in solubilities can be used as a simple method for separating mixtures of solids. Compounds with different types of structure will show different solubilities. For example, metallic structures are insoluble in all solvents. Ionic compounds, such as sodium and potassium salts, are often very soluble in water (dissociating into their ions). Simple molecular substances such as the halogens are soluble in hexane and water but others, particularly organic compounds, dissolve in hexane but not water. Giant covalent compounds, such as diamond and silicon dioxide, are insoluble.

When separating a mixture of two solids, the method is to select a solvent in which one is soluble and the other is insoluble. After adding the solvent, the mixture should be filtered (see next section). This gives a **residue**, which can be dried (the insoluble material) and a solution of the soluble compound. The soluble compound can be recovered by evaporating the solvent. For example, to separate a mixture of sulfur and iron filings, toluene is added as the solvent. The mixture is then filtered and the solvent evaporated. The iron filings are insoluble and so are left behind on the filter paper. The sulfur dissolved into the toluene and is produced after evaporation of the toluene.

KEY WORDS

solution: formed when a substance (solute) dissolves into another substance (solvent)

residue: the solid left behind in the filter paper after **filtration** has taken place

filtrate: the liquid that passes through the filter paper during filtration

Filtration

Filtration is used to separate an insoluble solid from a liquid (Figure 21.8a). The liquid part that passes through the filter is called the **filtrate** and the solid part that remains after filtration is called the residue. The useful product could be either the filtrate or the residue and care is needed to avoid disposing of the wrong part. For very fine solids, filtration can be carried out using a Buchner funnel and vacuum flask (Figure 21.8b).

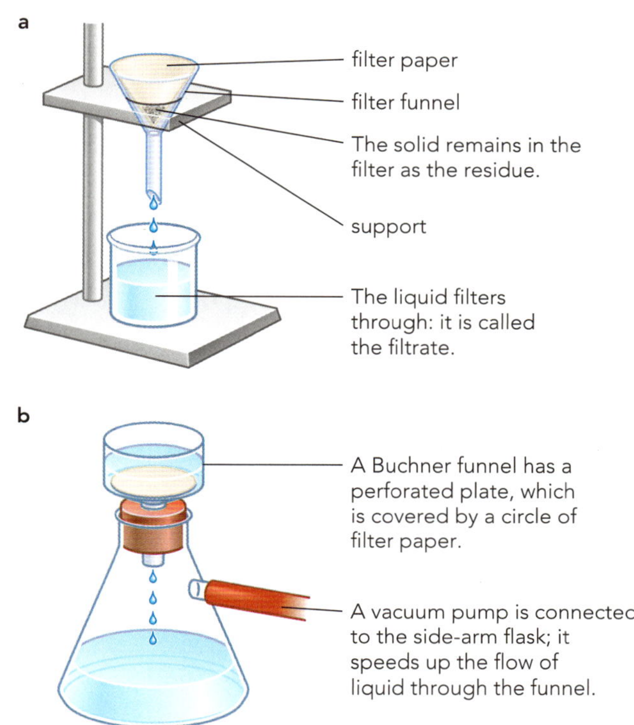

Figure 21.8: Filtration separates an insoluble solid from a liquid.

Crystallisation

Crystallisation is used to remove water from a solution containing an ionic compound (salt). This process works because the water has a lower boiling point than the salt. Evaporation can be carried out by carefully heating the sample in an evaporating basin or by placing the basin over a beaker of boiling water. Using boiling water avoids overheating the sample and prevents material from being lost because of rapid boiling of the liquid. To obtain larger crystals, a little water can be evaporated and then the solution left to cool and evaporate at room temperature. The crystals can be dried between filter papers.

EXPERIMENTAL SKILLS 21.1

Filtration and crystallisation of sodium chloride

This experiment uses a simple two-part method for obtaining pure crystals of sodium chloride (salt) from rock salt. You will learn how to use the techniques of filtration and crystallisation.

You will need:

- sample of rock salt
- spatula
- mortar and pestle
- glass rod
- 50 cm^3 measuring cylinder
- 100 cm^3 beaker
- filter funnel and filter paper
- 100 cm^3 conical flask
- evaporating basin
- Bunsen burner
- tripod
- gauze
- heat-resistant mat.

Safety

Wear eye protection throughout. Ensure long hair is tied back when using a Bunsen burner. Care must be taken when evaporating the water to avoid the salt from spitting.

Getting started

You should check the different solubilities of sand and sodium chloride by placing one spatula of each into a beaker containing 10 cm^3 of water. Does the sand dissolve? Does the salt dissolve? What does this tell you about the solubilities of these two substances? Can you link the difference in solubility to the bonding in the substances?

Method

1 Grind two heaped spatulas of rock salt using the mortar and pestle.

2 Carefully transfer the ground up salt to a beaker and add 50 cm^3 of cold water. Stir the mixture with a glass rod.

3 Carefully set up the equipment as shown in Figure 21.9 and filter the mixture into a conical flask. This will remove the insoluble materials and leave a solution of sodium chloride and water (the filtrate).

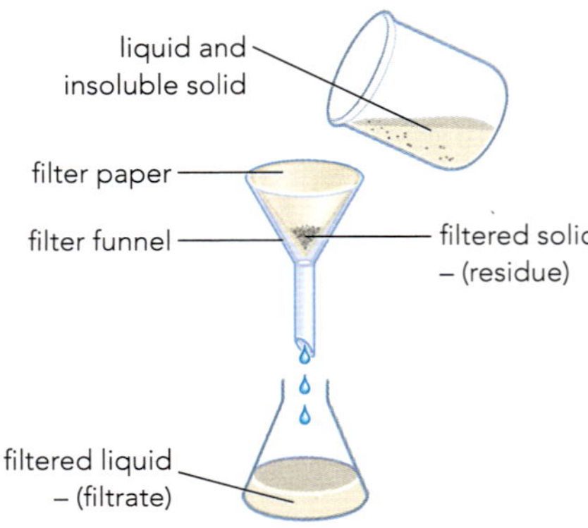

Figure 21.9: Filtration to remove the insoluble components of the rock salt mixture.

4 Place the filtrate into an evaporating basin and heat over a water-bath (Figure 21.10). Allow water to evaporate until crystals of salt start to form when a glass rod is placed in the solution.

Figure 21.10: Evaporation method to produce the salt crystals.

CONTINUED

5 Leave the evaporating basin overnight so that the rest of the water evaporates.

6 Pat dry the crystals between filter papers.

Questions

1 In filtration, the liquid that passes through the filter paper is called the filtrate. What general name is given to the insoluble part?

2 What advantage is there to the rock salt being ground up using a mortar and pestle rather than leaving it in larger pieces?

3 Why is it possible to obtain salt crystals by crystallisation? Would this method work for preparing iodine crystals from an aqueous solution of iodine?

Distillation

Simple distillation is used to separate the solvent from a solution containing a dissolved solid. The boiling point of the liquid is usually very much lower than that of the dissolved solid. The liquid solvent is more volatile than the dissolved solid and can easily be evaporated off in a distillation flask. **Anti-bumping granules** are often added to the liquid being distilled to prevent the formation of large bubbles in the liquid during boiling (Chapter 1). The solvent vapour is condensed by passing it down a water-cooled condenser and then collected as the **distillate**.

Seawater is a solution of various salts in water and Figure 21.11 shows how pure water can be obtained by simple distillation. Simple distillation can be used to obtain drinking water from salt water on a large scale (**desalination**).

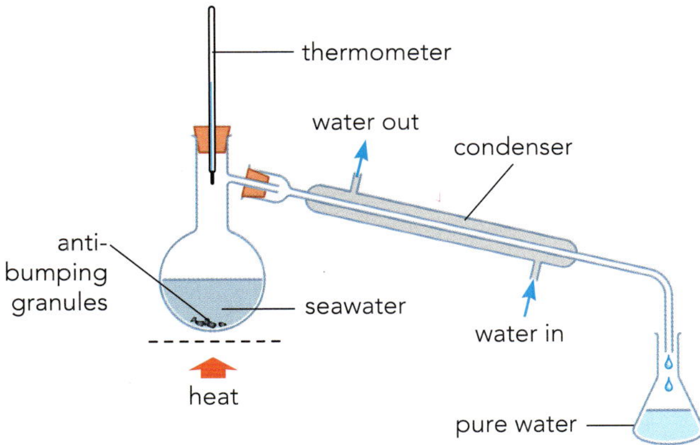

Figure 21.11: Simple distillation of seawater.

KEY WORDS

anti-bumping granules: small granules that help reduce the size of bubbles formed when a liquid boils; used for safety to stop the flask shaking

distillate: the liquid collected in the receiving flask during distillation

desalination: the removal of dissolved salts and minerals from seawater to produce drinking water

Separating the liquids from a mixture of two (or more) miscible liquids is again based on the fact that the liquids will have different boiling points. However, the boiling points are closer together than for a solid-in-liquid solution and fractional distillation must be used (Figure 21.12a). In fractional distillation, the most volatile liquid in the mixture distils first and the least volatile liquid distils last.

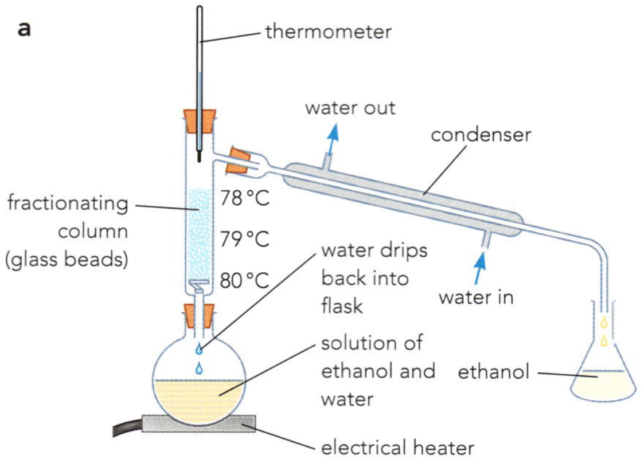

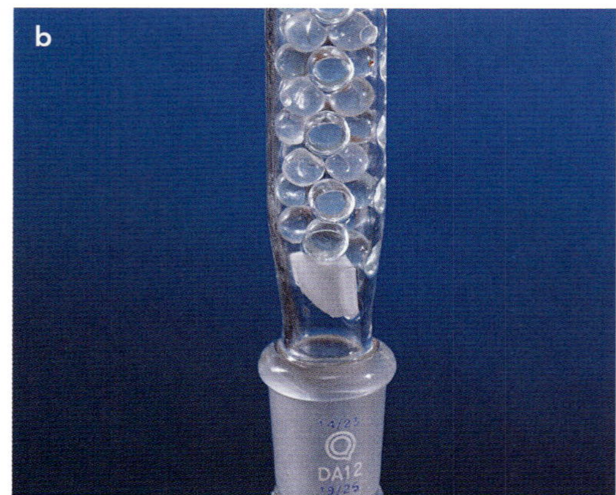

Figure 21.12 a: Separating a mixture of ethanol (alcohol) and water by fractional distillation. **b:** A close-up of the glass beads in the fractionating column.

For example, ethanol boils at 78 °C whereas water boils at 100 °C. When a mixture of the two is heated, ethanol and water vapours enter the fractionating column. Glass beads in the column provide a large surface area for condensation (Figure 21.12b). Evaporation and condensation take place many times as the vapours rise up the column. There is a range of temperatures in the column; higher at the bottom and lower at the top. Ethanol passes through the condenser first as the temperature of the column is raised above its boiling point. Water condenses in the column and flows back into the flask because the temperature of the column is below its boiling point of 100 °C.

The temperature on the thermometer stays at 78 °C until all the ethanol has distilled over. Only then does the temperature on the thermometer rise to 100 °C and the water distils over. By watching the temperature carefully, the two liquids (fractions) can be collected separately. Fractional distillation is a very important industrial process and is used in the separation of petroleum fractions (Chapter 20).

EXPERIMENTAL SKILLS 21.2

Distillation: pure water from inky water

In this experiment a mixture of ink and water will be separated to give pure water. The experiment uses simple distillation and is based on the different boiling points of the water and ink (water has a lower boiling point than ink).

You will need:

- inky water (mixture of ink and water)
- round-bottomed flask
- thermometer
- Liebig condenser
- Bunsen burner or electric heater (heating mantle)
- anti-bumping granules
- conical flask.

Safety

Wear eye protect throughout. Ensure long hair is tied back when using a Bunsen burner.

Getting started

Distillation is used to produce petrol from petroleum and in the purification of ethanol (Chapter 19). What two processes take place during distillation?

Method

1 Set the equipment up as shown in Figure 21.11.

2 Place the inky water into the round-bottomed flask and add a few anti-bumping granules to the round-bottomed flask (these are unevenly shaped pieces of an insoluble material added to a liquid to make them boil more calmly).

3 Gently heat the inky water.

CONTINUED

4 Record the temperature at which the liquid starts to boil using the thermometer.

5 Collect the distillate in a conical flask. Record its appearance in comparison to the inky water.

Questions

1 What property of the components in a mixture of liquids is used to bring about separation?

2 What temperature does steam condense at?

3 The technique used in this experiment is simple distillation. Explain how distillation is used to separate a mixture of liquids.

Component 1	Component 2	Technique	Example
Solid	Solid	Use of suitable **solvent** – separates on basis of different solubilities	Sand (giant covalent) and sodium chloride (ionic).
Insoluble solid	Liquid	**Filtration** – separates the solid (residue) and the liquid (filtrate)	Copper(II) oxide (residue) and water (filtrate).
Soluble solid	Liquid	**Crystallisation** – the liquid is evaporated to leave the solid	Copper(II) sulfate from water.
Liquid	Liquid (miscible)	**Distillation**/fractional distillation – separates on basis of different boiling points	A mixture of different hydrocarbons, e.g. crude oil.

Table 21.1: Overview of separation techniques.

Selecting the correct separation technique

We have looked at a range of simple separation techniques. When selecting which technique to use, the most important consideration is the nature of the mixture to be separated. Table 22.1 provides an overview of which technique to use for a range of two-component mixtures.

Testing purity by melting point analysis

The physical separation techniques we have looked at provide methods to produce **pure substances** from a **mixture**. A simple test for purity is to measure a melting point or boiling point and compare this to a standard reference value. For example, pure water (ice) melts at 0 °C, pure copper melts at 1085 °C, pure aspirin (acetylsalicylic acid) melts at 135 °C and pure oxygen melts at −218.8 °C. As with other physical properties, melting and boiling points can be used to identify a substance.

KEY WORDS

pure substance: a single chemical element or compound – it melts and boils at definite temperatures

mixture: two or more substances mixed together but not chemically combined – the substances can be separated by physical means

When a substance is impure, the melting point will change. Whereas a pure substance melts at a defined, specific temperature, a sample containing impurities (i.e. a mixture) will melt at a lower temperature (Chapter 1). Seawater is impure water and it freezes at a temperature well below 0 °C. In addition, the impurity reduces the sharpness of the melting point. An impure substance melts over a range of temperatures, not at a particular temperature.

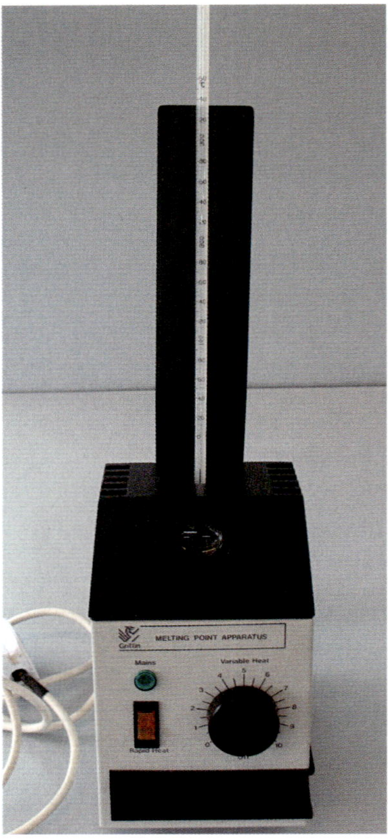

Figure 21.13 a: An electrically heated melting-point apparatus.

Measuring the melting point of a sample therefore provides a simple check for sample purity. Melting points are measured accurately using digital meters or an electrically heated melting-point apparatus. In the melting-point apparatus, a capillary tube is filled with a small amount of the solid and is placed in a heating block (Figure 21.13 a, b). The melting is viewed through a magnifying lens. Alternatively, the sample in the capillary tube can be attached to a thermometer and melting achieved using a water or oil bath (Chapter 1).

The presence of impurities will also alter the temperature at which a substance boils. As with melting points, pure substances have fixed, well-defined boiling points (e.g. pure water boils at 100 °C). The presence of an impurity will lead to the boiling point being raised above this value.

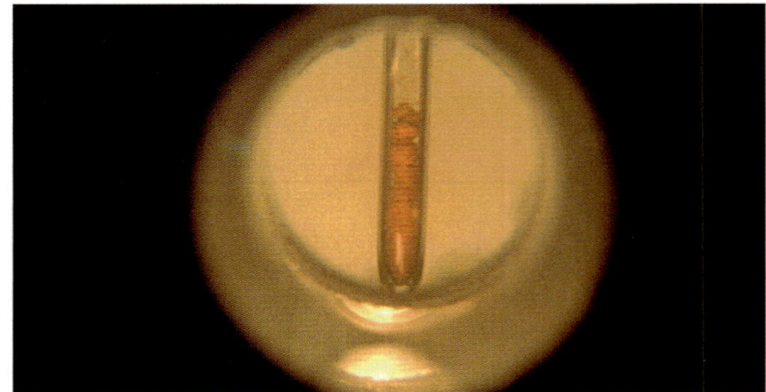

Figure 21.13 b: Magnified view through a lens of a sample being analysed for its melting point in a melting point apparatus.

ACTIVITY 21.1

A game of separation

Use a simple game to test other students' understanding of separation techniques. The options include:

1 *Missing labels*. On a piece of card produce a neat diagram of the apparatus used for a particular type of separation technique. On a separate piece of card produce a set of labels to go with the diagram you have just drawn and cut the labels out from the card. Test another student on their ability to place the labels in the correct places. To extend this game design use more than one set of equipment and have the labels jumbled up. If you want to add an extra level of challenge to the game, then remove labels from your selection and replace them with blank ones. If a blank is picked up, the player needs to correctly complete it.

2 *Chemical pairs*. Produce a set of 16 playing cards that match up a word or technique to its description (they could include diagrams, e.g. on one card a diagram of a condenser and on its pair the word condenser). Shuffle the cards and lay out face down in a 4×4 grid. Take it in turns to turn cards over in order to correctly identify pairs.

3 *A game of your choice*. You can design a completely different game to help you to learn the different types of separation and purification techniques.

CONTINUED

Peer assessment

In pairs, first discuss the game design. Think about how easy it was to start playing the game. How clear were the rules, was there any confusion? How engaging was the game? For example, would people want to play it more than once?

Secondly, think about the chemistry that was being learnt by playing the game. Were the questions too easy/too hard? How could the design be improved to make it more challenging? Are there other topics that you have learnt about that could also be taught/revised using the game format you used?

Finally, you could be the first to write a short product review for when this game becomes a commercial success.

21.3 Chromatography

Separating two or more dissolved solids in solution can be carried out by chromatography. There are several types of chromatography, but they all follow the same basic principles. **Paper chromatography** was originally developed as a method for separating soluble pigments (coloured substances such as dyes and inks) using filter paper.

The colour substances separate if:

- the pigments have different solubilities in the solvent
- the pigments have different degrees of attraction for the filter paper.

These two factors determine how fast the pigments move across the filter paper.

A drop of concentrated solution is usually placed on a pencil line (the baseline or origin) near the bottom edge of a strip of chromatography paper. A pencil is used to draw the line as it does not dissolve and interfere with the separation. The paper is then dipped in the solvent. The level of the solvent must start below the sample. Figure 21.14 shows stages of separation using paper chromatography.

The substances separate according to their solubility in the solvent. As the solvent moves up the paper, the substances are carried with it and begin to separate. The substance that is most soluble moves fastest up the paper. An insoluble substance would remain at the baseline. The chromatography run is stopped just before the **solvent front** reaches the top of the paper. Figure 21.15 shows the separation of some inks in which you can see the solvent front rising up the paper. The final product of a chromatography experiment can be dried and is known as a **chromatogram**.

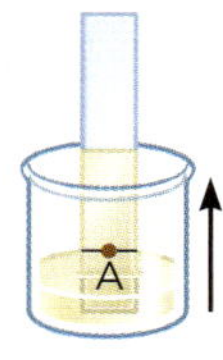

Stage 1
- The solution is spotted and allowed to dry. The original spot is identified as A.
- The solvent begins to move up the paper by capillary action.

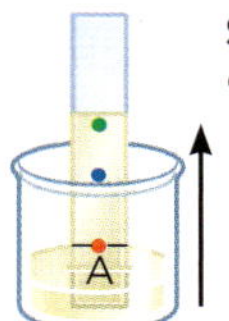

Stage 2
- The solvent moves up the paper, taking different components along at different rates.

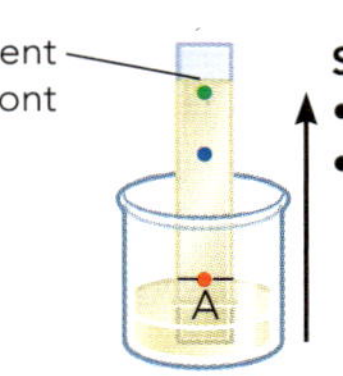

Stage 3
- The separation of the mixture is complete.
- The different components string out along the paper.

Figure 21.14: Stages of separation in paper chromatography.

KEY WORDS

paper chromatography: a simple type of chromatography used to separate the components of soluble substances based on their rate of migration in a solvent (mobile phase) on sheets of filter paper (stationary phase)

solvent front: the moving boundary of the liquid solvent that moves up the paper during chromatography

chromatogram: the result of a paper chromatography run, showing where the spots of the samples have moved to

Figure 21.15: Separation of mixtures of inks by paper chromatography.

Many different solvents are used in chromatography. Water and organic solvents (carbon-containing solvents) such as ethanol, ethanoic acid solution and propanone are commonly used. Organic solvents are useful because they dissolve many substances that are insoluble in water. When an organic solvent is used, the process is carried out in a tank with a lid to stop the solvent evaporating.

Using chromatography, we can get more information about the substances present in a mixture. Figure 21.16 shows a chromatogram of a mixture of dyes (labelled X). The three dyes present in the mixture can be seen as the spots vertically above where the sample was spotted on the baseline. By running pure samples of known dyes A, B and C (sometimes known as reference standards), we can determine whether any of these dyes were present in the mixture X.

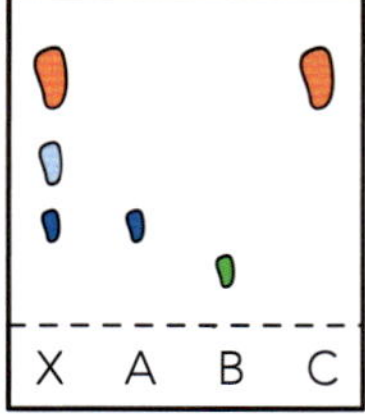

Figure 21.16: A chromatogram of a mixture of dyes X.

By looking horizontally across the chromatogram, we can see that mixture X contains dyes A and C, but not dye B. The chromatogram also shows there is a dye in X that has not been identified in this experiment.

Paper chromatography can be used to check for the purity of a substance. If the sample is pure, it should only give one spot when run in several different solvents. The identity of the sample can also be checked by comparing its movement to that of a sample we know to be pure.

The distance moved by a particular spot can be measured and related to the position of the solvent front. The ratio of these distances is called the **R_f value** (Figure 21.17).

$$R_f = \frac{\text{distance moved by the substance}}{\text{distance moved by the solvent front}}$$

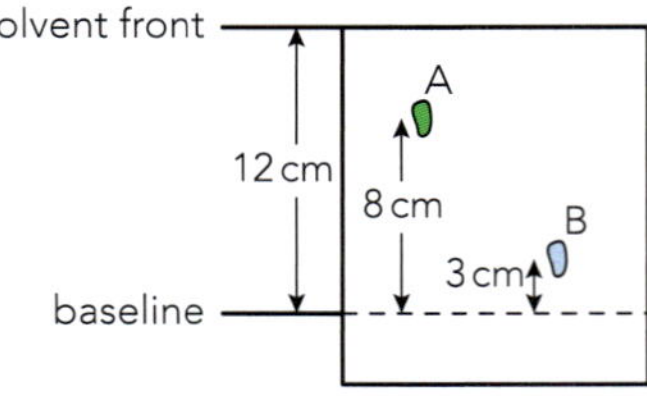

Figure 21.17: Measuring R_f values for substances A and B on a chromatogram.

The R_f value is always the same for a particular substance in a specific solvent at a given temperature, and can be used to identify unknown substances by comparison with reference values. From the chromatogram in Figure 21.17 it can be seen that the R_f values for substances A and B are 0.67 and 0.25, respectively. As an R_f value is a ratio, it always ranges between 0 and 1 and does not have any units.

As mentioned earlier, paper chromatography was originally used to separate solutions of coloured substances (dyes and pigments) since they could be seen as they moved up the paper. However, the usefulness of chromatography has been greatly increased by the use of **locating agents** (Figure 21.18) to separate mixtures of soluble *colourless* substances. A locating agent reacts with a colourless substance to produce coloured spots or spots that glow under ultraviolet light. The chromatography paper is treated with the locating agent after the chromatography run.

KEY WORDS

R_f value: in chromatography, the ratio of the distance travelled by the solute to the distance travelled by the solvent front

locating agent: a compound that reacts with invisible, colourless spots separated by chromatography to produce a coloured product that can be seen

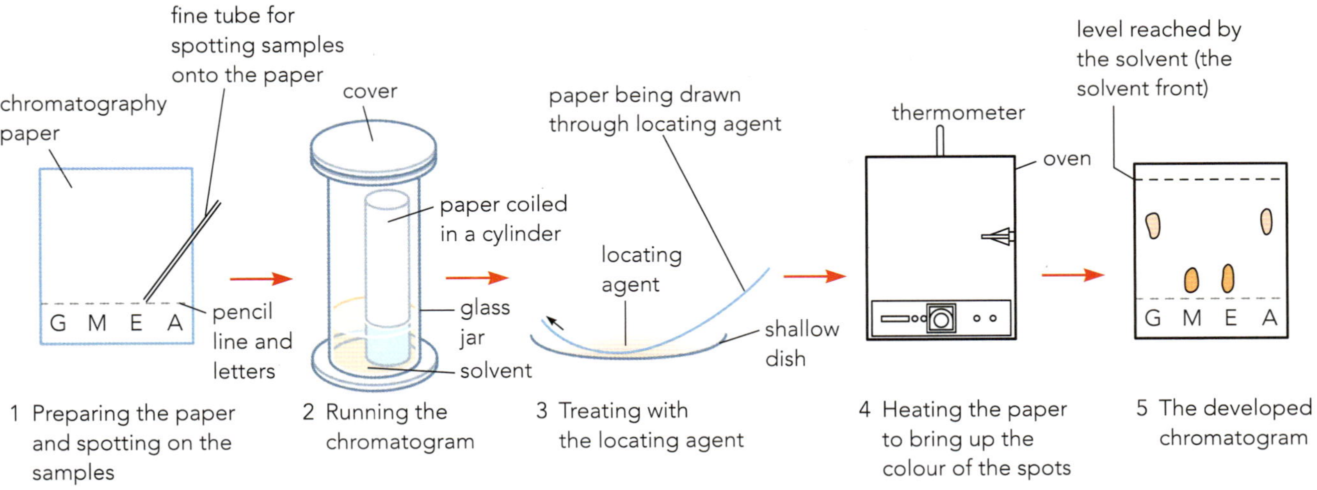

Figure 21.18: Chromatography using a locating agent to detect the spots on the paper. Alternatively, the locating agent can be sprayed on the paper.

Chromatography has proved very useful in the analysis of biologically important molecules such as sugars, amino acids and nucleotide bases. The technique can also be used to detect illegal substances in blood or urine (Figure 21.19) and to identify contaminants in food or drinking water.

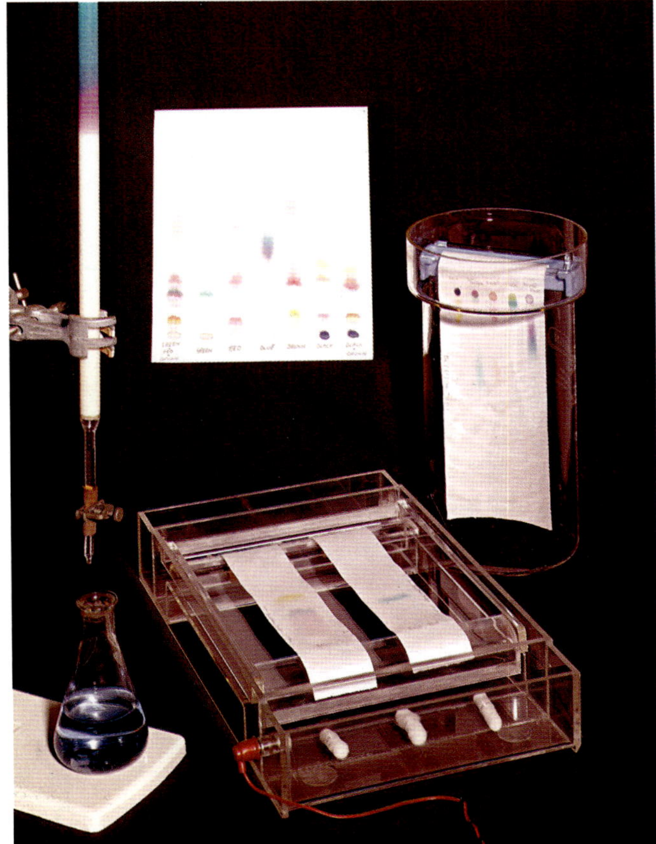

Figure 21.19: Paper chromatography is one of several different types of chromatography that work on the same principle.

Questions

4 A simple chromatogram from a red marker pen showed two dots in a vertical line. What does this tell you about the ink used in this pen?

5 A student tried to separate some inks from a black marker pen using paper chromatography but the ink remained on the baseline. Why did the ink not move and what should the student do to improve the experiment?

6 When analysing a chromatogram, the distance moved by the solvent from the baseline was 2.8 cm and the distance travelled by the sample was 2.3 cm. Calculate the R_f value for this compound to 2 decimal places. Using the same ink, but a larger piece of chromatography paper, the solvent front had travelled 9.7 cm. Predict, to 1 decimal place, the distance moved by the sample.

EXPERIMENTAL SKILLS 21.3

Paper chromatography

This experiment provides a simple method for analysing the colours in different food dyes or inks. You will learn how to produce a paper chromatogram.

You will also calculate R_f values.

You will need:

- three different inks or liquid food dyes labelled A, B and C
- 250 cm^3 beaker
- water or ethanol as the solvent
- rectangular piece of chromatography paper (approximately 8 cm × 4 cm)
- capillary tube or dropping pipette
- pencil
- ruler
- splint or pencil and bulldog clip to suspend the paper in the solvent
- hairdryer.

Safety

Wear eye protection throughout. Ethanol is toxic and so you must wash hands after use. Ethanol is also highly flammable and so this experiment must not be conducted near lit Bunsen burners.

Getting started

Selecting the correct solvent for a chromatography experiment is important. If the samples are too soluble, they will travel with the solvent front and not separate; if the samples are insoluble, they will remain on the baseline. Test the solubility of an ink by placing a small amount onto a piece of paper and then submerge the paper in a small amount of solvent. Does the ink remain on the paper or does it start to dissolve into the solvent? Does the ink dissolve more rapidly in different solvents?

Method

1. Draw a pencil line (baseline) horizontally on your piece of chromatography paper and add three crosses (X) at regular intervals. Under each cross write a letter A, B or C.
2. Add a drop of dye A carefully in the centre of the first cross, this should be done with the capillary tube. If using a pipette just touch the end onto the paper, do not squeeze. To get the best results use small amounts of sample as larger amounts will give very streaky results.
3. Repeat with the other dyes adding to the crosses labelled B and C.
4. Attach the paper to a splint or pencil using a clip and suspend the paper in the solvent (water). It is essential that the height of the solvent is between the bottom of the paper and the baseline.
5. The solvent should start to travel up the chromatography paper and any soluble dyes will start to move from the baseline.
6. Remove the paper from the water after the solvent has travelled about 80% of the way up the paper.
7. Mark the solvent front with a pencil to show how far the solvent moved.
8. Dry the chromatogram carefully with the hairdryer.
9. Deduce the R_f values for the different dyes.

The experiment can be repeated using ethanol as an alternative solvent. If using ethanol, a lid should be added to the equipment.

Questions

1. A student repeated the experiment using inks from a selection of permanent pens and the following method:
   a. Draw a baseline in pen horizontally on your piece of chromatography paper. The baseline should be about 0.5 cm from the bottom of the paper.

CONTINUED

b Add a drop of ink at regular intervals along the baseline.

c Attach the paper to a pencil using a bulldog clip and suspend the paper in water so that the baseline is just covered by the water.

d Remove from the water once the solvent has travelled all the way up the paper.

e Mark the solvent front in pencil.

Can you spot the errors in their method? Explain what the impact of each mistake they have made would be.

2 Why would a lid be required if ethanol was used as a solvent?

ACTIVITY 21.2

Mixing it up

1 Write down five facts linked to the different separation techniques, e.g. the liquid that passes through the filter paper is called the filtrate (Figure 21.20).

2 Write down two detailed explanations as to how two of the different separation techniques work, e.g. fractional distillation involves heating a mixture to its boiling point, different fractions have different boiling points and so evaporate at different temperatures, etc.

3 Identify one question you still have about separation techniques or one technique you are not yet confident with.

In pairs, compare your facts and then work together to identify which three facts you think are the most important.

Figure 21.20: Filtering flower essence at a perfume factory in France.

Peer assessment

Look at your partner's explanations of two different separation techniques. What are the strengths in your partner's explanations? How could their explanation be improved?

Finally, find an answer or explanation for your partner's question or try to explain to them the technique they are not yet confident with.

REFLECTION

One of the key skills in chemistry is the ability to design an experiment. In this chapter, we have considered not only the way in which apparatus is selected but also the techniques used to purify the products of chemical reactions. How will you apply your understanding of experimental design to suggest improvements in future practical experiments? Thinking back to the flow diagram given in Figure 21.2, what aspects of your experimental work do you still need to develop? How could you do this?

SUMMARY

There are different factors involved in the design and planning of an experiment, including choice of apparatus, level of accuracy, variables, sources of error and safety.
Different methods or pieces of apparatus each have advantages and disadvantages, and these should be accounted for when choosing which to use.
The correct apparatus to select when measuring time is a stopwatch or stopclock; temperature, a thermometer; mass, a balance; volume, the appropriate glassware and pH is Universal Indicator or a pH meter.
There is a range of different separation and purification techniques, including use of a solvent (to separate two solids), filtration (to separate an insoluble solid from a liquid), crystallisation (to separate an soluble solid from a solution) and simple or fractional distillation (to separate a liquid from a solution or a mixture of liquids).
The correct separation technique needs to be selected for a given mixture.
Melting point and boiling point data can be used to identify substances and assess their purity.
Paper chromatography is a simple separation technique and a chromatogram can be used to determine whether a sample is a pure substance or a mixture.
A reference sample is used to compare samples in chromatography.
An R_f value is a ratio calculated from a chromatogram and is the distance moved by the substance divided by the distance moved by the solvent front.
A locating agent can be used to highlight the position of soluble colourless substances in a chromatogram.

PROJECT

Separations Are Us

'Separations Are Us' is a new business that aims to sell technical know-how to chemical companies around the world. Their aim is to help companies purify products through distillation, filtration, crystallisation and chromatography (Figure 21.21).

You have been asked to produce an advertising brochure for the company that summarises each of these processes and that gives some possible applications of each.

As you work through your leaflet/brochure, ensure you meet all of the following criteria:

1 You have included at least four different separation techniques.

2 For each technique you have given a simple diagram to show the apparatus needed.

3 You have explained the chemistry involved in each method.

4 You have outlined what sort of mixture each method can be used to separate.

You may want to extend your brochure by including information on the possible strengths and weaknesses of the different methods being offered.

You may want to extend your brochure to include information about using melting points as a check for product purity.

Figure 21.21: A technician adjusting the valves on industrial filtration tanks.

EXAM-STYLE QUESTIONS

1 Coloured sweets contain edible dyes. These dyes can be separated by chromatography. The figure shows chromatograms obtained from three different orange sweets.

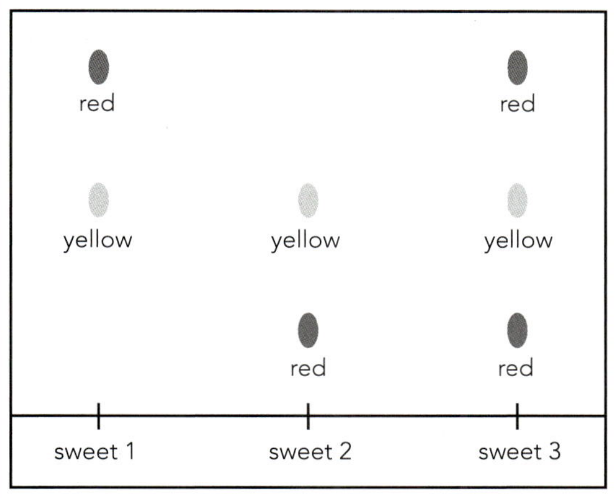

How many different red dyes were present in these orange sweets?

A 4 **B** 3 **C** 2 **D** 7 **[1]**

2 The diagrams A–F in the figure show the stages in an experiment to separate a mixture of copper(II) sulfate and copper(II) oxide and prepare clean dry crystals of copper(II) sulfate.

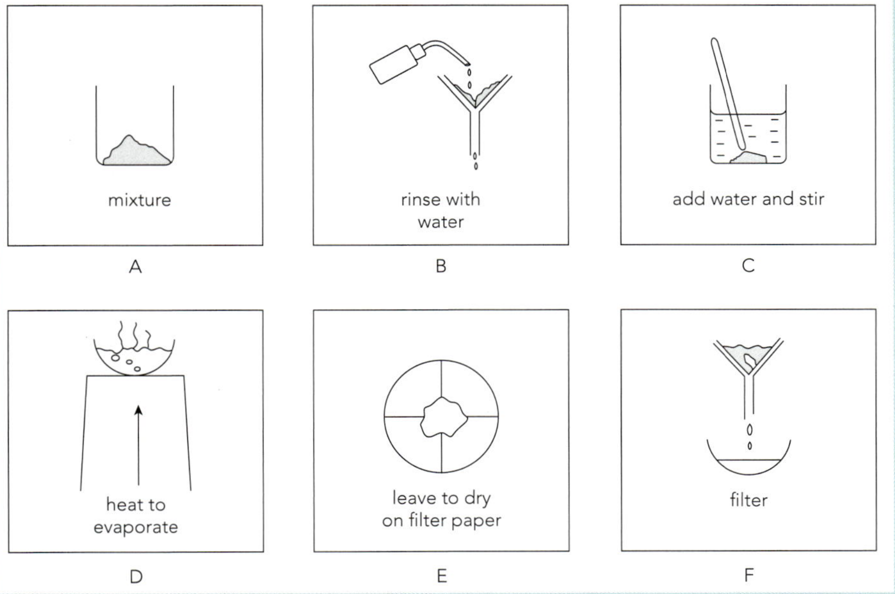

a Name the two pieces of apparatus drawn in **D**. **[2]**

b The experiment would not work if the different steps were in this order. Using the letters **A–F**, place the diagrams in the correct order. **[2]**

CONTINUED

c What colour would the solution be in the beaker in **C**? [1]

d Why was the mixture stirred with water and not acid? [1]

[Total: 6]

3 Malachite is an ore of copper which contains copper carbonate, $CuCO_3$. You are provided with a large lump of malachite. **Describe** an experiment to obtain a sample of pure copper from this ore.

Your answer should include:

- details of all apparatus used
- details of any chemicals used
- how you would separate the copper formed from any mixture. [6]

4 The following questions refer to different items of common laboratory apparatus.

a Why might a measuring cylinder be used instead of a gas jar to collect a gas over water? [1]

b Why is a measuring cylinder not used to measure 13.6 cm^3 of a liquid? [1]

c How can a Bunsen burner be adjusted to produce the hottest flame? [1]

d Why is a thermometer needed when ethanol is separated from water by distillation? [1]

e When crystallising from a solution, how do you know when to stop heating? [1]

[Total: 5]

5 Proteins are long chains of different amino acids joined together. Design an experiment to discover which amino acids make up a particular protein.

You are provided with a concentrated solution of the protein that has been split into its individual amino acids. You are also provided with chromatography apparatus. All amino acids are colourless compounds that are soluble in water. [6]

COMMAND WORD

describe: state the points of a topic / give characteristics and main features

SELF-EVALUATION CHECKLIST

After studying this chapter, think about how confident you are with the different topics. This will help you see any gaps in your knowledge and help you to learn more effectively.

I can	See Topic...	Needs more work	Almost there	Confident to move on
understand the different factors involved in the design and planning of an experiment	21.1			
name appropriate apparatus for the measurement of time, temperature, mass, volume and pH	21.1			
state possible advantages/disadvantages for selecting a particular method or piece of apparatus	21.1			
explain the terms solvent, solute, solution, saturated solution, residue and filtrate	21.2			
describe how a mixture of solids can be separated by using different solvents	21.2			
explain how filtration can be used to separate an insoluble compound from a solvent or solution	21.2			
describe how to crystallise a soluble salt from a solution	21.2			
explain how simple and fractional distillation can be used to separate mixtures of liquids	21.2			
identify the most suitable separation technique when given information about the substances present	21.2			
describe the use of melting and boiling points to determine purity	21.2			
explain how to carry out a simple paper chromatography to separate coloured compounds	21.3			
use a chromatogram to compare an unknown to a known sample and deduce whether a substance is pure or a mixture	21.3			
recall and use the equation to calculate R_f values	21.3			
describe how locating agents can be used to identify soluble colourless substances in chromatography	21.3			

Chapter 22
Chemical analysis

IN THIS CHAPTER YOU WILL:

- learn how characteristic colours produced by flame tests can be used to identify metal cations
- explore how precipitation reactions with solutions of sodium hydroxide and ammonia can be used to identify aqueous cations
- discover how chemical analysis is used to determine the presence of anions
- develop simple practical techniques for proving the identity of different gases
- describe acid–base titration as a form of quantitative analysis
- learn how to identify the end-point of a titration using an indicator.

GETTING STARTED

Chemists analyse samples of water, soil and air. When looking at these samples, chemists need to detect the presence of different substances quickly and accurately, particularly those that may be harmful to life. Can you think of the criteria for a chemical test that can be used to identify the presence of a particular substance? Think about reactions you have carried out or seen demonstrated where there are noticeable changes when a certain substance is present. Discuss in groups the possible reactions that could be used as tests and the changes that would need to be monitored. Who would need access to the findings of these tests and why? What would the analytical chemists need to do to ensure their conclusions were valid?

BOTTLED WATER – FINDING A SOLUTION TO WHAT IONS IT CONTAINS

Analytical tests are routinely used across all areas of Chemistry, including in the analysis of bottled water. The tests can help scientists to understand what chemicals are either present (positive test) or not present (negative test). The results of these tests can be qualitative, giving a yes/no answer, or quantitative, giving the actual amount of something present. For example, silver nitrate is added to domestic drinking water as a test for chloride ions. This is a qualitative test and the formation of a white precipitate (a positive result) would mean chloride ions were present in the water.

Bottled mineral water is sold in high and increasing volumes across the world (Figure 22.1). It is a mixture containing water and several soluble metal compounds, known as minerals. These minerals dissolve as the water passes through different types of rock. Minerals are made up of a cation (the metal ion) and an anion (a non-metal ion). Some minerals are beneficial to health, but care is also needed as in high concentration some can cause illness. As examples, very high levels of potassium can cause problems such as the life-threatening condition known as hyperkalaemia and high levels of chloride ions are toxic. To ensure product quality, legally each company that sells mineral water needs to display both the names and quantities of the ions that are present. Chemical tests are necessary to determine the quality and safety of mineral water in order to ensure potentially harmful or 'fake' drinking water does not reach the consumer.

Figure 22.1: Supermarkets carry a large range of different bottled waters.

Discussion questions

1 Some people might describe mineral water as being pure. Why do you think this could be? Can you explain why chemically this is not true?

2 Bottled mineral water is not allowed to be treated other than by filtration to remove sand particles and needs to have a guaranteed consistent chemical composition. What problems might this create for the companies selling bottled mineral water?

22.1 Tests to identify common cations

We all require certain levels of metal ions (cations) such as sodium, potassium, copper, iron and zinc to maintain a healthy body. However, in larger concentrations, some metal ions can become dangerous to human health, e.g. high levels of lead ions are linked to Alzheimer's Disease.

It is important for analytical chemists to detect the presence of metal ions in solution and in the following sections we will look at two **qualitative** approaches to identifying a range of cations. These are the flame test and precipitation reactions with aqueous sodium hydroxide or aqueous ammonia.

Flame tests to identify metal cations

As we have seen in Chapter 2, the distinctive colours produced in a firework display result from the different metal compounds that are used in the manufacture of the fireworks. The colours produced are due to electrons changing energy level in the different metal ions. This same principle is used as the basis for the flame test to test for the presence of some metal cations. In particular, the flame test is widely used for metal ions of the elements found in Group I (alkali metals) and Group II (Figure 22.2).

KEY WORDS

qualitative (analysis): the process used to determine the presence or absence of a substance in a given sample

cation: a positive ion that would be attracted to the cathode in electrolysis

The test for the presence of metal **cations** involves taking a nichrome wire probe and first heating it in a roaring Bunsen flame. The wire is then dipped into hydrochloric acid. These two steps are repeated several times until no residual colour is evident in the flame. This is done to clean the probe by removing traces of metal ions from previous experiments. The probe is then dipped into the acid again to wet it before placing the probe into the test sample. The wet probe will pick up a few crystals. The probe is then placed into a roaring flame and the colour is recorded. These steps are outlined in Figure 22.3 (and form the basis for Experimental Skills 2.1 in Chapter 2).

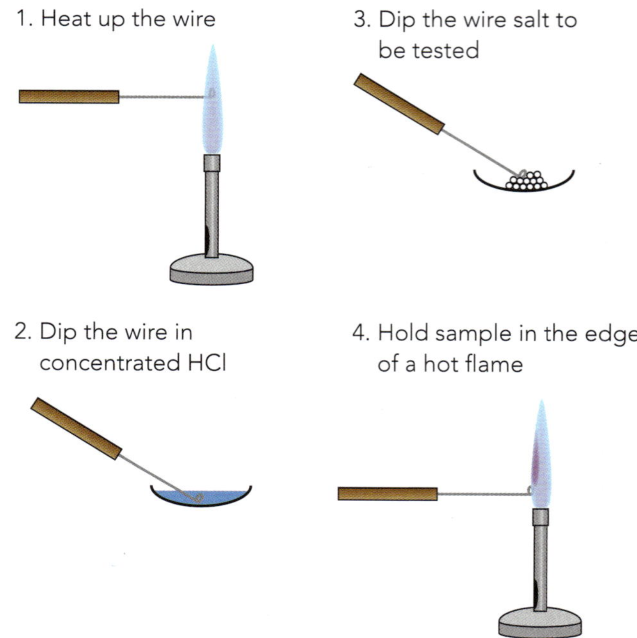

Figure 22.3: Method for the flame test.

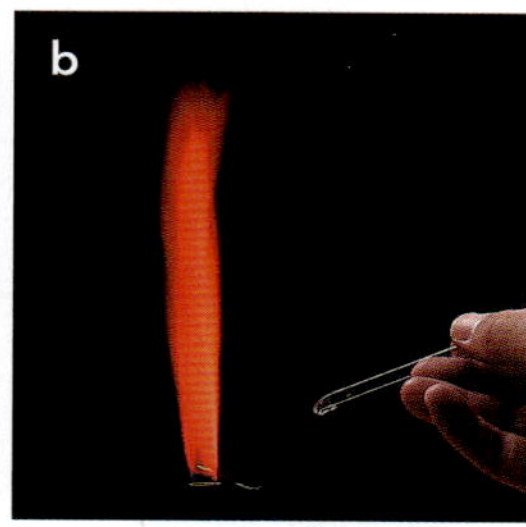

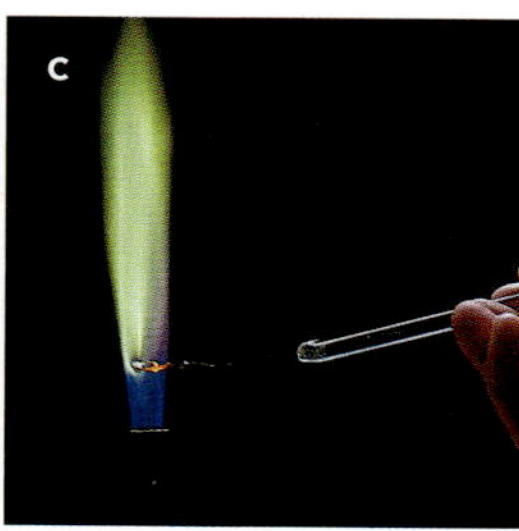

Figure 22.2: Some Group II metals give characteristic colours in the flame test: **a:** calcium, **b:** strontium and **c:** barium.

	Lithium, Li^+	Sodium, Na^+	Potassium, K^+	Calcium, Ca^{2+}	Barium, Ba^{2+}	Copper(II), Cu^{2+}
Flame colour	Red	Yellow	Lilac	Orange–red	Light green	Blue–green

Table 22.1: Standard colours obtained with the flame test for different metal ions.

Different metal ions produce characteristic colours (Table 22.1).

An alternative method is to take a wooden splint and dip it in a solution of the metal ion for an extended period of time. The wooden splint will absorb the test chemical and can then be placed into a roaring Bunsen burner flame (Figure 22.4). This will give similar results to those recorded in Table 22.1. You must take care not to set the splint on fire. If this does happen the flame will turn yellow and the results will be void as the yellow flame of the burning wood will dominate the colour of the metal ion.

Figure 22.4: Using a splint, or rolled paper, as an alternative method for the flame test: the yellow flame shows the presence of sodium.

Questions

1 Explain how a nichrome wire probe and Bunsen burner can be used to show the presence of sodium ions.

2 Flame tests on three different metal sulfates (SO_4^{2-}) gave three different coloured flames: lilac, red and light green. What metal ion was present in each test? Deduce the formulae of the three sulfates that were tested.

3 A flame test was carried out using a wooden splint soaked in a salt solution prepared with tap water. The flame test produced an orange–red colour. Why might a conclusion that this sample contained calcium ions not be valid? How could the experimental design be improved to ensure a valid result?

Using aqueous sodium hydroxide or aqueous ammonia to identify aqueous cations

It is possible to predict observations just by looking at a balanced chemical equation that includes the **state symbols**. For example, the reaction between copper(II) carbonate and sulfuric acid can be represented using the following symbol equation:

$$CuCO_3(s) + H_2SO_4(aq) \rightarrow CuSO_4(aq) + CO_2(g) + H_2O(l)$$

KEY WORDS

state symbols: symbols used to show the physical state of the reactants and products in a chemical reaction: they are s (solid), l (liquid), g (gas) and aq (in solution in water)

The state symbols enable us to say that if we observed this reaction, we would see the disappearance of the solid $CuCO_3$ and effervescence (bubbling) as CO_2 gas was produced. In this section, we will see reactions where a solid (a **precipitate**) is formed when two solutions are added together. The symbol equations will include aqueous (aq) state symbols in the reactants and a solid (s) as one of the products.

Not all aqueous ions produce characteristic flame colours and so alternative tests are needed. A commonly used test is to add aqueous sodium hydroxide NaOH(aq). When sodium hydroxide solution is added to a solution containing a metal cation (usually a transition metal ion), an insoluble metal hydroxide is formed in a **precipitation reaction** (Chapter 4). The transition metals are particularly well suited to this test as they form compounds with characteristic colours.

Precipitation reactions are less useful for salts containing Group I or Group II metal ions. Salts of the Group I metals tend to be soluble (Chapter 12) while those of Group II, if they do produce a precipitate, will produce a white precipitate that requires additional tests to identify the specific metal ion.

KEY WORDS

precipitate: an insoluble salt formed during a precipitation reaction

precipitation reaction: a reaction in which an insoluble salt is prepared from solutions of two suitable soluble salts

Precipitation with aqueous sodium hydroxide

Several aqueous cations react with a solution containing hydroxide ions to produce a characteristic precipitate. For example, aqueous iron(II) ions will initially produce a green precipitate when dilute sodium hydroxide is added (Figure 22.5).

iron(II) chloride + sodium hydroxide → iron(II) hydroxide + sodium chloride

$$FeCl_2(aq) + 2NaOH(aq) \rightarrow Fe(OH)_2(s) + 2NaCl(aq)$$

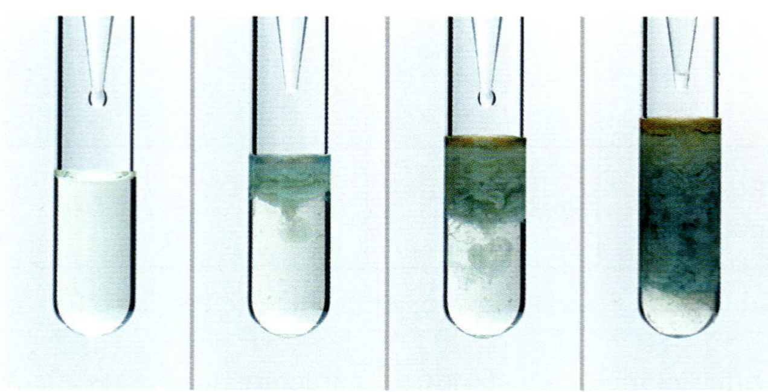

Figure 22.5: Precipitation reaction as sodium hydroxide solution is added to a solution containing aqueous iron(II) ions.

If left to stand, iron(II) hydroxide will slowly oxidise to iron(III) hydroxide at the surface of the reaction leading to a change in colour (green to brown) (Figure 22.5).

You can use a solution of aqueous sodium hydroxide to test for cations including aluminium, ammonium, calcium, chromium(III), copper(II), iron(II), iron(III) and zinc. The result of the precipitation reaction for each ion is given in Table 22.2.

Care needs to be taken when trying to identify aluminium, zinc and calcium ions as all of these ions produce a white precipitate. This means that a further stage in the analysis is needed. You can identify the calcium ions as they do not redissolve when an excess of sodium hydroxide is added; both zinc and aluminium ions will redissolve. As an alternative, to show the presence of calcium ions you could also do a flame test (the calcium ions would produce a characteristic orange–red coloured flame). To distinguish between the zinc and aluminium ions you need to repeat the test but using ammonia solution (the results for this test are shown in the next section).

Cation	Positive result when NaOH(aq) added
Aluminium, Al^{3+}	White precipitate, this redissolves when an excess of NaOH(aq) is added to give a clear, colourless solution
Ammonium, NH_4^+	No precipitate but produces ammonia on warming
Calcium, Ca^{2+}	White precipitate, this is insoluble in excess NaOH(aq)
Chromium (III), Cr^{3+}	Green precipitate, this will redissolve in excess NaOH(aq)
Copper(II), Cu^{2+}	Light blue precipitate, this is insoluble in excess NaOH(aq)
Iron(II), Fe^{2+}	Green precipitate, this is insoluble in excess NaOH(aq) but starts to turn brown near the surface if left to stand
Iron(III), Fe^{3+}	Red–brown precipitate, this is insoluble in excess NaOH(aq)
Zinc, Zn^{2+}	White precipitate, this will redissolve in excess NaOH(aq) to give a clear, colourless solution

Table 22.2: Results of precipitation reactions using NaOH(aq).

Questions

4 Dilute sodium hydroxide solution was added dropwise to copper(II) sulfate solution to form a light blue precipitate of copper(II) hydroxide. Which of these symbol equations is/are correct?

A $CuSO_4 + NaOH \rightarrow CuOH + NaSO_4$

B $CuSO_4(aq) + 2NaOH(l) \rightarrow Cu(OH)_2(s) + Na_2SO_4(aq)$

C $CuSO_4(aq) + 2NaOH(aq) \rightarrow Cu(OH)_2(s) + Na_2SO_4(aq)$

D $CuSO_4(aq) + NaOH(aq) \rightarrow Cu(OH)_2(s) + Na_2SO_4(s)$

E $CuSO_4 + 2NaOH \rightarrow Cu(OH)_2 + Na_2SO_4$

5 Which equation from question **4** reveals that the reaction involves precipitation?

6 A few drops of dilute sodium hydroxide solution (NaOH(aq)) were added to a sample of aluminium chloride ($AlCl_3$(aq)), and this produced a white precipitate. Write the balanced equation for this reaction, including state symbols.

Precipitation with aqueous ammonia

A solution of ammonia (NH_3) will also lead to precipitation reactions with solutions containing aqueous metal cations (Table 22.3). The initial changes with dilute ammonia solution are generally similar to those when using aqueous sodium hydroxide. This is because ammonia forms a weakly alkaline solution in water and only produces a low concentration of aqueous hydroxide ions.

Ammonia solution produces slightly different results when it is added in excess. For solutions containing zinc or aluminium ions, a few drops of ammonia solution will produce a white precipitate. When an excess is added it is possible to distinguish between the two metal cations as the aluminium precipitate is insoluble, but the zinc precipitate dissolves to give a colourless solution. The other difference to note is that copper(II) ions will initially give a characteristic pale blue precipitate but this will redissolve when excess ammonia is added, producing a dark blue solution (Figure 22.6).

Figure 22.6: Precipitation reaction as ammonia solution is added to a solution containing aqueous copper(II) ions.

Cation	Positive result when $NH_3(aq)$ added
Aluminium, Al^{3+}	White precipitate; insoluble in excess $NH_3(aq)$
Ammonium, NH_4^+	No reaction
Calcium, Ca^{2+}	No precipitate or very slight white precipitate
Chromium (III), Cr^{3+}	Green precipitate; insoluble in excess $NH_3(aq)$
Copper(II), Cu^{2+}	Light blue precipitate; soluble in excess $NH_3(aq)$ producing a dark blue solution
Iron(II), Fe^{2+}	Green precipitate; insoluble in excess $NH_3(aq)$ but starts to turn brown near the surface if left to stand
Iron(III), Fe^{3+}	Red–brown precipitate; insoluble in excess $NH_3(aq)$
Zinc, Zn^{2+}	White precipitate; redissolve in excess $NH_3(aq)$ to give a clear, colourless solution

Table 22.3: Results of precipitation reaction using $NH_3(aq)$.

EXPERIMENTAL SKILLS 22.1

Testing for aqueous cations

This experiment provides methods for the identification of common metal ions and the ammonium ion (NH_4^+) using sodium and ammonium hydroxide solutions. The method can be used in parallel with the tests for common anions discussed in the next section to confirm the identity of an ionic compound.

You will need:

- test-tubes
- test-tube rack
- Bunsen burner and heat-resistant mat
- 10 cm^3 measuring cylinder
- dropping pipette
- 0.5 mol/dm^3 sodium hydroxide solution
- 0.5 mol/dm^3 ammonia solution
- damp red litmus paper
- test samples including 0.2 mol/dm^3 solutions of the following compounds: copper(II) sulfate, iron(II) sulfate, iron(III) chloride, ammonium chloride, zinc sulfate, chromium(III) sulfate, aluminium chloride and calcium chloride.

Safety

Wear eye protection throughout. Be careful with chemicals. Never ingest them and always wash your hands after handling them. Avoid the chemicals coming into direct contact with skin; if they do then rinse the affected area under running water.

Getting started

Take 2 cm^3 of a solution of iron(II) sulfate and place it in a clean, dry test-tube. Then use the dropping pipette to add a few drops of 0.5 mol/dm^3 sodium hydroxide solution to the tube. Note the formation of a precipitate and in particular its colour.
Leave the test-tube in a rack and look at it again after a further ten minutes. Has there been any change in the colour of the precipitate?

Method

Part 1: Using sodium hydroxide

1 Place 2 cm^3 of a solution of the test sample into a clean test-tube.

2 Slowly add a few drops of dilute 0.5 mol/dm^3 sodium hydroxide solution and record the colour of any precipitate formed.

3 Continue to add dilute sodium hydroxide until it is in excess. This is necessary as some of the precipitates initially formed will redissolve.

4 *Additional step*: to test for the ammonium ion (NH_4^+) repeat steps 1 and 2 but then warm gently using a Bunsen burner. If a gas is produced then it should turn damp red litmus paper blue (see also test for ammonia gas in Topic 22.3).

Part 2: Using ammonia solution

1 Place 2 cm^3 of a solution of the test sample into a clean test-tube.

2 Slowly add a few drops of dilute ammonia solution (NH_3(aq)) and record the colour of any precipitate formed.

3 Continue to add dilute ammonia solution until it is in excess. This is necessary as some of the precipitates initially formed will redissolve.

Questions

1 The reaction between transition metal ions and sodium hydroxide produces coloured precipitates. What is a precipitate?

2 Transition metals can form more than one ion. What difference did you note between the precipitates of iron(II), Fe^{2+}(aq) and iron(III), Fe^{3+}(aq)?

3 Ammonia turns red litmus blue. What does this tell you about ammonia?

Summary of tests for aqueous cations

The flame test and the simple precipitation reactions using aqueous sodium hydroxide or aqueous ammonia provide qualitative methods of analysis for identifying aqueous cations. It is important when planning an analytical experiment to use a stepwise approach. This could include initially always starting with a flame test as this is a very quick and easy method. If the flame test does not yield a positive test result then sodium hydroxide solution should be added. A simple overview of how to analyse the results of the precipitate test is shown in Figure 22.7.

When carrying out these tests it is important that the aqueous solutions are prepared using distilled water. This is because tap water can contain dissolved minerals (metal compounds). These minerals may include metal ions such as calcium and zinc that would potentially interfere with the tests (Chapter 17). Also note that not all metal salts form soluble compounds, e.g. many metal carbonates show low levels of solubility (Chapter 12).

Questions

7 $5\,cm^3$ of iron(II) and iron(III) chloride were placed into separate clean test-tubes and a few drops of aqueous sodium hydroxide solution were added to each tube. Describe the observations for the positive test results expected for each salt.

8 $5\,cm^3$ of copper(II) sulfate solution was placed into a clean test-tube and aqueous sodium hydroxide was added dropwise. In another experiment, $5\,cm^3$ of copper(II) sulfate solution was placed into a clean test-tube and dilute ammonia solution was added dropwise. Describe fully what would be seen during each experiment.

9 A few drops of NaOH(aq) were added to a sample containing an unknown aqueous metal ion. The result was a green precipitate. A student concluded that the sample contained chromium(III) ions. Was their conclusion valid? If not, what should be done to provide a valid result?

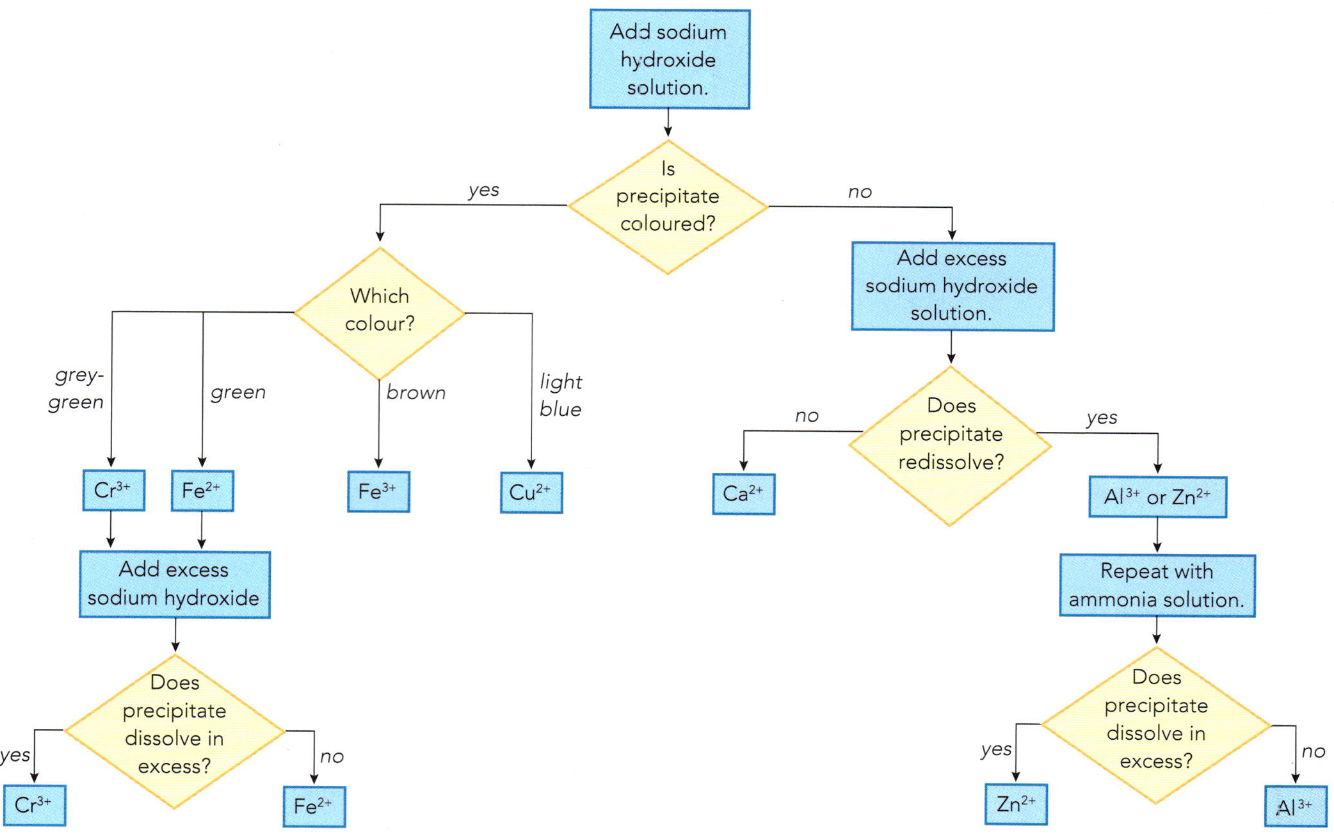

Figure 22.7: A strategy for identifying metal cations.

10 A solution of a metal ion produced white precipitates with dilute ammonia and aqueous sodium hydroxide. The solution is known to contain either calcium, zinc or aluminium ions. During a class discussion, student A said they would do a flame test to prove whether the solution contained calcium ions. Student B said they would add an excess of sodium hydroxide to test for zinc or aluminium ions. Why would student A's approach give valid results? What was wrong with the suggestion made by student B and what change in reagent would lead to a valid conclusion?

22.2 Tests to identify common anions

The previous sections have considered the main tests for the presence of common cations (positive ions), but as ionic compounds contain both a positive and a negative ion (anion), analytical techniques are needed for both anions and cations. The common anions include single elements (monatomic ions), e.g. chloride (Cl^-), bromide (Br^-) and iodide (I^-), and compound ions, which as the name suggests contain more than one element, one of which is oxygen, e.g. carbonate (CO_3^{2-}) and nitrate (NO_3^-). Some elements form more than one compound ion and an example of this is sulfur, which exists as sulfate (SO_4^{2-}) and sulfite (SO_3^{2-}).

Carbonate ions

The test for carbonate ions (CO_3^{2-}) relies on the fact that carbonates react with dilute acids to release carbon dioxide. As this reaction takes place in solution, the gas formed is visible as effervescence. Proof that the gas formed is carbon dioxide is determined by bubbling the gas through limewater. If carbon dioxide is present, the limewater changes from clear to cloudy/milky (Figure 22.8).

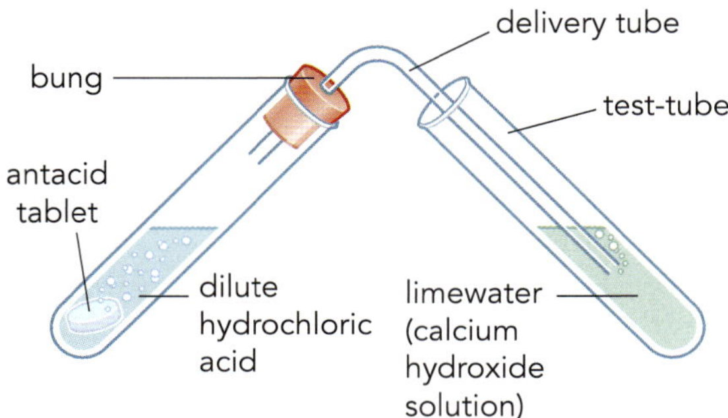

Figure 22.8: Testing for carbonate ions.

The reaction of solid magnesium carbonate with dilute hydrochloric is:

magnesium carbonate + hydrochloric acid → magnesium chloride + water + carbon dioxide

$$MgCO_3(s) + 2HCl(aq) \rightarrow MgCl_2(aq) + H_2O(l) + CO_2(g)$$

Halide ions

The test for halide ions (chloride (Cl^-), bromide (Br^-) and iodide (I^-)) is the formation of a precipitate when a solution of acidified silver nitrate is added. The test produces different colours of precipitate: a chloride forms a white precipitate; a bromide forms a cream precipitate and an iodide forms a yellow precipitate (Figure 22.9). The presence of carbonate ions in the test sample can interfere with this reaction and so they need to be removed. This is done by the addition of an acid, which reacts with carbonates to produce carbon dioxide, water and salt. The silver nitrate solution is acidified by adding a small amount of nitric acid.

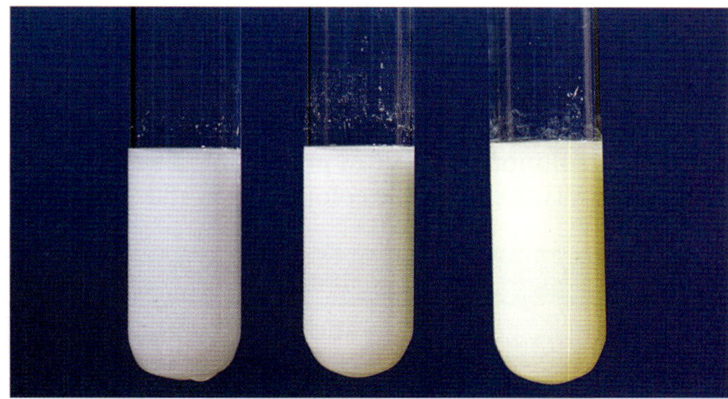

Figure 22.9: Precipitates produced in the tests for halide ions using acidified silver nitrate. The precipitates are silver chloride (white), silver bromide (cream) and silver iodide (yellow).

A solution containing sodium chloride reacts with silver nitrate solution to produce a white precipitate of silver chloride.

sodium chloride + silver nitrate → sodium nitrate + silver chloride

$$NaCl(aq) + AgNO_3(aq) \rightarrow NaNO_3(aq) + AgCl(s)$$

Sulfate ions

The test for sulfate ions (SO_4^{2-}) is the addition of acidified barium nitrate solution. As with the test for halide ions, the presence of carbonate ions can interfere with the results. The barium nitrate is therefore acidified using a small volume of nitric acid (or hydrochloric acid). If carbonate ions are present, then a white precipitate of barium carbonate will form. In the presence of sulfate ions, barium nitrate solution reacts to produce a white precipitate of barium sulfate.

A solution containing sodium sulfate reacts with barium nitrate solution to produce a white precipitate of barium sulfate and a solution of sodium nitrate.

barium nitrate + sodium sulfate → barium sulfate + sodium nitrate

$$Ba(NO_3)_2(aq) + Na_2SO_4(aq) \rightarrow BaSO_4(s) + 2NaNO_3(aq)$$

Sulfite ions

The most commonly occurring anion of sulfur is sulfate (SO_4^{2-}); however, sulfite ions (SO_3^{2-}) also exists. The test for sulfite uses its oxidation to sulfate by a transition metal ion. Some transition metal compounds are excellent oxidising agents, e.g. acidified potassium manganate(VII), $KMnO_4$. A solution of acidified potassium manganate(VII) has a purple colour. If it is added to a solution containing sulfite ions, a redox reaction takes place (Chapter 10) and the resulting solution is colourless.

Nitrate ions

The standard test for nitrate ions (NO_3^-) involves adding aqueous sodium hydroxide, aluminium foil and then heating gently. As with the test for sulfite ions, this involves a redox reaction. The nitrate ions are reduced to produce ammonia (NH_3) and the aluminium metal is oxidised to give Al^{3+}. The ammonia produced is a gas and can be tested by placing damp red litmus paper above the neck of the reaction vessel. The positive result for ammonia is a colour change of the damp red litmus from red to blue.

Summary of tests for anions

The tests for the anions are unique to individual ions, except for the Group VII halide ions. The presence of carbonate ions can lead to false results if not removed because most carbonates are insoluble. This need to acidify reagents to remove carbonate requires care to ensure the correct acid/reagent pair are selected. For example, using sulfuric acid to acidify barium nitrate solution would introduce sulfate ions. This would lead to the immediate precipitation of barium sulfate before adding it to any test sample.

Table 22.4 summarises the tests for anions.

Anion	Test	Test result
carbonate (CO_3^{2-})	add dilute hydrochloric acid; test for carbon dioxide gas	effervescence (fizzing), carbon dioxide produced (test with limewater)
chloride (Cl^-) (in solution)	acidify with dilute nitric acid, then add aqueous silver nitrate	white precipitate of silver chloride formed
bromide (Br^-) (in solution)	acidify with dilute nitric acid, then add aqueous silver nitrate	cream precipitate of silver bromide formed
iodide (I^-) (in solution)	acidify with dilute nitric acid, then add aqueous silver nitrate	yellow precipitate of silver iodide formed
sulfate (SO_4^{2-}) (in solution)	acidify with dilute nitric acid (or hydrochloric acid), then add barium nitrate solution	white precipitate of barium sulfate formed
sulfite (SO_3^{2-})	add dilute hydrochloric acid, then add aqueous potassium manganate(VII) solution	decolourises the purple potassium manganate(VII) solution
nitrate (NO_3^-) (in solution)	add aqueous sodium hydroxide, then add aluminium foil and warm carefully; test for ammonia gas	ammonia gas given off (test with moist red litmus)

Table 22.4: Tests for anions.

Questions

11 The test for carbonate ions can use any dilute acid. Hydrochloric acid is most commonly used but dilute sulfuric acid or nitric acid would also work. Explain why the type of acid used does not matter. Your answer should refer to the equations for each acid when added to sodium carbonate solution.

12 Two solutions were prepared, one of magnesium iodide (MgI_2) and the other of magnesium chloride ($MgCl_2$). How could acidified silver nitrate ($AgNO_3$) be used to show which solution contained the magnesium iodide? Give equations for the two reactions.

13 A student testing for sulfate ions used barium nitrate acidified with dilute sulfuric acid instead of the recommended hydrochloric acid. Would this change in acid have any impact on the student's results? Explain your answer fully.

14 In the test for nitrate ions (NO_3^-), the nitrate ion is reduced to ammonia (NH_3) and the aluminium is oxidised. Write a half equation for the oxidation of aluminium metal to aluminium ions.

ACTIVITY 22.1

Memorising the standard tests for ions

Work with a partner to produce a set of flash cards for the standard tests for ions (Figure 22.10). One person could focus on the aqueous anions and the other person could focus on the aqueous cations.

On the front side of each flash card, write the formula and name of the ion being tested. The cations should include Li^+, Na^+, K^+, Ba^{2+}, Ca^{2+}, Cu^{2+}, Fe^{2+}, Fe^{3+}, Cr^{2+}, Zn^{2+} and NH_4^+. The anions should include Cl^-, Br^-, I^-, SO_4^{2-}, SO_3^{2-}, NO_3^- and CO_3^{2-}. To help, you might use two different colours for the cards, one colour for cations and the other colour for anions.

On the reverse of each flash card give the test/reagent and result, e.g. Li^+ would be 'Test/reagent = flame test', 'Result = red'. Some ions would require two tests, e.g. Al^{3+} would be 'Test/reagent = NaOH(aq)', 'Result = white precipitate that will redissolve' and 'Test/reagent = NH_3(aq)', 'Result = white precipitate'

Once you have produced a set of flash cards, covering both the anions and cations, follow the instructions that follow:

1 Each person selects five cards at random from the set of flash cards and spends one minute learning what is on them.

2 After one minute, pass your cards to your partner.

3 Take it in turns to test each other by stating the standard test and result and asking for the ion. For example, 'Which ion could be tested for by using the flame test and gives a blue-green flame?' If your partner gives the correct answer (Cu^{2+}), give their card back, if not you can keep it.

4 Keep taking alternate turns to test each other until all five cards have been used.

5 At the end of the round, count up the number of cards each of you have. The player with the most cards wins that round.

You can use your set of flash cards to play further rounds of the game by:

- swapping the original set of five cards between you and your partner
- shuffle all of the cards and then deal each player five new cards
- test by stating the ion and asking for the standard test
- test by stating the ion and asking for the standard test and its result.

Self-assessment

How many cards did you win? What tests do you know really well? Which tests do you think you could learn more about? What other information could you have included on the cards to help you remember the tests? Did you think about using online software for creating virtual flash cards?

22.3 Tests to identify common gases

Animals and birds have been used to detect the presence of low levels of chemicals in a wide range of situations. Detection ('sniffer') dogs are trained to use their sense of smell to search for illegal drugs and canaries have been used in coal mines to detect the build-up of carbon monoxide and other toxic gases before they affect humans. Analytical chemists still use very simple tests to detect common gases such as hydrogen, oxygen, carbon dioxide, chlorine, ammonia and sulfur dioxide.

Hydrogen

Hydrogen (H_2) is a colourless and odourless gas. The standard test for hydrogen is to place a lighted splint at the neck of a test-tube. If hydrogen is present, there will be a characteristic 'squeaky pop' sound due to a small explosion as the hydrogen and oxygen rapidly react. The test is often referred to as 'the squeaky pop test' and the reaction involved is the rapid combustion of hydrogen to produce water.

Oxygen

Oxygen (O_2) is a colourless, odourless gas. The standard test for oxygen is to place a glowing splint into a test-tube of the gas. If oxygen is present the splint will relight (Figure 22.10). This is because substances burn much more rapidly in pure oxygen.

Figure 22.10: A glowing splint being relit by oxygen in a test-tube.

Carbon dioxide

Carbon dioxide (CO_2) is a colourless, odourless gas. To test for carbon dioxide the gas is bubbled through limewater (a dilute aqueous solution of calcium hydroxide) (Figure 22.11). If carbon dioxide is present the limewater turns from clear and colourless to cloudy/milky due to the formation of insoluble calcium carbonate.

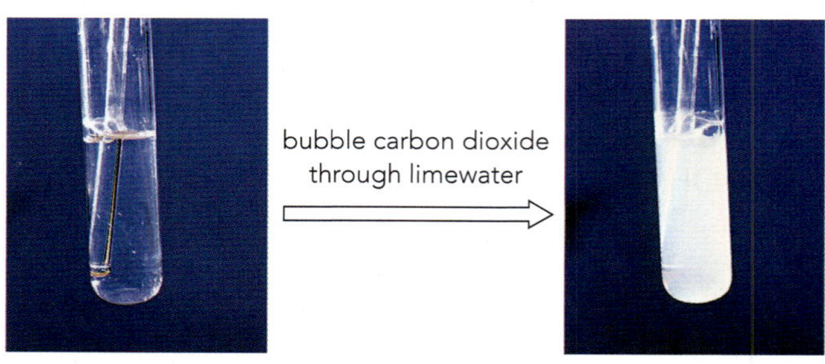

Figure 22.11: The limewater test for carbon dioxide.

Chlorine

Chlorine (Cl_2) is a toxic, green gas with a distinctive odour. The standard test for chlorine is to use damp blue litmus paper. The damp blue litmus paper will initially turn red before being bleached white (Figure 22.12). The chlorine gas reacts with the water to form hydrochloric acid (HCl) and hydrogen chlorate(I) (hypochlorous acid, HClO). The presence of hydrochloric acidic results in the blue litmus paper turning red. The hypochlorous acid is a bleaching agent and causes the litmus paper to turn white.

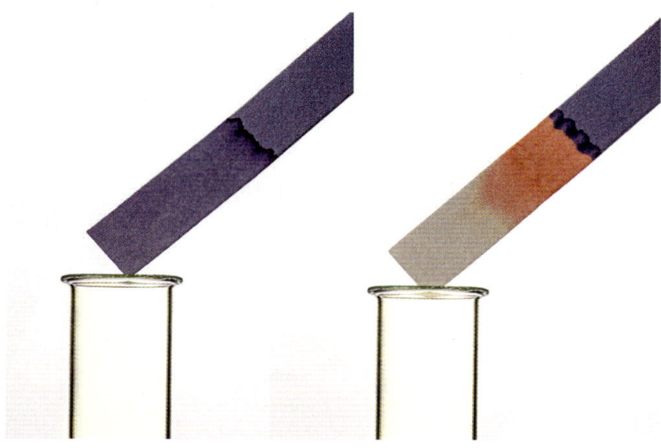

Figure 22.12: Testing for chlorine. Damp blue litmus paper is held over a test-tube of chlorine. The chlorine vapour turns the strip red and then bleaches it white.

If damp red litmus is used for the test for chlorine than it is simply bleached white.

Ammonia

Ammonia (NH_3) is a colourless gas with a distinctive, pungent odour. The standard test for ammonia gas is to use damp red litmus paper. The damp red litmus paper will turn blue. This is because ammonia will dissolve in water to form a weak base (ammonium hydroxide).

Sulfur dioxide

Sulfur dioxide (SO_2) is a colourless, toxic gas with an unpleasant odour. The standard test for sulfur dioxide is to pass it over a piece of filter paper dampened with a solution of acidified potassium manganate(VII). The purple colour of potassium manganate(VII) is removed and the solution becomes colourless. The colour change is due to the manganate(VII) ions that are deep purple being reduced to form manganese(II) sulfate, which is very pale pink (almost colourless). During this reaction the sulfur dioxide is oxidised to form sulfuric acid.

Summary of tests for common gases

Chemists need to check for the presence of different gases. For example, in industrial settings being able to detect leaks of chlorine, ammonia and sulfur dioxide can reduce the risk of serious accidents. Table 22.5 summarises the test used for a particular gas and the positive result given. Chlorine is acutely toxic, oxidising and hazardous to the aquatic environment and sulfur dioxide is corrosive and acutely toxic so tests involving these gases should be carried out in a well-ventillated laboratory or a fume cupboard.

Gas	Test	Positive result
Hydrogen (H_2)	Lit splint	Produces a squeaky pop
Oxygen (O_2)	Glowing splint	The splint is relit
Carbon dioxide (CO_2)	Bubble through limewater	Turns from clear to cloudy/milky
Chlorine (Cl_2)	Damp litmus paper	Bleached white
Ammonia (NH_3)	Damp red litmus paper	Turns from red to blue
Sulfur dioxide (SO_2)	Solution of acidified potassium manganate(VII)	Decolourised (purple to colourless)

Table 22.5: Summary of tests for gases.

Questions

15 A colourless, odourless gas produced a white precipitate when bubbled through limewater.

- **a** What gas was present?
- **b** Give the balanced symbol equation for the reaction between the gas and the limewater (calcium hydroxide).

16 A colourless, odourless gas did not relight a glowing splint and did not turn limewater cloudy. What conclusions can you make about this gas?

17 The test for sulfur dioxide involves the reaction with acidified potassium manganate(VII).

- **a** Why is there a colour change?
- **b** Given that this is a redox reaction (one which involves oxidation and reduction), using the equation, identify which element is oxidised and which is reduced.

$$2KMnO_4(aq) + 5SO_2(g) + 2H_2O$$
$$\rightarrow K_2SO_4 + 2MnSO_4(aq) + 2H_2SO_4(aq)$$

22.4 Quantitative analysis: acid–base titrations

Titration, a form of **quantitative** analysis, is used to determine the concentration of a wide range of substances. Titrations are used in many industries, including agriculture, food production, cosmetic manufacturing, pharmaceutical companies, water works and mining organisations. Here we will consider the use of titrations in determining the concentrations in an **acid–base titration**.

In an acid–base titration, a specified, fixed volume of base is placed into a conical flask by means of a volumetric pipette and filler. Typical pipette volumes are either $10.0\,cm^3$ or $25.0\,cm^3$ and the pipettes used are highly accurate.

The acid is dispensed using a burette, a piece of glassware that gives an accurate but variable volume of liquid. After carefully recording the start volume, the acid is added to the base by opening the tap at the bottom of the burette (Figure 22.13). The burette should be clamped vertically above the conical flask and a white tile may be placed under the conical flask to help see any colour change. The volume added from the burette during a titration is called the **titre** volume.

Figure 22.13: Performing an acid–base titration using methyl orange as indicator.

KEY WORDS

quantitative (analysis): the process used to determine the amount or percentage of a substance in a given sample

acid–base titration: a method of quantitative chemical analysis where an acid is added slowly to a base until it has been neutralised

titre: the volume of solution added from the burette during a titration

To determine the point at which the alkali has been neutralised, acid–base titrations use an **indicator**. An indicator is a chemical that gives a clear colour change at the end-point for the reaction. Care must be taken to add the acid dropwise from the burette as the end-point is approached. The conical flask must also be swirled to ensure everything is mixed and the reaction is complete. There are many different indicators including methyl orange and thymolphthalein. Methyl orange turns from yellow in basic conditions to red in acidic conditions, whereas thymolphthalein is blue in basic conditions and colourless in acidic conditions (Chapter 11).

When measuring the volume added from the burette it is important to be very precise. Read carefully from the base of the meniscus and to an accuracy of $\pm 0.05\,cm^3$. You should read with your eyes positioned horizontal to the meniscus (Figure 22.14). Two readings need to be taken from the burette: the start volume and end volume. The volume of liquid added from the burette is calculated as the difference between these two values.

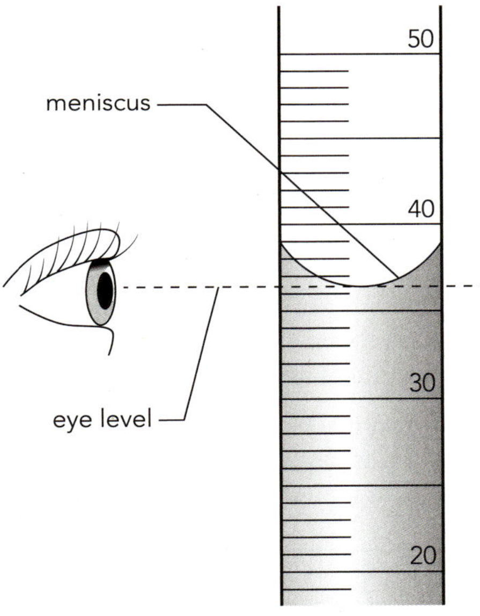

Figure 22.14: How to take an accurate burette reading.

KEY WORD

indicator: a substance that changes colour when added to acidic or alkaline solutions, e.g. litmus or thymolphthalein

Titrations are repeated until three consecutive results within $\pm 0.10\,cm^3$ are obtained. The first run in a titration experiment is a trial or pilot run. As the end-point is unknown, this first run is used to give an approximate titre volume. Any subsequent titrations will be added dropwise as the end-point is approached and then a mean titre should be calculated. See Chapter 5 for details of the calculations required to work out concentration in an acid–base titration.

The titration described here is for a titration involving solutions of an acid and an alkali (e.g. the titration method for preparing a soluble salt in Chapter 12). Titrations can also be carried out using an acid solution and an insoluble base (see Experimental Skills 11.1 in Chapter 11).

Questions

18 A titration uses two accurate pieces of glassware. What are they and how do they differ?

19 Why should methyl orange or thymolphthalein be used to detect the end-point of an acid–base titration rather than using Universal Indicator?

20 When carrying out a titration the method states 'the conical flask must also be swirled'. Why is this important?

21 A titration was repeated five times and the titre volumes were: $12.70\,cm^3$, $12.15\,cm^3$, $12.35\,cm^3$, $12.10\,cm^3$ and $12.10\,cm^3$. If these results were to be used to calculate a concentration, what was the mean titre?

REFLECTION

To identify different cations, anions and gases you will need to know the results of many different tests. There are different approaches to remembering this type of information:

- You could spend time repeating the information for each test and result or ask your peers to ask you questions.
- You could write a mnemonic (a sentence that helps you to recall a list of words). For example, you may already use **OIL RIG** as a mnemonic to remind yourself that **O**xidation **I**s **L**oss (of electrons) **R**eduction **I**s **G**ain (of electrons).
- You might even consider writing a short story: In the **Li**thuanian forest was a **red**wood cabin which **Na**talie entered, removing her **yellow** cap. **K**ristoff in his **lilac** jumper sat with a **Ca**t by the **orange-red** glow of the fire. Hidden here are the first four flame colours: Li (red), Na (yellow), K (lilac) and Ca (orange-red).

Which of these approaches do you think works best for you? How could you assess which approach is most effective in learning the standard tests for ions and gases?

SUMMARY

Flame tests can be used to identify the presence of metal cations based on the characteristic colour of the flames produced: Li^+(red), Na^+(yellow), K^+(lilac), Ca^{2+}(orange–red), Ba^{2+}(light green) and Cu^{2+}(blue–green).
Precipitation reactions with aqueous solutions of sodium hydroxide or ammonia can be used to test for the metal aqueous cations Al^{3+}, Ca^{2+}, Cu^{2+}, Fe^{2+}, Fe^{3+}, Cr^{3+} and Zn^2.
Sodium hydroxide can also be used to test for ammonium ions, which are reduced to ammonia.
To test for anions: carbonate ions are detected by effervescence with dilute acid, which produces carbon dioxide. Halide ions are detected by precipitation reactions with acidified silver nitrate. Sulfate ions are detected by a precipitation reaction with acidified barium nitrate. Sulfite ions are detected by a redox reaction with acidified potassium manganate(VII) which produces a visible colour change (purple to colourless). Nitrate ions are detected by a redox reaction using aluminium, which produces ammonia gas.
The tests for common gases are: hydrogen gives a squeaky pop with a lit splint, oxygen relights a glowing splint, carbon dioxide turns limewater milky, chlorine bleaches damp litmus paper. ammonia turns damp red litmus paper blue and sulfur dioxide turns acidified aqueous potassium manganate(VII) from purple to colourless.
Quantitative analysis provides information about the amount of substance present.
Acid–base titrations are a commonly used form of quantitative analysis, which use highly accurate glassware (the pipette and burette).
The end-point of a titration is detected by a colour change in an indicator: methyl orange turns from yellow to red (basic to acidic conditions), whereas thymolphthalein turns from blue to colourless (basic to acidic conditions).

PROJECT

The chemical detective

A very busy technician had taken five bottles containing white crystalline solids out of the store cupboard for use by the school's chemistry teacher (Figure 22.15). The samples were old and unfortunately the labels fell off. The teacher ended up with a tray of five bottles each with a different white solid and five labels: sodium carbonate, sodium chloride, zinc chloride, ammonium sulfate and ammonium carbonate.

Rather than disposing of the chemicals and buying new ones, working with a partner can you identify the lowest number of tests the teacher would need to positively identify all five compounds?

Figure 22.15: Correct labelling of reagent bottles is important to ensure laboratory safety.

You will need to write an experimental plan which your classmates can then evaluate. The plan should include: a list of chemicals, a list of tests and the methods for each experiment, and the positive test results that would be expected.

To evaluate another group's plan, consider the following points:

1 Have all the chemical solutions required been stated?

2 Are the methods given clear/logical?

3 Does their plan lead to valid results?

EXAM-STYLE QUESTIONS

1 A solution of an unknown salt is tested in two ways. First, excess ammonia solution is added to one portion of the solution and a white precipitate is formed that then re-dissolves. A second portion of the unknown salt is tested with excess sodium hydroxide solution. Again, a white precipitate is formed that re-dissolves.

What metal ion is present in the unknown solution?

A Ca^{2+}
B Al^{3+}
C Cu^{2+}
D Zn^{2} **[1]**

2 A mixture of two solid salts (**X** and **Y**) was tested. The mixture was a white powder. The mixture was added to water in a boiling tube and the mixture was shaken. Some of the solid dissolved and the resulting solution was filtered.

a Why was the solution filtered? **[1]**

Tests were carried out separately on the filtrate and the residue.

Tests on filtrate

Aqueous sodium hydroxide was added slowly until in excess	A white precipitate formed that re-dissolved
Aqueous ammonia was added slowly until in excess	A white precipitate formed that remained in excess
An acidified solution of silver nitrate was added	No react on was observed
An acidified solution of barium nitrate was added	A white precipitate was formed

Tests on residue

Dilute hydrochloric acid was added	Effervescence occurred producing a gas that turned lime water milky
Aqueous sodium hydroxide was added slowly until in excess	A white precipitate that did not redissolve was formed

b **Identify** the salt present in the:

i Filtrate (X) **[2]**

ii Residue (Y) **[2]**

c Which salt was soluble in water? **[1]**

[Total: 6]

COMMAND WORD

identify: name/select/recognise

CONTINUED

3 The table shows some salts and the results of tests carried out on the salts.

Complete the table.

salt	flame test	slowly adding excess aqueous sodium hydroxide	adding acidified silver nitrate
potassium sulfate		no reaction	
	green colour	no reaction	cream precipitate
lithium iodide		no reaction	
	no colour produced	green precipitate, re-dissolves	white precipitate

[6]

4 a What two tests can you do to prove that a sample of pale green crystals is iron(II) sulfate? **State** the tests and give the observed result of each test.

i Test I [2]

ii Test II [2]

b A white crystalline salt of sodium produces a gas when dilute hydrochloric acid is added to it. What two tests can you do to show whether the substance is sodium carbonate, sodium sulfite or neither? State the tests and **give** the observation expected with a positive test.

i Test A [2]

ii Test B [2]

[Total: 8]

5 Vinegar contains ethanoic acid and can be made from wine or malt (germinated and dried cereal grains). You are provided with two bottles of vinegar, one bottle of vinegar made from wine and one bottle of vinegar made from malt. Design an experiment to discover which of these vinegars contains the more concentrated solution of ethanoic acid.

Your answer should include.

- details of the apparatus used
- any measurements made
- how you decide which vinegar contains the most acid. [6]

COMMAND WORDS

state: express in clear terms

give: produce an answer from a given source or recall/memory

SELF-EVALUATION CHECKLIST

After studying this chapter, think about how confident you are with the different topics. This will help you to see any gaps in your knowledge and help you to learn more effectively.

I can	See Topic...	Needs more work	Almost there	Confident to move on
give the characteristic colours for the flame test for the metal cations: Li^+, Na^+, K^+, Ca^{2+}, Ba^{2+} and Cu^{2+}	22.1			
describe how aqueous solutions of sodium hydroxide or ammonia can be used to identify the cations Al^{3+}, Ca^{2+}, Cu^{2+}, Fe^{2+}, Fe^{3+}, Cr^{3+} and Zn^{2+}	22.1			
describe a simple test for the presence of ammonium ions	22.1			
describe the tests to identify carbonate, halide, sulfate, sulfite and nitrate anions	22.2			
describe the tests to identify different gases (hydrogen, oxygen, carbon dioxide, chlorine, ammonia and sulfur dioxide)	22.3			
state the difference between quantitative and qualitative analysis	22.4			
explain what a burette and a pipette are used for in a titration experiment	22.4			
describe how to perform an acid–base titration	22.4			
give the colour changes for methyl orange and thymolphthalein used to identify the end-point of an acid–base titration	22.4			

Glossary

Command Words

Below are the Cambridge International definitions for command words which may be used in exams. The information in this section is taken from the Cambridge International syllabus (0620/0971) for examination from 2023. You should always refer to the appropriate syllabus document for the year of your examination to confirm the details and for more information. The syllabus document is available on the Cambridge International website www.cambridgeinternational.org.

analyse: examine in detail to show meaning, identify elements and the relationship between them

calculate: work out from given facts, figures or information

compare: identify/comment on similarities and/or differences

consider: review and respond to given information

contrast: identify/comment on differences

deduce: conclude from available information

define: give precise meaning

demonstrate: show how or give an example

describe: state the points of a topic / give characteristics and main features

determine: establish an answer using the information available

evaluate: judge or calculate the quality, importance, amount, or value of something

examine: investigate closely, in detail

explain: set out purposes or reasons/make the relationships between things evident/provide why and/or how and support with relevant evidence

give: produce an answer from a given source or recall/memory

identify: name/select/recognise

justify: support a case with evidence/argument

predict: suggest what may happen based on available information

show (that): provide structured evidence that leads to a given result

sketch: make a simple freehand drawing showing the key features, taking care over proportions

state: express in clear terms

suggest: apply knowledge and understanding to situations where there are a range of valid responses in order to make proposals / put forward considerations

Key Words

accuracy: how close a value is to the true value

acid: a substance that dissolves in water, producing $H^+(aq)$ ions – a solution of an acid turns litmus red and has a pH below 7.
Acids act as proton donors

acid base reaction: (see **neutralisation**)

acid-base titration: a method of quantitative chemical analysis where an acid is added slowly to a base until it has been neutralised

acid rain: rain that has been made more acidic than normal by the presence of dissolved pollutants such as sulfur dioxide (SO_2) and oxides of nitrogen (nitrogen oxides, NO_x)

acidic oxides: oxides of non-metals that will react with bases and dissolve in water to produce acid solutions

activation energy (E_a): the minimum energy required to start a chemical reaction – for a reaction to take place the colliding particles must possess at least this amount of energy

addition polymer: a polymer formed by an addition reaction – the monomer molecules contain a C=C double bond

addition reaction: a reaction in which a simple molecule adds across the carbon–carbon double bond of an alkene

adsorption: the attachment of molecules to a solid surface

alcohols: a series of organic compounds containing the functional group –OH and with the general formula $C_nH_{2n+1}OH$

alkali metals: elements in Group I of the Periodic Table; they are the most reactive group of metals

alkalis: soluble bases that produce $OH^-(aq)$ ions in water – a solution of an alkali turns litmus blue and has a pH above 7

alkanes: a series of hydrocarbons with the general formula C_nH_{2n+2}; they are saturated compounds as they have only single bonds between carbon atoms in their structure

alkenes: a series of hydrocarbons with the general formula C_nH_{2n}; they are unsaturated molecules as they have a C=C double bond somewhere in the chain

alloys: mixtures of elements (usually metals) designed to have the properties useful for a particular purpose, e.g. solder (an alloy of tin and lead) has a low melting point

amide link (or **peptide link**)**:** the link between monomers in a protein or nylon, formed by a condensation reaction between a carboxylic acid group on one monomer and an amine group on the next monomer

amino acids: naturally occurring organic compounds that possess both an amino ($-NH_2$) group and an acid (–COOH) group in the molecule; there are 20 naturally occurring amino acids and they are polymerised in cells to make proteins

amphoteric compound: a compound (hydroxide or metal oxide) that reacts with both an acid and an alkali to give a salt and water

anion: a negative ion which would be attracted to the anode in electrolysis

anhydrous: an adjective to describe a substance without water combined with it

anode: the electrode in any type of cell at which oxidation (the loss of electrons) takes place – in electrolysis it is the positive electrode

anomalous: something that is unusual or unexpected and deviates from the normal; one of a series of repeated experimental results that is much larger or smaller than the others is an anomalous result

antacids: compounds used medically to treat indigestion by neutralising excess stomach acid

anti-bumping granules: small granules that help reduce the size of bubbles formed when a liquid boils; used for safety to stop the flask shaking

atmosphere: the layer of air and water vapour surrounding the Earth

atom: the smallest particle of an element that can take part in a chemical reaction

atomic number (or **proton number**) **(*Z*):** the number of protons in the nucleus of an atom

atomic theory: a model of the atom in which electrons can only occupy certain electron shells (energy levels) moving outwards from the nucleus of an atom

Avogadro constant: the number (6×10^{23}) of characteristic particles in 1 mole of a substance

Avogadro's law: equal volumes of any gas, under the same conditions of temperature and pressure, contain the same number of particles

balanced chemical (symbol) equation: a summary of a chemical reaction using chemical formulae – the total number of any of the atoms involved is the same on both the reactant and product sides of the equation

base: a substance that neutralises an acid, producing a salt and water as the only products.
Bases act as proton acceptors

basic oxide: oxide of a metal that will react with acids to neutralise the acid

bauxite: the major ore of aluminium; a form of aluminium oxide, Al_2O_3

biodegradable: a substance that can be broken down, or decomposed, by microorganisms

blast furnace: a furnace for extracting metals (particularly iron) by reduction with carbon that uses hot air blasted in at the base of the furnace to raise the temperature

boiling: the process of change from liquid to gas at the boiling point of the substance; a condition under which gas bubbles are able to form within a liquid – gas molecules escape from the body of a liquid, not just from its surface

boiling point: the temperature at which a liquid boils, when the pressure of the gas created above the liquid equals atmospheric pressure

bond energy: the energy required to break a particular type of covalent bond

brass: an alloy of copper and zinc; this alloy is hard

burette: a piece of glass apparatus used for delivering a variable volume of liquid accurately

calibration: the process of checking that the device gives accurate values by using it to read samples with known values

carboxylic acids: a homologous series of organic compounds containing the functional group –COOH (–CO_2H), with the general formula $C_nH_{2n+1}COOH$

catalyst: a substance that increases the rate of a chemical reaction but itself remains unchanged at the end of the reaction

catalytic converter: a device for converting polluting exhaust gases from cars into less dangerous emissions

catalytic cracking: the decomposition of long-chain alkanes into alkenes and alkanes of lower relative molecular mass; involves passing the larger alkane molecules over a catalyst heated to 500 °C

cathode: the electrode in any type of cell at which reduction (the gain of electrons) takes place; in electrolysis it is the negative electrode

cation: a positive ion which would be attracted to the cathode in electrolysis

ceramic: material such as pottery made from inorganic chemicals by high-temperature processing

chemical bonding: the strong forces that hold atoms (or ions) together in the various structures that chemical substances can form – metallic bonding, covalent bonding and ionic bonding

chemical feedstock: a chemical element or compound which can be used as a raw material for an industrial process making useful chemical products

chemical formula: a shorthand method of representing chemical elements and compounds using the symbols of the elements

chemical reaction (change): a change in which a new substance is formed

chemical symbol: a letter or group of letters representing an element in a chemical formula

chemiluminescence: light given out by certain chemical reactions

chlorophyll: a green pigment in plants which traps energy from the sun in photosynthesis

chromatogram: the result of a paper chromatography run, showing where the spots of the samples have moved to

chromatography: a technique employed for the separation of mixtures of dissolved substances, which was originally used to separate coloured dyes

clean dry air: containing no water vapour and only the gases which are always present in the air

climate change: changes in weather patterns brought about by global warming

closed system: a system where none of the reactants or products can escape the reaction mixture or the container where the reaction is taking place

coal: a black, solid fossil fuel formed underground over geological periods of time by conditions of high pressure and temperature acting on decayed vegetation

collision theory: a theory which states that a chemical reaction takes place when particles of the reactants collide with sufficient energy to initiate the reaction

combustion: a chemical reaction in which a substance reacts with oxygen – the reaction is exothermic

complete combustion: (*see also* **incomplete combustion**) a type of combustion reaction in which a fuel is burned in a plentiful supply of oxygen; the complete combustion of hydrocarbon fuels produces only carbon dioxide and water

compound: a substance formed by the chemical combination of two or more elements in fixed proportions

compound fertiliser: a fertiliser such as a NPK fertilizer or nitrochalk that contains more than one compound to provide elements to the soil

compound ion: an ion made up of several different atoms covalently bonded together and with an overall charge (can also be called a molecular ion; negatively charged compound ions containing oxygen can be called oxyanions)

compromise temperature: a temperature that gives sufficient product and a reasonable and economic rate of reaction

condensation: the change of a vapour or a gas into a liquid; during this process heat is given out to the surroundings

condensation polymer: a polymer formed by a condensation reaction, e.g. nylon is produced by the condensation reaction between 1,6-diaminohexane and hexanedioic acid; this is the type of polymerisation used in biological systems to produce proteins, nucleic acids and polysaccharides

condensation reaction: a reaction where two or more substances combine together to make a larger compound, and a small molecule is eliminated (given off)

concentration: a measure of how much solute is dissolved in a solvent to make a solution. Solutions can be dilute (with a high proportion of solvent), or concentrated (with a high proportion of solute)

Contact process: the industrial manufacture of sulfuric acid using the raw materials sulfur and air

control variable: a variable that is kept the same during an investigation

corrosion: the process that takes place when metals and alloys are chemically attacked by oxygen, water or any other substances found in their immediate environment

corrosive: a corrosive substance (e.g. an acid) is one that can dissolve or 'eat away' at other materials (e.g. wood, metals or human skin)

covalent bonding: chemical bonding formed by the sharing of one or more pairs of electrons between two atoms

cryolite: sodium aluminium fluoride (Na_3AlF_6), an ore of aluminium used in the extraction of aluminium to lower the operating temperature of the electrolytic cell. Now replaced by synthetic sodium aluminium fluoride produced from the common mineral fluorite.

crystallisation: the process of forming crystals from a saturated solution

decanting: the process of removing a liquid from a solid which has settled or from an immiscible heavier liquid by careful pouring

decomposition: (*see also* **thermal decomposition**) a type of chemical reaction where a compound breaks down into simpler substances

dehydration: a chemical reaction in which water is removed from a compound

density: expresses the relationship between the mass of a substance and the volume it occupies: density = mass/volume

dependent variable: the variable which is measured during a scientific investigation

desalination: the removal of dissolved salts and minerals from seawater to produce drinking water

desulfurisation: an industrial process for removing contaminating sulfur from fossil fuels such as petrol (gasoline) or diesel

diatomic molecules: molecules containing two atoms, e.g. hydrogen, H_2

diffusion: the process by which different fluids mix as a result of the random motions of their particles

displacement reaction: a reaction in which a more reactive element displaces a less reactive element from a solution of its salt

displayed formula: a representation of the structure of a compound which shows all the atoms and bonds in the molecule

dissociation: the separation of a covalent molecule into ions when dissolved in water

distillate: the liquid collected in the receiving flask during distillation

distillation: the process of boiling a liquid and then condensing the vapour produced back into a liquid: used to purify liquids and to separate liquids from solutions

dot-and-cross diagram: a diagram drawn to represent the bonding in a molecule, or the electrons in an ion; usually only the outer electrons are shown and they are represented by dots or crosses depending on which atom they are from

downward delivery: a method of collecting a gas which is denser than air by passing it downwards into a gas jar

ductile: a word used to describe the property that metals can be drawn out and stretched into wires

ductility: the ability of a substance to be drawn out into a wire

dynamic (chemical) equilibrium: two chemical reactions, one the reverse of the other, taking place at the same time, where the concentrations of the reactants and products remain constant because the rate at which the forward reaction occurs is the same as that of the reverse reaction

electrical conductivity: the ability to conduct electricity

electrical conductor: a substance that conducts electricity but is not chemically changed in the process

electrodes: the points where the electric current enters or leaves a battery or electrolytic cell

electrolysis: the breakdown of an ionic compound, molten or in aqueous solution, by the use of electricity

electrolyte: an ionic compound that will conduct electricity when it is molten or dissolved in water; electrolytes will not conduct electricity when solid

electrolytic cell: a cell consisting of an electrolyte and two electrodes (anode and cathode) connected to an external DC power source where positive and negative ions in the electrolyte are separated and discharged

electron: a subatomic particle with negligible mass and a charge of −1; electrons are present in all atoms, located in the shells (energy levels) outside the nucleus

electronic configuration: a shorthand method of describing the arrangement of electrons within the electron shells (or energy levels) of an atom; also referred to as electronic structure

electron shells (energy levels): (of electrons) the allowed energies of electrons in atoms – electrons fill these shells (or levels) starting with the one closest to the nucleus

electroplating: a process of electrolysis in which a metal object is coated (plated) with a layer of another metal

electrostatic forces: strong forces of attraction between particles with opposite charges – such forces are involved in ionic bonding

element: a substance which cannot be further divided into simpler substances by chemical methods; all the atoms of an element contain the same number of protons

empirical formula: a formula for a compound which shows the simplest ratio of atoms present

end point: the point in a titration when the indicator just changes colour showing that the reaction is complete

endothermic changes: a process or chemical reaction which takes in heat from the surroundings. ΔH for an endothermic change has a positive value.

energy level diagram: *see* **reaction pathway diagram**

enthalpy (H): the thermal (heat) content of a system

enthalpy change (ΔH): the heat change during the course of a reaction (also known as heat of reaction); can be either exothermic (a negative value) or endothermic (a positive value)

enzymes: protein molecules that act as biological catalysts

ester link: the link produced when an ester is formed from a carboxylic acid and an alcohol; also found in polyesters and in the esters present in fats and vegetable oils

esterification: the chemical reaction between an alcohol and a carboxylic acid that produces an ester; the other product is water

esters: a family of organic compounds formed by esterification, characterised by strong and pleasant tastes and smells

evaporation: a process occurring at the surface of a liquid, involving the change of state from a liquid into a vapour at a temperature below the boiling point

exothermic changes: a process or chemical reaction in which heat energy is produced and released to the surroundings. ΔH for an exothermic change has a negative value.

fermentation: a reaction carried out using a living organism, usually a yeast or bacteria, to produce a useful chemical compound; most usually refers to the production of ethanol

fertiliser: a substance added to the soil to replace essential elements lost when crops are harvested, which enables crops to grow faster and increases the yield

filtrate: the liquid that passes through the filter paper during filtration

filtration: the separation of a solid from a liquid, using a fine filter paper which does not allow the solid to pass through

fluid: a gas or a liquid; they are able to flow

formula unit: this unit of an element or compound is the molecule or group of ions defined by the chemical formula of the substance

fossil fuels: fuels, such as coal, oil and natural gas, formed underground over geological periods of time from the remains of plants and animals

fractional distillation: a method of distillation using a fractionating column, used to separate liquids with different boiling points

fractionating column: the vertical column which is used to bring about the separation of liquids in fractional distillation

fractions (from distillation): the different mixtures that distil over at different temperatures during fractional distillation

freezing point: the temperature at which a liquid turns into a solid – it has the same value as the melting point; a pure substance has a sharp freezing point

fuel: a substance that can be used as a source of energy, usually by burning (combustion)

fuel cell: a device for continuously converting chemical energy into electrical energy using a combustion reaction; a hydrogen fuel cell uses the reaction between hydrogen and oxygen

functional group: the atom or group of atoms responsible for the characteristic reactions of a compound

galvanising: the protection of iron and steel objects by coating with a layer of zinc

giant covalent structures: a substance where large numbers of atoms are held together by covalent bonds forming a strong lattice structure

giant ionic lattice (structure): a lattice held together by the electrostatic forces of attraction between positive and negative ions

giant metallic lattice: a regular arrangement of positive metal ions held together by the mobile 'sea' of electrons moving between the ions

giant molecular lattice (structure): substance where large numbers of atoms are joined by covalent bonds forming a strong lattice structure

giant structures: these are lattices where the structure repeats itself in all directions; the forces involved are the same in all directions holding the whole structure together

global warming: a long-term increase in the average temperature of the Earth's surface, which may be caused in part by human activities

greenhouse effect: the natural phenomenon in which thermal energy from the Sun is 'trapped' at the Earth's surface by certain gases in the atmosphere (greenhouse gases)

greenhouse gas: a gas that absorbs thermal energy reflected from the surface of the Earth, stopping it escaping the atmosphere

group number: the number of the vertical column that an element is in on the Periodic Table

groups: vertical columns of the Periodic Table containing elements with similar chemical properties; atoms of elements in the same group have the same number of electrons in their outer energy levels

Haber process: the industrial manufacture of ammonia by the reaction of nitrogen with hydrogen in the presence of an iron catalyst

half-equations: ionic equations showing the reactions at the anode (oxidation) and cathode (reduction) in an electrolytic cell

halides: compounds formed between an element and a halogen, e.g. sodium iodide

halogen displacement reactions: reactions in which a more reactive halogen displaces a less reactive halogen from a solution of its salt

halogens: elements in Group VII of the Periodic Table – generally the most reactive group of non-metals

hematite: the major ore of iron, iron(III) oxide

homologous series: a family of similar compounds with similar chemical properties due to the presence of the same functional group

hydrated salts: salts whose crystals contain combined water (*water of crystallisation*) as part of the structure

hydrated substance: a substance that is chemically combined with water; hydrated salts are an important group of such substances

hydration: the addition of the elements of water across a carbon–carbon double bond; H- adds to one carbon, and -OH to the other

hydrocarbons: organic compounds which contain carbon and hydrogen only; the alkanes and alkenes are two series of hydrocarbons

hydrogenation: an addition reaction in which hydrogen is added across the double bond in an alkene

hydrolysis: a chemical reaction between a covalent compound and water; covalent bonds are broken during the reaction and the elements of water are added to the fragments; can be carried out with acids or alkalis, or by using enzymes

immiscible: if two liquids form two layers when they are mixed together, they are said to be immiscible

incomplete combustion: a type of combustion reaction in which a fuel is burned in a limited supply of oxygen; the incomplete combustion of hydrocarbon fuels produces carbon, carbon monoxide and water (see also **complete combustion**).

independent variable: the variable that is altered during a scientific investigation

indicator: a substance which changes colour when added to acidic or alkaline solutions, e.g. litmus or thymolphthalein

inert: term that describes substances that do not produce a chemical reaction when another substance is added

insoluble: a substance that does not dissolve in a particular solvent

insulator: a substance that does not conduct electricity

intermolecular forces: the weak attractive forces which act between molecules

intermolecular space: the space between atoms or molecules in a liquid or gas. The intermolecular space is small in a liquid, but relatively very large in a gas.

ionic bonding: a strong electrostatic force of attraction between oppositely charged ions

ionic equation: the simplified equation for a reaction involving ionic substances: only those ions which actually take part in the reaction are shown

ions: charged particles made from an atom, or groups of atoms (compound ions), by the loss or gain of electrons

isomerism: the property shown by molecules which have the same molecular formula but different structures

isomers: compounds which have the same molecular formula but different structural arrangements of the atoms – they have different structural formulae

isotopes: atoms of the same element which have the same proton number but a different nucleon number; they have different numbers of neutrons in their nuclei; some isotopes are radioactive because their nuclei are unstable (radioisotopes)

kinetic particle theory: a theory which accounts for the bulk properties of the different states of matter in terms of the movement of particles (atoms or molecules) – the theory explains what happens during changes in physical state

lattice: a regular three-dimensional arrangement of atoms, molecules or ions in a crystalline solid

law of conservation of mass: matter cannot be lost or gained in a chemical reaction – the total mass of the reactants equals the total mass of the products

lime: a white solid known chemically as calcium oxide (CaO), produced by heating limestone; it can be used to counteract soil acidity, to manufacture calcium hydroxide (slaked lime) and also as a drying agent

limestone: a form of calcium carbonate ($CaCO_3$)

limewater: a solution of calcium hydroxide in water; it is an alkali and is used in the test for carbon dioxide gas

limiting reactant: the reactant that is not in excess

litmus: the most common indicator; turns red in acid and blue in alkali

locating agent: a compound that reacts with invisible, colourless spots separated by chromatography to produce a coloured product which can be seen

main-group elements: the elements in the outer groups of the Periodic Table, excluding the transition elements (Groups I–VIII)

malleability: the ability of a substance to be bent or beaten into shape

malleable: a word used to describe the property that metals can be bent and beaten into sheets

mass: practical measure of quantity of a sample found by weighing on a balance

mass concentration: the measure of the concentration of a solution in terms of the mass of the solute, in grams, dissolved per cubic decimetre of solution (g/dm^3)

mass number (or **nucleon number**) **(*A*):** the total number of protons and neutrons in the nucleus of an atom

mass spectrometer: an instrument in which atoms or molecules are ionised and then accelerated; the ions are then separated according to their mass

matter: anything that occupies space and has mass

melting point (m.p): the temperature at which a solid turns into a liquid – it has the same value as the freezing point; a pure substance has a sharp melting point

metallic bonding: an electrostatic force of attraction between the mobile 'sea' of electrons and the regular array of positive metal ions within a solid metal

metalloid (semi-metal): element which shows some of the properties of metals and some of non-metals, e.g. boron and silicon

metals: a class of chemical elements (and alloys) which have a characteristic shiny appearance and are good conductors of heat and electricity

methyl orange: an acid–base indicator that is red in acidic and yellow in alkaline solutions

microplastics: small pieces of plastic less than 5mm in length that enter natural ecosystems from a variety of sources including cosmetics, clothing and industrial processes: nurdles and microbeads are different types of microplastic

mineral: a naturally occurring rock containing a particular compound

miscible: if two liquids form a completely uniform mixture when added together, they are said to be miscible

mixture: two or more substances mixed together but not chemically combined – the substances can be separated by physical means

molar concentration: the measure of the concentration of a solution in terms of the number of moles of the solute dissolved per cubic decimetre of solution (mol/dm^3)

molar gas volume: 1 mole of any gas has the same volume under the same conditions of temperature and pressure (24 dm^3 at r.t.p.)

molar mass: the mass, in grams, of 1 mole of a substance

mole: the measure of amount of substance in chemistry; 1 mole of a substance has a mass equal to its relative formula mass in grams – that amount of substance contains 6.02×10^{23} (the Avogadro constant) atoms, molecules or formula units depending on the substance considered

molecular formula: a formula that shows the actual number of atoms of each element present in a molecule of the compound

molecule: a group of atoms held together by covalent bonds

monomer: a small molecule, such as ethene, which can be polymerised to make a polymer

nanotechnology: the study and control of matter on an atomic and molecular scale; it is aimed at engineering working systems at this microscopic level

natural gas: a fossil fuel formed underground over geological periods of time by conditions of high pressure and temperature acting on the remains of dead sea creatures; natural gas is more than 90% methane

neutralisation: a chemical reaction between an acid and a base to produce a salt and water only; summarised by the ionic equation $H^+(aq) + OH^-(aq) \rightarrow H_2O(l)$

neutron: an uncharged subatomic particle present in the nuclei of atoms – a neutron has a mass of 1 relative to a proton

nitrogen fixation: the direct use of atmospheric nitrogen in the formation of important compounds of nitrogen; most plants cannot fix nitrogen directly, but bacteria present in the root nodules of certain plants are able to take nitrogen from the atmosphere to form essential protein molecules

noble gases: elements in Group VIII – a group of stable, very unreactive gases

non-electrolytes: liquids or solutions that do not take part in electrolysis: they do not contain ions

non-metals: a class of chemical elements that are typically poor conductors of heat and electricity

non-renewable (finite) resources: sources of energy, such as fossil fuels, and other resources formed in the Earth over many millions of years, which we are now using up at a rapid rate and cannot replace

NPK fertiliser: fertilisers to provide the elements nitrogen, phosphorus and potassium for improved plant growth

nucleus: (of an atom) the central region of an atom that is made up of the protons and neutrons of the atom; the electrons orbit around the nucleus in different 'shells' or 'energy levels'

ore: a naturally occurring mineral from which a metal can be extracted

organic chemistry: studies on the structure, properties and reactions of organic compounds, which contain carbon in covalent bonding

oxidation: there are three definitions of oxidation:
i a reaction in which oxygen is added to an element or compound
ii a reaction involving the loss of electrons from an atom, molecule or ion
iii a reaction in which the oxidation state of an element is increased

oxidation number: a number given to show whether an element has been oxidised or reduced; the oxidation number of a simple ion is simply the charge on the ion

oxidising agent: a substance which oxidises another substance during a redox reaction

paper chromatography: a simple type of chromatography used to separate the components of soluble substances based on their rate of migration in a solvent (mobile phase) on sheets of filter paper (stationary phase)

particulates: very tiny solid particles produced during the combustion of fuels

percentage composition: the percentage by mass of each element in a compound

percentage purity: a measure of the purity of the product from a reaction carried out experimentally:

$$\text{percentage purity} = \frac{\text{mass of pure product}}{\text{mass of impure product}} \times 100$$

percentage yield: a measure of the actual yield of a reaction when carried out experimentally compared to the theoretical yield calculated from the equation:

$$\text{percentage yield} = \frac{\text{actual yield}}{\text{predicted yield}} \times 100$$

period: a horizontal row of the Periodic Table

period (row) number: the horizontal row of the Periodic Table that an element is in

periodic property: a property of the elements that shows a repeating pattern when plotted against proton number (***Z***)

Periodic Table: a table of elements arranged in order of increasing proton number (atomic number) to show the similarities of the chemical elements with related electronic configurations

petrol (or **gasoline**)**:** a clear flammable liquid derived from petroleum used primarily as a fuel in most combustion engines; it is obtained by the fractional distillation of petroleum

petroleum (or **crude oil**)**:** a fossil fuel formed underground over many millions of years by conditions of high pressure and temperature acting on the remains of dead sea creatures

pH scale: a scale running from below 0 to 14, used for expressing the acidity or alkalinity of a solution; a neutral solution has a pH of 7

photochemical reaction: a chemical reaction where the activation energy required to start the reaction is provided by light, usually of a particular wavelength, falling on the reactants

photochemical smog: a form of local atmospheric pollution found in large cities in which several gases react with each other to produce harmful products

photosynthesis: the chemical process by which plants synthesise glucose from atmospheric carbon dioxide and water giving off oxygen as a by-product: the energy required for the process is captured from sunlight by chlorophyll molecules in the green leaves of the plants

physical change: a change in the physical state of a substance or the physical nature of a situation that does not involve a change in the chemical substance(s) present

plastics: polymers that can be moulded or shaped by the action of heat and pressure

pollutants: substances, often harmful, which are added to another substance

polyamide: a polymer where the monomer units are joined together by amide (peptide) links, e.g. nylon and proteins

polyester: a polymer where the monomer units are joined together by ester links, e.g. PET

polymer: a substance consisting of very large molecules made by polymerising a large number of repeating units or monomers

polymerisation: the chemical reaction in which molecules (monomers) join together to form a long-chain polymer

porous pot: an unglazed pot that has channels (pores) through which gases can pass

position of equilibrium: the mixture of reactants and products at which a reversible reaction is in equilibrium under a particular set of physical conditions of temperature and pressure

precipitate: an insoluble salt formed during a precipitation reaction

precipitation: the sudden formation of a solid when either two solutions are mixed or a gas is bubbled into a solution

precipitation reaction: a reaction in which an insoluble salt is prepared from solutions of two suitable soluble salts

precision: the degree by which repeat measurements are consistent (close to each other)

products: (in a chemical reaction) the substance(s) produced by a chemical reaction

proteins: polymers of amino acids formed by a condensation reaction; they have a wide variety of biological functions

proton: a subatomic particle with a relative mass of 1 and a charge of +1 found in the nucleus of all atoms

proton number (or **atomic number**) **(Z):** the number of protons in the nucleus of an atom (*see also* **atomic number**)

pure substance: a single chemical element or compound – it melts and boils at definite precise temperatures

qualitative (analysis): the process used to determine the presence or absence of a substance in a given sample

quantitative: the ability to put numerical values to the properties being studied

quantitative (analysis): the process used to determine the amount or percentage of a substance in a given sample

rancid: a term used to describe oxidised organic material (food) – usually involving a bad smell

random errors: these are unpredictable variations in results caused by factors such as human errors

reactants: (in a chemical reaction) the chemical substances that react together in a chemical reaction

reaction pathway diagram (energy level diagram): a diagram that shows the energy levels of the reactants and products in a chemical reaction and shows whether the reaction is exothermic or endothermic

reaction rate: a measure of how fast a reaction takes place

reactivity: the ease with which a chemical substance takes part in a chemical reaction

reactivity series of metals: an order of reactivity, giving the most reactive metal first, based on results from a range of experiments involving metals reacting with oxygen, water, dilute hydrochloric acid and metal salt solutions

redox reaction: a reaction involving both reduction and oxidation

reducing agent: a substance which reduces another substance during a redox reaction

reduction: there are three definitions of reduction:
i a reaction in which oxygen is removed from a compound
ii a reaction involving the gain of electrons by an atom, molecule or ion
iii a reaction in which the oxidation state of an element is decreased

refluxing: a practical technique using a (reflux) condenser fitted vertically to condense vapours from an experiment back into a flask

relative atomic mass (A_r): the average mass of naturally occurring atoms of an element on a scale where the carbon-12 atom has a mass of exactly 12 units

relative formula mass (M_r): the sum of all the relative atomic masses of the atoms present in a 'formula unit' of a substance (*see also* **relative molecular mass**)

relative molecular mass (M_r): the sum of all the relative atomic masses of the atoms present in a molecule (*see also* **relative formula mass**)

renewable resources: sources of energy and other resources which cannot run out provided they are managed sustainably, or which can be made at a rate faster than our current rate of use

repeatability: where an experiment (or series of experiments) can be repeated using the same method and obtain **reproducible** (similar) results

residue: the solid left behind in the filter paper after filtration has taken place

resolution: the smallest division on the instrument, e.g. this could be 1mm on a ruler or 0.1g on a balance

respiration: the chemical reaction (a combustion reaction) by which biological cells release the energy stored in glucose for use by the cell or the body; the reaction is exothermic and produces carbon dioxide and water as the chemical by-products

reversible reaction: a chemical reaction that can go either forwards or backwards, depending on the conditions

R_f value: in chromatography, the ratio of the distance travelled by the solute to the distance travelled by the solvent front

r.t.p.: room temperature and pressure: the standard values are 25 °C/298 K and 101.3 kPa/1 atmosphere pressure

run-off: water which travels over the surface of the land before entering waterways such as rivers and lakes; runoff from farmland may contain dissolved substances such as fertilisers

rust: a loose, orange–brown, flaky layer of hydrated iron(III) oxide, $Fe_2O_3{\cdot}xH_2O$, found on the surface of iron or steel

rusting: the corrosion of iron and steel to form rust (hydrated iron(III) oxide)

sacrificial protection: a method of rust protection involving the attachment of blocks of a metal more reactive than iron to a structure; this metal is corroded rather than the iron or steel structure

salts: ionic compounds made by the neutralisation of an acid with a base (or alkali), e.g. copper(II) sulfate and potassium nitrate

saturated hydrocarbons: hydrocarbons molecules in which all the carbon–carbon bonds are single covalent bonds

saturated solution: a solution that contains as much dissolved solute as possible at a particular temperature

'sea' of delocalised electrons: term used for the free, mobile electrons between the positive ions in a metallic lattice

significant figures: the number of digits in a number, not including any zeros at the beginning; for example the number of significant figures in 0.0682 is three

simple distillation: a distillation method for separating the liquid solvent from a solution containing dissolved solids

simple molecular substances: substances made up of individual molecules held together by covalent bonds: there are only weak forces between the molecules

slag: a molten mixture of impurities, mainly calcium silicate, formed in the blast furnace

solubility: a measure of how much of a solute dissolves in a solvent at a particular temperature

soluble: a solute that dissolves in a particular solvent

solute: the solid substance that has dissolved in a liquid (the solvent) to form a solution

solution: formed when a substance (solute) dissolves into another substance (solvent)

solvent: the liquid that dissolves the solid solute to form a solution; water is the most common solvent but liquids in organic chemistry that can act as solvents are called *organic solvents*

solvent front: the moving boundary of the liquid solvent that moves up the paper during chromatography

sonorous: a word to describe a metallic substance that rings like a bell when hit with a hammer

stainless steel: an alloy of iron that resists corrosion; this steel contains a significant proportion of chromium which results in the alloy being resistant to rusting

standard solution: a solution whose concentration is known precisely – this solution is then used to find the concentration of another solution by titration

state symbols: symbols used to show the physical state of the reactants and products in a chemical reaction: they are s (solid), l (liquid), g (gas) and aq (in solution in water)

states of matter: solid, liquid and gas are the three states of matter in which any substance can exist, depending on the conditions of temperature and pressure

stoichiometry: the ratio of the reactants and products in a balanced symbol equation

strong acid: an acid that is completely ionised when dissolved in water – this produces the highest possible concentration of $H^+(aq)$ ions in solution, e.g. hydrochloric acid

structural formula: the structural formula of an organic molecule shows how all the groups of atoms are arranged in the structure; ethanol, CH_3CH_2OH, for example

structural isomerism: a property of compounds that have the same molecular formula but different structural formulae; the individual compounds are known as structural isomers

subatomic particles: very small particles – protons, neutrons and electrons – from which all atoms are made

sublimation: the direct change of state from solid to gas or gas to solid: the liquid phase is bypassed

substitution reaction: a reaction in which an atom (or atoms) of a molecule is (are) replaced by different atom(s), without changing the molecule's general structure

suspension: a mixture containing small particles of an insoluble solid, or droplets of an insoluble liquid, spread (suspended) throughout a liquid

systematic errors: these are consistent errors which may arise due to a problem with the experimental design or in a piece of equipment being used

thermal conductivity: the ability to conduct heat

thymolphthalein: an acid–base indicator that is colourless in acidic solutions and blue in alkaline solutions

titration: a method of quantitative analysis using solutions: one solution is slowly added to a known volume of another solution using a burette until an end point is reached

titre: the volume of solution added from the burette during a titration

transition metals (transition elements): elements from the central region of the Periodic Table – they are hard, strong, dense metals that form compounds that are often coloured

universal indicator: a mixture of indicators that has different colours in solutions of different pH

unsaturated hydrocarbons: hydrocarbons whose molecules contain at least one carbon–carbon double or triple bond

volatile: term that describes a liquid that evaporates easily; it is a liquid with a low boiling point because there are only weak intermolecular forces between the molecules in the liquid

volatility: the property of how easily a liquid evaporates

volumetric pipette: a pipette used to measure out a volume of solution accurately

water of crystallisation: water included in the structure of certain salts as they crystallise, e.g. copper(II) sulfate pentahydrate ($CuSO_4{\cdot}5H_2O$) contains five molecules of water of crystallisation per molecule of copper(II) sulfate

weak acid: an acid that is only partially dissociated into ions in water – usually this produces a low concentration of $H^+(aq)$ in the solution, e.g. ethanoic acid

word equation: a summary of a chemical reaction using the chemical names of the reactants and products

zero error: a type of systematic error in a measuring instrument, e.g. the reading on a balance may not reset to zero when there is nothing on the balance

The Periodic Table of Elements

Group																	
I	II											III	IV	V	VI	VII	VIII
							1 H hydrogen 1										2 He helium 4
3 Li lithium 7	4 Be beryllium 9		**Key** atomic number atomic symbol name relative atomic mass									5 B boron 11	6 C carbon 12	7 N nitrogen 14	8 O oxygen 16	9 F fluorine 19	10 Ne neon 20
11 Na sodium 23	12 Mg magnesium 24											13 Al aluminium 27	14 Si silicon 28	15 P phosphorus 31	16 S sulfur 32	17 Cl chlorine 35.5	18 Ar argon 40
19 K potassium 39	20 Ca calcium 40	21 Sc scandium 45	22 Ti titanium 48	23 V vanadium 51	24 Cr chromium 52	25 Mn manganese 55	26 Fe iron 56	27 Co cobalt 59	28 Ni nickel 59	29 Cu copper 64	30 Zn zinc 65	31 Ga gallium 70	32 Ge germanium 73	33 As arsenic 75	34 Se selenium 79	35 Br bromine 80	36 Kr krypton 84
37 Rb rubidium 85	38 Sr strontium 88	39 Y yttrium 89	40 Zr zirconium 91	41 Nb niobium 93	42 Mo molybdenum 96	43 Tc technetium –	44 Ru ruthenium 101	45 Rh rhodium 103	46 Pd palladium 106	47 Ag silver 108	48 Cd cadmium 112	49 In indium 115	50 Sn tin 119	51 Sb antimony 122	52 Te tellurium 128	53 I iodine 127	54 Xe xenon 131
55 Cs caesium 133	56 Ba barium 137	57–71 lanthanoids	72 Hf hafnium 178	73 Ta tantalum 181	74 W tungsten 184	75 Re rhenium 186	76 Os osmium 190	77 Ir iridium 192	78 Pt platinum 195	79 Au gold 197	80 Hg mercury 201	81 Tl thallium 204	82 Pb lead 207	83 Bi bismuth 209	84 Po polonium –	85 At astatine –	86 Rn radon –
87 Fr francium –	88 Ra radium –	89–103 actinoids	104 Rf rutherfordium –	105 Db dubnium –	106 Sg seaborgium –	107 Bh bohrium –	108 Hs hassium –	109 Mt meitnerium –	110 Ds darmstadtium –	111 Rg roentgenium –	112 Cn copernicium –	113 Nh nihonium –	114 Fl flerovium –	115 Mc moscovium –	116 Lv livermorium –	117 Ts tennessine –	118 Og oganesson –

lanthanoids	57 La lanthanum 139	58 Ce cerium 140	59 Pr praseodymium 141	60 Nd neodymium 144	61 Pm promethium –	62 Sm samarium 150	63 Eu europium 152	64 Gd gadolinium 157	65 Tb terbium 159	66 Dy dysprosium 163	67 Ho holmium 165	68 Er erbium 167	69 Tm thulium 169	70 Yb ytterbium 173	71 Lu lutetium 175
actinoids	89 Ac actinium –	90 Th thorium 232	91 Pa protactinium 231	92 U uranium 238	93 Np neptunium –	94 Pu plutonium –	95 Am americium –	96 Cm curium –	97 Bk berkelium –	98 Cf californium –	99 Es einsteinium –	100 Fm fermium –	101 Md mendelevium –	102 No nobelium –	103 Lr lawrencium –

The volume of one mole of any gas is $24\,dm^3$ at room temperature and pressure (r.t.p.).

> Index

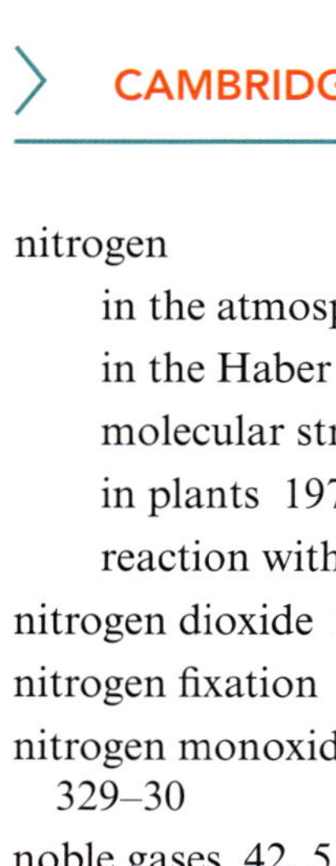

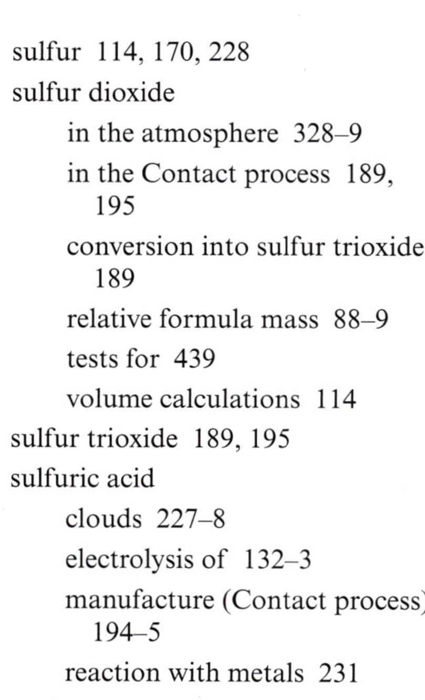

> Acknowledgements

The authors and publishers would like to thank the following for reviewing this coursebook: Farhat Deeba, Joan Hope-Jones and Jarina Iqbal

The authors and publishers acknowledge the following sources of copyright material and are grateful for the permissions granted. While every effort has been made, it has not always been possible to identify the sources of all the material used, or to trace all copyright holders. If any omissions are brought to our notice, we will be happy to include the appropriate acknowledgements on reprinting.

Thanks to the following for permission to reproduce images:

Illustrations by Tech-Set Ltd

Unit 1: Nikkytok/GI; NASA, ESA and Erich Karkoschka (University of Arizona); Charles D. Winters/SPL; Instants/GI; Menahem Kahana/GI; Adam Gault/GI; Simon Murrell/GI; Andrew Lambert Photography/SPL; Andrew Mcclenaghan/SPL; Andrew Lambert Photography/SPL; NASA/JPL Photojournal; Schon/GI; **Unit 2:** Paper Boat Creative/GI; Don Eigler, IBM Almaden Research Center; Image originally created by IBM Corporation; Stocktrek Images/GI; Charles D.Winters/SPL; GIPhotoStock/SPL; Ray Bradshaw/GI;
Unit 3: Mina De La O/GI; Maciej Frolow/GI; Bet_Noire/GI; GIPhotoStock/SPL; Giphotostock/SPL; Nobelmuseet/Wikimedia; Martyn F. Chillmaid/SPL; Adam Gault/GI; INchendio/GI; Wikimedia; **Unit 4:** Charles D.Winters/SPL; Andrew Lambert Photography/SPL; MvH/GI; GIPhotoStock/SPL; Steven Puetzer/GI; Sergio Amiti/GI; **Unit 5:** Caracterdesign/GI; IP Galanternik D.U./GI; Asafta/GI; Cover of Green Chemistry, Vol. 17, No. 6, June 2015 published by the Royal Society of Chemists; **Unit 6:** Monty Rakusen/GI; Paul Kane/GI; Vwpics/Nano Calvo/Universal Images Group/GI; Trevor Clifford Photography/SPL; Monty Rakusen/GI; Robert Houser/GI; David Becker/GI;
Unit 7: Andre Schoenherr/GI; Westend61/GI; GIPhotoStock/SPL; DuncanL/GI; ViewStock/GI; Geography Photos/Universal Images Group/GI; PeopleImages/GI; **Unit 8:** Jose A. Bernat Bacete/GI; Marylexa/GI; Khaled Desouki/GI; Al Fenn/GI; Martyn F. Chillmaid/SPL; Charles D.Winters/SPL; Charles D.Winters/SPL; Roger Harris/SPL; Martyn F. Chillmaid/SPL; J.C. Revy, ISM/SPL; Astrid & Hanns-Frieder Michler/SPL; H. Armstrong Roberts/ClassicStock/GI; Gary Yeowell/GI; **Unit 9:** Westend61/GI; Universal History Archive/Universal Images Group/GI; Wikimedia; Martyn F. Chillmaid/SPL; Andrew Lambert Photography/SPL (x3); Deutches Museum von Meisterwerken der Naurwissenschaft und Technik; Darryl Dyck/Bloomberg/GI; Martyn F. Chillmaid/SPL; Zorazhuang/GI; MirageC/GI; Paul Hennessy/NurPhoto/GI; **Unit 10:** Andrew Lambert Photography/SPL (x3); Trevor Clifford Photography/SPL (x2); Martyn F.Chillmaid/SPL; Eclipse_images/GI; **Unit 11:** Napaporn Leadprathom/GI; Kola Sulaimon/GI; Lacaosa/GI; Andrew Lambert Photography/SPL; Andrew Lambert Photography/SPL; Andrew Lambert Photography/SPL; European Space Agency/SPL; Leslie Garland Picture Library; Richard Packwood/GI; Martyn F.Chillmaid/SPL; Charles D.Winters/SPL; GustoImages/SPL; Arnold Fisher/SPL; Andrew Lambert Photography/SPL; Turtle Rock Scientific/SPL; Iryna Veklich/GI; **Unit 12:** Abstract Aerial Art/GI; Tuul & Bruno Morandi/GI; Jamain/Wikimedia; Biris Paul Silviu/GI; Richard Harwood; Martyn F.Chillmaid/SPL; Martyn F.Chillmaid/SPL; Andrew Lambert Photography/SPL; GIPhotoStock/SPL; Jami Tarris/GI; Beautifulchemistry.net/SPL; **Unit 13:** Image Source/GI; Carlos Jones, Oak Ridge National Lab; MC Talbot; Andrew Lambert Photography/SPL; Andrew Lambert Photography/SPL; Atlantide Phototravel/GI; Richard Harwood; Martyn F.Chillmaid/SPL; Siriwat Nakha/GI; **Unit 14:** Peter Dazeley/GI; Turtle Rock Scientific/Science Source; SPL; Charles D.Winters/SPL; Universal History Archive/GI; Dea/S.Vannini/GI; James King-Holmes/SPL; Volker Pape/GI; Ullstein bild/GI; DuKai photographer/GI; Masako Ishida/GI; Mongkol Nitirojsakul/GI; Photo_Concepts/GI; Francesco Carta fotografo/GI; Monty Rakusen/GI; Tuomas Lehtinen/GI; Jsolie/GI; Sean Gladwell/GI; **Unit 15:** Alextov/GI; Jeffrey Coolidge/GI; Werner Forman/GI; Tim Graham/GI; Charles D.Winters/GI; Turtle Rock Scientific/SPL; Martyn F.Chillmaid/SPL; Charles D. Winters/SPL; Charles D. Winters/SPL; Joseph Scherschel/GI; Recep-bg/GI; **Unit 16:** Marc Friederich/GI; Ribeiroantonio/GI; Matthew Horwood/GI; Rosenfeld Images Ltd/SPL; Ben Johnson/SPL; George Karbus Photography/GI; MC Talbot; laughingmango/GI; **Unit 17:** Rapeepong Puttakumwong/GI; Sarote Pruksachat/GI; Wikimedia; Andrew Holt/GI; Byronsdad/GI; Martyn F.Chillmaid/SPL; Douglas Sacha/GI; Tristan Savatier/GI; Si-Gal/GI; Ruben Ramos/GI; **Unit 18:** Andrew Brookes/GI; Christian Darkin/SPL; Andrew Lambert Photography/SPL; Richard Harwood; Martyn F. Chillmaid/SPL; Andrew Lambert Photography/SPL;
Unit 19: Hirkophoto/GI; Ljeoma Uchegbu/Wikimedia; Sonu Mehta/Hindustan Times/GI; Martyn F. Chillmaid/SPL; Haidar Mohammed Ali/GI; David Taylor/SPL; Paul Rapson/SPL; Paul Rapson/SPL; David R. Frazier/SPL; Trevor Clifford Photography/SPL; Friedrich Saurer/SPL; Adam Brackenbury/SPL; **Unit 20:** Paranyu Pithayarungsarit/GI; Wikimedia; Sputnik/SPL; Anthony Brawley/GI; VCG/GI; David Talbot/Richard Harwood; Bess Adler/Bloomberg/GI; Science&Society Picture Library/GI; Adrian Bicker/SPL; Charles D.Winters/SPL; Kyodo News/GI; William Campbell /GI; Lisa Maree Williams/GI; Stefano Montesi/GI; Bloomberg Creative Photos/GI; NOAA via Flickr; Mikkel Juul Jensen/SPL; Education Images/GI; Steve Gschmeissner/SPL; Wladimir Bulgar/SPL; Brent Durand/GI; **Unit 21:** Anchalee Phanmaha/GI; Sultan Al-Hasani/GI; Tdub303/GI; Andrew Lambert/SPL; Richard Harwood; Andrew Lambert/SPL; Andrew Lambert/SPL; Science & Society Picture Library/GI; Giraudou Laurent/GI; Jock Fistick/Bloomberg/GI; **Unit 22:** SPL; Felix Wong/South China Morning Post/GI; Andrew Lambert/SPL; Andrew Lambert/SPL; Andrew Lambert/SPL; Andrew Lambert/SPL; Turtle Rock Scientific/Science Source/SPL; Turtle Rock Scientific/SPL; Richard Harwood; Andrew Lambert Photography/SPL; Richard Harwood; Martyn F.Chillmaid/SPL; Martyn F.Chillmaid/SPL; BraunS/GI.

Key: GI= Getty Images; SPL = Science Photo Library